Nanomaterials in tissue engineering

Related titles:

Tissue engineering using ceramics and polymers
(ISBN 978-1-84569-176-9)

Electrospinning for tissue regeneration
(ISBN 978-1-84569-741-9)

Nanomedicine: Technologies and applications
(ISBN 978-0-85709-233-5)

Details of these books and a complete list of titles from Woodhead Publishing can be obtained by:

- visiting our web site at www.woodheadpublishing.com
- contacting Customer Services (e-mail: sales@woodheadpublishing.com; fax: +44 (0) 1223 832819; tel.: +44 (0) 1223 499140 ext. 130; address: Woodhead Publishing Limited, 80 High Street, Sawston, Cambridge CB22 3HJ, UK)
- in North America, contacting our US office (e-mail: usmarketing@ woodheadpublishing.com; tel.: (215) 928 9112; address: Woodhead Publishing, 1518 Walnut Street, Suite 1100, Philadelphia, PA 19102–3406, USA)

If you would like e-versions of our content, please visit our online platform: www. woodheadpublishingonline.com. Please recommend it to your librarian so that everyone in your institution can benefit from the wealth of content on the site.

We are always happy to receive suggestions for new books from potential editors. To enquire about contributing to our Biomaterials series, please send your name, contact address and details of the topic/s you are interested in to laura.overend@ woodheadpublishing.com. We look forward to hearing from you.

The team responsible for publishing this book:

Commissioning Editor: Laura Overend
Publications Coordinator: Lucy Beg
Project Editor: Kate Hardcastle
Editorial and Production Manager: Mary Campbell
Production Editor: Adam Hooper
Project Manager: Newgen Knowledge Works Pvt Ltd
Copyeditor: Newgen Knowledge Works Pvt Ltd
Proofreader: Newgen Knowledge Works Pvt Ltd
Cover Designer: Terry Callanan

Woodhead Publishing Series in Biomaterials: Number 56

Nanomaterials in tissue engineering

Fabrication and applications

Edited by
A. K. Gaharwar, S. Sant,
M. J. Hancock and S. A. Hacking

WOODHEAD
PUBLISHING

Oxford   Cambridge   Philadelphia   New Delhi

Published by Woodhead Publishing Limited,
80 High Street, Sawston, Cambridge CB22 3HJ, UK
www.woodheadpublishing.com
www.woodheadpublishingonline.com

Woodhead Publishing, 1518 Walnut Street, Suite 1100, Philadelphia,
PA 19102-3406, USA

Woodhead Publishing India Private Limited, 303 Vardaan House, 7/28 Ansari Road,
Daryaganj, New Delhi – 110002, India
www.woodheadpublishingindia.com

First published 2013, Woodhead Publishing Limited
British Library Cataloguing in Publication Data
A catalogue record for this book is available from the British Library.

Library of Congress Control Number: 2013937804

ISBN 978-0-85709-596-1 (print)
ISBN 978-0-85709-723-1 (online)
ISSN 2049-9485 Woodhead Publishing Series in Biomaterials (print)
ISSN 2049-9493 Woodhead Publishing Series in Biomaterials (online)

The publisher's policy is to use permanent paper from mills that operate a
sustainable forestry policy, and which has been manufactured from pulp which is
processed using acid-free and elemental chlorine-free practices. Furthermore, the
publisher ensures that the text paper and cover board used have met acceptable
environmental accreditation standards.

Typeset by Newgen Knowledge Works Pvt Ltd
Printed by Lightning Source

Cover image courtesy of M. Bucaro and B. Hatton, Wyss Institute for Bio-inspired
Engineering, Harvard University, Cambridge, MA, USA.

Contents

v

Contributor contact details

(* = main contact)

Editors

A. K. Gaharwar
David H. Koch Institute for
    Integrative Cancer Research
Massachusetts Institute of
    Technology
500 Main Street, 76–661
Cambridge
MA 02139
USA

E-mail: gaharwar@mit.edu

S. Sant
Assistant Professor
School of Pharmacy
Department of Pharmaceutical
    Sciences
Department of Bioengineering
McGowan Institute for
    Regenerative Medicine
University of Pittsburgh
3501 Terrace Street
527 Salk Hall
Pittsburgh
PA 15261
USA

E-mail: shs149@pitt.edu

M. J. Hancock
Broad Institute
320 Charles Street
Cambridge
MA 02141
USA

E-mail: hancock@alum.mit.edu

S. A. Hacking
Department of Orthopaedic
    Surgery
Massachusetts General Hospital
    and Harvard Medical School

and

Director, Laboratory for
    Musculoskeletal Research &
    Innovation
Massachusetts General Hospital
55 Fruit Street, GRJ1120
Boston
MA 02114
USA

E-mail: adam.hacking@gmail.com

Chapter 1

NaJuang Kim, Sang Jin Lee* and
    Anthony Atala
Wake Forest Institute for
    Regenerative Medicine
Wake Forest School of Medicine
Medical Center Boulevard
Winston-Salem
NC 27157
USA

E-mail: nakim@wakehealth.edu;
    sjlee@wakehealth.edu; aatala@
    wakehealth.edu

Chapter 2

Ashish A. Kulkarni* and P. S. Rao
Division of Biomedical Engineering
Department of Medicine and
  Center for Regenerative
  Therapeutics
Brigham and Women's Hospital
Cambridge
MA 02139
USA

E-mail: aakulkarni@rics.bwh.
  harvard.edu; kulkarni.poornima@
  gmail.com

Chapter 3

Benjamin D. Hatton
Materials Science and Engineering
University of Toronto
Toronto
ON M5S 3E4
Canada

E-mail: benjamin.hatton@utoronto.
  ca

Chapter 4

Rui R. Costa and João F. Mano*
3B's Research Group –
  Biomaterials, Biodegradables and
  Biomimetics
University of Minho
Headquarters of the European
  Institute of Excellence on Tissue
  Engineering and Regenerative
  Medicine
AvePark
Zona Industrial da Gandra
S. Claudio do Barco
4806–909 Caldas das Taipas
  Guimarães
Portugal

and

ICVS/3B's
PT Government Associated
  Laboratory
Braga/Guimarães
Portugal

E-mail: rui.costa@dep.uminho.pt;
  jmano@dep.uminho.pt

Chapter 5

David A. Stout and N. Gozde
  Durmus
School of Engineering and Center
  of Biomedical Engineering
Brown University
182 Hope Street, Box D
Providence
RI 02912
USA

E-mail: David_Stout@brown.edu;
  Naside_Durmus@brown.edu

Thomas J. Webster*
Department of Chemical
  Engineering
Northeastern University
360 Huntington Avenue
313A Snell Engineering Center
Boston
MA 02115
USA

E-mail: th.webster@neu.edu

Chapter 6

Honglin Chen, Roman
  Truckenmüller, Clemens van
  Blitterswijk and Lorenzo
  Moroni*
Department of Tissue Regeneration

MIRA Institute for Biomedical
  Technology and Technical
  Medicine
Faculty of Science and Technology
University of Twente
P.O. Box 217
7500AE Enschede
The Netherlands

E-mail: h.chen@utwente.nl;
  r.k.truckenmuller@utwente.nl;
  c.a.vanblitterswijk@utwente.nl;
  l.moroni@utwente.nl

Chapter 7

R. R. Sehgal and R. Banerjee*
Department of Biosciences &
  Bioengineering
Indian Institute of Technology
  Bombay
Powai
Mumbai
India

E-mail: rekha@iitb.ac.in; rinti@iitb.
  ac.in

Chapter 8

Daniela F. Coutinho, Manuela
  M. Estima Gomes and Rui L.
  Gonçalves dos Reis*
3B's Research Group –
  Biomaterials, Biodegradables and
  Biomimetics
University of Minho
Headquarters of the European
  Institute of Excellence on Tissue
  Engineering and Regenerative
  Medicine
AvePark
Zona Industrial da Gandra

S. Claudio do Barco
4806–909 Caldas das Taipas
  Guimarães
Portugal

and

ICVS/3B's
PT Government Associated
  Laboratory
Braga/Guimarães
Portugal

E-mail: d.coutinho@dep.uminho.
  pt; megomes@dep.uminho.pt;
  rgreis@dep.uminho.pt

Chapter 9

Yulia Sapir
The Avram and Stella
  Goldstein-Goren Department of
  Biotechnology Engineering
Ben-Gurion University of the
  Negev
Beer-Sheva 84105
Israel

E-mail: yuliasapir@gmail.com

Boris Polyak
Department of Surgery
Drexel University College of
  Medicine
Philadelphia
PA 19102
USA

and

Department of Pharmacology and
  Physiology
Drexel University
Philadelphia
PA 19102
USA

Smadar Cohen*
The Avram and Stella
Goldstein-Goren Department of
Biotechnology Engineering
Ben-Gurion University of the
Negev
Beer-Sheva 84105
Israel

and

Ilse Katz Institute for Nanoscale
    Science and Technology
Ben-Gurion University of the
    Negev
Beer-Sheva 84105
Israel

E-mail: scohen@bgu.ac.il

Chapter 10

Mustafa E. Marti and Anup D.
    Sharma
Department of Chemical and
    Biological Engineering
Iowa State University
Ames
IA 50011–2230
USA

E-mail: mustafaesenmarti@gmail.
    com; adsharma@iastate.edu

Donald S. Sakaguchi
Department of Genetics,
    Development, and Cell Biology
Iowa State University
Ames
IA 50011–2230
USA

E-mail: dssakagu@iastate.edu

Surya K. Mallapragada*
Department of Chemical and
    Biological Engineering

Iowa State University
Ames
IA 50011–2230
USA

E-mail: suryakm@iastate.edu

Chapter 11

Eun Ji Chung
Institute of BioNanotechnology in
    Medicine
Northwestern University
303 E. Superior Street, 11th Floor
Chicago
IL 60611–3015
USA

E-mail: eunchung2008@u.
    northwestern.edu

Nirav Shah
Parkview Musculoskeletal Institute
7600 West College Drive
Palos Heights
IL 60463
USA

Ramille N. Shah*
Institute of BioNanotechnology in
    Medicine
Northwestern University
303 E. Superior Street, 11th Floor
Chicago
IL 60611–3015
USA

and

Department of Materials Science
    and Engineering
McCormick School of Engineering
Northwestern University
2220 Campus Drive
Evanston

IL 60208-3108
USA

and

Department of Surgery
Feinberg School of Medicine
  Northwestern University
251 East Huron St., Galter 3-150
Chicago
IL 60611
USA

E-mail: ramille-shah@northwestern.
  edu

Chapter 12

Aaron Cipriano
Department of Bioengineering
University of California at
  Riverside
900 University Avenue
Riverside
CA 92521
USA

E-mail: acipr001@ucr.edu

Huinan Liu*
Department of Bioengineering
  and The Materials Science and
  Engineering Program
University of California at
  Riverside
900 University Avenue, MSE 217
Riverside
CA 92521
USA

E-mail: huinan.liu@ucr.edu

Chapter 13

Emily C. Beck
University of Kansas
Bioengineering Program
4132 Learned Hall, 1530 W 15th St
Lawrence
KS 66045
USA

E-mail: beckem@ku.edu

Michael S. Detamore*
University of Kansas
Department of Chemical and
  Petroleum Engineering
4132 Learned Hall, 1530 W 15th St
Lawrence
KS 66045
USA

E-mail: detamore@ku.edu

Chapter 14

Benedetto Marelli, Chiara E. Ghezzi
  and Showan N. Nazhat*
Department of Mining and
  Materials Engineering
MH Wong Building
McGill University
3610 University Street
Montreal
Qc H3A 2B2
Canada

E-mail: benedetto.marelli@mail.
  mcgill.ca; chiara.ghezzi@mail.
  mcgill.ca; showan.nazhat@mcgill.
  ca

Chapter 15

Samer H. Zaky, S. Yoshizawa and
C. Sfeir*
Center for Craniofacial
  Regeneration
Department of Oral Biology
University of Pittsburgh School of
  Dental Medicine

University of Pittsburgh
598 Salk Hall
3501 Terrace St.
Pittsburgh
PA 15261–1964
USA

E-mail: shz33@pitt.edu; say15@pitt.
  edu; csfeir@pitt.edu

Woodhead Publishing Series in Biomaterials

Foreword

Tissue engineering promises to revolutionize the way millions of patients with organ failure or tissue damage are treated. Yet the challenges are significant. Materials synthesis, characterization, design, and application are a central part of tissue engineering. This book provides guidance on these important issues, particularly at the nano scale.

The book is divided into three parts following an introduction to biomedical nanomaterials in tissue engineering. Firstly, there are six chapters involving the fabrication of such nanomaterials including novel techniques such as layer-by-layer self-assembly, synthesis of carbon based systems, creation of nanofibrous scaffolds, and delivery of growth factors. The following section examines the application of nanomaterials in 'soft' tissue engineering – in particular vascularized tissue, heart neural tissue, and cartilage and ligaments. The final section considers the application of nanomaterials in 'hard' tissue engineering, in particular bone, dental, and craniofacial tissues.

This book should be a useful guide for scientists and engineers interested in nanomaterials and their role in tissue engineering. Its value will only grow as this field continues to mature, in part based on the very advances discussed in these chapters.

Robert S. Langer Sc.D.
David H. Koch Institute Professor
Massachusetts Institute of Technology (MIT)
USA

Introduction

This book presents the latest groundbreaking research and reviews exciting, emerging trends at the intersection of nanomaterials and tissue engineering. Written from a biological perspective with biological applications in mind, this book provides the necessary insight and rationale to facilitate the incorporation of nanomaterials into your work.

This book is divided into three parts that collectively provide an important and succinct overview of nanomaterials in tissue engineering. The first part provides an essential review of the fabrication methods necessary to produce tissue scaffolds incorporating nanomaterials or nanofeatures. Part 1 also explores the various means by which nanomaterials can direct cell behavior. The sub-micron surface structures present on ordered nanoporous surfaces can affect cell adhesion, morphology, viability, genetic regulation, apoptosis, and motility. Nanofabrication can also be used to exert control over the local environment. Biomaterials with precisely controlled pore size, porosity, and interconnectivity at the nano- and micro-scale, are ideal drug delivery matrices or separation membranes. Stimuli-responsive materials with nanofeatures are leading a revolution in smart coatings. For example, nanofibrous cellular scaffolds synthesized by molecular self-assembly, phase separation, or electrospinning can be engineered with chemical and biological cues to provide stimuli-responsive and dynamic behavior. Finally, nanomaterials that deliver bioactive agents to control cellular behavior for recombinant applications are discussed.

The second and third parts of the book summarize the latest advances in nanomaterial applications for the tissue engineering of soft and hard tissues, respectively. For example, electrically conductive polymer scaffolds containing graphene or carbon nanotubes promote axonal outgrowth and neural regeneration. Injectable supramolecular nanostructures from peptide-based self-assembling materials have shown promise for cartilage regeneration. For hard tissues, biological building blocks such as collagen and hydroxyapatite nanoparticles have been used with micro- and nano-fabrication technologies to synthesize nanostructured scaffolds that mimic the remarkable hierarchical architectures found in bone and enamel. Bioactive ceramic, β-tricalcium phosphate, calcium phosphate, and bio-glass nanoparticles have been added to polymeric networks to enhance their strength, guide biomineralization, and facilitate their integration within the skeleton. The engineering of interfacial tissues that mimic the bone–cartilage or bone–tendon interface present special challenges that are being addressed by gradients of bioactive ceramic nanoparticles.

Tissue engineering has become an extremely active research field. The application of nanomaterials in tissue engineering is enabling exciting advances with significant and tangible benefits for public health.

1

Biomedical nanomaterials in tissue engineering

N. J. KIM, S. J. LEE and A. ATALA,
Wake Forest School of Medicine, USA

DOI: 10.1533/9780857097231.1

Abstract: Tissue engineering aims to develop functional tissue substitutes that can be used for reconstructing damaged tissues or organs. To engineer a tissue construct, cells are generally seeded on biomaterial scaffolds that recapitulate the extracellular matrix (ECM) and microenvironment in order to enhance tissue development. Recently, it has been recognized that biomedical nanomaterials play a central role in tissue engineering as they may better support tissue regeneration. Here, we address various biomedical nanomaterials and their effects on cells and tissues. The considerations needed to create a scaffold material for tissue engineering applications are discussed. We will also review the current progress and future directions for nanomaterials in tissue engineering.

Key words: nanomaterial, biomaterial, tissue engineering, regenerative medicine.

1.1 Introduction

Tissue engineering strategies combine the principles of cell transplantation, material science, and engineering to develop biological substitutes that can restore and maintain the normal function of diseased or injured tissues/ organs (Atala, 2007). Despite the technical advance over the past decades, the use of these approaches has been restricted to research applications and few have been used in clinic. Many clinicians still employ biodegradable polyesters that were first approved for use in humans over 30 years ago. This is a serious concern because morphogenesis is strongly affected by the inter-actions between the cells and the extracellular environment during normal tissue development. While the simple synthetic polymers that have been in use provide support for neo-tissue development, they do not successfully mimic the complex interactions between the tissue-specific cells and the tissue-specific extracellular matrices (ECMs) that promote functional tissue regeneration. Therefore, 'smart biomaterials' that actively participate in

1

functional tissue regeneration (Furth *et al.*, 2007) must be developed and used for future applications.

A biomedical nanomaterial is a nanostructured material that is used to construct a device designed to interface with tissues or tissue components in performance of its function (Williams, 2009). Currently, a biomedical nanomaterial in tissue engineering is considered a smart biomaterial due to its multi-functionality. Even though nanoscale is usually defined as being smaller than one-tenth of a micrometer in at least one dimension, this definition is also used to describe materials that are smaller than 1 μm in tissue engineering applications. These materials include nanoparticles, nanofibers, nanotubes, nanowires, and any other nanoscaled materials or devices. A nanomaterial has been commonly described and recognized as a nanostructured building block of biological replacements used to restore or improve physiological function of bioengineered tissues/organs. Nanomaterials are increasingly studied for tissue engineering and regenerative medicine because they can closely mimic natural biological environments. Using these materials, nanostructures and other properties of native tissues can be constructed. In addition, nanomaterials have better functional, mechanical, and physical characteristics as well as superior biocompatibility compared to macroscaled materials. These properties can be modified by varying methods of fabrication and assembly. The use of nanomaterials has opened up immense possibilities and facilitated significant progress in tissue engineering to date.

Since tissue engineering involves developing functional substitutes for damaged tissues/organs, materials need to be inert or biocompatible when in contact with living cells or tissues. Moreover, the interaction between a material and cells influences cellular behaviors such as migration, adhesion, proliferation, and differentiation. Molecular signals from the extracellular environment affect the functional properties and the physical integrity of a material. Thus, the interactions among materials, cells, and the neighboring environment determine the performance and outcome of the tissue engineering scaffolds. In this regard, nanostructured biomaterials play important roles in boosting cell–material or cell–cell interactions (Huppertz, 2011). This chapter will review biomedical nanomaterials that provide nanoscaled environments for tissue engineering applications and will describe the design considerations for these materials. The current progress and future directions for nanomaterials in tissue engineering will also be discussed.

1.2   Overview of nanomaterials in tissue engineering

This section will introduce fundamentals of nanomaterials in tissue engineering. Basic principles and utilization of nanomaterials in tissue engineering will be discussed.

1.2.1   Basic principles of nanomaterials

The basic requirements of scaffold materials for use in tissue engineering applications are biocompatibility, biodegradability, and low immunogenicity in order to assimilate cellular function and tissue formation with little adverse effect. Incompatible biomaterials are destined for inflammatory response or foreign-body reaction that eventually leads to rejection and/or necrosis (Atala, 2008). While classic tissue-engineered scaffolds provide temporary mechanical support during the spatial tissue organization of cells, a smart biomaterial scaffold can maintain adequate mechanical integrity and simultaneously accelerate tissue formation during the early stages of development. Principally, a scaffold should facilitate the localization and the delivery of tissue-specific cells to designated sites in the body, maintain a three-dimensional architecture that permits the formation of neo-tissues, and guide the development of neo-tissues with appropriate function (Kim and Mooney, 1998). However, it has been recognized that most classic tissue-engineered scaffolds do not properly recapitulate the extracellular environment, because this environment is a dynamic and hierarchically organized nanoscaled composite that regulates essential cellular functions such as morphogenesis, migration, adhesion, proliferation, and differentiation (Dvir *et al.*, 2011b). The extracellular environment encompasses the ECM proteins, soluble factors, and cytokines secreted from cells. In addition, ECM proteins consist of proteoglycans, structural proteins (e.g. collagens and elastin), and adhesive proteins (e.g. fibronectin and laminin).

In order to provide an effective extracellular environment using biomaterial scaffolds, there are critical factors that need to be taken into account including the biological, mechanical, and the physical properties of a target tissue/organ. Moreover, the significance of the nanoscaled configuration of the extracellular environment in promoting essential cellular interactions has been recognized and can be employed to engineer functional tissue constructs. For example, a scaffold composed of nanoscaled fibers provides excellent cellular interactions as well as tissue compatibility. There has been remarkable interest in exploring the potency of biomimetic fibers in tissue engineering. One important feature of these fibrous biomaterials is the vastly increased ratio of surface area to volume. These attractive aspects of advanced biomaterials enable improved mimicking of the extracellular environment and ultimately being used to engineer more complex tissues/organs when compared to traditional biomaterials.

1.2.2   Utilization of nanomaterials in tissue engineering

There is a need in tissue engineering to develop advanced biomaterial scaffolding systems, which could provide a proper extracellular environment for a regenerating tissue including biophysical (topography) and biochemical (delivery of bioactive molecules) cues. Such a material could mimic natural ECMs to regulate cellular behaviors for example, stem cell differentiation (McMurray *et al.*, 2011). One major determinant of success or failure in tissue engineering is the physicochemical interaction between cells and materials. Cells interact with the neighboring environment by nanoscaled extracellular signals. The purpose of nanoscaled tissue engineering is to channel these interactions through nanostructured biomaterials and to guide cellular behaviors towards regeneration. To improve cell–material and cell–cell interactions for comprehensive tissue regeneration, nanomaterials have been fabricated and modified by various methods (Wheeldon *et al.*, 2011).

Topographical cues generated by the ECM have significant effects upon cellular interactions. Studies have shown that substratum topography has direct effects on the ability of cells to orient, migrate, and produce an organized cytoskeletal arrangement (Flemming *et al.*, 1999). The topography of a native tissue matrix is a complex structure that comprises pores, fibers, ridges, and other nanoscaled features. Fundamental understanding of cell–substrate interactions is important for tissue engineering applications and development of medical implant devices. In addition, various designs have applied to construct scaffolds in terms of the nanotopography of the biomaterial surface. For example, nanoscaled electrospun scaffolds composed of poly(ε-caprolactone) (PCL) and type I collagen have enhanced the cell attachment and the cytoskeletal orientation of human aortic endothelial cells when compared to micro-scaled fibers due to the increased contact guidance of the smaller fibers (Plate I, see colour section between pages 264 and 265) (Ju *et al.*, 2010). It has also been shown that increasing the surface area of nanofibrous scaffolds enhanced protein adsorption to the scaffold. Scaffolds with nanofibrous-pore walls have selectively adsorbed more serum proteins, fibronectin, and vitronectin than solid-pore wall scaffolds. As a result, this structure has allowed greater osteoblast attachment (Woo *et al.*, 2003). Nanopillar substrate topography has been reported to promote bovine corneal endothelial cell responses. Corneal endothelial cells in nature interact with the nanotopography of the underlying Descemet's membrane, which consists of a densely packed interwoven mesh of nanoscaled fibers and pores. These cell–substratum and cell–cell interactions play crucial roles *in vivo*. This nanopillar substrate has been shown to help cells develop a healthy corneal endothelium-like monolayer structure and function (Teo *et al.*, 2012). Nanoscale morphology provides initial structural integrity for tissue and control cellular behaviors, which lead to the restoration of functional tissue. Mimicking the surrounding environment of target cells through nanostructure design has significant effects on tissue regeneration (Goldberg *et al.*, 2007). Proper manipulation of nanomaterials can boost the performance of scaffolds, leading to successful applications in tissue engineering.

On the other hand, an increasing number of studies have incorporated therapeutic components via drug/protein delivery systems to accelerate tissue maturation/formation and improve the function of damaged tissues/organs. The development of nanoencapsulation technology as a delivery tool can play an important role in tissue engineering application. Nanoencapsulation can achieve significantly improved delivery of bioactive molecules such as growth factors, genes, and small molecules, as well as providing targeted delivery of molecules in a cell or tissue specific manner. In addition, it can enhance the transcytosis of molecules across tight epithelial and endothelial barriers and the delivery of large macromolecules to the intracellular sites of action. Nanoencapsulation has also been used for the co-delivery of two or more molecules in combination therapy, for the visualization of drug target sites by combining therapeutic agents with imaging modalities, and for the real time readouts of the *in vivo* efficacy of a therapeutic agent (Farokhzad and Langer, 2009).

Growth factors are signaling polypeptides that guide cells towards development and direct specific cellular responses (Cross and Dexter, 1991). A growth factor can trigger cell migration, differentiation, or proliferation. The specific effect of a growth factor can be manipulated through selective control of the tissue environment (Lee *et al.*, 2011). As an example, a poly(L-lactide) (PLLA)/gelatin scaffold containing stromal-derived factor-1 alpha (SDF-1α) has significantly enhanced the recruitment of host stem cell-like cells into an implanted construct. The recruited stem cell-like cells have shown the potential to differentiate into multiple lineages (Ko *et al.*, 2012).

Nucleic acids can be very potent tools for tissue engineering applications due to their ability to selectively control cellular responses. Both DNA and RNA have been utilized as cargo within scaffolds and as the building blocks of the scaffolds themselves (Seeman, 2010). A porous scaffold fabricated from poly(L-lactic-*co*-glycolic acid) (PLGA) seeded with cells containing a vascular endothelial growth factor (VEGF)-encoding plasmid (pVEGF) has exhibited long-term sustained release of pVEGF for 105 days *in vivo*. This VEGF expression has increased the blood vessel density in and around the scaffold. Transfected cells have been located at the periphery of the scaffolds initially, and then slowly spread within the pores of the scaffold. Eventually, these cells have been found throughout the scaffold interior after about 4 months (Jang *et al.*, 2005). In another study, muscle specific microRNAs (miRNA), including miR-1, miR-133, and miR-206, have been mixed with collagen and injected into injured rat skeletal muscle. The miRNAs have promoted myotube differentiation and up-regulated myogenic regulatory factors (Nakasa *et al.*, 2010). In addition to the restorative benefits of the molecules, the incorporation method and the release profile of molecules from the scaffold need to be considered in regard to the use of biological or therapeutic factors in tissue engineering applications. Bioactive molecules must be stably included and protected until they can be released to a specific target. To control the release of these molecules, the material degradation rate or porosity need to be taken into consideration so that biological factors can be released at the proper rate to coordinate the recovery of tissue.

1.3   Biomedical nanomaterials in tissue engineering applications

In order to discuss on biomedical nanomaterials that are used in tissue engineering purposes, various types of nanomaterials, their constructions, and the applications will be considered. This section will present examples of each usage that give rise to detailed investigations.

1.3.1   Basic design considerations of nanomaterials

Biomedical nanomaterials are the building blocks of scaffolds in tissue engineering applications. Architectural properties, such as porosity and surface features, also contribute to tissue repair in terms of cell–material interactions. From a molecular standpoint, biodegradability and biocompatibility are desirable to increase the success rate of implantation of a construct as well as to avoid additional surgical procedures and adverse side effects. Degradation of these materials must also be considered, since it can be employed as an additional infiltration path for cells or as a release mechanism for bioactive molecules stored within the scaffold. Proper cell infiltration is vital, because it leads to cell migration, proliferation, and differentiation within the scaffold. The scaffold should also support cells and regulate the extracellular environment to enhance tissue alignment and cell–cell interactions. Furthermore, these features lead to sufficient neovascularization, adequate oxygen supply, and a mechanism for waste disposal. Without these features, the regenerated tissue will not function properly. Thus, the selection of an appropriate nanomaterial has a significant impact on functional tissue regeneration (Powers et al., 2002; Paszek and Weaver, 2004; Georges and Janmey, 2005; Yeung et al., 2005; Engler et al., 2006; Xu et al., 2006; De Laporte and Shea, 2007).

In the following sections, we will discuss current biomedical nanomaterials used in tissue engineering applications, as well as their advantages and disadvantages. These materials are categorized as metallic nanomaterials, carbon nanotubes and nanoceramics, nanoparticles for biomedical imaging, and nanofibers, according to their structural nature and applications. In each section, the use of the nanomaterials in different applications will be discussed.

1.3.2   Metallic nanomaterials

Metallic nanostructures such as iron oxide, copper, titanium, silver, and gold are used for cancer diagnostics and therapeutics, as well as in the engineering of 'hard' tissues such as bone (Table 1.1). They are shaped in various ways, including spheres, shells, rods, and wires. The main advantages of using metals include their optical adjustability, electrical conductivity, surface chemistry, and ease of fabrication. They have been used as diagnostic and therapeutic tools for cancer due to their optical and surface functional diversity (Sperling et al., 2008; Sau et al., 2010; Bhattacharyya et al., 2011; Conde et al., 2012). Because of their highly tunable optical properties, they are easily fabricated to desirable electromagnetic wavelengths based on their shape, size, and composition. Their capacity to convert light or radio frequencies into heat enables thermal ablation of target tissues. It also allows simultaneous diagnostic and photothermal therapy inside patients' bodies (Lee and El-Sayed, 2006; Jain et al., 2008; Day et al., 2009; Chen et al., 2010). The surface chemistry of metallic nanomaterials makes functional modification possible. Different moieties including antibodies, peptides, nucleic acids, and polymers, have been attached to metal implants. This functional versatility gives metal nanostructures target-specificity and a long half-life in serum depending on the specific moieties bound to them (Nishiyama, 2007; Ghosh et al., 2008; Sperling and Parak, 2010).

Nanostructured gold is a good candidate for use in medical applications due to its low toxicity and good electrical conductivity. Gold nanowires incorporated within alginate scaffolds have been shown to bridge the electrically resistant pore walls of alginate and improve electrical communication between adjacent cardiac cells. The gold nanowires have also enhanced

Table 1.1 Metallic nanomaterials in tissue engineering

Material	Application	Effect
Gold	Nanorods for chemical and biological sensor (Lee and El-Sayed, 2006)	Enhanced sensing resolution
Gold	Nanowires in alginate scaffold for cardiac patches (Dvir et al., 2011a)	Improved electrical communication, tissue thickness and alignment, enabled cells contract along with electrical stimuli
Silver	Silver nanoparticle-containing gelatin nanofibers for wound healing (Rujitanaroj et al., 2008)	Increased antimicrobacterial activity
Titanium and Ti6Al4V	Nanopowder in compacts for bone regeneration (Miller et al., 2004)	Increased osteoblast adhesion

the tissue thickness and alignment. Moreover, the cells in this tissue have shown the ability to contract synchronously corresponding to electrical stimuli (Plate II) (Dvir *et al.*, 2011a). Another application of nanometals is in bone and cartilage regeneration. Nanophase metals, synthesized with Ti, Ti6Al4V, and CoCrMo alloys, have increased osteoblast adhesion to the surface compared to conventional metals due to the increased surface particle boundaries in nanophase materials (Plate II). These nanometal surfaces have similar chemistry to natural tissues with the exception of surface roughness. This increased cell adhesion can lead to increased deposition of calcium-containing minerals, which is crucial in bone regeneration (Webster and Ejiofor, 2004). Despite the functional diversity of metallic nanomaterials, they have limited application due to their non-biodegradable features, which restrict high-dose applications. Many non-biodegradable metals can cause toxicity in adjacent tissue, which in turn may result in undesirable outcomes. The potential toxicity from nanometals can be reduced by coating the metal with a more biocompatible material or by surface modification (Ai *et al.*, 2011).

1.3.3   Carbon nanotubes and nanoceramics

Carbon nanotubes (CNTs) are the most widely used organic nanomaterials in tissue engineering and regenerative medicine (Table 1.2). Superior biocompatibility based on surface chemistry, mechanical strength, and chemical inertness makes them fit for use as a building block of various tissue scaffolds. CNTs have very high surface/volume ratio due to their diameter at the nanoscale (Veetil and Ye, 2009; Dvir *et al.*, 2011b). CNTs are one of the strongest materials in existence; thus, they have been used to coat bone scaffolds and improve their mechanical properties (Yu *et al.*, 2000). The effects of a thin film of poly(ethylene glycol) (PEG) conjugated multiwalled carbon nanotubes (MWCNT) spray dried onto preheated coverslips have been investigated for their ability to influence human mesenchymal stem cell (hMSC) proliferation, morphology, and final differentiation into osteoblasts (Nayak *et al.*, 2010). CNTs are used in cardiopulmonary applications as catheters, because they are highly compatible with blood. They are chemically inert and do not wear out. For these reasons, integrating CNTs into polyurethane has significantly enhanced its tensile strength and elongation capability. CNTs have significantly improved the anticoagulant properties of polyurethane as well, which suggests that there is potential for the use of carbon nanotube-based materials in other blood-contacting applications (Meng *et al.*, 2005). In addition, due to their superior electrical conductivity, CNTs have been applied in neuronal tissue regeneration. As a substrate for cultured neurons, CNTs have promoted the neurite outgrowth and the branching of neurons. They have been reported to promote neuron attachment, growth, differentiation, and long-term survival (Mattson *et al.*, 2000; Hu *et al.*, 2004, 2005; Lovat *et al.*, 2005). CNTs have also enhanced the signal transmission efficiency of neurons by forming tight contacts with the cell membranes. This tight contact is thought to favor electrical shortcuts between the proximal and distal compartments of the neuron. As a result, neurons cultured on CNTs have improved responsiveness and integrative properties (Lovat *et al.*, 2005; Mazzatenta *et al.*, 2007; Massobrio *et al.*, 2008; Cellot *et al.*, 2009). The surface of a CNT can be functionally modified for drug delivery applications to increase delivery efficiency and target specificity. Various therapeutic elements including cells, proteins, nucleic acids, and any other molecules can be integrated into CNTs (Veetil and Ye, 2009). However, toxicity of CNTs has been reported in lung tissue after being used in pulmonary applications. Toxicity to other organs or systemic toxicity has not yet been determined. There is much controversy surrounding the potential carcinogenicity of CNTs as well. Hence, more extensive studies are required to get a comprehensive picture of CNTs as a biocompatible nanomaterial (van der Zande *et al.*, 2011).

Nanoceramics have comparable mechanical and chemical compatibilities to many biological materials (Table 1.2). This makes nanoceramics good candidates for hard tissue engineering and applications in which the scaffold is to be in specific contact with biological fluids. The natural mineral

Table 1.2 Carbon nanotubes and nanoceramics in tissue engineering

Material	Application	Effect
Carbon nanotubes (CNT)	CNT ropes to coat bone scaffold (Yu *et al.*, 2000)	Improved mechanical strength
	CNT integrated polyurethane fillers for cardiovascular surgery (Meng *et al.*, 2005)	Enhanced tensile strength and elongation capability, improved anticoagulant property
	CNT for neural tissue regeneration (Mattson *et al.*, 2000; Hu *et al.*, 2004; Hu *et al.*, 2005; Lovat *et al.*, 2005; Mazzatenta *et al.*, 2007; Massobrio *et al.*, 2008; Cellot *et al.*, 2009)	Promoted neuron attachment, growth, differentiation and long-term survival, improved neuronal responsiveness and integrative properties
Alumina and titania	Nanopowder in compacts for bone regeneration (Webster *et al.*, 1999)	Increased osteoblast adhesion
Hydroxyapatite	Nanocrystalline HA in PLLA scaffold for bone scaffold (Kim *et al.*, 2006)	Accelerated osteoblast growth and mineralization

phase of bones and dentin is calcium phosphate (hydroxyapatite, HA), which makes ceramics suitable as replacements for defective bone (Park and Lakes, 2007). For these reasons, ceramic nanoparticles have been studied for orthopedic and dental applications. These ceramics also have surface chemistry that is virtually inert in biological fluid, enabling cardiovascular applications. Nanophase alumina and titania have also been reported to increase osteoblast adhesion compared to non-nanophase materials *in vitro* (Webster *et al.*, 1999). The enhanced adhesion of osteoblasts has been attributed to the grain size of the ceramic, indicating that the critical grain size for osteoblast adhesion was 49–67 nm for alumina and 32–56 nm for titania. The result suggests the potential of nanoceramics to simulate surface grain size of natural bone, facilitating protein interactions and thus, enhancing osteoblast adhesion.

1.3.4   Nanoparticles for biomedical imaging

Various types of nanoparticles are now under investigation for use in biomedical imaging including solid lipid nanoparticles, liposomes, micelles, nanotubes, metallic nanoparticles, quantum dots, dendrimers, polymeric nanoparticles, and iodinated nanoparticles (Choi and Frangioni, 2010). Biomedical imaging supports the development of enabling technologies including real-time, non-invasive tools for assessing the function of engineered tissues and real-time assays that monitor the interaction of cells and their environment at the molecular and organelle level. For instance, implanted scaffolds are designed to eventually degrade and be replaced by cells generating an essentially normal tissue over time. Thus, the rate of scaffold degradation and ECM production by cells must be equivalent for successful outcomes. In addition, scaffold degradation can affect cell viability, cell growth, and the host response to the implanted construct (Burg *et al.*, 2000; Hutmacher, 2000; Middleton and Tipton, 2000; Wu and Ding, 2004). Many efforts have been focused on investigating *in vitro* scaffold degradation behavior by measuring changes in scaffold weight (Agrawal *et al.*, 2000; Lu *et al.*, 2000; Wu and Ding, 2004), morphological changes (Wang and Hon, 2003; Oh *et al.*, 2011), viscosity (Agnihotri *et al.*, 2006), mechanical properties (Lee *et al.*, 2008; Oh *et al.*, 2011), and molecular weight (Cho and An, 2006; Oh *et al.*, 2011), all of which can be dependent on culture conditions. However, *in vivo* scaffold degradation cannot be fully measured by traditional *in vitro* methods, because body conditions cannot be mimicked *in vitro*. Thus, investigation of *in vivo* scaffold degradation currently requires implantation into animal models. However, this involves sacrificing numerous animals at various time points after implantation, which often leads to inaccurate conclusions.

In order to avoid these pitfalls, non-invasive optical imaging is currently being investigated for real-time monitoring of *in situ* degradation of implanted scaffolds, as well as that of cellular behaviors, using a combination of nanoparticles and tissue-engineered scaffolds (Choi and Frangioni, 2010). The feasibility of using ultrasound elasticity imaging has been demonstrated as a non-invasive monitoring tool for the mechanical performance and degradation of scaffolds *in vitro* and *in vivo* (Kim *et al.*, 2008). Saldanha *et al.* (2008) have used a magnetic resonance imaging (MRI)-based method of tracking cells that had been labeled with ferumoxides for non-invasive *in vivo* detection and longitudinal assessment of implanted cells. Several studies have used fluorescence imaging techniques for the *in situ* monitoring of implanted scaffold materials (Murphy and Lever, 2002; Yang *et al.*, 2005). Most recently, non-invasive fluorescence imaging has been performed to sequentially follow *in vivo* material-mass loss to model the degradation of materials hydrolytically and enzymatically (Artzi *et al.*, 2011). These advanced imaging techniques can provide real-time monitoring of *in vivo* degradation of scaffold materials and the ability to track implanted cells *in vivo*. For instance, inorganic nanoparticles such as quantum dots offer several advantages (Cho *et al.*, 2010): their optical and magnetic properties can be tailored by controlling composition, structure, size, and shape; their surfaces can be modified with ligands to target specific molecules of tissue/organ; the contrast enhancement provided can be equivalent to millions of molecular counterparts; and they can be integrated with a combination of different functions for multimodal imaging. Nanoparticles for biomedical imaging are highly sensitive, target-specific, and safe to use for improving the future tissue engineering applications.

1.3.5   Nanofibers

Synthetic biodegradable polymers, including various polyesters, are the most widely used scaffold materials in tissue engineering. They are built by linking monomers through primary covalent bonding. For this reason, the size and composition of repeating monomers are precisely controllable. The malleability in synthesis, fabrication, and modification enables nanopolymer fabrication in virtually any shape, size, and functionality. On the other hand, natural polymers, including collagen, elastin, chitosan, and polypeptides, have also been utilized to build therapeutic scaffolds. Since they are natural materials, they have the advantage of biocompatibility and no toxicity. These synthetic or natural materials can be used to fabricate nanofibrous scaffolds using various fabrication methods.

As mentioned above, the interactions between cells and the ECM are fundamental processes that govern signaling for cell adhesion, alignment, migration, and differentiation. In the past decade, there has been tremendous interest in exploring the potency of biomimetic nanofibrous structures in tissue engineering, and several techniques originally developed for different purposes have been used for this application. For instance, electrospinning and molecular self-assembly are common nanofabrication methods to create three-dimensional (3D) scaffolds composed of sub-micron fibrous structures. Among these, electrospinning technology has been widely employed to fabricate tissue-engineered scaffolds that mimic native ECM architecture (Table 1.3). Thus, understanding of the interactions between electrospun fibrous scaffolds and mammalian cells is crucial to the successful production of target tissues and organs.

Focal adhesions of cells can be regulated by the topography of biomaterials. Focal adhesion lies at the convergence of integrin adhesion, signaling, and the actin cytoskeleton. It is also known to control signaling complexes and integrin function. For this reason electrospinning is especially attractive, because the diameters of electrospun fibers can be easily and quickly controlled by varying the fabrication parameters. This provides different topographical cues for cells seeded on the scaffolds. Plate I shows the cytoskeletal organization and focal adhesion of endothelial cells on an electrospun PCL/collagen scaffold with different fiber diameters. Endothelial cells on nanoscaled fibers (0.5 µm, Plate I) show a better-developed cytoskeletal organization and improved focal adhesion compared to other fiber diameters (Ju et al., 2010). Because the fiber diameters of electrospun nanofiber scaffolds are orders of magnitude smaller than the size of most cells, the cells are able to organize around the fibers or spread and attach to adsorbed proteins at multiple focal points (Elias et al., 2002; Xu et al., 2004).

Furthermore, the electrospun nanofibrous mats have been shown to enhance adhesion and proliferation of mesenchymal stem cells (MSCs) when compared to a flat smooth surface (Finne-Wistrand et al., 2008). Several studies have demonstrated that electrospun fibrous scaffolds can enhance cellular responses, including cell adhesion and maintenance of cell phenotype (Chua et al., 2005; Lee et al., 2005; Badami et al., 2006; Luong-Van et al., 2006). Shih et al. (2006) demonstrated the adhesion, proliferation, motility, and differentiation of human MSCs on electrospun type I collagen nanofibers of different diameters. These indicate that the nanoscaled fibers can support initial cell adhesion, which then affects further cell proliferation and differentiation.

Electrospun fiber orientation can influence cell proliferation in addition to controlling cell orientation and tissue growth. It is known that musculoskeletal tissues exhibit significant anisotropic mechanical properties and highly oriented cells underneath the ECM. An essential step in engineering functional skeletal muscle tissue is to mimic the structure of native tissue, which comprises highly oriented myofibers developed

Table 1.3 Cellular interaction of various cell types with electrospun nanofibrous substrates

Cell types	Materials	Fiber configuration	Outcomes	Reference
Mesenchymal stem cells (MSCs)	PEUU	Fiber diameter and alignment	Ligament-like cell differentiation	Bashur *et al.*, 2009
	PLA	Fiber alignment	Cell alignment and *in vivo* differentiation	Hashi *et al.*, 2007
	PCL	Fiber alignment	Chondrogenic differentiation	Wise *et al.*, 2009
	Collagen	Fiber diameter	Cell adhesion and differentiation	Shih *et al.*, 2006
	PCL, collagen, PES	Randomly oriented fibers	Hepatocyte-like cell differentiation	Kazemnejad *et al.*, 2009
Osteoprogenitor cell (MC3T3-E1)	PLLDO	Randomly oriented fibers	Cell adhesion and proliferation	Finne-Wistrand *et al.*, 2008
	PLA	Fiber diameter	Cell spreading, proliferation, and differentiation	Badami *et al.*, 2006
Embryonic stem cells	PCL	Fiber alignment	Neural cell differentiation	Xie *et al.*, 2009
Skeletal muscle cell	PCL/collagen	Fiber alignment	Cell alignment and myotube formation	Choi *et al.*, 2008
C2C12	PU	Fiber alignment, mechanical and electrical stimulation	Cell alignment and maturation	Liao *et al.*, 2008
DRG	PEU	Fiber alignment	Cell adhesion and alignment	Riboldi *et al.* 2005, 2008
	PDO	Fiber alignment	Neurite outgrowth	Chow *et al.*, 2007
	PLA	Fiber alignment	Neurite outgrowth	Wang *et al.*, 2009
DRG, Schwann cells, OEC, fibroblasts	PCL/collagen	Fiber alignment	Neurite outgrowth, cell adhesion and migration	Schnell *et al.*, 2007
Schwann cells	PCL	Fiber alignment	Cell maturation	Chew *et al.*, 2008
Neural stem cells	PES	Fiber diameter	Cell differentiation and proliferation	Christopherson *et al.*, 2009
Fibroblasts	PLGA	Fiber diameter and orientation	Cell adhesion and proliferation	Bashur *et al.*, 2006
	PMMA	Fiber alignment	Cell migration	Liu *et al.*, 2009
3T3 fibroblasts	PGA/collagen	Fiber diameter and alignment	Cell adhesion	Tian *et al.*, 2008
Vascular endothelial cells	PCL-PU	Fiber alignment	Cell attachment and proliferation	Williamson *et al.*, 2006
	PLCL	Fiber alignment	Phenotypic maintenance	He *et al.*, 2006

DRG: dorsal root ganglia, OEC: olfactory ensheathing cell, PCL: polycaprolactone, PDO: polydioxanone, PES: polyethersulfone, PEU: polyesterurethane, PEUU: poly(ester urethane)urea, PLA: polylactide, PLCL: poly(lactide-co-caprolactone), PLGA: poly(lactide-co-glicolide), PLLDO: poly(L-lactide)-*co*-(1,5,-dioxepan-2-one), PMMA: poly(methyl methacrylate), PU: polyurethane.

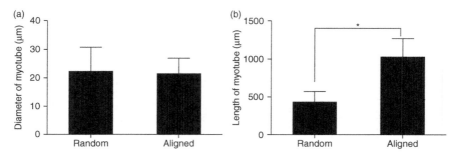

1.1 Human skeletal muscle cell morphologies at 7 days after cell differentiation on the electrospun PCL/collagen meshes: (a) diameter and (b) length of myotubes (*$P < 0.05$). (*Source*: Reproduced with permission from Choi *et al.*, *Biomaterials*, **29**, pp. 2899–906 © Elsevier (Choi *et al.*, 2008).)

from fused mononucleated muscle cells. It is well known that structure and organization of muscle fibers dictate tissue function. The ability to efficiently organize muscle cells to form aligned myotubes *in vitro* would greatly benefit skeletal muscle tissue engineering. There are many factors that can guide cellular growth and orientation, including uniaxial mechanical stimulation generated by a bioreactor. This stimulation facilitates muscle cell orientation and accelerates muscle tissue formation. It has been demonstrated that aligned nanofibers greatly influence muscle cell organization and enhance myotube formation. The myotubes formed on aligned nanofibrous scaffolds have highly organized and significantly longer morphology than myotubes formed on randomly oriented scaffolds, while the diameter of the myotubes has shown no significant difference (Fig. 1.1, Plate III) (Choi *et al.*, 2008). The aligned nanofibrous scaffolds are able to promote skeletal muscle morphogenesis into parallel-oriented myotubes. In combination with proper cell seeding, it may provide implantable functional muscle tissues that can closely mimic native ones for large muscle defects.

Electrospinning has evolved as a powerful tool in tissue engineering. This fabrication technology provides the ability to control biomaterial composition, fiber diameter, fiber alignment, geometry, and drug/protein incorporation into a scaffold. Nanofibers generated by electrospinning are able to support the adhesion and proliferation of a wide variety of cell types (Table 1.3); moreover, the cells are able to maintain their phenotypic and functional characteristics on nanofibrous scaffolds. Additionally, a growing body of evidence demonstrates that micro- to nano-scaled topography plays an important role in controlling the adhesion, proliferation, and survival of adult and embryonic stem cells in culture (Xie *et al.*, 2009). Therefore, electrospun nanofibrous scaffold can serve as a tool for studying the topographical aspects of cellular interactions that would lead to improved tissue formation. Furthermore, these scaffolds can be functionalized by adding biochemical and mechanical cues to enhance cellular interactions for tissue engineering applications.

Self-assembly methods involve the organization of molecules into highly ordered structures through specific and local interactions among the components without any external direction (Whitesides *et al.*, 1991). Biomolecules such as peptides, DNA, and RNA have been modified to stabilize into a structure without involving other materials. They can form structures such as nanoparticles or layered sheets to operate as diagnostic tools or self-releasing drug delivery scaffolds (Guo, 2010; Seeman, 2010; Liu *et al.*, 2011; Lee *et al.*, 2012). A peptide nanofibrous scaffold has been constructed through β (beta)-sheet self-assembly of interweaving nanofibers. Then chondrocytes have been encapsulated within the scaffold for cartilage repair. The scaffold has promoted morphogenesis and cartilage-like ECM formation rich in proteoglycans and type II collagen, indicating a stable chondrocyte phenotype. As ECM accumulated, mechanically functional new tissue has been deposited leading to increased material stiffness (Kisiday *et al.*, 2002). A short peptide (rabies virus glycoprotein (RVG)) has been used to deliver small interfering RNA (siRNA) through the blood-brain barrier to the brain, because of the specific binding to the acetylcholine receptor on the surface of neuronal cells. The RVG peptide has delivered siRNA to the neuronal cells, resulting in robust protection against fatal viral encephalitis in mice (Kumar *et al.*, 2007). The synthesis technique and functional efficiencies of biomolecule-based scaffolds requires further investigation for stable and optimized outcome.

1.4   Future trends

Biomedical nanomaterials have a central role and great potential as smart biomaterials in tissue engineering and regenerative medicine. The structural similarities to natural tissue in nanoscale create a familiar environment for adjacent cells and tissues, allowing them to behave as if they are interacting with natural tissues. Significant developments have been achieved by using nanostructured materials in both the diagnostic and therapeutic tissue engineering fields. Numerous types of synthetic and natural biomaterials offer functional and structural diversity. When used with proper fabrication techniques and in combination with appropriate cells, the potential of nanoscale materials can be infinite – from drug delivery to tissue scaffold applications.

The majority of biomedical nanomaterials is synthesized with nano-scale features. The size and shape can be controlled by differing fabrication methods, which can lead to a change in cell–material interaction. If the material has a biodegradable nature (e.g., aliphatic polyesters of collagens), such manipulation can contribute to tissue regeneration. However, if non-biodegradable materials (e.g. inorganic nanomaterials or carbon nanotubes) interact with cells or tissues, the aftermath is not predictable. In fact, several local and systemic toxicity issues of non-biodegradable nanomaterials have been reported including inflammation, fibrosis, oxidative stress, immune responses, and mutagenesis (van der Zande *et al.*, 2011). This can potentially lead to complications and failure in tissue restoration. To overcome these challenges and successfully proceed to clinical translation, the fundamental cause and the toxic mechanisms need to be elucidated. Proper modifications can then be made based on the source of toxicity. Many non-biodegradable materials are designed to be secreted through the circulatory, digestive, or integumentary system. The effect of modifications may be further assessed *in vivo* to avoid any potential risk.

Tissue engineering is an interdisciplinary field that embraces chemistry, engineering, biology, and medicine. To achieve successful tissue regeneration for clinical application, many collaborative studies on the synthesis, characterization, and application of many tissue engineering strategies are yet be completed. The main goal of using biomedical nanomaterials in tissue engineering is to accomplish structural and functional restoration of defective body parts by creating engineered constructs that are as close to native tissue as possible. Developing nanomaterials suitable for clinical use is critical in accomplishing this goal.

1.5   References

Agnihotri, S. A., Kulkarni, V. D., Kulkarni, A. R. and Aminabhavi, T. M. (2006). Degradation of chitosan and chemically modified chitosan by viscosity measurements. *J Appl Polym Sci*, **102**, 3255–8.

Agrawal, C. M., McKinney, J. S., Lanctot, D. and Athanasiou, K. A. (2000). Effects of fluid flow on the in vitro degradation kinetics of biodegradable scaffolds for tissue engineering. *Biomaterials*, **21**, 2443–52.

Ai, J., Biazar, E., Jafarpour, M., Montazeri, M., Majdi, A., Aminifard, S., Zafari, M., Akbari, H. R. and Rad, H. G. (2011). Nanotoxicology and nanoparticle safety in biomedical designs. *Int J Nanomedicine*, **6**, 1117–27.

Artzi, N., Oliva, N., Puron, C., Shitreet, S., Artzi, S., Bon Ramos, A., Groothuis, A., Sahagian, G. and Edelman, E. R. (2011). In vivo and in vitro tracking of erosion in biodegradable materials using non-invasive fluorescence imaging. *Nat Mater*, **10**, 704–9.

Atala, A. (2007). Engineering tissues, organs and cells. *J Tissue Eng Regen Med*, **1**, 83–96.

Atala, A. (2008). Bioengineered tissues for urogenital repair in children. *Pediatr Res*, **63**, 569–75.

Badami, A. S., Kreke, M. R., Thompson, M. S., Riffle, J. S. and Goldstein, A. S. (2006). Effect of fiber diameter on spreading, proliferation, and differentiation of osteoblastic cells on electrospun poly(lactic acid) substrates. *Biomaterials*, **27**, 596–606.

Bashur, C. A., Dahlgren, L. A. and Goldstein, A. S. (2006). Effect of fiber diameter and orientation on fibroblast morphology and proliferation on electrospun poly(D,L-lactic-co-glycolic acid) meshes. *Biomaterials*, **27**, 5681–8.

Bhattacharyya, S., Kudgus, R. A., Bhattacharya, R. and Mukherjee, P. (2011). Inorganic nanoparticles in cancer therapy. *Pharm Res*, **28**, 237–59.

Burg, K. J., Porter, S. and Kellam, J. F. (2000). Biomaterial developments for bone tissue engineering. *Biomaterials*, **21**, 2347–59.

Cellot, G., Cilia, E., Cipollone, S., Rancic, V., Sucapane, A., Giordani, S., Gambazzi, L., Markram, H., Grandolfo, M., Scaini, D., Gelain, F., Casalis, L., Prato, M., Giugliano, M. and Ballerini, L. (2009). Carbon nanotubes might improve neuronal performance by favouring electrical shortcuts. *Nat Nanotechnol*, **4**, 126–33.

Chen, H., Shao, L., Ming, T., Sun, Z., Zhao, C., Yang, B. and Wang, J. (2010). Understanding the photothermal conversion efficiency of gold nanocrystals. *Small*, **6**, 2272–80.

Chew, S. Y., Mi, R., Hoke, A. and Leong, K. W. (2008). The effect of the alignment of electrospun fibrous scaffolds on Schwann cell maturation. *Biomaterials*, **29**, 653–61.

Cho, E. C., Glaus, C., Chen, J., Welch, M. J. and Xia, Y. (2010). Inorganic nanoparticle-based contrast agents for molecular imaging. *Trends Mol Med*, **16**, 561–73.

Cho, H. and An, J. (2006). The effect of epsilon-caproyl/D,L-lactyl unit composition on the hydrolytic degradation of poly(D,L-lactide-ran-epsilon-caprolactone)-poly(ethylene glycol)-poly(D,L-lactide-ran-epsilon-caprolactone). *Biomaterials*, **27**, 544–52.

Choi, H. S. and Frangioni, J. V. (2010). Nanoparticles for biomedical imaging: Fundamentals of clinical translation. *Mol Imaging*, **9**, 291–310.

Choi, J. S., Lee, S. J., Christ, G. J., Atala, A. and Yoo, J. J. (2008). The influence of electrospun aligned poly(epsilon-caprolactone)/collagen nanofiber meshes on the formation of self-aligned skeletal muscle myotubes. *Biomaterials*, **29**, 2899–906.

Chow, W. N., Simpson, D. G., Bigbee, J. W. and Colello, R. J. (2007). Evaluating neuronal and glial growth on electrospun polarized matrices: Bridging the gap in percussive spinal cord injuries. *Neuron Glia Biol*, **3**, 119–126.

Christopherson, G. T., Song, H. and Mao, H. Q. (2009). The influence of fiber diameter of electrospun substrates on neural stem cell differentiation and proliferation. *Biomaterials*, **30**, 556–64.

Chua, K. N., Lim, W. S., Zhang, P., Lu, H., Wen, J., Ramakrishna, S., Leong, K. W. and Mao, H. Q. (2005). Stable immobilization of rat hepatocyte spheroids on galactosylated nanofiber scaffold. *Biomaterials*, **26**, 2537–47.

Conde, J., Doria, G. and Baptista, P. (2012). Noble metal nanoparticles applications in cancer. *J Drug Deliv*, **2012**, 751075.

Cross, M. and Dexter, T. M. (1991). Growth factors in development, transformation, and tumorigenesis. *Cell*, **64**, 271–80.

Day, E. S., Morton, J. G. and West, J. L. (2009). Nanoparticles for thermal cancer therapy. *J Biomech Eng*, **131**, 074001.

De Laporte, L. and Shea, L. D. (2007). Matrices and scaffolds for DNA delivery in tissue engineering. *Adv Drug Deliv Rev*, **59**, 292–307.

Dvir, T., Timko, B. P., Brigham, M. D., Naik, S. R., Karajanagi, S. S., Levy, O., Jin, H., Parker, K. K., Langer, R. and Kohane, D. S. (2011a). Nanowired three-dimensional cardiac patches. *Nat Nanotechnol*, **6**, 720–5.

Dvir, T., Timko, B. P., Kohane, D. S. and Langer, R. (2011b). Nanotechnological strategies for engineering complex tissues. *Nat Nanotechnol*, **6**, 13–22.

Elias, K. L., Price, R. L. and Webster, T. J. (2002). Enhanced functions of osteoblasts on nanometer diameter carbon fibers. *Biomaterials*, **23**, 3279–87.

Engler, A. J., Sen, S., Sweeney, H. L. and Discher, D. E. (2006). Matrix elasticity directs stem cell lineage specification. *Cell*, **126**, 677–89.

Farokhzad, O. C. and Langer, R. (2009). Impact of nanotechnology on drug delivery. *ACS Nano*, **3**, 16–20.

Finne-Wistrand, A., Albertsson, A. C., Kwon, O. H., Kawazoe, N., Chen, G., Kang, I. K., Hasuda, H., Gong, J. and Ito, Y. (2008). Resorbable scaffolds from three different techniques: Electrospun fabrics, salt-leaching porous films, and smooth flat surfaces. *Macromol Biosci*, **8**, 951–9.

Flemming, R. G., Murphy, C. J., Abrams, G. A., Goodman, S. L. and Nealey, P. F. (1999). Effects of synthetic micro- and nano-structured surfaces on cell behavior. *Biomaterials*, **20**, 573–88.

Furth, M. E., Atala, A. and Van Dyke, M. E. (2007). Smart biomaterials design for tissue engineering and regenerative medicine. *Biomaterials*, **28**, 5068–73.

Georges, P. C. and Janmey, P. A. (2005). Cell type-specific response to growth on soft materials. *J Appl Physiol*, **98**, 1547–53.

Ghosh, P., Han, G., De, M., Kim, C. K. and Rotello, V. M. (2008). Gold nanoparticles in delivery applications. *Adv Drug Deliv Rev*, **60**, 1307–15.

Goldberg, M., Langer, R. and Jia, X. (2007). Nanostructured materials for applications in drug delivery and tissue engineering. *J Biomater Sci Polym Ed*, **18**, 241–68.

Guo, P. (2010). The emerging field of RNA nanotechnology. *Nat Nanotechnol*, **5**, 833–42.

Hashi, C. K., Zhu, Y., Yang, G. Y., Young, W. L., Hsiao, B. S., Wang, K., Chu, B. and Li, S. (2007). Antithrombogenic property of bone marrow mesenchymal stem cells in nanofibrous vascular grafts. *Proc Natl Acad Sci U S A*, **104**, 11915–20.

He, W., Yong, T., Ma, Z. W., Inai, R., Teo, W. E. and Ramakrishna, S. (2006). Biodegradable polymer nanofiber mesh to maintain functions of endothelial cells. *Tissue Eng*, **12**, 2457–66.

Hu, H., Ni, Y., Mandal, S. K., Montana, V., Zhao, B., Haddon, R. C. and Parpura, V. (2005). Polyethyleneimine functionalized single-walled carbon nanotubes as a substrate for neuronal growth. *J Phys Chem B*, **109**, 4285–9.

Hu, H., Ni, Y., Montana, V., Haddon, R. C. and Parpura, V. (2004). Chemically functionalized carbon nanotubes as substrates for neuronal growth. *Nano Lett*, **4**, 507–11.

Huppertz, B. (2011). Nanoparticles: Barrier thickness matters. *Nat Nanotechnol*, **6**, 758–9.

Hutmacher, D. W. (2000). Scaffolds in tissue engineering bone and cartilage. *Biomaterials*, **21**, 2529–43.

Jain, P. K., Huang, X., El-Sayed, I. H. and El-Sayed, M. A. (2008). Noble metals on the nanoscale: Optical and photothermal properties and some applications in imaging, sensing, biology, and medicine. *Acc Chem Res*, **41**, 1578–86.

Jang, J. H., Rives, C. B. and Shea, L. D. (2005). Plasmid delivery in vivo from porous tissue-engineering scaffolds: Transgene expression and cellular transfection. *Mol Ther*, **12**, 475–83.

Ju, Y. M., Choi, J. S., Atala, A., Yoo, J. J. and Lee, S. J. (2010). Bilayered scaffold for engineering cellularized blood vessels. *Biomaterials*, **31**, 4313–21.

Kazemnejad, S., Allameh, A., Soleimani, M., Gharehbaghian, A., Mohammadi, Y., Amirizadeh, N. and Jazayery, M. (2009). Biochemical and molecular character-ization of hepatocyte-like cells derived from human bone marrow mesenchy-mal stem cells on a novel three-dimensional biocompatible nanofibrous scaf-fold. *J Gastroenterol Hepatol*, **24**, 278–87.

Kim, B. S. and Mooney, D. J. (1998). Development of biocompatible synthetic extra-cellular matrices for tissue engineering. *Trends Biotechnol*, **16**, 224–30.

Kim, K., Jeong, C. G. and Hollister, S. J. (2008). Non-invasive monitoring of tissue scaffold degradation using ultrasound elasticity imaging. *Acta Biomater*, **4**, 783–90.

Kim, S. S., Park, M. S., Gwak, S. J., Choi, C. Y. and Kim, B. S. (2006). Accelerated bone-like apatite growth on porous polymer/ceramic composite scaffolds in vitro. *Tissue Eng*, **12**, 2997–3006.

Kisiday, J., Jin, M., Kurz, B., Hung, H., Semino, C., Zhang, S. and Grodzinsky, A. J. (2002). Self-assembling peptide hydrogel fosters chondrocyte extracellular matrix production and cell division: Implications for cartilage tissue repair. *Proc Natl Acad Sci USA*, **99**, 9996–10001.

Ko, I. K., Ju, Y. M., Chen, T., Atala, A., Yoo, J. J. and Lee, S. J. (2012). Combined sys-temic and local delivery of stem cell inducing/recruiting factors for in situ tissue regeneration. *FASEB J*, **26**, 158–68.

Kumar, P., Wu, H., McBride, J. L., Jung, K. E., Kim, M. H., Davidson, B. L., Lee, S. K., Shankar, P. and Manjunath, N. (2007). Transvascular delivery of small interfer-ing RNA to the central nervous system. *Nature*, **448**, 39–43.

Lee, C. H., Shin, H. J., Cho, I. H., Kang, Y. M., Kim, I. A., Park, K. D. and Shin, J. W. (2005). Nanofiber alignment and direction of mechanical strain affect the ECM production of human ACL fibroblast. *Biomaterials*, **26**, 1261–70.

Lee, J. B., Hong, J., Bonner, D. K., Poon, Z. and Hammond, P. T. (2012). Self-assembled RNA interference microsponges for efficient siRNA delivery. *Nat Mater*, **11**, 316–22.

Lee, K., Silva, E. A. and Mooney, D. J. (2011). Growth factor delivery-based tissue engineering: general approaches and a review of recent developments. *J R Soc Interface*, **8**, 153–70.

Lee, K. S. and El-Sayed, M. A. (2006). Gold and silver nanoparticles in sensing and imaging: Sensitivity of plasmon response to size, shape, and metal composition. *J Phys Chem B*, **110**, 19220–5.

Lee, S. J., Liu, J., Oh, S. H., Soker, S., Atala, A. and Yoo, J. J. (2008). Development of a composite vascular scaffolding system that withstands physiological vascular conditions. *Biomaterials*, **29**, 2891–8.

Liao, I.-C., Liu, J. B., Bursac, N. and Leong, K. W. (2008). Effect of electromechanical stimulation on the maturation of myotubes on aligned electrospun fibers. *Cell Mol Bioeng*, **1**, 133–145.

Liu, L., Busuttil, K., Zhang, S., Yang, Y., Wang, C., Besenbacher, F. and Dong, M. (2011). The role of self-assembling polypeptides in building nanomaterials. *Phys Chem Chem Phys*, **13**, 17435–44.

Liu, Y., Franco, A., Huang, L., Gersappe, D., Clark, R. A. and Rafailovich, M. H. (2009). Control of cell migration in two and three dimensions using substrate morphology. *Exp Cell Res*, **315**(15), 2544–2557.

Lovat, V., Pantarotto, D., Lagostena, L., Cacciari, B., Grandolfo, M., Righi, M., Spalluto, G., Prato, M. and Ballerini, L. (2005). Carbon nanotube substrates boost neuronal electrical signaling. *Nano Lett*, **5**, 1107–10.

Lu, L., Peter, S. J., Lyman, M. D., Lai, H. L., Leite, S. M., Tamada, J. A., Vacanti, J. P., Langer, R. and Mikos, A. G. (2000). In vitro degradation of porous poly(L-lactic acid) foams. *Biomaterials*, **21**, 1595–605.

Luong-Van, E., Grondahl, L., Chua, K. N., Leong, K. W., Nurcombe, V. and Cool, S. M. (2006). Controlled release of heparin from poly(epsilon-caprolactone) electrospun fibers. *Biomaterials*, **27**, 2042–50.

Massobrio, G., Massobrio, P. and Martinoia, S. (2008). Modeling the neuron-carbon nanotube-ISFET junction to investigate the electrophysiological neuronal activity. *Nano Lett*, **8**, 4433–40.

Mattson, M. P., Haddon, R. C. and Rao, A. M. (2000). Molecular functionalization of carbon nanotubes and use as substrates for neuronal growth. *J Mol Neurosci*, **14**, 175–82.

Mazzatenta, A., Giugliano, M., Campidelli, S., Gambazzi, L., Businaro, L., Markram, H., Prato, M. and Ballerini, L. (2007). Interfacing neurons with carbon nanotubes: Electrical signal transfer and synaptic stimulation in cultured brain circuits. *J Neurosci*, **27**, 6931–6.

McMurray, R. J., Gadegaard, N., Tsimbouri, P. M., Burgess, K. V., McNamara, L. E., Tare, R., Murawski, K., Kingham, E., Oreffo, R. O. and Dalby, M. J. (2011). Nanoscale surfaces for the long-term maintenance of mesenchymal stem cell phenotype and multipotency. *Nat Mater*, **10**, 637–44.

Meng, J., Kong, H., Xu, H. Y., Song, L., Wang, C. Y. and Xie, S. S. (2005). Improving the blood compatibility of polyurethane using carbon nanotubes as fillers and its implications to cardiovascular surgery. *J Biomed Mater Res A*, **74**, 208–14.

Middleton, J. C. and Tipton, A. J. (2000). Synthetic biodegradable polymers as orthopedic devices. *Biomaterials*, **21**, 2335–46.

Murphy, C. L. and Lever, M. J. (2002). A ratiometric method of autofluorescence correction used for the quantification of Evans blue dye fluorescence in rabbit arterial tissues. *Exp Physiol*, **87**, 163–70.

Nakasa, T., Ishikawa, M., Shi, M., Shibuya, H., Adachi, N. and Ochi, M. (2010). Acceleration of muscle regeneration by local injection of muscle-specific microRNAs in rat skeletal muscle injury model. *J Cell Mol Med*, **14**, 2495–505.

Nayak, T. R., Jian, L., Phua, L. C., Ho, H. K., Ren, Y. and Pastorin, G. (2010). Thin films of functionalized multiwalled carbon nanotubes as suitable scaffold materials for stem cells proliferation and bone formation. *ACS Nano*, **4**, 7717–25.

Nishiyama, N. (2007). Nanomedicine: nanocarriers shape up for long life. *Nat Nanotechnol*, **2**, 203–4.

Oh, S. H., Park, S. C., Kim, H. K., Koh, Y. J., Lee, J. H. and Lee, M. C. (2011). Degradation behavior of 3D porous polydioxanone-b-polycaprolactone scaffolds fabricated using the melt-molding particulate-leaching method. *J Biomater Sci Polym Ed*, **22**, 225–237.

Park, J. B. and Lakes, R. S. (2007). *Biomaterials: An introduction*, New York, Springer.

Paszek, M. J. and Weaver, V. M. (2004). The tension mounts: mechanics meets morphogenesis and malignancy. *J Mammary Gland Biol Neoplasia*, **9**, 325–42.

Powers, M. J., Domansky, K., Kaazempur-Mofrad, M. R., Kalezi, A., Capitano, A., Upadhyaya, A., Kurzawski, P., Wack, K. E., Stolz, D. B., Kamm, R. and Griffith, L. G. (2002). A microfabricated array bioreactor for perfused 3D liver culture. *Biotechnol Bioeng*, **78**, 257–69.

Riboldi, S. A., Sadr, N., Pigini, L., Neuenschwander, P., Simonet, M., Mognol, P., Sampaolesi, M., Cossu, G. and Mantero, S. (2008). Skeletal myogenesis on highly orientated microfibrous polyesterurethane scaffolds. *J Biomed Mater Res A*, **84**, 1094–101.

Riboldi, S. A., Sampaolesi, M., Neuenschwander, P., Cossu, G. and Mantero, S. (2005). Electrospun degradable polyesterurethane membranes: potential scaffolds for skeletal muscle tissue engineering. *Biomaterials*, **26**, 4606–15.

Rujitanaroj, P.-O., Pimpha, N. and Supaphol, P. (2008). Wound-dressing materials with antibacterial activity from electrospun gelatin fiber mats containing silver nanoparticles. *Polymer*, **49**, 4723–32.

Saldanha, K. J., Piper, S. L., Ainslie, K. M., Kim, H. T. and Majumdar, S. (2008). Magnetic resonance imaging of iron oxide labelled stem cells: Applications to tissue engineering based regeneration of the intervertebral disc. *Eur Cell Mater*, **16**, 17–25.

Sau, T. K., Rogach, A. L., Jackel, F., Klar, T. A. and Feldmann, J. (2010). Properties and applications of colloidal nonspherical noble metal nanoparticles. *Adv Mater*, **22**, 1805–25.

Schnell, E., Klinkhammer, K., Balzer, S., Brook, G., Klee, D., Dalton, P. and Mey, J. (2007). Guidance of glial cell migration and axonal growth on electrospun nanofibers of poly-epsilon-caprolactone and a collagen/poly-epsilon-caprolactone blend. *Biomaterials*, **28**, 3012–25.

Seeman, N. C. (2010). Nanomaterials based on DNA. *Annu Rev Biochem*, **79**, 65–87.

Shih, Y. R., Chen, C. N., Tsai, S. W., Wang, Y. J. and Lee, O. K. (2006). Growth of mesenchymal stem cells on electrospun type I collagen nanofibers. *Stem Cells*, **24**, 2391–7.

Sperling, R. A. and Parak, W. J. (2010). Surface modification, functionalization and bioconjugation of colloidal inorganic nanoparticles. *Philos Transact A Math Phys Eng Sci*, **368**, 1333–83.

Sperling, R. A., Rivera Gil, P., Zhang, F., Zanella, M. and Parak, W. J. (2008). Biological applications of gold nanoparticles. *Chem Soc Rev*, **37**, 1896–908.

Teo, B. K., Goh, K. J., Ng, Z. J., Koo, S. and Yim, E. K. (2012). Functional reconstruction of corneal endothelium using nanotopography for tissue-engineering applications. *Acta Biomater*, **8**(8), 2941–52.

Tian, F., Hosseinkhani, H., Hosseinkhani, M., Khademhosseini, A., Yokoyama, Y., Estrada, G. G. and Kobayashi, H. (2008). Quantitative analysis of cell adhesion on aligned micro- and nanofibers. *J Biomed Mater Res A*, **84**, 291–9.

van der Zande, M., Junker, R., Walboomers, X. F. and Jansen, J. A. (2011). Carbon nanotubes in animal models: a systematic review on toxic potential. *Tissue Eng Part B Rev*, **17**, 57–69.

Veetil, J. V. and Ye, K. (2009). Tailored carbon nanotubes for tissue engineering applications. *Biotechnol Prog*, **25**, 709–21.

Wang, H. B., Mullins, M. E., Cregg, J. M., Hurtado, A., Oudega, M., Trombley, M. T. and Gilbert, R. J. (2009). Creation of highly aligned electrospun poly-L-lactic acid fibers for nerve regeneration applications. *J Neural Eng*, **6**, 016001.

Wang, J. W. and Hon, M. H. (2003). Preparation and characterization of pH sensitive sugar mediated (polyethylene glycol/chitosan) membrane. *J Mater Sci Mater Med*, **14**, 1079–88.

Webster, T. J. and Ejiofor, J. U. (2004). Increased osteoblast adhesion on nanophase metals: Ti, Ti6Al4V, and CoCrMo. *Biomaterials*, **25**, 4731–9.

Webster, T. J., Siegel, R. W. and Bizios, R. (1999). Osteoblast adhesion on nanophase ceramics. *Biomaterials*, **20**, 1221–7.

Wheeldon, I., Farhadi, A., Bick, A. G., Jabbari, E. and Khademhosseini, A. (2011). Nanoscale tissue engineering: Spatial control over cell-materials interactions. *Nanotechnology*, **22**, 212001.

Whitesides, G. M., Mathias, J. P. and Seto, C. T. (1991). Molecular self-assembly and nanochemistry: A chemical strategy for the synthesis of nanostructures. *Science*, **254**, 1312–9.

Williams, D. F. (2009). On the nature of biomaterials. *Biomaterials*, **30**, 5897–909.

Williamson, M. R., Black, R. and Kielty, C. (2006). PCL-PU composite vascular scaffold production for vascular tissue engineering: Attachment, proliferation and bioactivity of human vascular endothelial cells. *Biomaterials*, **27**, 3608–16.

Wise, J. K., Yarin, A. L., Megaridis, C. M. and Cho, M. (2009). Chondrogenic differentiation of human mesenchymal stem cells on oriented nanofibrous scaffolds: Engineering the superficial zone of articular cartilage. *Tissue Eng Part A*, **15**, 913–21.

Woo, K. M., Chen, V. J. and Ma, P. X. (2003). Nano-fibrous scaffolding architecture selectively enhances protein adsorption contributing to cell attachment. *J Biomed Mater Res A*, **67**, 531–7.

Wu, L. and Ding, J. (2004). In vitro degradation of three-dimensional porous poly(D,L-lactide-co-glycolide) scaffolds for tissue engineering. *Biomaterials*, **25**, 5821–30.

Xie, J., Willerth, S. M., Li, X., Macewan, M. R., Rader, A., Sakiyama-Elbert, S. E. and Xia, Y. (2009). The differentiation of embryonic stem cells seeded on electrospun nanofibers into neural lineages. *Biomaterials*, **30**, 354–62.

Xu, C., Inai, R., Kotaki, M. and Ramakrishna, S. (2004). Electrospun nanofiber fabrication as synthetic extracellular matrix and its potential for vascular tissue engineering. *Tissue Eng*, **10**, 1160–8.

Xu, M., West, E., Shea, L. D. and Woodruff, T. K. (2006). Identification of a stage-specific permissive in vitro culture environment for follicle growth and oocyte development. *Biol Reprod*, **75**, 916–23.

Yang, Y., Yiu, H. H. and El Haj, A. J. (2005). On-line fluorescent monitoring of the degradation of polymeric scaffolds for tissue engineering. *Analyst*, **130**, 1502–6.

Yeung, T., Georges, P. C., Flanagan, L. A., Marg, B., Ortiz, M., Funaki, M., Zahir, N., Ming, W., Weaver, V. and Janmey, P. A. (2005). Effects of substrate stiffness on cell morphology, cytoskeletal structure, and adhesion. *Cell Motil Cytoskeleton*, **60**, 24–34.

Yu, M. F., Files, B. S., Arepalli, S. and Ruoff, R. S. (2000). Tensile loading of ropes of single wall carbon nanotubes and their mechanical properties. *Phys Rev Lett*, **84**, 5552–5.

Part I
Fabrication of nanomaterials for tissue engineering applications

2

Synthesis of polymeric nanomaterials for biomedical applications

A. A. KULKARNI and P. S. RAO,
Brigham and Women's Hospital, USA

DOI: 10.1533/9780857097231.1.27

Abstract: This chapter focuses on two types of polymeric nanomaterials, nanoparticles and nano scaffolds. The first section will discuss various polymers that have been used to synthesize nanomaterials, their structures, properties, advantages and applications. The second part of this chapter discusses different methods of polymer nanoparticle synthesis such as solvent evaporation, emulsification-diffusion, salting out, supercritical fluid technology, polymerization, ion gelation and nanoprecipitation. Also it further describes the most commonly used methods for the preparation of polymeric nano scaffolds, including electrospinning, molecular self-assembly, phase separation and solvent casting methods. In the final part of the chapter, the techniques used for the detailed characterization of these nanomaterials will be discussed.

Key words: nanotechnology, polymers, polymeric nanomaterials, nanoparticles, polymeric scaffolds, drug delivery, tissue engineering.

2.1   Introduction

Comprehensive research over the last two decades has shown that significant advances in the nanotechnology field has resulted in paradigm shifting changes in healthcare and it continues as a major driving force in achieving a variety of evolutionary milestones in the field of medicine (Langer, 1990; LaVan et al., 2003; Shi et al., 2010, 2011; Tanner et al., 2011; Parveen et al., 2012). Polymer nanomaterials represent the largest and most versatile class of biomaterials that has been studied extensively and utilized in an increasing number of fields that include drug delivery (Hadjiargyrou and Chiu, 2008; Elsabahy and Wooley, 2012; Santo et al., 2012), imaging (Lu, 2010; Shokeen et al., 2011), and regenerative medicine and tissue engineering (Sokolsky-Papkov et al., 2007; Kohane and Langer, 2008). Effective drug delivery systems require that the active drug is transported to the desired location bypassing the biological barriers, thereby improving its pharmacodynamic and pharmacokinetic profiles without the need to alter the drug

27

entity (Davis *et al.*, 2008). In order to achieve this, several characteristics of the carrier system such as size, shape, stability, surface charge, composition, etc., need to be optimized since it plays a very important role (Petros and DeSimone, 2010). Significant advances in polymer chemistry and the development of sophisticated synthetic techniques have enabled the engineering of polymeric nanomaterials with a high degree of uniformity and precise control of their physicochemical properties (O'Reilly *et al.*, 2006; Iha *et al.*, 2009; Elsabahy and Wooley, 2012). The ease with which polymer structures can be modified has made possible the development of polymeric nanomaterials with the ability to incorporate targeting moiety, the ability to develop multicomponent systems that can encapsulate multiple drugs with different physical and chemical properties and the ability to deliver imaging as well as therapeutic modality and with desired biologically relevant characteristics such as degradation rates, *in vivo* stability, etc. (Abeylath *et al.*, 2011).

Tissue engineering is an interdisciplinary field that works at the interface of engineering, biology and medicine and deals with the development of biological mimics to restore, replace or enhance tissue or whole organ function (Langer and Vacanti, 1993). This field is built on the principle that cells seeded or recruited into three dimensional biocompatible scaffolds can grow into fully functional tissues under a controlled biomimetic environment (Shi *et al.*, 2010). In animal tissue, the cells are surrounded by a three dimensional nanofibrous support called the extracellular matrix (ECM) which functions not only as a cell support but also directs cell behavior via cell–ECM interactions (Stevens and George, 2005; Shi *et al.*, 2010). Furthermore, it also plays an important role in storing, releasing and activating a wide range of biological factors that help in cell–cell and cell–soluble factor interactions (Taipale and Keski-Oja, 1997; Shi *et al.*, 2010). The success of tissue engineering depends on the development of the scaffold system that resembles the complexity and functionality of the ECM, to function as a suitable substitute (Goldberg *et al.*, 2007). Biodegradable polymer materials have been explored for the design of scaffold systems due to the higher degree of synthetic flexibility that can be utilized to develop a tissue specific and complex support system. Several synthetic procedures were developed to engineer the three dimensional polymeric scaffolds.

In this chapter, we will review the techniques for synthesis of polymeric nanosized materials which have been utilized in drug delivery and tissue engineering applications. The first section will discuss various polymers that have been commonly used to synthesize nanomaterials, their structures, properties, advantages and applications. In the second part, we will describe the different methods for the synthesis of well-defined nanoparticles and nano scaffolds. Various methods for polymer nanoparticles synthesis include solvent evaporation, emulsification-diffusion, salting out, supercritical fluid technology, surfactant-free emulsion, polymerization, ion gelation and nanoprecipitation. The different methods for the preparation of scaffolds include electrospinning, molecular self-assembly, phase separation and the solvent casting method. The final section will describe techniques used for the detailed characterization of these nanomaterials.

2.2   Types of polymers used in nanomaterials

Polymers are most widely tested as biomaterials in fields that include drug delivery, tissue engineering scaffolds, medical device fabrication and diagnostics. Polymeric nanomaterials are becoming increasingly important in the biomedical field because they offer several advantages compared with other materials, such as high synthetic flexibility, high surface to volume ratio, high porosity with very small pore size, biodegradation, biocompatibility and suitable mechanical properties. There are different types of polymers that are used for designing nanomaterials based on the need and advantages they offer. The three main types of polymers that are used for nanomaterial synthesis are natural polymers, degradable synthetic polymers and non-degradable synthetic polymers. In Table 2.1 the most widely used polymers for nanomaterials' synthesis are listed.

2.3   Synthesis of polymeric nanoparticles

Polymeric nanoparticles are classified into nanosphere and nanocapsules. The drugs can be either encapsulated by or bound to the surface of the nanoparticle. In nanocapsules, the solid material surrounds the liquid or the semiliquid core. The core can be composed of oil encapsulating the lipophilic drugs or an aqueous phase encapsulating hydrophilic drugs (Fig. 2.1). In the nanosphere the entire mass is solid; this can be a lipophilic nanosphere, polyelectrolyte complex or nanogel.

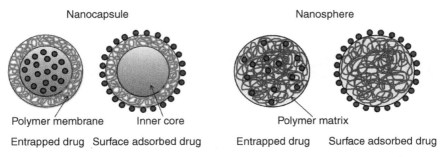

2.1 Types of polymeric nanoparticles: nanocapsules and nanospheres.

Table 2.1 Polymers used in nanomaterials

Type	Name	Structure	Properties	Applications
Natural polymers	Chitosan		Composed of β-1,4-linked-2amino-2-deoxy-D-glucopyranose, biocompatible. Good adhesion, coagulation and immunostimulating activity (Muzzarelli et al., 2005; Bravo-Osuna et al., 2007)	Transdermal drug delivery (Agnihotri and Aminabhavi, 2004; Agnihotri et al., 2004; Eroglu et al., 2006)
	Alginate		Linear polymers of (1–4) β-D-mannuronic acid and α-L-guluronic acid, isolated from bacteria and brown seaweeds. Gelling in water depends on carboxylic acid and bivalent counter ions interactions (Pawar and Edgar, 2012)	Delivery of drugs and proteins (Pawar and Edgar, 2012); cell-based therapies (Zimmermann et al., 2007)
	Hyaluronic acid		High molecular weight glycosaminoglycan, contains repeating units of N-acetylglucosamine and glucuronic acid (Chen and Abatangelo, 1999)	Long term delivery of peptides and proteins (Pouyani and Prestwich, 1994); wound dressing applications (Chen and Abatangelo, 1999); synthetic bone graft (Campoccia et al., 1998)

Material	Type	Description	Applications
Albumin	Protein	The pure form of human serum albumin is 66 kDa protein. Biodegradable, non-toxic, non-immunogenic (Chuang et al., 2002)	Stabilizing component in pharmaceuticals, coatings for medical devices, blood-volume expanders, drug carriers. The release kinetics of drug depends on the loading method used and degree of cross linking (Merodio et al., 2001; Arnedo et al., 2002)
Collagen	Protein	Abundantly available from living organisms and can be purified in pure form. Non-antigenic, biocompatible, biodegradable. Rod shaped protein with molecular weight of 300 kDa. Functional groups can be modified to tune biodegradation rates in vivo (Traub, 1971)	Tissue engineering applications such as cardiovascular, musculoskeletal and nerve tissue engineering; drug delivery applications (Lee et al., 2001; Buehler, 2006)
Gelatin	Protein	Derived from collagen, biocompatible, biodegradable (Chiellini et al., 2008)	Drug delivery applications (Young et al., 2005); cosmetic applications

(Continued)

Table 2.1 Continued

Type	Name	Structure	Properties	Applications
Degradable synthetic polymers	Poly(glycolic acid)		Linear aliphatic polyester, insoluble in water. Hydrolytically instable, non-toxic, good mechanical properties (Sokolsky-Papkov et al., 2007)	Implantable medical devices, tissue engineering and drug delivery applications (Kumari et al., 2010)
	Poly(D or L lactic acid)		Thermoplastic aliphatic polyester, chiral (Sokolsky-Papkov et al., 2007)	Medical implants, drug delivery and tissue engineering applications (Kumari et al., 2010)
	Poly(L,L-lactic acid)		Good mechanical properties such as tensile strength, improved suture (Sokolsky-Papkov et al., 2007)	Tissue engineering applications, orthopaedic fixation devices, high-strength fibers (Kumari et al., 2010)
	Poly(lactide-co-glycolide)		Biodegradable, biocompatible, FDA approved. Degradation rate and mechanical properties can be varied easily by changing monomer ratio and controlling degree of polymerization and molecular weight (Sokolsky-Papkov et al., 2007)	Drug delivery and tissue engineering applications (Kumari et al., 2010)

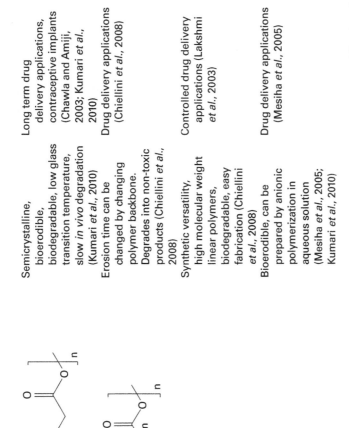

Poly(epsilon-caprolactone)	Semicrystalline, bioerodible, biodegradable, low glass transition temperature, slow *in vivo* degradation (Kumari et al., 2010)	Long term drug delivery applications, contraceptive implants (Chawla and Amiji, 2003; Kumari et al., 2010)
Polyanhydrides	Erosion time can be changed by changing polymer backbone. Degrades into non-toxic products (Chiellini et al., 2008)	Drug delivery applications (Chiellini et al., 2008)
Polyphosphazenes	Synthetic versatility, high molecular weight linear polymers, biodegradable, easy fabrication (Chiellini et al., 2008)	Controlled drug delivery applications (Lakshmi et al., 2003)
Poly(alkyl cyanoacrylate)	Bioerodible, can be prepared by anionic polymerization in aqueous solution (Mesiha et al., 2005; Kumari et al., 2010)	Drug delivery applications (Mesiha et al., 2005)

(*Continued*)

Table 2.1 Continued

Type	Name	Structure	Properties	Applications
Block co-polymers	Poly(lactide)-poly(ethylene glycol)		PEG is used to install stealth property in nanomaterials. In aqueous media self assembles into nanostructures (Sokolsky-Papkov et al., 2007)	Drug delivery applications
	Poly(lactide-co-glycolide)-poly(ethylene glycol)		Biocompatible, biodegradable, self assembles into nanostructures in water (Sokolsky-Papkov et al., 2007)	Most widely used for drug delivery applications
	Poly(epsilon-caprolactone)-poly(ethylene glycol)		Self assembles into nanostructures in water (Sokolsky-Papkov et al., 2007)	Drug delivery applications
Non-degradable synthetic polymers	Polyethylene (PE) High density polyethylene (HDPE)		High strength and lubricity, chemically resistant (Shastri, 2003)	Orthopaedic implants and catheters
	Polypropylene (PP)		Chemically resistant, rigid	Drug delivery; meshes and sutures

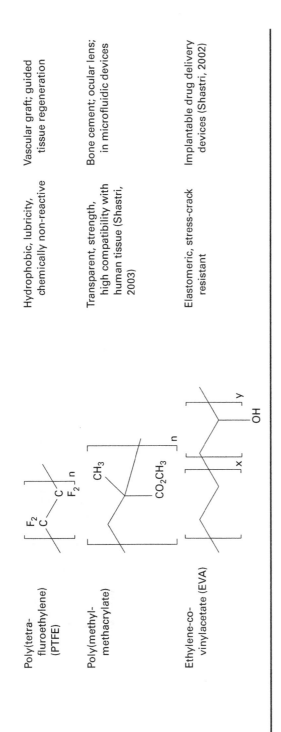

Poly(tetra-fluroethylene) (PTFE)	Hydrophobic, lubricity, chemically non-reactive	Vascular graft; guided tissue regeneration
Poly(methyl-methacrylate)	Transparent, strength, high compatibility with human tissue (Shastri, 2003)	Bone cement; ocular lens; in microfluidic devices
Ethylene-co-vinylacetate (EVA)	Elastomeric, stress-crack resistant	Implantable drug delivery devices (Shastri, 2002)

2.3.1   Solvent emulsion evaporation method

This method was developed in the late 1970s and is extensively used for nanoparticle preparation. This is a two-step method in which the first step includes the formation of an emulsion of polymer solution in an aqueous phase (Fig. 2.2). The second step is the evaporation of a solvent from the emulsified solution which results in the precipitation of the nanosphere. The polymer is dissolved in an organic solvent such as dichloromethane (DCM), chloroform or ethyl acetate, to which the drug is added. The resulting solution is emulsified in the aqueous solution containing the dispersing agent using a high speed homogenizer. The high speed stirring is necessary to reduce the size of emulsion droplets as this is directly related to the final size of the nanoparticle. The solvent is then evaporated at room temperature under pressure or by continuous stirring which results in nanoparticle formation (Zambaux *et al.*, 1998). Raghuvanshi *et al.* (2002) used the solvent evaporation method to obtain polylactic-co-glycolic acid (PLGA)–polylactic acids (PLA) entrapped tetanus toxoid nanoparticles which resulted in long lasting antibody response. Similarly, rhodium-loaded nanoparticles were prepared using a 1:1 (weight ratio) blend of PLA and polyethylene glycol (PEG)-g-PLA copolymer with varying PEG density ratios using the o/w emulsion solvent evaporation method. They found that a PEG density of 4–7% was necessary for PLA-g-PEG nanoparticle for an optimum drug delivery application (Essa *et al.*, 2011). Le Thi Mai Hoa *et al.* (2009) prepared a nanoparticle using the drug ketoprofen and co-polymer Eudragit E 10 using the solvent emulsion evaporation method. The drug and the Eudragit E 100, which is a stabilization agent, are dissolved in chloroform

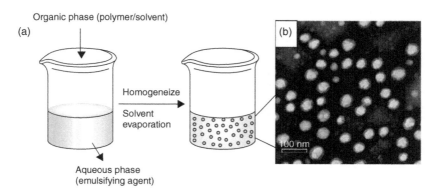

2.2 (a) Schematic representation of the emulsion solvent evaporation method. (b) TEM image of nanoparticles prepared by emulsion solvent evaporation method. (*Source*: TEM image reprinted from Jiang *et al.* (2011) with permission from Elsevier.)

and mixed along with sodium dodecyl sulfate and stirred until complete evaporation of solvent is achieved. Then the nanoparticle dispersions containing protective excipients glucose and lactose are frozen at −32°C for a minimum of 12 h and freeze dried at −55°C and 0.5 kPa for 24 h. Important parameters to consider are polymer molecular weight and concentration, drug and emulsifier concentration, sheer rate, organic to aqueous phase ratio, solvent nature and evaporation rate. Rahman *et al.* (2010) studied the effect of formulation and process variables on the characteristics of cyclosporine A (CyA)-PLGA nanoparticles. They found that emulsifier concentration, drug concentration and the type of organic solvent influenced the entrapment efficiency and drug concentration, and stirring rate affected the nanoparticle size. Julienne *et al.* (1992) found that different concentrations of PLGA (0.79%, 2.5%, 5% and 7.5% w/v) resulted in different mean nanoparticle sizes (220, 178, 177 and 236 nm, respectively). They also showed the effect of different organic to aqueous phase ratios on nanoparticle size; they found that a lower organic to aqueous phase resulted in smaller nanoparticle size (Julienne *et al.*, 1992).

2.3.2   Double emulsification solvent evaporation method

In this method, the aqueous solution containing a drug is first emulsified in the polymer-containing organic solution with sonication. The oil–water emulsion is then added to the aqueous phase containing a surfactant and emulsified using homogenization. The resulting mixture is then stirred at room temperature or heated under pressure for the evaporation of the solvent. The nanoparticles are obtained after separation and lyophilization (Fig. 2.3). The w/o/w emulsion is mainly used to entrap the hydrophilic compound. The important parameters for stable nanoparticle formation are drug concentration, sheer force, solvent and polymer molecular weight. Bilati *et al.* (2005) studied the effect of different process parameters on size and entrapment efficiency of peptide encapsulation in PLGA nanoparticles. They found that a higher molecular weight of polymer, increasing drug loading and presence of uncapped carboxylic acid increased the entrapment efficiency, while the higher polymer molecular weight and shorter mixing times at the second stage of emulsification increased the nanoparticle size. Lamprechet *et al.* (2000) found that an increase in polymer concentration led to increase in PLGA and poly(epsilon-caprolactone) (PCL) encapsulated protein nanoparticle size whereas they found that an increase in polyvinyl alcohol (PVA) concentration led to a smaller emulsion droplet and hence smaller nanoparticle size. High shear stress is a basic requirement for reducing the emulsion droplet size and hence making small nanoparticles

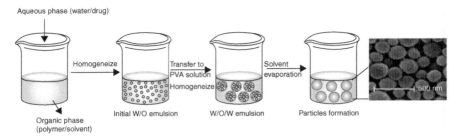

2.3 Schematic showing preparation of nanoparticles using the double emulsification solvent evaporation method. TEM showing spherical nanoparticles prepared by double emulsification solvent evaporation method. (*Source*: TEM image reprinted from McNeer *et al.* (2011) with permission from Elsevier.)

by double emulsion. Prabha and Labhasetwar (2004) prepared PLGA and PLA nanoparticle entrapping plasmid DNA using different polymer compositions and molecular weights: the molecular weight of 12 kDa and 143 kDa resulted in 563 + 6 nm and 375 + 22 nm nanoparticles whereas PLGA (molecular wt of 53 kDa) copolymer ratio of 75:25 and 50:50 resulted in 485 ± 11 nm and 685 ± 40 nm nanoparticles.

2.3.3   Emulsion diffusion method

This is a two-step method. The organic solution containing the polymer and drug is emulsified in an aqueous solution containing solvent and stabilizer using a homogenizer. An excess amount of water is added to the resulting mixture which enables the solvent to diffuse from the droplet, resulting in the formation of nanoparticles. The solvent and excess water is removed by lyophilization. The oil, drug and the polymer are usually soluble in the organic solvent whereas the water should be partially soluble to facilitate the diffusion followed by the dilution (Fig. 2.4). Checot *et al.* (2007) found that the final nanocapsule properties depended mainly on the oil to polymer ratio that fixed the shell thickness. The size of the primary emulsion and, in turn, the size of the nanoparticles depended on the mixing parameters (mixing device, shear rate). They also found that the molar mass of stabilizers did not influence the nanoparticle size but the increased stabilizer concentration resulted in an increase in nanoparticle size. Wanxue Ma and team (2011) used the emulsion diffusion method to prepare PLGA nanoparticles to encapsulate antigens for cancer treatment (Ma *et al.*, 2011). They found the encapsulating antigens in PLGA increased antigen loading, shelf life of antigens, stability of peptide by preventing its degradation, and made the delivery and targeting more feasible. Dongmei Cun and team (2010) encapsulated small

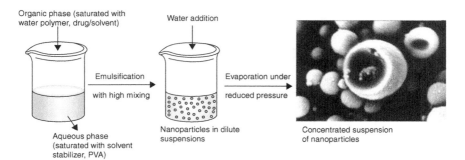

Organic phase (saturated with water polymer, drug/solvent)

Water addition

Emulsification with high mixing

Evaporation under reduced pressure

Aqueous phase (saturated with solvent stabilizer, PVA)

Nanoparticles in dilute suspensions

Concentrated suspension of nanoparticles

2.4 Schematic representation of the emulsion diffusion method. SEM showing nanocapsules synthesized by emulsion diffusion method. (*Source*: SEM figure reprinted from Olvera-Martinez *et al.* (2005) with permission from American Pharmaceutical Association.)

interfering RNA (siRNA) in PLGA using the double emulsion evaporation method which resulted in stable nanoparticles (Cun *et al.*, 2010). They found that the loading efficiency could be increased by changing the inner water phase volume, the concentration of polymer and sonication time.

2.3.4   Emulsion-reverse salting out method

This method is similar to the emulsion diffusion method except for the emulsion composition. The solvent used to dissolve the drug and polymer is usually a water miscible solvent like acetone (Fig. 2.5). The aqueous phase contains a salting out agent of electrolytes, typically magnesium chloride, calcium chloride and magnesium acetate, or non-electrolytes such as sucrose and colloidal stabilizers. The salting agent aids in the salting out of the organic solvent. The oil–water emulsion formed from the combination of aqueous and organic phase is further diluted with water or aqueous solution which enables solvent diffusion and thus induces the formation of nanoparticles. The addition of water reduces the ionic strength which aids in organic solvent diffusion and nanoparticle formation (Astete and Sabliov, 2006). Zheng Zhang and team (2006) synthesized dexamethasone loaded in poly(trimethylene carbonate) (PTMC) and monomethoxy poly(ethylene   glycol)-block-poly(trimethylene   carbonate)   (mPEG– PTMC) nanoparticles using single emulsion and the salting out method (Zhang *et al.*, 2006). They found that the size of the nanoparticle depended on polymer concentration and stirring speed. The diffusion coefficient of dexamethasone loaded in PTMC using the salting out method was higher than the single emulsion method, as the fast extraction of solvent during the salting out method resulted in a porous nanoparticle. Song *et al.* (2008)

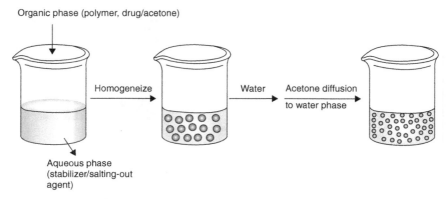

2.5 Schematic showing the emulsion-reverse salting out method.

optimized the encapsulation of two hydrophilic drugs, Vincristine sulfate (VCR) and Verapamil hydrochloride (VRP), in PLGA nanoparticles prepared through combining the single emulsion and salting out method. They studied the influence of different process parameters on nanoparticle size and entrapment efficiency. They found that the nanoparticle size increased with an increase in polymer molecular weight and concentration and decreased with an increased PVA concentration and increased acetone/dichloromethane (A/D) volume. The organic solvent removal rate and the salt concentration had a significant effect on nanoparticle size. The increase in PLGA concentration and molecular weight, salt concentration and organic solvent removal rate resulted in increased drug entrapment efficiency and an increase in PVA concentration. The initial concentrations of drug, w/o volume ratio and sonication time decreased the encapsulation efficiency. They found that combining the two methods of emulsion solvent evaporation and salting out resulted in higher drug entrapment efficiency. Song *et al.* (2009) found that co-encapsulating Vincristine and Verapamil in PLGA nanoparticles using the combined emulsion solvent evaporation and salting out methods resulted in less normal tissue toxicity and fewer drug to drug interactions.

2.3.5   Nanoprecipitation

Nanoprecipitation is also referred to as solvent displacement or the interfacial deposition method. It is a less extensive, less energy consuming and widely used procedure for nanoparticle formation. In this method, the particle formation is spontaneous and does not require the formation of emulsion or high speed mixing. The water miscible solvent, such

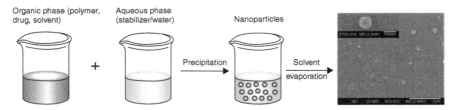

2.6 Schematic representation of the nanoprecipitation method. TEM image showing drug encapsulated PLGA–PEG nanoparticles. (*Source*: TEM image is reprinted from Graf *et al.* (2012) with permission from American Chemical Society.)

as acetone, containing dissolved polymer and the drug is added to a surfactant-containing aqueous phase, which results in instant formation of small drops of polymer containing drugs (Fig. 2.6). The nanoparticles are formed by the Marangoni effect which occurs at the interface of a solvent and non-solvent due to interfacial tension. The rapid diffusion of solvent results in polymer precipitation into nanoparticles. Usually surfactant or stabilizers are included to control the size and to ensure the stability of nanoparticles (Rong *et al.*, 2011). The advantages of nanoprecipitation are that it does not require high speed mixing, sonication or a very high temperature. The conventional nanoprecipitation method is more suitable for hydrophobic drugs but has less efficiency to encapsulate water-soluble drugs. Niu *et al.* (2009) encapsulated the plasmid DNA in PLGA nanoparticles using the modified nanoprecipitation method and double emulsion solvent evaporation method. In this method, they used dimethyl sulfoxide (DMSO) as a solvent instead of acetone to dissolve DNA and PLGA which was then injected into an aqueous stabilizer medium. They found that the DNA encapsulation in PLGA nanoparticles was higher in the modified nanoprecipitation method (97%) as compared to the double emulsion solvent evaporation method. Xie and Smith (2010) used fluidic nanoprecipitation system (FNPS) for synthesis of PLGA nanoparticles. They found that nanoparticles obtained using the FNPS method were highly uniform with the particle size of 148 ± 14 nm as compared to 211 ± 70 nm using the conventional nanoprecipitation method.

2.3.6   Emulsion-coacervation method

In this procedure, the emulsion is obtained by addition of an organic phase containing polymer and active substance with an aqueous phase containing stabilizers using mechanical stirring or ultrasound. Simple coacervation is done either by changing temperature or pH, by addition of electrolytes

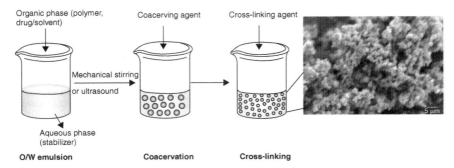

2.7 Schematic representation of the emulsion-coacervation method. SEM image showing poly-[bis(p-carboxyphenoxy)propane (PCPP) particles prepared using emulsion-coacervation method. (*Source*: SEM image is reprinted from Garlapati *et al.* (2009) with permission from Elsevier.)

or by adding materials that are not compatible with the polymer solution, resulting in polymer precipitation forming nanocapsules (Fig. 2.7). The cross linking agents are added to obtain the rigid shell structure of the nanocapsules. The coacervating agents lower the solvation of dissolved polymers and induce a thin solvated shell (Mora-Huertas *et al.*, 2010). Deda *et al.* (2009) encapsulated hydrophobic 5,10,15-triphenyl-20-(3-N-methylpyridinium-yl) porphyrin (3MMe porphyrin) in marine atelocollagen/xanthane gum using the coacervation method. They used Tween-20 in the oily phase to increase the dispersion of 3MMe porphyrin and resulted in size reduction of the nanoparticles. Anhydrous $NaSO_4$ was used to promote the salting out process which leads to coacervation. Gallarate *et al.* (2011) encapsulated insulin and leuprolide as model peptides within solid lipid nanoparticle (SLN) using coacervation. A known amount of peptide ethanol mixture was added to the aqueous mixture containing stearic acid. An acidifying solution (0.1 M HCl) was then added to induce nanoparticle formation using coacervation (Gallarate *et al.*, 2011). A slightly higher temperature was used to form a clear solution that did not hamper the chemical stability of peptides. Hao (2012) also used the above procedure to generate baicalin-loaded solid lipid nanoparticles which resulted in increased bioavailability of baicalin after oral administration (Hao *et al.*, 2012).

2.3.7   Polymerization method

This is one of the first methods used for the synthesis of polymeric nanoparticles. In this method, the nanoparticles are prepared by the polymerization of monomers (Fig. 2.8). The monomer is added into an emulsion and

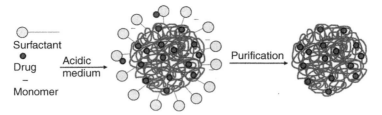

2.8 Schematic representation of nanoparticle production by anion polymerization.

the polymer is obtained by polymerization which, in turn, results in nano-particles being formed. A drug is dissolved in the polymerization medium either before the addition of the monomer or at the end of the polymer-ization reaction. The drug can also be attached or adsorbed into the fully formed nanoparticles. The nanoparticles can be purified to remove the sta-bilizers. Reese and Asher (2002) used small monodisperse colloidal parti-cles as a seed along with a monomer solution which was then polymerized to obtain nanoparticles. They used the emulsifier-free emulsion technique to synthesize latex consisting of poly(styrene-(*co*-2-hydroxyethyl meth-acrylate)) spherical particles with a surface functionalized with sulfate and carboxylic acid groups. Several monomers have been used to form the shell surrounding a magnetic nanoparticle by polymerization tech-niques. Chen *et al.* (2009) coated a biocompatible polysiloxane containing amphiphilic diblock copolymer, poly(ethylene oxide)-block-poly(γ-met hacryloxypropyltrimethoxysilane) (PEO-b-PγMPS) on iron oxide nano-particles (IONPs) to make the nanoparticles very stable for biomedical applications, using reversible addition fragmentation chain transfer poly-merization. Purushotham *et al.* (2009) coated poly-n-isopropylacrylamide (PNIPAM) on magnetic nanoparticles using dispersing free radical poly-merization methods.

2.3.8   Ionic gelation method

Some natural macromolecules can be used for the formation of nanopar-ticles. They are usually water-soluble polymers which are used for the syn-thesis of biodegradable nanoparticles employing the ionic gelation method. This involves the transition of liquid to gel due to ionic interaction. Some of the examples of nanoparticles prepared by this method include agarose nanoparticles, chitosan nanoparticles, albumin nanoparticles and gelatin nanoparticles. Gelatin nanoparticles are formed from gelatin emulsion by cooling the mixture below gelation point in an ice bath. Nagarwal *et al.*

(2012) synthesized chitosan-alginate-chitosan nanoparticles for entrapping 5-FU using the gelation method for ophthalmic delivery. In the first step, a solution of alginate containing 5-FU was prepared to which the chitosan solution was added dropwise. The colloidal solution, separated and redispersed in water, was freeze dried to get chitosan-sodium alginate (SA-CH) nanoparticles. In the second step chitosan is physically adsorbed on SA-CH nanoparticles. They found that the chitosan coated on CH-SA nanoparticles resulted in an increased bioavailability of drugs as compared to an uncoated one. Zhu *et al.* (2012) prepared thiolated CH-SA nanoparticles using the ionic gelation method for ocular drug delivery. They found that the thiolated CH-SA nanoparticles were more stable and had higher mucoadhesive properties and drug deliveries compared to CH-SA nanoparticles. Low molecular weight chitosan nanoparticles prepared using sodium tripolyphosphate (TPP) as a cross-linking agent were found to be highly stable and monodisperse for drug delivery applications (Trapani *et al.*, 2011; Fan *et al.*, 2012). An example is Methotrextrate (MTX) loaded chitosan or glycolchitosan nanoparticles was synthesized using ionic gelation method in the presence and absence of Tween-80 for brain treatment applications (Trapani *et al.*, 2011). The MTX-loaded chitosan nanoparticles were prepared using pentasodium tripolyphosphate (TPP) as a cross linking agent. The Tween-80 was added to the chitosan solution before the addition of MTX/TPP to maintain the Tween-80 concentration in the dispersion of 0.1% (v/v). Trapani *et al.* (2011) found that the presence of a low concentration of Tween-80 resulted in enhanced transport of MTX from the nanoparticles across the cells in *in vitro* studies.

2.3.9   Nanoparticles from supercritical fluid

In this procedure, the drug and polymer is mixed in a saturation vessel to form supercritical fluid and the resulting mixture is passed through an expansion nozzle (Fig. 2.9). The polymer solution expands, the supercritical fluid evaporates and the drug-containing polymer precipitates. This technique is clean because the precipitated solute is free of solvent. It also provides advantages such as suitable technological and biopharmaceutical properties and high quality products. It has been demonstrated for numerous applications involving protein drug delivery systems. However, this new process requires a high initial capital investment for equipment and elevated operating pressures require high pressure equipment. In addition, compressed supercritical fluids require elaborate recycling measures to reduce energy costs. Another disadvantage of this method is that most of the polymers have low solubility in supercritical fluid (Randolph *et al.*, 1993; Sun *et al.*, 2005).

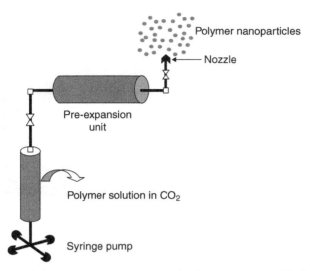

2.9 Schematic representation of polymer nanoparticle formation using supercritical fluid. (*Source*: Reprinted from J. Prasad Rao (2011) with permission from Elsevier.)

2.4   Synthesis of polymeric scaffolds

Polymers are the primary materials for various tissue engineering scaffolds such as skin, cartilage and bone. The synthesis of polymeric scaffolds that can mimic the structure and functions of natural ECMs are of great importance in tissue engineering applications. Many different methods have been introduced to synthesize the scaffolds, including electrospinning, molecular self-assembly, phase separation and the solvent casting or porogen leaching method.

2.4.1   Electrospinning

Electrospinning is widely used for the generation of nanofibers because of its simplicity, versatility and low cost. The high surface area and porosity of electrospun nanofibers makes them a potential candidate for tissue engineering applications. The typical set up of electrospinning is shown in Fig. 2.10. It usually contains a high voltage supply, a syringe with a small-diameter needle, a polymer solution or melt system and a melt collector.

In electrospinning the nanofibers are produced by the application of an electric field to the polymer solution. A strong electric field is applied between the needle tip containing the polymer solution and metallic collector which results in the charging of droplets at the capillary tip. Mutual

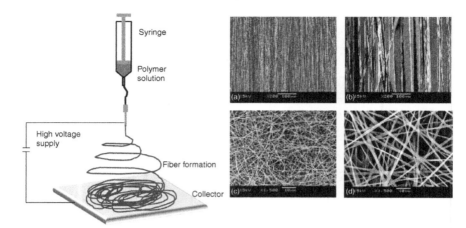

2.10 Schematic showing experimental set-up of the electrospinning method. SEM showing PLLA nanofibers: (a) aligned nanofibers, (b) aligned microfibers, (c) random nanofibers and (d) random microfibers. (*Source*: SEM images reprinted from Yang *et al.* (2005) with permission from Elsevier.)

repulsion and attraction between the charged droplets and collector can cause a force which acts against the surface tension. When the applied voltage exceeds the critical voltage, the repulsion electrostatic forces overcome the surface tension and the polymer solution jet is ejected towards the metallic collector which then forms nanofibers as the solvent evaporates. Yang *et al.* (2005) prepared a highly efficiently aligned poly (L-lactic acid) (PLLA) nanofibrous scaffold using the electrospinning method for neural tissue engineering applications. The solvents DCM and dimethylformamide (DMF) in the ratio 70:30 were used to dissolve the polymer. The concentration of the polymer solution was varied to optimize the diameter of the electrospun fibers. Stankus *et al.* (2008) synthesized a hybrid nanofibrous scaffold of a poly(ester-urethane) urea (PEUU) and urinary bladder matrix (UBM) using electrospinning for soft tissue engineering applications. They found the tensile strength and breaking strain increased with increased concentrations of PEUU. The incorporation of UBM increased the bioactivity of the scaffold which resulted in better cytocompatibility, cell adhesion and proliferation. Shafiee *et al.* (2011) used the electrospun method to synthesize a hybrid scaffold containing biomaterials (PVA/PCL) for seeding bone marrow mesenchymal stem cells (MSC) for cartilage regeneration applications. Two separate syringes and pumps were used to synthesize the hybrid PVA/PCL scaffold. They found that the hybrid PVA/PCL scaffold led to a higher viability, proliferation and initial attachment of MSC than the PCL nanofibers. Li *et al.* (2012) grafted chitosan onto the PLGA surface using cross-linking agents which were then used to obtain nanofibers using the electrospinning technique. They found that chitosan-graft-PLGA nanofibers improved the hydrophilicity, protein adsorption capacity and mechanical properties of the scaffold.

2.4.2   Molecular self-assembly

Self-assembling systems present attractive platforms for engineering materials with controlled nanostructures. The key requirement for molecular self-assembly is to design molecular building blocks which can assemble from the bottom up through the formation of non-covalent bonds such as hydrogen bonds, Van der Waals forces, electrostatic and hydrophobic interaction. Though the non-covalent bonds are individually weak, when they are working together they can form a strong bond which can hold an entire structure together. This method relies on intermolecular forces to assemble small units into fibers with diameters of approximately 5–10 nm, arranged into networks with a very high water content (99%). Self-assembly is a rather complex laboratory procedure that is limited to only a few polymer configurations (diblock copolymers, triblock copolymers, triblocks from peptide-amphiphile and dendrimers).

There are generally two classes of self-assembling construction subunits. The first class belongs to amphiphilic molecules from nature containing hydrophobic and hydrophilic segments which can form nanofibrous structures. The natural amphiphilic molecules are phospholipids, amphiphilic proteins, etc. The second class of peptides has a hydrophilic head and a hydrophobic tail, much like lipids or detergents. They sequester their hydrophobic tail inside micelles, vesicles or nanotube structures, and their hydrophilic heads are exposed to water (Liu and Zhao, 2011).

Several peptide scaffolds have been shown to support cell attachment, enhance cell survival and to induce cell differentiation for a variety of mammalian primary and tissue culture cells. Gelain *et al.* (2006) fabricated the 3-D nano scaffolds using the self-assembling peptide RADA 16 (Ac-RADARADARADARADA-COHN$_2$) to embed adult neural stem cells. It is easy to incorporate the desired cellular motifs in the scaffolds in order to enhance the neural cell survival. Zhao *et al.* (2010) studied the effect of differing the position of aromatic residue phenylalanine on the scaffold nanostructures using two self-assembly β-sheet peptides RADAFI ([CH$_3$CONH]–RADARADFRADARADA–[CONH$_2$]) and RADAFII ([CH$_3$CONH]–RADFRADARADARADA–[CONH$_2$]). He concluded that the significant change in self-assembly behaviors was observed by simple modification of the position of the phenylalanine group. The presence of π–π stacking resulted in a twisted morphology of nanofibers and hence a stronger network. Liuyun *et al.* (2009) prepared the nanocomposite of chitosan (CS)/carboxymethyl cellulose (CMC)/nano-hydroxyapatite (n-HA) using self-assembly through superficial static electricity interaction among CS, n-HA and CMC. The n-HA concentration was varied from 0% to 60% and an n-HA concentration of 40% was found to have the highest mechanical property for bone tissue regeneration.

2.4.3   Phase separation

Phase separation is a method for creating biocompatible scaffold matrices by precipitation of polymers from a polymer-poor phase and a polymer-rich phase. The advantage of the phase separation process is that it is a relatively simple procedure and requires minimal apparatus. In this process, the polymer is dissolved in solution and the phase separation is induced, either thermally or through the addition of a non-solvent to the polymer solution, to create a gel. The polymer solution under this condition become thermodynamically unstable and tends to separate into two phases. Water is then used to extract the solvent from the gel; the polymer-rich phase then solidifies on reducing the temperature to a 3-D porous composite scaffold. However, this method is limited to being effective with only a select number of polymers and with a low yield. Cheng and Kisaalita (2010) has fabricated the nanofibrous and microfibrous combined scaffold using thermally induced phase separation and particulate leaching techniques. They found that the combination of nano/micro scaffold had greater cellular activity than micro-only or nano-only scaffolds. The nanofibers improved neural differentiation while the microfibers facilitated cell infiltration. Wang *et al.* (2010) used silane modified nano-hydroxyapatite (MNHAP) and poly-L-lactic acid (PLLA) (80:20 w/w) to fabricate nanocomposites using thermally induced phase separation. They found that the PLLA/MNHAP composite nano scaffold had high porosity (>90%), compressive modulus (4.2 fold) and protein adsorption (2.8 fold) compared to the pure PLLA nano scaffold. Yang *et al.* (2004) fabricated the PLLA scaffold using the liquid–liquid phase separation method for nerve tissue engineering applications. They found that the nano scaffold prepared had high porosity, nano-diameter fibers and a high surface area. Vaquette and Cooper-White (2012) used the combination of electrospinning and thermally induced phase separation (TIPS) to obtain a thick 3-D scaffold of PCL and PLGA. The use of electrospinning and TIPS together enabled them to obtain a scaffold which was thicker (as compared to that obtained by electrospinning only) and with higher tensile strength (as compared to that obtained by TIPS only).

2.4.4   Solvent casting or porogen leaching method

This method is used to obtain scaffolds of uniform pore size, pore structure and porosity. In this method the organic solvent containing the polymer is poured into a 3-D mold containing a porogen such as pre-synthesized microspheres, salt or sugar particles. The solvent is then evaporated which results in polymer solidification around the porogen. The porogen is then leached away by rinsing with water or organic solvent to generate the 3-D scaffold. The advantage of this method is the ease of fabrication. The disadvantage is the use of organic solvent and the long period of rinsing in water which may result in loss of drug payloads from the controlled-release scaffold. Huang *et al.* (2007) synthesized nanoHA-PHEMA/PCL nanocomposites by incorporating nano-sized hydroxyapatite (nanoHA) into poly-2-hydroxyethylmethacrylate (PHEMA)/polycaprolactone (PCL) using the porogen leaching method to mimic natural bone. The result obtained was bioactive and supported the growth and proliferation of primary human osteoblast cells. Johari *et al.* (2012) synthesized the poly(ε-caprolactone) (PCL)/nano fluoridated hydroxyapatite (FHA) using the solvent casting method employing sodium chloride as the porogen for bone tissue engineering applications. They varied the ratio of the FHA and found that the biodegradation of the scaffold increased with increased FHA concentrations. They also found that cell viability increased with increasing porosity of the scaffold. The salient features of different methods used for polymeric nanoparticles is summarized in Table 2.2.

2.5   Characterization of the nanomaterials

The advances in polymer nanotechnology have enabled the development of a plethora of nanomaterials with varying composition, size, shape and surface characteristics. In order to understand the biological fate of each nanomaterial it is very important to know the effect of their physicochemical characteristics on the interactions with biological components. Furthermore, characterization of mechanical properties and the degradation rates of the polymer scaffold is essential for a range of tissue engineering applications. Several methods are used to characterize nanomaterial constructs after production and drug loading and are discussed below.

2.5.1   Structural characterization

The size distribution profiles of polymeric nanoparticles are determined by the dynamic light scattering technique (Mora-Huertas *et al.*, 2010). This technique measures the hydrodynamic radius of a spherical particle by shining

Table 2.2 Synthesis of polymeric nanoparticles

Method	Polymers used	Salient features	Active agent (Only selected)
Solvent emulsion evaporation method	PLA, PGA, PLGA	Oil-in-water (o/w) emulsification process, used primarily to encapsulate hydrophobic drugs, high speed homogenization or sonication required (Park et al., 2005)	Anticancer drugs (Fonseca et al., 2002; Kim and Martin, 2006), narcotic agents, local anesthetics, steroids, fertility control agents (McGinity and O'Donnell, 1997)
Double emulsification solvent evaporation method	PCL, PLA, PLGA	Water-in-oil-in-water (w/o/w) double emulsification method, can be used to encapsulate water soluble drugs, vigorous stirring is required (Lamprecht et al., 2000; Rosca et al., 2004; Bilati et al., 2005; McNeer et al., 2011)	Insulin, antibacterial agent (Ciprofloxacin.HCl, Penicilin G), protein (BSA), plasmid DNA, tetanus toxoid (Mora-Huertas et al., 2010)
Emulsion diffusion method	PCL, PLA	Three phases are required, organic, aqueous and dilution can be used to encapsulate lipophilic and hydrophilic drugs, PVA is generally used as stabilizing agent, easy scale up, nanoparticle size can be controlled (Quintanar-Guerrero et al., 1998, Checot et al., 2007)	Indomethacine, antibacterial agent (Hinokitinol), analgesic agent (Eugenol), antineoplastic agent (Chlorambucil Chlofibrate Vitamin E), estradiol, insulin (Csaba et al., 2004; Mora-Huertas et al., 2010; Sun et al., 2011)
Emulsion-reverse salting out method	PLA, PLGA	Can be used to encapsulate lipophilic drugs, high loading efficiency can be achieved, scale up is easy, purification is necessary to remove salting agents	Vincristine, Dexamethasone (Zhang et al., 2006; Song et al., 2008)

Method	Polymer/material	Advantages	Applications
Nanoprecipitation	PLA, PCL, PCL-PEG, PLGA, PLGA-PEG, Poly(N-acryolamide), Poly(methyl-methacrylate)	Most suitable for encapsulation of hydrophobic drugs, single step, easy procedure, reproducible, fast, easy to scale up, low concentration of surfactants (Song et al., 2008; Xie and Smith, 2010)	Paclitaxel (Chan et al., 2011), Docetaxel (Kolishetti et al., 2010), Gemcitabine, Indomethacin
Emulsion-coacervation method	Sodium alginate, Chitosan acetic acid, Gelatin	Nanomaterial from natural polymers, uses mechanical stirring or ultrasound, stabilization can be achieved by physical or covalent cross linking or by addition of cross linking agent (Hao et al., 2012)	Turmeric oil, Triamcinolone acetonide (Mora-Huertas et al., 2010)
Polymerization method	Cyanoacrylate monomer	Used for polymerization of alkylcyanoacrylates, easy method to synthesize core-shell tuned nanoparticles, size of the nanoparticle can be controlled by using surfactant, purification is required (Chen et al., 2009)	Magnetic nanoparticles, Radiolabeling agents, fluorescent dyes, peptides, siRNA (Reese and Asher, 2002; Purushotham et al., 2009)
Nanoparticle from supercritical fluid	PLA	No toxic organic solvent is required, so environmentally safe procedure, use supercritical fluids, very few polymers are soluble in supercritical fluids, cannot be used for high molecular weight (MW) polymers (Sun et al., 2005)	Phosporous, magnetic nanoparticles, carbon nanotubes (Shekunov et al., 2006; Byrappa et al., 2008)

the monochromatic light beam onto a solution with particles in Brownian motion, which changes the wavelength of incoming light in relation to the size of the particle. This method is widely used because of its several advantages such as short measurement duration, easy and reproducible procedure, fully automated instrumentation, simple sample preparation or no sample preparation required and a non-destructive, non-invasive method of measurement. Polymer nanoparticle morphology is determined by electron microscopy techniques, such as scanning electron microscopy (SEM), transmission electron microscopy (TEM) and atomic forced microscopy (AFM). Morphologies of the polymer scaffolds such as diameter, alignment and geometry that are used for tissue engineering applications are determined by SEM, TEM or AFM. Scanning electron microscopy is usually used for this purpose due to ease of use and easy accessibility, but since the image is generated by using an electron beam of 1–40 keV and could interact up to a depth of 1 μm, it has the potential to damage samples with diameters < 200 nm. Also, non-conductive samples that include polymers often suffer from limitations such as variations in surface potential which introduce astigmatism, instabilities and false X-ray signals. Additionally, obtaining good quality images using SEM is sometimes difficult due to a condition called charging during which charge accumulates on the surface of a nonconducting specimen and causes excessive brightness. To overcome these challenges, a thin coating of a conductive metal, such as fine gold, is often required which can lead to questionable accuracy for very thin fibers. TEM is generally used for analysis of smaller sized samples. Furthermore, TEM enables the visualization of the internal structure of samples such as pore size and distribution and provides two dimensional images magnified as much as 100 000 times by use of transmitted electrons. Average pore size, size distribution, pore volume, pore interconnectivity, pore shape, pore throat size and pore wall roughness are important parameters in polymeric scaffolds since they determine the amount of surface area that will allow cell growth and uniform cell distribution. To analyze the detailed pore geometry, the mercury porosimetry technique is used. The crystal structure of polymer fibers is determined by molecular structure X-ray diffraction. Fourier transform infrared spectroscopy (FTIR) or nuclear magnetic resonance (NMR) spectroscopy are commonly used to determine the functional groups if the chemical modifications are performed on the drug while synthesizing the nanoparticles.

2.5.2   Drug/active agent loading

The amount of drug encapsulated in polymer nanoparticles is determined by using spectroscopic techniques such as UV-spectroscopy or fluorescence spectroscopy, depending on the spectral signature of the active drugs. Often high performance liquid chromatography is used to determine the amount of drug encapsulated in the polymeric nanomaterial. Fluorescently labeled molecules can be used to determine the loading. In the case of polymeric scaffolds, the distribution of drugs or proteins can be determined by using fluorescently labeled molecules and observed with fluorescent microscopy or with attenuated total reflectance FTIR (Mora-Huertas *et al.*, 2010).

2.5.3 Surface characteristics (surface area, surface charge)

In polymeric nanoparticles which have been engineered for drug delivery applications, the surface properties in combination with size determine the way that the nanoparticles interact with their local environment, which ultimately determines the fate of the nanoparticles in the body. Since nanoparticles have high surface area to volume ratios compared to larger particles, the control of their surface properties is important in controlling their behavior *in vivo*. The interactions between nanoparticles and macrophages through non-specific interactions increase with a larger surface charge (either positive or negative) which leads to rapid and greater clearance by the reticuloendothelial system. This loss of nanoparticles to the undesired locations could be minimized by sterically stabilizing the surface with polymers, such as polyethylene glycol, or by controlling the surface charge. In tissue engineering, polymeric scaffolds provide the first point of contact and facilitate the attachment of cells in the surrounding tissue. Consequently, optimum surface properties are crucial. A high surface area to volume ratio is essential to accommodate the number of cells required for regeneration or restoration of the tissue or organ. The optimization of surface characteristics could lead to the enhancement of the biomaterials' performance.

Several techniques have been utilized to determine the surface properties of the nanoparticles which can be measured either by measuring the surface charge, quantifying the functional groups on the surface or changes in hydrophilicity of surface. The most widely used method for understanding the surface charges of the nanoparticles is to measure zeta potential of nanoparticles in aqueous suspension. The zeta potential may be positively charged or negatively charged depending on the type of polymer used or the type of functional modifications on the nanoparticle surface. X-ray photoelectron spectroscopy (XPS) or electron spectroscopy for chemical analysis (ESCA) is a quantitative spectroscopic technique which is widely used for surface elemental analysis. The analysis is performed by irradiating the material with the beam of X-rays causing the emission of electrons from the material surface (up to 100 Å) which is specific to the type of atom available on the surface of the nanoparticle (Soppimath *et al.*, 2001).

2.5.4 Drug release kinetics and degradation characteristics

Controlled release of the drug from the polymeric nanoparticles and subsequent degradation of the polymer into biocompatible components impacts the efficacy of nanoparticle formulation significantly in drug delivery applications. The drugs are released from the nanoparticles through one or a combination of processes, such as desorption of adsorbed drug, diffusion through the polymer matrix and/or erosion of the polymer matrix. Several methods were developed to understand the release of drugs from the polymer nanoparticles and to correlate their efficacy *in vivo* with release kinetics. One of the most widely used methods is the diffusion through dialysis membrane method. In this technique the nanoparticles are incubated in different buffer solutions or biologically relevant fluids and the drugs released are measured at predetermined time intervals. Other methods include the reverse dialysis sac technique, ultracentrifugation or ultrafiltration.

Biodegradable polymeric scaffolds are often used in tissue engineering as implants to support the tissue during the regeneration phase and are designed to degrade slowly so that the support is completely removed when the tissue is regenerated. The ideal scaffold degradation rate is the one which matches the growth of tissue. Scaffold degradation can occur through one or more processes that are mediated by physical or chemical parameters, such as pH or biological agents such as enzymes. The rate of polymer degradation is affected by several factors such as the chemical structure of the polymer, the presence of cleavable bonds and the copolymer ratio.

2.5.5 Biocompatibility

Biocompatibility is defined differently in different contexts. In broad terms,

> Biocompatibility is defined as the ability of a biomaterial to perform its desired function with respect to a medical therapy, without eliciting any undesirable or systemic effects in the recipient or beneficiary of that therapy, but generating the most appropriate beneficial cellular or tissue response in that specific situation, and optimizing the clinically relevant performance of the therapy. (Black, 2005)

'The biocompatibility of tissue engineering products such as scaffold or matrix refers to the ability to perform as a substrate that will support the appropriate cellular activity including the facilitation of molecular and mechanical signaling systems' (Onuki *et al.*, 2008). The factors that influence biocompatibility of polymeric nanomaterials are polymer compositions such as co-polymer ratio, chemical structure of the polymer, functional groups of the polymer, morphologies and synthetic processes of nanomaterials. Widely used polymers such as PLA, PGA and PLGA are commonly used as biodegradable materials for tissue engineering applications.

2.5.6   Mechanical characterization

The factors that govern the mechanical stability of the polymeric scaffolds are mechanical strength, degradation rate, elasticity and absorption at the material interface. Successful implantation therefore depends on the optimal mechanical properties of the polymer scaffold. The degradation rate of the polymer and the mechanical stability of the scaffold should be balanced such that the scaffold should endure the load of regenerated tissues but should degrade at a steady rate. The mechanical strength of the polymer scaffold is measured by tensile strength, Young's modulus of elasticity and maximum strain. The tensile strength of the material is the maximum stress it can withstand before suffering permanent deformation. The elastic modulus measures the strain in response to an applied tensile stress along the force. The factors that affect the mechanical properties of the polymer scaffolds depend on the polymer characteristics such as polymer length and cross-linking, as well as the porosity properties such as pore size, shape and orientation (Dhandayuthapani *et al.*, 2011).

2.6   Future trends

Polymeric nanomaterials have emerged as highly versatile biomaterials that hold great promise in revolutionizing various biomedical fields, including drug delivery and tissue engineering. The biggest advantage of polymeric nanomaterials is their synthetic versatility, as described in various synthetic procedures, which enables precise control of the size, shape, surface properties, geometry, composition, stability, degradation rates and mechanical strength. These factors have been shown to play a very significant role in predicting the biological outcome of the nanoparticle therapy which has increased our understanding of biological interactions of nanomaterials, and is a key to developing improved nanomaterials. Numerous preclinical studies underscore the potential of polymer nanomaterials in revolutionizing drug delivery by altering the pharmacokinetic and pharmacodynamic profile of the active agent, thereby enhancing the therapeutic efficacy while simultaneously reducing systemic toxicity.

Although large number of nanoparticle formulations have been reported in the preclinical studies in animal models, very few have translated into clinics. The challenges that need to be overcome in order to expedite the transition of nanomaterial from bench-side to bed-side include: (a) the bio-compatibility and biodegradability of polymers, (b) development of easy and reproducible synthetic procedures or bioconjugation techniques for the attachment of active agents, (c) optimization of nanoparticle surface chemistries in order to achieve longer circulation times, desired biodistribution profiles and tunable drug release kinetics, (d) development of highly stable formulations with increased shelf-life, (e) optimization of process chemistry parameters to develop easily scalable procedures and (f) development of better chemical and biological characterization techniques in order to avoid the discrepancies arising from batch to batch variations.

Tissue engineering is one of the fastest growing interdisciplinary fields and polymeric scaffolds play an important role in the advancement of this field. The studies on the design of polymeric scaffolds emphasize the development of materials which can act as biologically active support systems to restore, replace or enhance tissue or whole organ function. Some of the functions that these scaffolds are designed to perform include: promoting cell–cell and cell–ECM interactions; allowing transfer of biological factors to permit cell survival; signaling and differentiation; degrading into biocompatible components with a rate that corresponds to tissue regeneration; and showing no toxicity *in vivo*. Several fabrication techniques have been developed in order to design scaffolds with the desired structural, mechanical and porosity properties. Despite the progress in the design and fabrication of scaffolds, there are many limitations, such as biocompatibility, surface topology and mechanical properties, which need to be overcome in order to fabricate implants with the best performance.

2.7   References

Abeylath, S. C., Ganta, S., Iyer, A. K. and Amiji, M. (2011). Combinatorial-designed multifunctional polymeric nanosystems for tumor-targeted therapeutic delivery. *Acc Chem Res*, **44**, 1009–17.

Agnihotri, S. A. and Aminabhavi, T. M. (2004). Controlled release of clozapine through chitosan microparticles prepared by a novel method. *J Control Release*, **96**, 245–59.

Agnihotri, S. A., Mallikarjuna, N. N. and Aminabhavi, T. M. (2004). Recent advances on chitosan-based micro- and nanoparticles in drug delivery. *J Control Release*, **100**, 5–28.

Arnedo, A., Espuelas, S. and Irache, J. M. (2002). Albumin nanoparticles as carriers for a phosphodiester oligonucleotide. *Int J Pharm*, **244**, 59–72.

Astete, C. E. and Sabliov, C. M. (2006). Synthesis and characterization of PLGA nanoparticles. *J Biomater Sci Polym Ed*, **17**, 247–89.

Bilati, U., Allemann, E. and Doelker, E. (2005). Poly(D,L-lactide-co-glycolide) protein-loaded nanoparticles prepared by the double emulsion method – processing and formulation issues for enhanced entrapment efficiency. *J Microencapsul*, **22**, 205–14.

Black, J. (2005). *Biological Performance of Materials: Fundamentals of Biocompatibility*, New York, NY, Marcel Dekker.

Bravo-Osuna, I., Vauthier, C., Farabollini, A., Palmieri, G. F. and Ponchel, G. (2007). Mucoadhesion mechanism of chitosan and thiolated chitosan-poly(isobutyl cyanoacrylate) core-shell nanoparticles. *Biomaterials*, **28**, 2233–43.

Buehler, M. J. (2006). Nature designs tough collagen: explaining the nanostructure of collagen fibrils. *Proc Natl Acad Sci U S A*, **103**, 12285–90.

Byrappa, K., Ohara, S. and Adschiri, T. (2008). Nanoparticles synthesis using super-critical fluid technology – towards biomedical applications. *Adv Drug Deliv Rev*, **60**, 299–327.

Campoccia, D., Doherty, P., Radice, M., Brun, P., Abatangelo, G. and Williams, D. F. (1998). Semisynthetic resorbable materials from hyaluronan esterification. *Biomaterials*, **19**, 2101–27.

Chan, J. M., Rhee, J. W., Drum, C. L., Bronson, R. T., Golomb, G., Langer, R. and Farokhzad, O. C. (2011). In vivo prevention of arterial restenosis with paclitaxel-encapsulated targeted lipid-polymeric nanoparticles. *Proc Natl Acad Sci U S A*, **108**, 19347–52.

Chawla, J. S. and Amiji, M. M. (2003). Cellular uptake and concentrations of tamox-ifen upon administration in poly(epsilon-caprolactone) nanoparticles. *AAPS PharmSci*, **5**, E3.

Checot, F., Rodriguez-Hernandez, J., Gnanou, Y. and Lecommandoux, S. (2007). pH-responsive micelles and vesicles nanocapsules based on polypeptide diblock copolymers. *Biomol Eng*, **24**, 81–5.

Chen, H., Wu, X., Duan, H., Wang, Y. A., Wang, L., Zhang, M. and Mao, H. (2009). Biocompatible polysiloxane-containing diblock copolymer PEO-b-PγMPS for coating magnetic nanoparticles. *ACS Appl Mater Interfaces*, **1**, 2134–40.

Chen, W. Y. and Abatangelo, G. (1999). Functions of hyaluronan in wound repair. *Wound Repair Regen*, **7**, 79–89.

Cheng, K. and Kisaalita, W. S. (2010). Exploring cellular adhesion and differentiation in a micro-/nano-hybrid polymer scaffold. *Biotechnol Prog*, **26**, 838–46.

Chiellini, F., Piras, A. M., Errico, C. and Chiellini, E. (2008). Micro/nanostruc-tured polymeric systems for biomedical and pharmaceutical applications. *Nanomedicine (Lond)*, **3**, 367–93.

Chuang, V. T., Kragh-Hansen, U. and Otagiri, M. (2002). Pharmaceutical strategies utilizing recombinant human serum albumin. *Pharm Res*, **19**, 569–77.

Csaba, N., Gonzalez, L., Sanchez, A. and Alonso, M. J. (2004). Design and characteri-sation of new nanoparticulate polymer blends for drug delivery. *J Biomater Sci Polym Ed*, **15**, 1137–51.

Cun, D., Foged, C., Yang, M., Frokjaer, S. and Nielsen, H. M. (2010). Preparation and characterization of poly(DL-lactide-co-glycolide) nanoparticles for siRNA delivery. *Int J Pharm*, **390**, 70–5.

Dhandayuthapani, B., Yoshida, Y., Maekawa, T. and Kumar, D. S. (2011). Polymeric scaffolds in tissue engineering application: A review. *Int J Polym Sci*, **2011**, 1–19.

Davis, M. E., Chen, Z. G. and Shin, D. M. (2008). Nanoparticle therapeutics: An emerging treatment modality for cancer. *Nat Rev Drug Discov*, **7**, 771–82.

Deda, D. K., Uchoa, A. F., Carita, E., Baptista, M. S., Toma, H. E. and Araki, K. (2009). A new micro/nanoencapsulated porphyrin formulation for PDT treatment. *Int J Pharm*, **376**, 76–83.

Elsabahy, M. and Wooley, K. L. (2012). Design of polymeric nanoparticles for biomedical delivery applications. *Chem Soc Rev*, **41**, 2545–61.

Eroglu, M., Kursaklioglu, H., Misirli, Y., Iyisoy, A., Acar, A., Isin Dogan, A. and Denkbas, E. B. (2006). Chitosan-coated alginate microspheres for embolization and/or chemoembolization: In vivo studies. *J Microencapsul*, **23**, 367–76.

Essa, S., Rabanel, J. M. and Hildgen, P. (2011). Characterization of rhodamine loaded PEG-g-PLA nanoparticles (NPs): Effect of poly(ethylene glycol) grafting density. *Int J Pharm*, **411**, 178–87.

Fan, W., Yan, W., Xu, Z. and Ni, H. (2012). Formation mechanism of monodisperse, low molecular weight chitosan nanoparticles by ionic gelation technique. *Colloids Surf B Biointerfaces*, **90**, 21–7.

Fonseca, C., Simoes, S. and Gaspar, R. (2002). Paclitaxel-loaded PLGA nanoparticles: Preparation, physicochemical characterization and in vitro anti-tumoral activity. *J Control Release*, **83**, 273–86.

Gallarate, M., Battaglia, L., Peira, E. and Trotta, M. (2011). Peptide-loaded solid lipid nanoparticles prepared through coacervation technique. *Int J Chem Eng*, **2011**, 132435.

Garlapati, S., Facci, M., Polewicz, M., Strom, S., Babiuk, L. A., Mutwiri, G., Hancock, R. E., Elliott, M. R. and Gerdts, V. (2009). Strategies to link innate and adaptive immunity when designing vaccine adjuvants. *Vet Immunol Immunopathol*, **128**, 184–91.

Gelain, F., Bottai, D., Vescovi, A. and Zhang, S. (2006). Designer self-assembling peptide nanofiber scaffolds for adult mouse neural stem cell 3-dimensional cultures. *PLoS ONE*, **1**, e119.

Goldberg, M., Langer, R. and Jia, X. (2007). Nanostructured materials for applications in drug delivery and tissue engineering. *J Biomater Sci Polym Ed*, **18**, 241–68.

Graf, N., Bielenberg, D. R., Kolishetti, N., Muus, C., Banyard, J., Farokhzad, O. C. and Lippard, S. J. (2012). αVβ3 integrin-targeted PLGA-PEG nanoparticles for enhanced anti-tumor efficacy of a Pt(IV) prodrug. *ACS Nano*, **6**, 4530–9.

Hadjiargyrou, M. and Chiu, J. B. (2008). Enhanced composite electrospun nanofiber scaffolds for use in drug delivery. *Expert Opin Drug Deliv*, **5**, 1093–106.

Hao, J., Wang, F., Wang, X., Zhang, D., Bi, Y., Gao, Y., Zhao, X. and Zhang, Q. (2012). Development and optimization of baicalin-loaded solid lipid nanoparticles prepared by coacervation method using central composite design. *Eur J Pharm Sci*, **47**, 497–505.

Huang, J., Lin, Y. W., Fu, X. W., Best, S. M., Brooks, R. A., Rushton, N. and Bonfield, W. (2007). Development of nano-sized hydroxyapatite reinforced composites for tissue engineering scaffolds. *J Mater Sci Mater Med*, **18**, 2151–7.

Iha, R. K., Wooley, K. L., Nystrom, A. M., Burke, D. J., Kade, M. J. and Hawker, C. J. (2009). Applications of orthogonal 'click' chemistries in the synthesis of functional soft materials. *Chem Rev*, **109**, 5620–86.

Jiang, X., Sha, X., Xin, H., Chen, L., Gao, X., Wang, X., Law, K., Gu, J., Chen, Y., Jiang, Y., Ren, X., Ren, Q. and Fang, X. (2011). Self-aggregated pegylated poly (trimethylene carbonate) nanoparticles decorated with c(RGDyK) peptide for targeted paclitaxel delivery to integrin-rich tumors. *Biomaterials*, **32**, 9457–69.

Johari, N., Fathi, M. H., Golozar, M. A., Erfani, E. and Samadikuchaksaraei, A. (2012). Poly(epsilon-caprolactone)/nano fluoridated hydroxyapatite scaffolds for bone tissue engineering: In vitro degradation and biocompatibility study. *J Mater Sci Mater Med*, **23**, 763–70.

Julienne, M. A., Gomez Amoza, J. and Benoit, J. (1992). Preparation of poly(D,L-Lactide/glycolide) nanoparticles of controlled particle size distribution: Application of experimental design. *Drug Dev Ind Pharm*, **18**, 1063–77.

Kim, D. H. and Martin, D. C. (2006). Sustained release of dexamethasone from hydrophilic matrices using PLGA nanoparticles for neural drug delivery. *Biomaterials*, **27**, 3031–7.

Kohane, D. S. and Langer, R. (2008). Polymeric biomaterials in tissue engineering. *Pediatr Res*, **63**, 487–91.

Kolishetti, N., Dhar, S., Valencia, P. M., Lin, L. Q., Karnik, R., Lippard, S. J., Langer, R. and Farokhzad, O. C. (2010). Engineering of self-assembled nanoparticle platform for precisely controlled combination drug therapy. *Proc Natl Acad Sci U S A*, **107**, 17939–44.

Kumari, A., Yadav, S. K. and Yadav, S. C. (2010). Biodegradable polymeric nanoparticles based drug delivery systems. *Colloids Surf B Biointerfaces*, **75**, 1–18.

Lakshmi, S., Katti, D. S. and Laurencin, C. T. (2003). Biodegradable polyphosphazenes for drug delivery applications. *Adv Drug Deliv Rev*, **55**, 467–82.

Lamprecht, A., Ubrich, N., Hombreiro Perez, M., Lehr, C., Hoffman, M. and Maincent, P. (2000). Influences of process parameters on nanoparticle preparation performed by a double emulsion pressure homogenization technique. *Int J Pharm*, **196**, 177–82.

Langer, R. (1990). New methods of drug delivery. *Science*, **249**, 1527–33.

Langer, R. and Vacanti, J. P. (1993). Tissue engineering. *Science*, **260**, 920–6.

Lavan, D. A., McGuire, T. and Langer, R. (2003). Small-scale systems for in vivo drug delivery. *Nat Biotechnol*, **21**, 1184–91.

Le Thi Mai Hoa, Nguyen Tai Chi, Nguyen Minh Triet, Le Ngoc Thanh Nhan and Dang Mau Chien (2009). Preparation of drug nanoparticles by emulsion evaporation method. *J Phys: Conf Ser*, **187**, 012047.

Lee, C. H., Singla, A. and Lee, Y. (2001). Biomedical applications of collagen. *Int J Pharm*, **221**, 1–22.

Li, A. D., Sun, Z. Z., Zhou, M., Xu, X. X., Ma, J. Y., Zheng, W., Zhou, H. M., Li, L. and Zheng, Y. F. (2012). Electrospun chitosan-graft-PLGA nanofibres with significantly enhanced hydrophilicity and improved mechanical property. *Colloids Surf B Biointerfaces*, **102C**, 674–81.

Liu, J. and Zhao, X. (2011). Design of self-assembling peptides and their biomedical applications. *Nanomedicine (Lond)*, **6**, 1621–43.

Liuyun, J., Yubao, L. and Chengdong, X. (2009). A novel composite membrane of chitosan-carboxymethyl cellulose polyelectrolyte complex membrane filled with nano-hydroxyapatite I. Preparation and properties. *J Mater Sci Mater Med*, **20**, 1645–52.

Lu, Z. R. (2010). Molecular imaging of HPMA copolymers: Visualizing drug delivery in cell, mouse and man. *Adv Drug Deliv Rev*, **62**, 246–57.

Ma, W., Smith, T., Bogin, V., Zhang, Y., Ozkan, C., Ozkan, M., Hayden, M., Schroter, S., Carrier, E., Messmer, D., Kumar, V. and Minev, B. (2011). Enhanced presentation of MHC class Ia, Ib and class II-restricted peptides encapsulated in biodegradable nanoparticles: a promising strategy for tumor immunotherapy. *J Transl Med*, **9**, 34.

McGinity, J. W. and O'Donnell, P. B. (1997). Preparation of microspheres by the solvent evaporation technique. *Adv Drug Deliv Rev*, **28**, 25–42.

McNeer, N. A., Schleifman, E. B., Glazer, P. M. and Saltzman, W. M. (2011). Polymer delivery systems for site-specific genome editing. *J Control Release*, **155**, 312–6.

Merodio, M., Arnedo, A., Renedo, M. J. and Irache, J. M. (2001). Ganciclovir-loaded albumin nanoparticles: Characterization and in vitro release properties. *Eur J Pharm Sci*, **12**, 251–9.

Mesiha, M. S., Sidhom, M. B. and Fasipe, B. (2005). Oral and subcutaneous absorption of insulin poly(isobutylcyanoacrylate) nanoparticles. *Int J Pharm*, **288**, 289–93.

Mora-Huertas, C. E., Fessi, H. and Elaissari, A. (2010). Polymer-based nanocapsules for drug delivery. *Int J Pharm*, **385**, 113–42.

Muzzarelli, R. A., Guerrieri, M., Goteri, G., Muzzarelli, C., Armeni, T., Ghiselli, R. and Cornelissen, M. (2005). The biocompatibility of dibutyryl chitin in the context of wound dressings. *Biomaterials*, **26**, 5844–54.

Nagarwal, R. C., Kumar, R. and Pandit, J. K. (2012). Chitosan coated sodium alginate-chitosan nanoparticles loaded with 5-FU for ocular delivery: In vitro characterization and in vivo study in rabbit eye. *Eur J Pharm Sci*, **47**, 678–85.

Niu, X., Zou, W., Liu, C., Zhang, N. and Fu, C. (2009). Modified nanoprecipitation method to fabricate DNA-loaded PLGA nanoparticles. *Drug Dev Ind Pharm*, **35**, 1375–83.

O'Reilly, R. K., Hawker, C. J. and Wooley, K. L. (2006). Cross-linked block copolymer micelles: Functional nanostructures of great potential and versatility. *Chem Soc Rev*, **35**, 1068–83.

Olvera-Martinez, B. I., Cazares-Delgadillo, J., Calderilla-Fajardo, S. B., Villalobos-Garcia, R., Ganem-Quintanar, A. and Quintanar-Guerrero, D. (2005). Preparation of polymeric nanocapsules containing octyl methoxycinnamate by the emulsification-diffusion technique: Penetration across the stratum corneum. *J Pharm Sci*, **94**, 1552–9.

Onuki, Y., Bhardwaj, U., Papadimitrakopoulos, F. and Burgess, D. J. (2008). A review of the biocompatibility of implantable devices: Current challenges to overcome foreign body response. *J Diabetes Sci Technol*, **2**, 1003–15.

Park, J. H., Ye, M. and Park, K. (2005). Biodegradable polymers for microencapsulation of drugs. *Molecules*, **10**, 146–61.

Parveen, S., Misra, R. and Sahoo, S. K. (2012). Nanoparticles: A boon to drug delivery, therapeutics, diagnostics and imaging. *Nanomedicine*, **8**, 147–66.

Pawar, S. N. and Edgar, K. J. (2012). Alginate derivatization: A review of chemistry, properties and applications. *Biomaterials*, **33**, 3279–305.

Petros, R. A. and Desimone, J. M. (2010). Strategies in the design of nanoparticles for therapeutic applications. *Nat Rev Drug Discov*, **9**, 615–27.

Pouyani, T. and Prestwich, G. D. (1994). Functionalized derivatives of hyaluronic acid oligosaccharides: Drug carriers and novel biomaterials. *Bioconjug Chem*, **5**, 339–47.

Prabha, S. and Labhasetwar, V. (2004). Critical determinants in PLGA/PLA nanoparticle-mediated gene expression. *Pharm Res*, **21**, 354–64.

Prasad Rao, J. and Geckler, K. E. (2011). Polymer nanoparticles: Preparation techniques and size-control parameters. *Prog Polym Sci*, **36**, 887–913.

Purushotham, S., Chang, P. E., Rumpel, H., Kee, I. H., Ng, R. T., Chow, P. K., Tan, C. K. and Ramanujan, R. V. (2009). Thermoresponsive core-shell magnetic nanoparticles for combined modalities of cancer therapy. *Nanotechnology*, **20**, 305101.

Quintanar-Guerrero, D., Allemann, E., Doelker, E. and Fessi, H. (1998). Preparation and characterization of nanocapsules from preformed polymers by a new process based on emulsification-diffusion technique. *Pharm Res*, **15**, 1056–62.
Raghuvanshi, R. S., Katare, Y. K., Lalwani, K., Ali, M. M., Singh, O. and Panda, A. K. (2002). Improved immune response from biodegradable polymer particles entrapping tetanus toxoid by use of different immunization protocol and adjuvants. *Int J Pharm*, **245**, 109–21.
Rahman, Z., Zidan, A. S., Habib, M. J. and Khan, M. A. (2010). Understanding the quality of protein loaded PLGA nanoparticles variability by Plackett-Burman design. *Int J Pharm*, **389**, 186–94.
Randolph,T.W.,Randolph,A.D.,Mebes,M.and Yeung,S.(1993).Sub-micrometer-sized biodegradable particles of poly(L-lactic acid) via the gas antisolvent spray precipitation process. *Biotechnol Prog*, **9**, 429–35.
Reese, C. E. and Asher, S. A. (2002). Emulsifier-free emulsion polymerization produces highly charged, monodisperse particles for near infrared photonic crystals. *J Colloid Interface Sci*, **248**, 41–6.
Rong, X., Xie, Y., Hao, X., Chen, T., Wang, Y. and Liu, Y. (2011). Applications of polymeric nanocapsules in field of drug delivery systems. *Curr Drug Discov Technol*, **8**, 173–87.
Rosca, I. D., Watari, F. and Uo, M. (2004). Microparticle formation and its mechanism in single and double emulsion solvent evaporation. *J Control Release*, **99**, 271–80.
Santo, V. E., Gomes, M. E., Mano, J. F. and Reis, R. L. (2012). From nano- to macro-scale: nanotechnology approaches for spatially controlled delivery of bioactive factors for bone and cartilage engineering. *Nanomedicine (Lond)*, **7**, 1045–66.
Shafiee, A., Soleimani, M., Chamheidari, G. A., Seyedjafari, E., Dodel, M., Atashi, A. and Gheisari, Y. (2011). Electrospun nanofiber-based regeneration of cartilage enhanced by mesenchymal stem cells. *J Biomed Mater Res A*, **99**, 467–78.
Shastri, P. V. (2002). Toxicology of polymers for implant contraceptives for women. *Contraception*, **65**, 9–13.
Shastri, V. P. (2003). Non-degradable biocompatible polymers in medicine: Past, present and future. *Curr Pharm Biotechnol*, **4**, 331–7.
Shekunov, B. Y., Chattopadhyay, P., Seitzinger, J. and Huff, R. (2006). Nanoparticles of poorly water-soluble drugs prepared by supercritical fluid extraction of emulsions. *Pharm Res*, **23**, 196–204.
Shi, J., Votruba, A. R., Farokhzad, O. C. and Langer, R. (2010). Nanotechnology in drug delivery and tissue engineering: From discovery to applications. *Nano Lett*, **10**, 3223–30.
Shi, J., Xiao, Z., Kamaly, N. and Farokhzad, O. C. (2011). Self-assembled targeted nanoparticles: Evolution of technologies and bench to bedside translation. *Acc Chem Res*, **44**, 1123–34.
Shokeen, M., Pressly, E. D., Hagooly, A., Zheleznyak, A., Ramos, N., Fiamengo, A. L., Welch, M. J., Hawker, C. J. and Anderson, C. J. (2011). Evaluation of multivalent, functional polymeric nanoparticles for imaging applications. *ACS Nano*, **5**, 738–47.
Sokolsky-Papkov, M., Agashi, K., Olaye, A., Shakesheff, K. and Domb, A. J. (2007). Polymer carriers for drug delivery in tissue engineering. *Adv Drug Deliv Rev*, **59**, 187–206.

Song, X., Zhao, Y., Wu, W., Bi, Y., Cai, Z., Chen, Q., Li, Y. and Hou, S. (2008). PLGA nanoparticles simultaneously loaded with vincristine sulfate and verapamil hydrochloride: Systematic study of particle size and drug entrapment efficiency. *Int J Pharm*, **350**, 320–9.

Song, X. R., Cai, Z., Zheng, Y., He, G., Cui, F. Y., Gong, D. Q., Hou, S. X., Xiong, S. J., Lei, X. J. and Wei, Y. Q. (2009). Reversion of multidrug resistance by co-encapsulation of vincristine and verapamil in PLGA nanoparticles. *Eur J Pharm Sci*, **37**, 300–5.

Soppimath, K. S., Aminabhavi, T. M., Kulkarni, A. R. and Rudzinski, W. E. (2001). Biodegradable polymeric nanoparticles as drug delivery devices. *J Control Release*, **70**, 1–20.

Stankus, J. J., Freytes, D. O., Badylak, S. F. and Wagner, W. R. (2008). Hybrid nano-fibrous scaffolds from electrospinning of a synthetic biodegradable elastomer and urinary bladder matrix. *J Biomater Sci Polym Ed*, **19**, 635–52.

Stevens, M. M. and George, J. H. (2005). Exploring and engineering the cell surface interface. *Science*, **310**, 1135–8.

Sun, S., Liang, N., Kawashima, Y., Xia, D. and Cui, F. (2011). Hydrophobic ion pairing of an insulin-sodium deoxycholate complex for oral delivery of insulin. *Int J Nanomedicine*, **6**, 3049–56.

Sun, Y. P., Meziani, M. J., Pathak, P. and Qu, L. (2005). Polymeric nanoparticles from rapid expansion of supercritical fluid solution. *Chemistry*, **11**, 1366–73.

Taipale, J. and Keski-Oja, J. (1997). Growth factors in the extracellular matrix. *FASEB J*, **11**, 51–9.

Tanner, P., Baumann, P., Enea, R., Onaca, O., Palivan, C. and Meier, W. (2011). Polymeric vesicles: From drug carriers to nanoreactors and artificial organelles. *Acc Chem Res*, **44**, 1039–49.

Trapani, A., Denora, N., Iacobellis, G., Sitterberg, J., Bakowsky, U. and Kissel, T. (2011). Methotrexate-loaded chitosan- and glycol chitosan-based nanopar-ticles: A promising strategy for the administration of the anticancer drug to brain tumors. *AAPS PharmSciTech*, **12**, 1302–11.

Traub, W. and Piez, K. A. (1971). *The Chemistry and Structure of Collagen*, New York, USA, Academic Press.

Vaquette, C. and Cooper-White, J. (2012). A simple method for fabricating 3-D mul-tilayered composite scaffolds. *Acta Biomater,* **9**(1), 4599–4608.

Wang, X., Song, G. and Lou, T. (2010). Fabrication and characterization of nano-composite scaffold of PLLA/silane modified hydroxyapatite. *Med Eng Phys*, **32**, 391–7.

Xie, H. and Smith, J. W. (2010). Fabrication of PLGA nanoparticles with a fluidic nanoprecipitation system. *J Nanobiotechnology*, **8**, 18.

Yang, F., Murugan, R., Ramakrishna, S., Wang, X., Ma, Y. X. and Wang, S. (2004). Fabrication of nano-structured porous PLLA scaffold intended for nerve tissue engineering. *Biomaterials*, **25**, 1891–900.

Yang, F., Murugan, R., Wang, S. and Ramakrishna, S. (2005). Electrospinning of nano/micro scale poly(L-lactic acid) aligned fibers and their potential in neural tissue engineering. *Biomaterials*, **26**, 2603–10.

Young, S., Wong, M., Tabata, Y. and Mikos, A. G. (2005). Gelatin as a delivery vehi-cle for the controlled release of bioactive molecules. *J Control Release*, **109**, 256–74.

Zambaux, M. F., Bonneaux, F., Gref, R., Maincent, P., Dellacherie, E., Alonso, M. J., Labrude, P. and Vigneron, C. (1998). Influence of experimental parameters on the characteristics of poly(lactic acid) nanoparticles prepared by a double emulsion method. *J Control Release*, **50**, 31–40.

Zhang, Z., Grijpma, D. W. and Feijen, J. (2006). Poly(trimethylene carbonate) and monomethoxy poly(ethylene glycol)-block-poly(trimethylene carbonate) nanoparticles for the controlled release of dexamethasone. *J Control Release*, **111**, 263–70.

Zhao, Y., Tanaka, M., Kinoshita, T., Higuchi, M. and Tan, T. (2010). Nanofibrous scaffold from self-assembly of beta-sheet peptides containing phenylalanine for controlled release. *J Control Release*, **142**, 354–60.

Zhu, X., Su, M., Tang, S., Wang, L., Liang, X., Meng, F., Hong, Y. and Xu, Z. (2012). Synthesis of thiolated chitosan and preparation nanoparticles with sodium alginate for ocular drug delivery. *Mol Vis*, **18**, 1973–82.

Zimmermann, H., Shirley, S. G. and Zimmermann, U. (2007). Alginate-based encapsulation of cells: Past, present, and future. *Curr Diab Rep*, **7**, 314–20.

3

Engineering nanoporous biomaterials

B. D. HATTON, University of Toronto, Canada

DOI: 10.1533/9780857097231.1.64

Abstract: A review is presented for the synthesis and engineering of ordered nanoporous oxide materials and surface topologies for biomedical applications. Techniques for bottom-up self-assembly and top-down etching can be used to define different pore sizes for different materials and in different applications, associated with drug delivery and the engineering of cell scaffolds.

Key words: nanoporous, self-assembly, template, biomedical, drug delivery, nanotube, scaffold.

3.1   Introduction

In recent years a convergence of top-down and bottom-up engineering approaches have led to an increasing sophistication in the synthesis and design of inorganic nanoporous materials. These developments have happened largely in the field of materials chemistry, but they have implications for a wide range of fields. Currently, various methods exist to synthesize highly ordered, nanoporous structures having uniform, monodisperse pore sizes of a few nanometers, to 10s or 100s of nanometers, and into the micron scale. The choice of technique depends on the size range, the material, and the macroscale morphology. Nanoporous materials and surface structures are finding important uses in biological and biomedical applications for a number of reasons. In addition to the pore size, the porous volume, porous connectivity, hierarchies, and structural gradients can be precisely controlled from the nanometer to the micron scale, and beyond. This precise engineering of surface area at the nanometer scale has important implications for drug delivery materials, separation membranes, cell scaffolds, and surface topologies for biomaterial implants. In each case, precision in the structural design can lead to a high degree of control in the biological response and biomaterial performance *in vitro* or *in vivo*. This review will focus on methods that have been developed for the synthesis of *inorganic* nanoporous layers or surface structures, with a particular emphasis on oxides, and having an ordered, monodisperse pore

64

size (as opposed to disordered, etched, or fibrous mesh structures). In this context, 'nanoporous' can refer very generally to sub-micron pore sizes. The biomedical applications of such materials are broad and increasing as the biomedical community further interacts with and develops these advances within materials science.

A number of excellent reviews summarize developments in different aspects of nano- and micro-structured materials and surface topographies for biomedical applications (Tirrell *et al.*, 2002; Kokubo *et al.*, 2004; Baker *et al.*, 2005; Khademhosseini *et al.*, 2006; Norman and Desai, 2006; Thomas *et al.*, 2006; Hertz and Bruce, 2007; Lim and Donahue, 2007; Mendonça *et al.*, 2008; Variola *et al.*, 2009; Anselme *et al.*, 2010a). However, there have been few reviews that summarize recent developments in nanoporous structures for biomedicine, and therefore this review will aim to provide a perspective on developments within this field more specifically. A particular focus is given to the synthesis methods for these materials, which are important to consider for any practical biomedical applications. The two main applications for nanoporous structures in biomedicine are for drug delivery and cell scaffold surface topologies, which are introduced below.

3.1.1 Drug delivery

Drug delivery remains a significant application for nanoporous structures, predominantly because of the very high surface areas that they can achieve. Surface areas of 500–1000 m^2/g can be achieved for many of the materials presented in this review, and there have been various methods developed to control specific surface chemistries (surface functionalization) with highly precise pore size control and interconnectivity. As of 2001 the US market for advanced drug delivery systems was already nearing \$20 billion (Langer, 2001). Part of this interest involves further enhancing the functionality of scaffolds for tissue engineering to give them the capability of releasing an array of biological signals with an adequate dose for a defined time frame (Langer, 2000, 2001; Tao and Desai, 2003; Langer and Tirrell, 2004; Biondi *et al.*, 2008; Farokhzad and Langer, 2009; Shi *et al.*, 2010). The release of a bolus dose of growth factor, for example, may not necessarily be effective since diffusion can happen rapidly from the target site and can quickly become enzymatically deactivated. Designing nanoporous structures to control diffusional release can also encourage better tissue regeneration over long time frames.

The design of the nanostructured delivery vehicle, or surface, can influence the effectiveness of a drug for a number of reasons (Farokhzad and Langer, 2009). While drug development costs remain very high (typically \$250–300 million, and 12+ years development time), the costs of designing novel and specific structures for drug delivery can be relatively low. Therefore, advances in the design, synthesis and engineering of nanoporous structures – even non-biodegradable, inert structures like inorganic oxides – represent a significant advancement for drug delivery. Such structures can incorporate highly precise pore size, gradient porosities, multiple biofunctionalities, and tailored release rates, depending on their design and method of manufacture.

Drug loading of nanoporous structures is typically achieved by soaking the porous structure in a concentrated drug solution and allowing capillary mechanisms to infiltrate the porous volume. Ordered nanoporous structures, the topic of this review, have certain advantages over disordered templates, whether inert or biodegradable. Primarily, if the pore size, distribution, density and interconnectivity are all precisely defined, then the drug loading and release rates can also be well defined. The pore size of these nanoporous structures is known to be highly important, especially when the pore dimensions are comparable to the size of the drug molecule itself. In these cases the diffusion rate is determined by the pore size, a process known as hindered or restricted diffusion. The diffusion of molecules through nanoporous structures and membranes has been well studied in theory and in practice (Pappenheimer, 1953; Deen, 1987; Gultepe *et al.*, 2010). In addition to the pore size, the surface chemical functionality, whether charged or hydrophobic, can be used to modify the release kinetics of the drug, depending on the drug characteristics.

3.1.2   Cell scaffolds

Nanoporous surfaces have been found to cause some interesting effects on cell growth and development. There are various examples showing the effects of nanoscale surface topologies on cell growth, particularly in the field of bone implant surface engineering (Shalabi *et al.*, 2006; Christenson *et al.*, 2007; Tran *et al.*, 2009; Ehrenfest *et al.*, 2010; Anselme *et al.*, 2010b). Porous scaffolds for cell growth typically have pore sizes on the order of 50–100 μm or larger, well above the nanoscale, which allows for cell growth and vascular network development. However, many studies have demonstrated that surface and scaffold features at the 1–50 μm scale, such as one-dimensional (1D) grooves and two-dimensional (2D) arrays of posts, can also greatly influence cell growth and function (Curtis and Wilkinson, 1997; Khademhosseini *et al.*, 2006; Norman, 2006; Hertz, 2007; Anselme *et al.*, 2010b). The mechanisms which relate cell morphology and differentiation to microscale topographic features are still not completely understood. However, length scales associated with certain cellular features, such as filopodia (normally 5–35 μm in length) (Wood and Martin, 2002) and integrin receptors, are known to play important roles in several morphogenetic processes and mechanical sensing.

Natural tissues have a hierarchical structure ranging from the macroscale (>1 mm), to microscale (1 µm–1 mm), and nanoscale (<1 µm). As a result, it is clear that individual cells (typically in the size range 10–50 µm) respond in different ways to structures at different length scales. Bone, for example, has a hierarchical structure and includes nanostructures that consist of collagen fibers and hydroxyapatite nanocrystals of 2–5 nm diameter, 20–50 nm length. There are many examples of surfaces structured topographically at the sub-micron scale which have been shown to provide important topographic cues for cells. In short, topographic features at the sub-micron scale have been shown to affect a wide variety of growth parameters, such as cell adhesion, morphology, viability, genetic regulation, apoptosis, and motility (Khademhosseini et al., 2006; Streicher et al., 2007; Anselme et al., 2010a). Recent results on nanoscale topographies suggest that there are features within this size range to which cells respond, specifically due to the size associated with integrin cluster formation, about 50–70 nm (Arnold et al., 2004).

There is an important biomedical interest in nanoporous structures for the surface engineering of bone implants (Shalabi et al., 2006; Christenson et al., 2007; Tran et al., 2009; Anselme et al., 2010b; Ehrenfest et al., 2010). Every year approximately 500 000 total hip and knee replacements are performed in the US (Streicher et al., 2007). Although most outcomes are successful, problems of implant loosening and failure do remain and incur significant cost and health effects. To improve fixation and mechanical bonding, there is great interest in developing surface structures and topographies that stimulate rapid bone regeneration and improve mechanical fixation to the adjacent bone (Popat et al., 2005; Streicher et al., 2007; Ehrenfest et al., 2010). Results in recent years have made it increasingly apparent that surface nanoscale topographies can enhance *in vitro* bone cell response and may provide better *in vivo* osseointegration than implant surfaces roughened by techniques such as etching and grit-blasting (Shalabi et al., 2006; Christenson et al., 2007; Tran et al., 2009; Anselme et al., 2010b; Ehrenfest et al., 2010). Methods to produce highly ordered, nanoporous surface structures have the potential to define completely new types of interactions between implant surfaces and cells. The surface area can be greatly increased and the surface topography can be engineered to better resemble native bone tissue, for example.

Methods to produce nanoporous (and nanostructured topography) materials and surfaces can be classified as either *subtractive* (i.e. etching) or *additive* (i.e. surface deposition). In this review, sections have been divided into general classes of nanoporous surfaces and their methods of synthesis: (1) nanotubes and etched nanoporous surfaces; (2) mesoporous surfaces (self-assembled and phase separated supramolecular templates); and (3) nanostructures from self-assembled colloidal templates. There are other techniques for generating nanoporous surface structures, such as electrospun nanoporous scaffolds, photolithography, and three-dimensional (3D) optical lithography, but these are either not ordered, involve small-scale, expensive conditions, or have not been highly developed for biomaterial applications.

3.2   Nanotubes and etched nanoporous surfaces

Many examples exist of fabricating nanoporous or nanotube surfaces using electrochemical etching, and the oxidation of Al and Ti metals and alloys are prime examples. Electrochemical anodic oxidation of aluminum is a well-established technique to control the oxide (Al_2O_3, or alumina) surface layer thickness and nanoporosity. It was reported as early as 1953 (Keller *et al.*, 1953), and later in the 1970s (O'Sullivan and Wood, 1970), that anodic etching of aluminum could produce vertically aligned pores, generally known as anodized aluminum oxide (AAO). Later work, largely by Masuda *et al.*, showed that control over the acid chemistry, pH, and voltage could be used to define ordered arrangements of monodisperse pores, vertically oriented relative to the surface (Masuda and Fukuda, 1995; Masuda *et al.*, 1997; Li *et al.*, 1998; Nielsch *et al.*, 2002; Lee *et al.*, 2006b).

Nanoporous AAO can be directly grown on any aluminum surfaces, such as implant materials, and the pores are oriented normal to the aluminum surface, making them highly accessible. The pore diameters in nanoporous AAO can be controlled within a wide range, from 5 nm to 10 μm, but typically they are generated between 50 and 300 nm, as shown in Fig. 3.1 (Masuda *et al.*, 1997) and Fig. 3.2a (Gultepe *et al.*, 2007). The pore diameter and wall thickness is determined by the anodization voltage and acidic conditions. The porous volume usually ranges from about 40% to 60%, and the pore depth is generally determined by the anodization time, but is typically around 10 to 50 μm. There are several examples of nanoporous AAO pores extending to mm-scale depths.

The conditions for long range ordering in nanoporous AAO also depend on the acid electrolyte, such as sulfuric, oxalic or phosphoric acid, and the anodizing voltage used. In the initial stages of anodic oxidation a random array of holes is formed in the aluminum surface. Various models have been developed to describe the formation, self-organization, and growth mechanisms, such as that by Parkhutik and Shershulsky (1992). As the oxide layer grows the pores can rearrange after a period of time to become a close-packed hexagonal structure. For example, anodizing at 25 V in 0.5 M sulfuric acid can cause the pores to become ordered after about 10 h (Masuda and Fukuda, 1995).

Masuda and Fukuda (1995) developed a two-step anodization process to allow highly ordered nanoporous Al_2O_3 structures to be fabricated over relatively large areas. A first anodization step is performed, typically in an

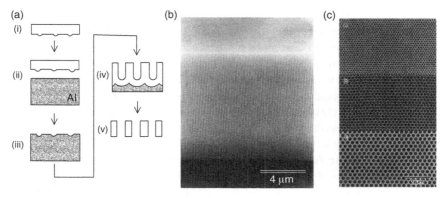

3.1 (a) Illustration of the molding process for an ordered hexagonal channel array in AAO; physically forming the Al, anodization and growth of the channels, removal of the backing Al and underlying Al₂O₃ layer; (b) cross-section of an AAO layer showing highly organized arrangement of straight channels; (c) control of pore sizes, from 100 nm (40V), 150 nm (60V) and 200 nm (80V), using anodization in oxalic acid. (*Source*: Reprinted with permission from Masuda *et al.* (1997). © 1997 American Institute of Physics.)

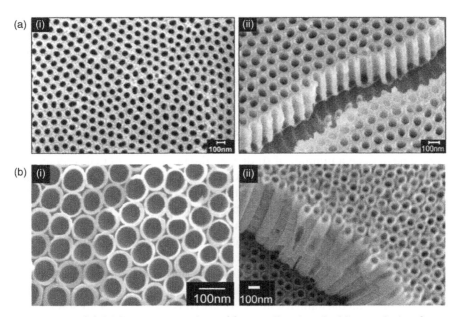

3.2 (a) AAO nanoporous layer (*Source*: Reprinted with permission from Gultepe *et al.* (2007). © 2007 American Institute of Physics.); (b) TiO₂ nanotube layer. (*Source*: Reprinted with permission from Gultepe *et al.* (2010). © (2010) Elsevier.)

oxalic acid electrolyte (e.g., 0.3 M, 58°C, stirred, 30V DC, 7 h). Then the Al_2O_3 layer on the unprotected side of the sheet is removed, using a mixture of phosphoric and chromic acid, exposing the textured substrate for a second anodization step. The second anodization can be performed under similar conditions to the first stage, which results in a regular, close-packed, hexagonal arrangement of nanopores, extending parallel throughout the layer. Chemical etching (such as phosphoric acid 5%) can be used to widen the pore size (Masuda and Fukuda, 1995; Parkinson, 2009). The pores can also be ordered by a 'pre-texturing' method by using an imprint of shallow indentations on the Al surface. Etching and pre-texturing each help to initiate the formation of individual pores (Masuda *et al.*, 1997).

Titanium and its alloys have a history since the 1970s in orthopedic and dental implant usage, owing to a combination of mechanical strength and good biocompatibility. Like Al, Ti surfaces always have an oxide layer (TiO_2 or titania), which is responsible for its biocompatibility (Streicher *et al.*, 2007; Mendonça *et al.*, 2008; Ehrenfest *et al.*, 2010). Nanoporous TiO_2, in the form of a vertically oriented array of nanotubes, can also be synthesized by anodization methods, as for Al (Macak *et al.*, 2005, 2007; Ghicov and Schmuki, 2009; Roy *et al.*, 2011). The oxidation conditions for synthesizing TiO_2 nanotube arrays were first developed by Zwilling *et al.* (1999) and Gong *et al.* (2001). Figure 3.2b shows a TiO_2 nanotube array (Gultepe *et al.*, 2010). Finally, there is a strong correlation between the anodization voltage and the pore size, which is also similar to Al anodization. Figure 3.3 shows an ordered array of TiO_2 nanotubes formed on the Ti surface using a phosphate-fluoride electrolyte, where the pore size is highly dependent on the voltage, from 1 to 20 V (Park *et al.*, 2007).

3.2.1   Nanotube surfaces as cell scaffolds

There have been a number of studies on the interaction of cells on nanoporous Al_2O_3 and TiO_2 surfaces. Karlsson *et al.* (2003) cultured primary human osteoblast-like cells *in vitro* on nanoporous Al_2O_3 membranes, having a 200 nm pore size. The cells exhibited normal growth patterns and morphology, with some indication of increased alkaline phosphatase (ALP) which is a measure of mineralization activity. Further, they demonstrated SEM evidence of filipodia penetration into the Al_2O_3 pores, as shown in Fig. 3.4.

Popat *et al.* (2005) and Swan *et al.* (2005) also investigated osteoblast response on nanoporous Al_2O_3 surfaces using a two-step anodization process. Swan *et al.* (2005) found that the human fetal osteoblast cells adopted a highly spread morphology on 77 nm pore Al_2O_3, after 4 days of culture. Popat *et al.* (2005) considered long-term osteoblast performance by measuring adhesion after 1 day and proliferation after 4 days of seeding cells on 100 nm pore nanoporous alumina membranes. Cells cultured on the nanoporous Al_2O_3 membranes showed

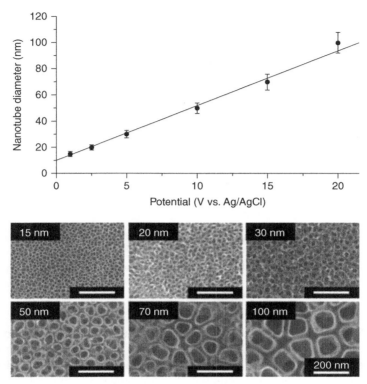

3.3 TiO$_2$ nanotube layers having different pore sizes. Top: nanotube diameter controlled by anodization potential, from 1 to 20 V. Bottom: images of nanotube layers with pore size from 15 to 100 nm indicated at top left of each image. Scale bars 200 nm. (*Source*: Reprinted with permission from Park *et al.* (2007). © 2007 American Chemical Society.)

a higher protein content as indicated by bicinchoninic acid (BCA) assay and X-ray photoelectron spectroscopy (XPS) results, compared to nonporous controls. Also, the osteoblasts on the nanoporous Al$_2$O$_3$ laid more matrix compared to cells that adhered to other control surfaces. LaFlamme *et al.* (2007) tested the biocompatibility and cytotoxicity of nanoporous AAO membranes, using the degree of complement activation as a measure of biocompatibility. Finally, Parkinson *et al.* (2009) investigated the influence of anodized Al$_2$O$_3$ nanoporous surfaces, having pore sizes between 40 and 500 nm, as tissue scaffolds for wound repair. They studied the effects of pore size on the attachment, growth, and migration of immortalized human skin cell lines: keratinocyte (epidermal) cells and fibroblast (dermal) cell lines.

The highly biocompatible properties of TiO$_2$ have made the fabrication and engineering of TiO$_2$ nanotube surfaces of particular interest for biomaterial applications (Burns *et al.*, 2009). Popat *et al.* (2007) synthesized TiO$_2$ nanotubes

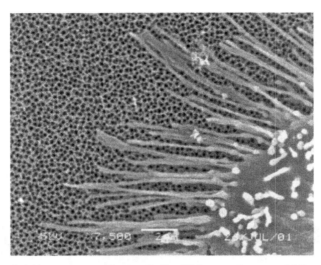

3.4 Human osteoblast cell after 24 h growth on an AAO surface. Scale bar 2 μm. (*Source*: Reprinted with permission from Karlsson *et al.* (2003). © 2003 Elsevier.)

with 80 nm pores (400 nm length), by anodization at 20 V for 20 min. The TiO$_2$ surfaces were seeded with mesenchymal stem cells (MSCs) from rats and compared to conventional Ti and polystyrene surfaces. After 7 days they found a 40% increase in the number of cells on the nanotubular TiO$_2$ surfaces compared to flat TiO$_2$, suggesting that nanoscale topographical cues promote cell adhesion and proliferation. Alkaline phosphatase activity was measured for up to 3 weeks of culture, showing higher ALP levels (approximately 50%) for the cells on the nanotubular surfaces compared to the Ti surfaces.

Park *et al.* (2007) tested the response of cells on TiO$_2$ nanotube surfaces having pore diameters between 15 and 100 nm, and found a maximum response at a lateral spacing of 15–30 nm. Further, they found that cells did not respond well (and often underwent apoptosis) on 100 nm nanotube surfaces. They suggested that the 15–30 nm spacing is close to the lateral spacing of integrin receptors in focal contacts on the extracellular matrix, a finding consistent with other published results (Arnold *et al.*, 2004). In any case, the adhesion, spreading, and growth at this scale were enhanced compared to nonporous TiO$_2$ surfaces. In comparison, von Wilmowsky *et al.* (2009) found that 30 nm TiO$_2$ nanotubes enhanced the expression of type-I collagen, but did not change the osteoblast proliferation, compared to untreated Ti.

Bjursten *et al.* (2010) found that 80 nm inner diameter TiO$_2$ nanotubes enhanced osteoblast adhesion by 300–400% *in vitro*. They also performed mechanical adhesion tests *in vivo* which showed that the tensile fracture force to remove TiO$_2$ nanotube implants from rabbit tibia was nine times greater compared to grit-blasted Ti implants (the control), after 4 weeks of bone growth. The authors suggest that the nanocrystalline structure and large surface area of the TiO_2 nanotubes allow a greater spreading and activation of proteins for osteoblast cell recruitment.

Brammer *et al.* (2008) investigated the rate of endothelialization for TiO_2 nanotube surfaces having an average 100 nm outer diameter, 70 nm inner diameter, 15 nm wall thickness, and 250 nm height, prepared by anodization. A major factor affecting late stent thrombosis is the delayed endothelialization of intracoronary stents, because the exposed stent surface can act as a nucleation site for thrombosis. As a result, the rate of cellular migration is a major requirement for the success of stent implantation. Brammer *et al.* (2008) compared the cellular response of primary bovine aortic endothelial cells (BAECs) on flat Ti and Ti nanotubes for cell viability and monolayer formation. The endothelial cells had enhanced interactions with the TiO_2 nanotube surface, allowing enhanced cellular migration and enhanced endothelial response *in vitro*, compared to the flat Ti surface. The nanotubes also caused an increase in extracelluar matrix (ECM) deposition, a unidirectional cytoskeleton of actin filaments, and organized filipodia on the nanotube surface. In an extension, Oh *et al.* (2009) determined that stem cell fate can be sensitive to TiO_2 nanotube dimensions, comparing small (30 nm) diameter to 70–100 nm diameter surfaces. The latter caused much higher cell elongation and osteoblast-like differentiation and demonstrates the critical effect of topological scale on cellular behavior. Hoess *et al.* (2012) used an Al_2O_3 membrane as a means of culturing hepatocytes, which are difficult to culture due to their sensitivity to culture conditions. They suggest that the porous membranes allowed nutrients and other molecules to be exchanged through the undersides of adherent cells, in addition to the exchange already occurring through the cell surface directly exposed to culture media.

3.2.2   Nanotube surfaces for drug delivery

There have been a number of studies that have tested nanoporous Al_2O_3 and TiO_2 surfaces for drug delivery applications (Losic and Simovic, 2009; Han *et al.*, 2011). Kipke and Schmid (2004) experimentally measured the diffusion rates of different species through Al_2O_3 membranes as a function of pore size and showed that pore size can influence the release of dye contained within sodium dodecyl sulfate (SDS) micelles. Kang *et al.* (2007) tested the design of drug-eluting stents using 316 stainless steel coronary stents coated with nanoporous Al_2O_3 (anodized directly on the stent surface). Using high-performance liquid chromatography (HPLC), they measured the rate of release of the drug 2-deoxyadenosine from AAO films having pore sizes from 13 nm to 27 nm and different pore depths. Song *et al.* (2009) addressed the problem

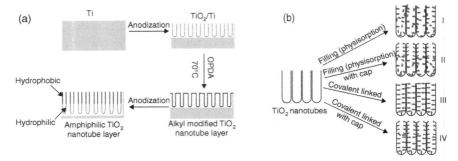

3.5 Example of drug release from a TiO$_2$ nanotube layer, modified to have a hydrophobic cap to control the drug release rate: (a) method of anodization of the TiO$_2$, followed by surface modification to make hydrophilic and hydrophobic regions on the channels; (b) method of drug loading within the modified channels. (*Source*: Reprinted (adapted) from Song *et al.* (2009). © 2009 American Chemical Society.)

of uncontrolled release of drugs or therapeutics from nanoporous surfaces by developing a hydrophobic cap on a hydrophilic TiO$_2$ nanotube surface (90 nm pore size) to test the effectiveness of controlled release kinetics. The cap prevents uncontrolled leaching of the hydrophilic drug into the surrounding aqueous environment. The photocatalytic nature of TiO$_2$ can cause a UV-induced cleavage of organic monolayers holding the cap in place, to allow drug release, as shown in Fig. 3.5 (Song *et al.*, 2009).

Nanoporous Al$_2$O$_3$ membranes having through-thickness porosity can function as substrates to isolate specific cell cultures or operate as biosensing platforms. Takoh *et al.* (2004) mounted a permeable nanoporous Al$_2$O$_3$ membrane (20 nm pores, 50% porosity) on a 40 μm thick polydimethylsiloxane (PDMS) support having a 50 μm diameter hole. HeLa cells were cultured on the membrane such that a chemical (ethanol) could be delivered from the opposite side, to selectively and locally affect cells. Scanning electrochemical microscopy (SECM) was used to simultaneously measure the electroactive species around the stimulated and non-stimulated cells using a 5.0 μm-radius Pt microelectrode.

3.3   Self-assembled supramolecular organic templates

An alternative to etching surfaces by anodic oxidation of aluminum, titanium, or silicon is the bottom-up self-assembly approach for nanoporous materials. The self-organization of soft matter at the nanoscale can be used as a template for inorganic (or organic) synthesis and polymerization. Recent research has shown that an enormous range of structures, porosities, and symmetries can be achieved with self-assembled organic templates at a scale from a few nanometers to tens of nanometers. A wide range of inorganic materials has been incorporated around such organic templates. Such approaches have led to a diversity of nanoporous inorganic structures, after removing the organic template.

Surfactants and block copolymers can form lyotropic liquid-crystal phases which can be used as templates for lamellar, hexagonal, and cubic structures for inorganic materials, such as SiO_2, Al_2O_3, and TiO_2. Such structures, generally known as mesoporous materials, have attracted attention for application in catalysis, membrane separation, and drug delivery (Soler-Illia *et al.*, 2002; Hamley, 2003; Wan and Zhao, 2007; Sanchez *et al.*, 2008). Kresge *et al.* (1992) first demonstrated the co-operative self-assembly of a surfactant template with silica, before dissolving out the organic template to leave a mesoporous silica scaffold.

The inorganic material often plays an important role in template self-assembly, where structuring occurs through a cooperative organization of inorganic and organic materials. In the formation of mesoporous silica, a common method is to mix a tetraalkoxy silane (i.e. TEOS) and surfactant in an aqueous solution. Both ionic and nonionic surfactants have been used to template structures (known as MCM-type materials), as have amphiphilic block copolymers, producing SBA materials. The cooperative self-assembly process generates a structure where the silica forms a shell around the hydrophilic exterior of the surfactant or block copolymer micelles, as illustrated in Plate IV (see colour section between pages 264 and 265). A common structure for mesoporous materials is a close-packed hexagonal arrangement of cylindrical micelles, producing a honeycomb arrangement of pores once the template is removed. Figure 3.6 shows some examples of mesoporous silica films imaged by transmission electron microscopy (TEM), for hexagonal honeycomb and cubic porous symmetries. The pore sizes are typically in the range 2–5 nm for surfactant templates (MCM), and from about 10 nm to 50 nm for block copolymer (SBA) templates (Hamley, 2003; Wan and Zhao, 2007).

Zhao *et al.* (1998a, 1998b) were one of the first groups to demonstrate the use of triblock copolymer templates to make nanoporous silica having pores in the range 5–30 nm (Fig. 3.6c). Block copolymers having hydrophobic and hydrophilic blocks can effectively behave as giant surfactants. Since that time block copolymers have been widely applied to generate a vast range of mesoporous structures for a broad range of inorganic materials. Block copolymers have some advantages over simple ionic surfactants. In particular, they can be engineered to have a wider range of pore sizes and a broader range of mesoporous symmetries, with a growing opportunity to chemically synthesize polymer chains having specific hydrophobic and hydrophilic blocks. Most of the novel block copolymer-templated nanoporous layers have been developed for novel photovoltaic structures for energy-related research. For example, nanoporous TiO_2 can be synthesized as a novel 'double gyroid' structure having interpenetrating porous networks (Nedelcu *et al.*, 2008).

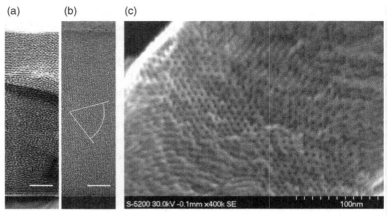

3.6 Examples of mesoporous silica films by transmission electron microscopy (TEM): (a) and (b) show cross-sections of hexagonal and cubic structures, respectively (50 nm scale bars) (*Source*: Reprinted with permission from Lu *et al.* (1997). © 1997 Nature Publishing Group.); (c) hexagonal structure of mesoporous organo silica from a triblock copolymer template. (*Source*: Courtesy of B. Hatton, W. Hunks.)

For the generation of mesoporous films, Lu *et al.* (1997) demonstrated a relatively simple method for the formation of mesoporous silica thin films by an evaporative deposition coating method, which has since been extended to a range of other oxides (Fig. 3.6a,b). In this approach, the self-assembly of a surfactant template occurs in the meniscus of a surfactant/silica solution as it moves across a (suitably wetting) substrate, to form a surfactant/silica nanocomposite layer. Typically such layers have a highly ordered hexagonal arrangement of channels parallel to the substrate surface. Subsequent thermal decomposition (calcination) of the layer burns out the surfactant template to leave behind a highly ordered nanoporous layer.

Nanoporous layers synthesized by self-assembled templated assembly have been extended to a wide range of materials beyond silica. For example, highly ordered, crystalline mesoporous TiO_2 layers have been synthesized for a range of pore sizes (Choi *et al.*, 2004). Various organosilica materials having terminal or bridging organic groups (Asefa *et al.*, 1999) have been synthesized in mesoporous form, which is important for incorporating chemical functionality. Hatton *et al.* (2005) showed that evaporative self-assembly could be used to deposit highly ordered layers of nanoporous organosilicas.

3.3.1   Drug release from mesoporous layers

There have been many studies and efforts to test and engineer mesoporous materials, primarily silicates, for drug release and drug delivery applications (Xia and Chang, 2006; Gultepe *et al.*, 2010; Manzano and Vallet-Regí, 2010). The very large surface area, uniform pore size, and chemical stability of mesoporous materials allows the drug release rate to be highly controlled and predictable. Furthermore, the biocompatibility of silica, and the relative ease of synthesis (in particle or layered morphologies) are attractive for drug release applications. Hudson *et al.* (2008) demonstrated some variation in the toxicity of mesoporous silica particles depending on the template chemistry and processing conditions. Bass *et al.* (2007) investigated the structural stability of mesoporous silica films under biologically relevant conditions, finding that the degradation times can vary widely, ranging from hours to days. As for nanoporous AAO or TiO$_2$ structures, there are important issues associated with the diffusion of molecules through nanopores, when the pore size becomes close to the molecular size (Beerdsen *et al.*, 2006).

Vallet-Regí *et al.* (2001) and Lai *et al.* (2003) were among the first to demonstrate drug release from mesoporous silica particles. Since then, there have been many studies on drug release from a range of mesoporous structures, such as gentamycin release from SBA mesoporous layers (Doadrio *et al.*, 2004), and a hydrophobic anticancer drug, camptothecin, from 100 nm-sized mesoporous silica spheres (Lu *et al.*, 2007) (Fig. 3.7, Plate V). Their results show the successful uptake of drug-loaded mesoporous particles by pancreatic cancer cells, among others. López-Noriega *et al.* (2006) demonstrated the application of mesoporous bioactive glasses for bone tissue regeneration. Active control of drug release from mesoporous materials can be achieved by pH (Yang *et al.*, 2005), temperature (Zhou *et al.*, 2007), or ultrasound-based (Kim *et al.*, 2006) stimulation. Izquierdo-Barba *et al.* (2008) and Yan *et al.* (2004) have demonstrated enhanced osteoblast

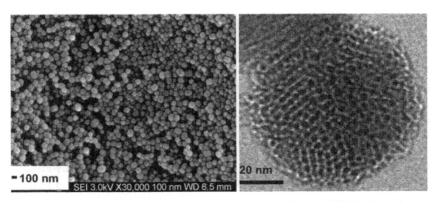

3.7 Mesoporous silica particles for drug delivery: SEM (left) and TEM (right) images of 100 nm mesoporous silica particles. (*Source*: Reprinted (adapted) from Lu *et al.* (2007). © 2007 Wiley-VCH.)

mineralization on mesoporous bioactive silica, which they attribute to the high surface area and potential for ion diffusional interchange.

3.4   Self-assembled colloidal templates

At a larger length scale from supramolecular organic templates (i.e. surfactants and block copolymers), a great deal of work has been done to develop self-assembled colloidal templates for nanoporous inorganic structures. Colloidal templates can be based on emulsions or solid colloidal particles. Emulsion templating of silica, using an oil/water emulsion (stabilized by surfactant), was first developed by Imhof and Pine (1997), and recently reviewed in general by Zhang and Cooper (2005). Fractionation and centrifuging the stabilized emulsion can be used to create a monodisperse pore size, though the pore sizes tend to be in the macroporous range (>10 μm). Lin *et al.* (2011) demonstrated the culture of epithelial and fibroblast cells within monodisperse gelatin porous scaffolds synthesized by emulsions generated by a microfluidic device. The average pore size was relatively large (60–90 μm), with 25 μm interconnecting pores.

Self-assembled colloidal crystals (also known as 'synthetic opal') consist of close-packed arrays of solid colloidal particles and have been widely used to make nanoporous 'inverse opal' structures (Lytle and Stein, 2006). First developed in the late 1990s (Velev *et al.*, 1997; Holland *et al.*, 1998), they can be made by the infiltration of inorganic precursor solutions into the interstices of closely-packed polymer spheres. The polymeric template can be removed by dissolution or thermal decomposition (as above, for mesoporous materials) after inorganic polymerization. Sol-gel solutions have been used as matrix precursors for the synthesis of various oxide inverse opals, such as SiO_2, TiO_2, and Al_2O_3. Matrices have also been deposited via infiltration of polymer precursors, salt precursors, nanoparticles, or vapor phase precursors. Inverse opals can exhibit a high degree of interconnected porosity (~75%) with extremely uniform size (average size normally in the range 100–1000 nm) and periodic distributions of pores, achieved through colloidal monodispersity. Such structures have been shown to be potentially useful in a wide range of fields, including photonics, sensing, and catalysis. For applications in tissue engineering and cell culture there have been some examples for macroporous inverse opal scaffolds having pore sizes >50 μm (Choi *et al.*, 2010), as discussed further below. The monodisperse pore sizes very precisely define the interconnecting pore size, porous volume, and interconnectivity, leading to better control over diffusion rates and cell/scaffold interactions. However, while conventional self-assembly has yielded highly ordered inverse opal materials over modest (<10 μm) length scales, such processes have been plagued by uncontrolled defect formation over larger (mm to cm) length scales, thus limiting their real-world applications.

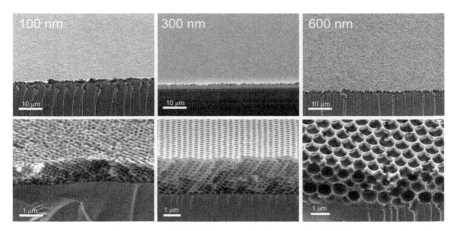

3.8 SiO$_2$ inverse opal films by colloidal co-assembly for 100, 300 and 600 nm pores. (*Source*: Courtesy of B. Hatton.)

Conventional colloidal self-assembly of inverse opal films has generally been conducted in three sequential steps: (1) assembly of a colloidal crystal template (e.g., comprising polymer latex spheres); (2) infiltration and deposition of a matrix phase, or a precursor to a solid matrix phase; and (3) selective removal of the colloidal crystal template to yield an inverse porous structure. In the first step, thin film colloidal crystal templates can be deposited by evaporative or 'flow-controlled' deposition. But these self-assembly processes usually result in the formation of cracks (as well as domain boundaries, colloid vacancies, and other defects).

In our lab, we have demonstrated the evaporative co-assembly of a sacrificial colloidal template with a matrix material in a single step to yield a colloidal composite, thereby avoiding the need for liquid infiltration into a pre-assembled porous structure. The secondary infiltration step tends to cause a range of defects such as cracks and non-uniform infiltration. We have demonstrated that colloidal co-assembly represents an effective and efficient method to deposit nanoporous oxide layers over large areas, and curved surfaces, without cracks (Hatton *et al.*, 2010). Figure 3.8 shows a cross-section of SiO$_2$ inverse opal films having 100, 300 and 600 nm pores, after calcination to remove the polymer template.

There has been surprisingly little application of ordered inverse opal structures for drug delivery, which may be due to the challenges in making continuous surfaces, or because the pore sizes tend to be larger than 50 nm which may lead to release rates too rapid for certain applications. Cherdhirankorn *et al.* (2010) have experimentally tested the diffusion of a tracer dye in 360 nm pore SiO$_2$ inverse opals. There have been certain examples of inverse opal layers used as sensors for biological species, such

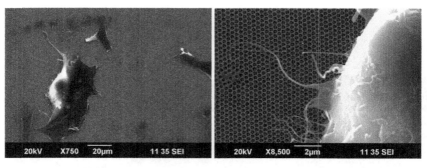

3.9 HMSCs cultured on SiO$_2$ inverse opal films; SEM image of a cell on a 300 nm pore SiO$_2$ inverse opal film. (*Source*: Courtesy M. Bucaro, B. Hatton.)

as glucose sensing (Nakayama *et al.*, 2003; Lee *et al.*, 2004), using a change in color. For cell culture on inverse opal structures, various groups have investigated macroporous (i.e. >50 μm) inverse opal materials (Yan *et al.*, 2001; Kotov *et al.*, 2004; Lee *et al.*, 2006a, 2009; Choi *et al.*, 2009; Nichols *et al.*, 2009; da Silva *et al.*, 2010; Kuo and Chiu, 2011). In our own research, we have been investigating the effects of inverse opal pore size on the growth and morphology of cells. Figure 3.9 shows images of human mesenchymal stem cells (HMSCs) cultured on 300 nm and 1.3 μm SiO$_2$ inverse opal surfaces (Plate VI).

3.5   Conclusion

This review has presented a wide range of synthesis techniques for the engineering of highly ordered nanoporous structures, from top-down etching methods to bottom-up self-assembly. These synthetic methods can have, and will have, an important impact on the design of implant surfaces and other biomaterials due to the highly precise and sophisticated control over surface area and porous connectivity at the nano- and micro-length scales. These new techniques were developed to control the porous structure itself and to engineer the surface chemistry within the pores.

A wealth of research indicates that topological features at the sub-micron scale can have a critically important effect on the biological response of cells and the biocompatibility of medical implant devices. Surface modification by nanoporous and nanotube layers is particularly interesting because it not only improves the biocompatibility of medical implants, by controlling the pore size and lateral spacing geometry, but it also offers the opportunity to regulate tissue engineering by filling nanopores with drugs and biologically active signaling molecules.

There are some important benefits to controlling the pore size and porous interconnectivity with such a high degree of precision. The diffusional release rate of drugs is much more easily controlled and defined when the porous networks through which they diffuse is uniform and directional. Also, nanophase materials in the sub-micron regime mimic more closely the natural constituents of bone (hydroxyapatite and collagen) than do disordered surfaces etched by coarsely controlled techniques, such as grit-blasting. Therefore, the topography of nanoporous AAO and TiO_2 nanotubes can closely resemble the porous structure of native bone tissue. As new methods of chemical synthesis and materials engineering emerge to define these nanoporous networks, we will learn and understand more about the complex relationship between nanoporous structure and biological processes. Further, new approaches to the active control of drug release from nanoporous scaffolds, as opposed to passive diffusional release, promise a new generation of responsive biomaterials.

3.6   References

Anselme, K., Davidson, P., Popa, A.M., Giazzon, M., Liley, M. and Ploux, L. (2010a). The interaction of cells and bacteria with surfaces structured at the nanometre scale. *Acta Biomaterialia* **6**, 3824–3846.

Anselme, K., Ponche, A. and Bigerelle, M. (2010b). Relative influence of surface topography and surface chemistry on cell response to bone implant materials. Part 2: biological aspects. *Proceedings of the Institution of Mechanical Engineers, Part H: Journal of Engineering in Medicine* **224**, 1487–1507.

Arnold, M., Cavalcanti-Adam, E.A., Glass, R., Blümmel, J., Eck, W., Kantlehner, M., Kessler, H. and Spatz, J.P. (2004). Activation of integrin function by nanopatterned adhesive interfaces. *ChemPhysChem* **5**, 383–388.

Asefa, T., MacLachlan, M.J., Coombs, N. and Ozin, G.A. (1999). Periodic mesoporous organosilicas with organic groups inside the channel walls. *Nature* **402**, 867–871.

Baker, L.A., Jin, P. and Martin, C.R. (2005). Biomaterials and biotechnologies based on nanotube membranes. *Critical Reviews in Solid State and Materials Sciences* **30**, 183–205.

Bass, J.D., Grosso, D., Boissiere, C., Belamie, E., Coradin, T. and Sanchez, C. (2007). Stability of mesoporous oxide and mixed metal oxide materials under biologically relevant conditions. *Chemistry of Materials* **19**, 4349–4356.

Beerdsen, E., Dubbeldam, D. and Smit, B. (2006). Understanding diffusion in nanoporous materials. *Physical Review Letters* **96**, 44501.

Biondi, M., Ungaro, F., Quaglia, F. and Netti, P.A. (2008). Controlled drug delivery in tissue engineering. *Advanced Drug Delivery Reviews* **60**, 229–242.

Bjursten, L.M., Rasmusson, L., Oh, S., Smith, G.C., Brammer, K.S. and Jin, S. (2010). Titanium dioxide nanotubes enhance bone bonding in vivo. *Journal of Biomedical Materials Research Part A* **92**, 1218–1224.

Brammer, K.S., Oh, S., Gallagher, J.O. and Jin, S. (2008). Enhanced cellular mobility guided by TiO₂ nanotube surfaces. *Nano Letters* **8**, 786–793.

Burns, K., Yao, C. and Webster, T.J. (2009). Increased chondrocyte adhesion on nanotubular anodized titanium. *Journal of Biomedical Materials Research Part A* **88**, 561–568.

Cherdhirankorn, T., Retsch, M., Jonas, U., Butt, H.J. and Koynov, K. (2010). Tracer diffusion in silica inverse opals. *Langmuir* **26**, 10141–10146.

Choi, S.Y., Mamak, M., Coombs, N., Chopra, N. and Ozin, G.A. (2004). Thermally stable two-dimensional hexagonal mesoporous nanocrystalline anatase, Meso-nc-TiO₂: Bulk and crack-free thin film morphologies. *Advanced Functional Materials* **14**, 335–344.

Choi, S.W., Xie, J. and Xia, Y. (2009). Chitosan-based inverse opals: Three-dimensional scaffolds with uniform pore structures for cell culture. *Advanced Materials* **21**, 2997–3001.

Choi, S.W., Zhang, Y. and Xia, Y. (2010). Three-dimensional scaffolds for tissue engineering: The importance of uniformity in pore size and structure. *Langmuir* **26**, 19001–19006.

Christenson, E.M., Anseth, K.S., van den Beucken, J.J.J.P., Chan, C.K., Ercan, B., Jansen, J.A., Laurencin, C.T., Li, W.J., Murugan, R., Nair, L.S., Ramakrishna, S., Tuan, R.S., Webster, T.J. and Mikos, A.G. (2007). Nanobiomaterial applications in orthopedics. *Journal of Orthopaedic Research* **25**, 11–22.

Curtis, A. and Wilkinson, C. (1997). Topographical control of cells. *Biomaterials* **18**, 1573–1583.

da Silva, J., Lautenschläger, F., Sivaniah, E. and Guck, J.R. (2010). The cavity-to-cavity migration of leukaemic cells through 3D honey-combed hydrogels with adjustable internal dimension and stiffness. *Biomaterials* **31**, 2201–2208.

Deen, W.M. (1987). Hindered transport of large molecules in liquid-filled pores. *AIChE Journal* **33**, 1409–1425.

Doadrio, A.L., Sousa, E.M.B., Doadrio, J.C., Pérez Pariente, J., Izquierdo-Barba, I. and Vallet-Regí, M. (2004). Mesoporous SBA-15 HPLC evaluation for controlled gentamicin drug delivery. *Journal of Controlled Release* **97**, 125–132.

Ehrenfest, D.M.D., Coelho, P.G., Kang, B.S., Sul, Y.T. and Albrektsson, T. (2010). Classification of osseointegrated implant surfaces: Materials, chemistry and topography. *Trends in Biotechnology* **28**, 198–206.

Farokhzad, O.C. and Langer, R. (2009). Impact of nanotechnology on drug delivery. *Acs Nano* **3**, 16–20.

Ghicov, A. and Schmuki, P. (2009). Self-ordering electrochemistry: A review on growth and functionality of TiO₂ nanotubes and other self-aligned MOx structures. *Chemical Communications*, 2791–2808.

Gong, D., Grimes, C.A., Varghese, O.K., Hu, W., Singh, R.S., Chen, Z. and Dickey, E.C. (2001). Titanium oxide nanotube arrays prepared by anodic oxidation. *Journal of Materials Research* **16**, 3331–3334.

Gultepe, E., Nagesha, D., Menon, L., Busnaina, A. and Sridhar, S. (2007). High-throughput assembly of nanoelements in nanoporous alumina templates. *Applied Physics Letters* **90**, 163119.

Gultepe, E., Nagesha, D., Sridhar, S. and Amiji, M. (2010). Nanoporous inorganic membranes or coatings for sustained drug delivery in implantable devices. *Advanced Drug Delivery Reviews* **62**, 305–315.

Hamley, I.W. (2003). Nanostructure fabrication using block copolymers. *Nanotechnology* **14**, R39–R54.

Han, C.M., Lee, E.J., Kim, H.E., Koh, Y.H. and Jang, J.H. (2011). Porous TiO$_2$ films on Ti implants for controlled release of tetracycline-hydrochloride (TCH). *Thin Solid Films* **519**, 8074–8076.

Hatton, B.D., Landskron, K., Whitnal, W., Perovic, D.D. and Ozin, G.A. (2005). Spin-coated periodic mesoporous organosilica thin films – towards a new generation of low-dielectric constant materials. *Advanced Functional Materials* **15**, 823–829.

Hatton, B., Mishchenko, L., Davis, S., Sandhage, K.H. and Aizenberg, J. (2010). Assembly of large-area, highly ordered, crack-free inverse opal films. *Proceedings of the National Academy of Sciences of the United States of America* **107**, 10354–10359.

Hertz, A. and Bruce, I.J. (2007). Inorganic materials for bone repair or replacement applications. *Nanomedicine* **2**, 899–918.

Hoess, A., Thormann, A., Friedmann, A. and Heilmann, A. (2012). Self-supporting nanoporous alumina membranes as substrates for hepatic cell cultures. *Journal of Biomedical Materials Research Part A* **100A**, 2230–2238.

Holland, B.T., Blanford, C.F. and Stein, A. (1998). Synthesis of macroporous minerals with highly ordered three-dimensional arrays of spheroidal voids. *Science* **281**, 538–540.

Hudson, S.P., Padera, R.F., Langer, R. and Kohane, D.S. (2008). The biocompatibility of mesoporous silicates. *Biomaterials* **29**, 4045–4055.

Imhof, A. and Pine, D.J. (1997). Ordered macroporous materials by emulsion templating. *Nature* **389**, 948–951.

Izquierdo-Barba, I., Arcos, D., Sakamoto, Y., Terasaki, O., López-Noriega, A. and Vallet-Regí, M. (2008). High-performance mesoporous bioceramics mimicking bone mineralization. *Chemistry of Materials* **20**, 3191–3198.

Kang, H.J., Kim, D.J., Park, S.J., Yoo, J.B. and Ryu, Y.S. (2007). Controlled drug release using nanoporous anodic aluminum oxide on stent. *Thin Solid Films* **515**, 5184–5187.

Karlsson, M., Pålsgård, E., Wilshaw, P.R. and Di Silvio, L. (2003). Initial in vitro interaction of osteoblasts with nano-porous alumina. *Biomaterials* **24**, 3039–3046.

Keller, F., Hunter, M.S. and Robinson, D.L. (1953). Structural features of oxide coatings on aluminum. *Journal of The Electrochemical Society* **100**, 411–419.

Kim, H.J., Matsuda, H., Zhou, H. and Honma, I. (2006). Ultrasound-triggered smart drug release from a poly(dimethylsiloxane)–mesoporous silica composite. *Advanced Materials* **18**, 3083–3088.

Kipke, S. and Schmid, G. (2004). Nanoporous alumina membranes as diffusion controlling systems. *Advanced Functional Materials* **14**, 1184–1188.

Khademhosseini, A., Langer, R., Borenstein, J. and Vacanti, J.P. (2006). Microscale technologies for tissue engineering and biology. *Proceedings of the National Academy of Sciences of the United States of America* **103**, 2480–2487.

Kokubo, T., Kim, H.M., Kawashita, M. and Nakamura, T. (2004). Bioactive metals: preparation and properties. *Journal of Materials Science: Materials in Medicine* **15**, 99–107.

Kotov, N.A., Liu, Y., Wang, S., Cumming, C., Eghtedari, M., Vargas, G., Motamedi, M., Nichols, J. and Cortiella, J. (2004). Inverted colloidal crystals as three-dimensional cell scaffolds. *Langmuir* **20**, 7887–7892.

Kresge, C.T., Leonowicz, M.E., Roth, W.J., Vartuli, J.C. and Beck, J.S. (1992). Ordered mesoporous molecular sieves synthesized by a liquid-crystal template mechanism. *Nature* **359**, 710–712.

Kuo, Y.C. and Chiu, K.H. (2011). Inverted colloidal crystal scaffolds with laminin-derived peptides for neuronal differentiation of bone marrow stromal cells. *Biomaterials* **32**, 819–831.

La Flamme, K.E., Popat, K.C., Leoni, L., Markiewicz, E., La Tempa, T.J., Roman, B.B., Grimes, C.A. and Desai, T.A. (2007). Biocompatibility of nanoporous alumina membranes for immunoisolation. *Biomaterials* **28**, 2638–2645.

Lai, C.Y., Trewyn, B.G., Jeftinija, D.M., Jeftinija, K., Xu, S., Jeftinija, S. and Lin, V.S.Y. (2003). A mesoporous silica nanosphere-based carrier system with chemically removable CdS nanoparticle caps for stimuli-responsive controlled release of neurotransmitters and drug molecules. *Journal of the American Chemical Society* **125**, 4451–4459.

Langer, R. (2000). Biomaterials in drug delivery and tissue engineering: One laboratory's experience. *Accounts of Chemical Research* **33**, 94–101.

Langer, R. (2001). Drugs on target. *Science* **293**, 58–59.

Langer, R. and Tirrell, D.A. (2004). Designing materials for biology and medicine. *Nature* **428**, 487–492.

Lee, J., Shanbhag, S. and Kotov, N.A. (2006a). Inverted colloidal crystals as three-dimensional microenvironments for cellular co-cultures. *Journal of Materials Chemistry* **16**, 3558–3564.

Lee, J., Cuddihy, M.J., Cater, G.M. and Kotov, N.A. (2009). Engineering liver tissue spheroids with inverted colloidal crystal scaffolds. *Biomaterials* **30**, 4687–4694.

Lee, W., Ji, R., Gösele, U. and Nielsch, K. (2006b). Fast fabrication of long-range ordered porous alumina membranes by hard anodization. *Nature Materials* **5**, 741–747.

Lee, Y.J., Pruzinsky, S.A. and Braun, P.V. (2004). Glucose-sensitive inverse opal hydrogels: Analysis of optical diffraction response. *Langmuir* **20**, 3096–3106.

Li, F., Zhang, L. and Metzger, R.M. (1998). On the growth of highly ordered pores in anodized aluminum oxide. *Chemistry of Materials* **10**, 2470–2480.

Lim, J.Y. and Donahue, H.J. (2007). Cell sensing and response to micro-and nanostructured surfaces produced by chemical and topographic patterning. *Tissue Engineering* **13**, 1879–1891.

Lin, J., Lin, W., Hong, W., Hung, W., Nowotarski, S.H., Gouveia, S.M., Cristo, I. and Lin, K. (2011). Morphology and organization of tissue cells in 3D microenvironment of monodisperse foam scaffolds. *Soft Matter* **7**, 10010–10016.

López-Noriega, A., Arcos, D., Izquierdo-Barba, I., Sakamoto, Y., Terasaki, O. and Vallet-Regí, M. (2006). Ordered mesoporous bioactive glasses for bone tissue regeneration. *Chemistry of Materials* **18**, 3137–3144.

Losic, D. and Simovic, S. (2009). Self-ordered nanopore and nanotube platforms for drug delivery applications. *Expert Opinion on Drug Delivery* **6**, 1363–1381.

Lu, Y., Ganguli, R., Drewien, C.A., Anderson, M.T., Brinker, C.J., Gong, W., Guo, Y., Soyez, H., Dunn, B., Huang, M.H. and Zink, J.I. (1997). Continuous formation of supported cubic and hexagonal mesoporous films by sol-gel dip-coating. *Nature* **389**, 364–368.

Lu, J., Liong, M., Zink, J.I. and Tamanoi, F. (2007). Mesoporous silica nanoparticles as a delivery system for hydrophobic anticancer drugs. *Small* **3**, 1341–1346.

Lytle, J.C. and Stein, A. (2006). Recent progress in synthesis and applications of inverse opals and related macroporous materials prepared by colloidal crystal templating. *Annual Reviews of Nano Research* **1**, 1–79.

Macák, J.M., Tsuchiya, H. and Schmuki, P. (2005). High-aspect-ratio TiO_2 nanotubes by anodization of titanium. *Angewandte Chemie International Edition* **44**, 2100–2102.

Macak, J.M., Tsuchiya, H., Ghicov, A., Yasuda, K., Hahn, R., Bauer, S. and Schmuki, P. (2007). TiO_2 nanotubes: Self-organized electrochemical formation, properties and applications. *Current Opinion in Solid State and Materials Science* **11**, 3–18.

Manzano, M. and Vallet-Regí, M. (2010). New developments in ordered mesoporous materials for drug delivery. *Journal of Materials Chemistry* **20**, 5593–5604.

Masuda, H. and Fukuda, K. (1995). Ordered metal nanohole arrays made by a two-step replication of honeycomb structures of anodic alumina. *Science* **268**, 1466–1468.

Masuda, H., Yamada, H., Satoh, M., Asoh, H., Nakao, M. and Tamamura, T. (1997). Highly ordered nanochannel-array architecture in anodic alumina. *Applied Physics Letters* **71**, 2770–2772.

Mendonça, G., Mendonça, D.B.S., Aragão, F.J.L. and Cooper, L.F. (2008). Advancing dental implant surface technology-from micron- to nanotopography. *Biomaterials* **29**, 3822–3835.

Nakayama, D., Takeoka, Y., Watanabe, M. and Kataoka, K. (2003). Simple and precise preparation of a porous gel for a colorimetric glucose sensor by a templating technique. *Angewandte Chemie* **115**, 4329–4332.

Nedelcu, M., Lee, J., Crossland, E.J.W., Warren, S.C., Orilall, M.C., Guldin, S., Hüttner, S., Ducati, C., Eder, D., Wiesner, U., Steiner, U. and Snaith, H. J. (2009). Block copolymer directed synthesis of mesoporous TiO_2 for dye-sensitized solar cells. *Soft Matter* **5**, 134–139.

Nichols, J.E., Cortiella, J., Lee, J., Niles, J.A., Cuddihy, M., Wang, S., Bielitzki, J., Cantu, A., Mlcak, R., Valdivia, E., Yancy, R., McClure, M. L. and Kotov, N. A. (2009). In vitro analog of human bone marrow from 3D scaffolds with biomimetic inverted colloidal crystal geometry. *Biomaterials* **30**, 1071–1079.

Nielsch, K., Choi, J., Schwirn, K., Wehrspohn, R.B. and Gösele, U. (2002). Self-ordering regimes of porous alumina: The 10 porosity rule. *Nano Letters* **2**, 677–680.

Norman, J.J. and Desai, T.A. (2006). Methods for fabrication of nanoscale topography for tissue engineering scaffolds. *Annals of Biomedical Engineering* **34**, 89–101.

Oh, S., Brammer, K.S., Li, Y.S., Teng, D., Engler, A.J., Chien, S. and Jin, S. (2009) Stem cell fate dictated solely by altered nanotube dimension. *Proceedings of the National Academy of Sciences of the United States of America* **106**, 2130–2135.

O'Sullivan, J. and Wood, G. (1970). The morphology and mechanism of formation of porous anodic films on aluminium. *Proceedings of the Royal Society of London. Series A, Mathematical and Physical Sciences*, **317**, 511–543.

Pappenheimer, J.R. (1953) Passage of molecules through capillary walls. *Physiological Reviews* **33**, 387–423.

Park, J., Bauer, S., von der Mark, K. and Schmuki, P. (2007) Nanosize and vitality: TiO_2 nanotube diameter directs cell fate. *Nano Letters* **7**, 1686–1691.

Parkhutik, V.P. and Shershulsky, V.I. (1992) Theoretical modelling of porous oxide growth on aluminium. *Journal of Physics D: Applied Physics* **25**, 1258.

Parkinson, L.G., Giles, N.L., Adcroft, K.F., Fear, M.W., Wood, F.M. and Poinern, G.E. (2009). The potential of nanoporous anodic aluminium oxide membranes to influence skin wound repair. *Tissue Engineering Part A* **15**, 3753–3763.

Popat, K.C., Leary Swan, E.E., Mukhatyar, V., Chatvanichkul, K.I., Mor, G.K., Grimes, C.A. and Desai, T.A. (2005). Influence of nanoporous alumina membranes on long-term osteoblast response. *Biomaterials* **26**, 4516–4522.

Popat, K.C., Leoni, L., Grimes, C.A. and Desai, T.A. (2007). Influence of engineered titania nanotubular surfaces on bone cells. *Biomaterials* **28**, 3188–3197.

Roy, P., Berger, S. and Schmuki, P. (2011). TiO$_2$ nanotubes: Synthesis and applications. *Angewandte Chemie International Edition* **50**, 2904–2939.

Sanchez, C., Boissière, C., Grosso, D., Laberty, C. and Nicole, L. (2008). Design, synthesis, and properties of inorganic and hybrid thin films having periodically organized nanoporosity. *Chemistry of Materials* **20**, 682–737.

Shalabi, M.M., Gortemaker, A., Van't Hof, M.A., Jansen, J.A. and Creugers, N.H.J. (2006). Implant surface roughness and bone healing: A systematic review. *Journal of Dental Research* **85**, 496–500.

Shi, J., Votruba, A.R., Farokhzad, O.C. and Langer, R. (2010). Nanotechnology in drug delivery and tissue engineering: From discovery to applications. *Nano Letters* **10**, 3223–3230.

Soler-Illia, G.J.A.A., Sanchez, C., Lebeau, B. and Patarin, J. (2002). Chemical strategies to design textured materials: from microporous and mesoporous oxides to nanonetworks and hierarchical structures. *Chemical Reviews* **102**, 4093–4138.

Song, Y.Y., Schmidt-Stein, F., Bauer, S. and Schmuki, P. (2009). Amphiphilic TiO$_2$ nanotube arrays: An actively controllable drug delivery system. *Journal of the American Chemical Society* **131**, 4230–4232.

Streicher, R.M., Schmidt, M. and Fiorito, S. (2007). Nanosurfaces and nanostructures for artificial orthopedic implants. *Nanomedicine* **2**, 861–874.

Swan, E.E.L., Popat, K.C., Grimes, C.A. and Desai, T.A. (2005). Fabrication and evaluation of nanoporous alumina membranes for osteoblast culture. *Journal of Biomedical Materials Research Part A* **72**, 288–295.

Takoh, K., Takahashi, A., Matsue, T. and Nishizawa, M. (2004). A porous membrane-based microelectroanalytical technique for evaluating locally stimulated culture cells. *Analytica Chimica Acta* **522**, 45–49.

Tao, S.L. and Desai, T.A. (2003). Microfabricated drug delivery systems: From particles to pores. *Advanced Drug Delivery Reviews* **55**, 315–328.

Thomas, V., Dean, D.R. and Vohra, Y.K. (2006). Nanostructured biomaterials for regenerative medicine. *Current Nanoscience* **2**, 155–177.

Tirrell, M., Kokkoli, E. and Biesalski, M. (2002). The role of surface science in bioengineered materials. *Surface Science* **500**, 61–83.

Tran, P.A., Sarin, L., Hurt, R.H. and Webster, T.J. (2009). Opportunities for nanotechnology-enabled bioactive bone implants. *Journal of Materials Chemistry* **19**, 2653–2659.

Vallet-Regí, M., Rámila, A., del Real, R.P. and Pérez-Pariente, J. (2001). A new property of MCM-41: Drug delivery system. *Chemistry of Materials* **13**, 308–311.

Variola, F., Vetrone, F., Richert, L., Jedrzejowski, P., Yi, J.H., Zalzal, S., Clair, S., Sarkissian, A., Perepichka, D.F., Wuest, J.D., Rosei, F. and Nanci, A. (2009). Improving biocompatibility of implantable metals by nanoscale modification of surfaces: An overview of strategies, fabrication methods, and challenges. *Small* **5**, 996–1006.

Velev, O.D., Jede, T.A., Lobo, R.F. and Lenhoff, A.M. (1997). Porous silica via colloidal crystallization. *Nature* **389**, 447–448.

von Wilmowsky, C., Bauer, S., Lutz, R., Meisel, M., Neukam, F.W., Toyoshima, T., Schmuki, P., Nkenke, E. and Schlegel, K.A. (2009). In vivo evaluation of anodic TiO$_2$ nanotubes: An experimental study in the pig. *Journal of Biomedical Materials Research Part B: Applied Biomaterials* **89**, 165–171.

Wan, Y. and Zhao, D.Y. (2007). On the controllable soft-templating approach to mesoporous silicates. *Chemical Reviews* **107**, 2821–2860.

Wood, W. and Martin, P. (2002). Structures in focus – filopodia. *The International Journal of Biochemistry and Cell Biology* **34**, 726–730.

Xia, W. and Chang, J. (2006). Well-ordered mesoporous bioactive glasses (MBG): A promising bioactive drug delivery system. *Journal of Controlled Release* **110**, 522–530.

Yan, H., Zhang, K., Blanford, C.F., Francis, L.F. and Stein, A. (2001). In vitro hydroxycarbonate apatite mineralization of CaO-SiO$_2$ sol-gel glasses with a three-dimensionally ordered macroporous structure. *Chemistry of Materials* **13**, 1374–1382.

Yan, X., Yu, C., Zhou, X., Tang, J. and Zhao, D. (2004). Highly ordered mesoporous bioactive glasses with superior in vitro bone-forming bioactivities. *Angewandte Chemie International Edition* **43**, 5980–5984.

Yang, Q., Wang, S., Fan, P., Wang, L., Di, Y., Lin, K. and Xiao, F.S. (2005). pH-responsive carrier system based on carboxylic acid modified mesoporous silica and polyelectrolyte for drug delivery. *Chemistry of Materials* **17**, 5999–6003.

Zhang, H. and Cooper, A.I. (2005). Synthesis and applications of emulsion-templated porous materials. *Soft Matter* **1**, 107–113.

Zhao, D., Yang, P., Melosh, N., Feng, J., Chmelka, B.F. and Stucky, G.D. (1998a). Continuous mesoporous silica films with highly ordered large pore structures. *Advanced Materials* **10**, 1380–1385.

Zhao, D., Feng, J., Huo, Q., Melosh, N., Fredrickson, G.H., Chmelka, B.F. and Stucky, G.D. (1998b) Triblock copolymer syntheses of mesoporous silica with periodic 50 to 300 angstrom pores. *Science* **279**, 548–552.

Zhou, Z., Zhu, S. and Zhang, D. (2007) Grafting of thermo-responsive polymer inside mesoporous silica with large pore size using ATRP and investigation of its use in drug release. *Journal of Materials Chemistry* **17**, 2428–2433.

Zwilling, V., Aucouturier, M. and Darque-Ceretti, E. (1999). Anodic oxidation of titanium and TA6V alloy in chromic media. An electrochemical approach. *Electrochimica Acta* **45**, 921–929.

4

Layer-by-layer self-assembly techniques for nanostructured devices in tissue engineering

R. R. COSTA and J. F. MANO, 3B's Research Group –
Biomaterials, Biodegradables and Biomimetics,
University of Minho, Portugal

DOI: 10.1533/9780857097231.1.88

Abstract: Devices prepared by bottom-up strategies have been proposed for several biomedical applications. Among various self-assembly techniques, layer-by-layer (LbL) offers ease of preparation, versatility, fine control over the materials structure, and robustness under physiological conditions. Due to its self-assembled nature and use of small masses of material, coatings can be produced using aqueous solutions under mild conditions of temperature, pH, and pressure, with minimal energy requirements and material waste. Herein, we review the multitude of LbL mechanisms for constructing nanostructured devices relevant to applications such as tunable cell adhesion and metabolism, drug delivery, and biomineralized and smart systems.

Key words: layer-by-layer, self-assembly, biomaterials, tissue engineering.

4.1   Introduction

The modification of surfaces has been a key aspect in biological and bio-technological applications, including cell expansion, biomaterials development, and preparation of substrates for regenerative medicine (Falconnet *et al.*, 2006; Hook *et al.*, 2009; Tirrell *et al.*, 2002). In the field of implantable biomaterials and tissue engineered constructs, the bulk material properties are generally considered to determine the overall properties of a biomaterial. Surface properties, however, are of utmost importance as the surface constitutes the first interface between an implant and the organism, and will drive subsequent tissue and cellular events including protein adsorption, cell adhesion and inflammatory response, all of which are necessary for tissue remodelling (Boudou *et al.*, 2010; Hubbell, 1999; Stevens and George, 2005). Thus, surface engineering is highly important for developing devices with improved biological performance.

Many processing techniques have been suggested and employed for the purpose of modifying substrates (Prakash *et al.*, 2009), but in the past

88

two decades alternatives to common surface engineering strategies have emerged. Inspired by nature, scientists and engineers have developed techniques based on biomimetic concepts. In nature, evolutionary processes have constantly improved the design and functionality of natural materials. A well-known example is the layered organization of nacre found in the shells of sea animals. This layered and hierarchical structure containing calcium minerals is capable of dissipating high amounts of mechanical energy (Luz and Mano, 2009; Mayer, 2005). Mimicking natural layered structures has been attempted in the laboratory. In particular, the alternate adsorption of polyanions and polycations onto solid substrates is a reliable, flexible, and simple approach to modify a substrate while allowing the integration of a wide range of molecules, including those with biological relevance. As a result, multilayered thin films can be constructed with nanometric precision. This technique was first introduced by Decher, Hong and Schmitt and has denominated the layer-by-layer (LbL) field (Decher *et al.*, 1992).

LbL is a self-assembly-driven surface modification strategy that allows the construction of nanostructured films onto substrates of any geometry, from simple bidimensional surfaces to more complex three-dimensional porous scaffolds. The underlying principle lies in the existence of multiple intermolecular interactions, such as electrostatic contacts, hydrophobic interactions and hydrogen bonding, where the cooperative effects of multipoint attractions play the most important role. It is a technique that is easy to perform, versatile, offers fine control over the materials structure, and produces materials that are robust in physiological environments (Clark and Hammond, 2000; Gribova *et al.*, 2012; Lyklema and Deschênes, 2011). Because this technique may be performed using various complementary substances, a vast choice of available materials may be added to a multilayer formulation, such as proteins, nucleic acids, polysaccharides, virus particles, organic polymers, molecular assemblies, and inorganic substances, allowing construction of LbL assemblies with very specific properties (Boudou *et al.*, 2010). The build-up can be performed at mild conditions, namely standard laboratory temperatures and pressures, and mild pH, and can be tuned by making small adjustments in such conditions. Moreover, LbL does not involve the use of organic solvents. The mild and non-aggressive processing conditions of LbL make it an appealing technique for tissue engineering applications. Molecules with biological relevance may be embedded in the multilayer materials, with bioactivity preserved, particularly important for drug delivery applications. Biomaterials and ligands with more specific functions, such as cell adhesion enhancers, may be used in order to mimic biological structures and functions, rendering a substrate more instructive for cell behaviour (Ariga *et al.*, 2008; Bertrand *et al.*, 2000; Tang *et al.*, 2006). Exploiting the advantages of LbL, multilayer systems have already been proposed for various biomedical applications. An example includes bio-mimetic composite coatings that regulate drug release or manipulate the adhesion, differentiation, proliferation, and even function of attached cells. In the following sections, these applications and associated multilayer formulations will be reviewed.

4.2   Interaction between biomaterials as ingredients for multilayer formulations

One of the key features of LbL adsorption is the ability to use ingredients from distinct classes of materials. Usually the most important requirement is that two distinct materials possess complementarity, that is, the properties of both materials result in attraction. A well-known and highly exploited example is the self-assembly driven by electrostatic interactions of oppositely charged components, in which polyanions and polycations interact and assemble (Hammond, 2004). Besides electrostatic self-assembly (ESA), other types of interactions that can drive the construction of multilayers include short-range interactions, such as van der Waals forces and hydrogen bonding (Clark and Hammond, 2000). The strength of these interactions may depend on several intrinsic factors, such as the molecular weight of polymers, the size of inorganic particles, charge density, polarity, and also external factors like temperature, the pH of the solvent, and ionic strength (see following sections). Importantly, it is now acknowledged that the driving force behind multilayer formation is not only mutual interaction, but also the gain in entropy due to the release of counter-ions (V. Klitzing, 2006; Lyklema and Deschênes, 2011). The variation of one or more of these parameters can lead to a more favourable interaction and consequent build-up of a nanostructured coating. The result is, however, dependent on the original properties of the materials: for instance, varying the pH of a weak polyelectrolyte will greatly affect its overall charge density. Several biomaterials of synthetic and natural origin have been used since LbL was first introduced, many of which are capable of being used for multilayered constructs while displaying biologically relevant properties.

4.2.1   Build-up mechanisms of synthetic and inorganic materials

Many types of charged molecules and nano-objects are suitable for deposition by ESA. So far, the most popular deposition materials have been synthetic polyelectrolytes. Synthetic materials are usually man-made and can be highly functional (Fetters et al., 1994). With the availability of thousands of monomers and their seemingly endless combinations, polymer science has developed a wide array of complex but well-defined structures with specific properties, such as molecular weight and charge density. Other materials suitable for deposition using ESA are inorganic (e.g., calcium-containing or clay mineral particles) or metallic materials, usually consisting of a single geometrical charged structure (e.g., spheres or platelet-like objects) with natural or artificial origins (Fendler, 1996; Zabet-Khosousi and Dhirani, 2008).

Synthetic polymers, though not often similar to biological structures, have been used to define and quantify parameters important for the construction of nanostructured multilayer coatings. The main advantages of synthetic polymers are that they are easily chemically modified and they enable certain buildup parameters to be precisely adjusted. The pioneers of LbL proved its concept by creating ultrathin homogeneous films on solid surfaces (Decher et al., 1992). Positively charged aminopropylsilanized fused quartz or silicon single crystal substrates served as supports for the ESA build-up. Poly(styrene sulfonate) (PSS) and poly(vinyl sulfate) (PVS) served as polyanions and poly(allylamine hydrochloride) (PAH) and poly-4-vinylbenzyl-(N,N-diethyl-N-methyl)-ammonium iodide as polycations. Immersing the positive support in an aqueous solution of one of the polyanions resulted in the adhesion of the polymeric molecules and reversal of charge density of the substrate. A rinsing step followed to remove loosely adsorbed molecules. Then the substrate was exposed to a solution of one of the polycations, again reversing the charge. By repeating this process, the stepwise construction of a coating was achieved. With 39 layers, Decher et al. were able to produce films as thin as 44 nm.

The striking simplicity of fabricating a wide range of LbL-modified substrates was rapidly recognized by the scientific community. Many studies with different materials collectively identified the key parameters influencing the fabrication of nanostructured multilayer films and determined how to optimize them. The pH of the polyelectrolyte solutions used during build-up has been studied by using poly(acrylic acid) (PAA) and PAH to fabricate polyelectrolyte multilayers (PEMs) (Shiratori and Rubner, 2000; Yoo et al., 1998). Varying the pH of the dipping solutions considerably changes the charge density of a weak polyelectrolyte when operating above or below the pK_a of the polymer. When adjusting the pH of PAA and of PAH above and below their pK_a, respectively, a higher degree of ionization could be exploited, favouring ESA-driven adsorption. In the opposite case, when operating near the acidity constant, the charge density is at its lowest. Consequently, fewer charged molecules are available to form a new top layer and more polyelectrolyte molecules are needed to interact with the underlying layer for ESA (Izumrudov et al., 1999; Schlenoff and Dubas,

2001; Tang *et al.*, 2006). This affects the internal structure of the film in terms of layer conformation since adjusting the pH modifies the charge density of the polyelectrolytes. An additional observation is that increasing the charge density of the adsorbing polymer will favour thinner polyelectrolyte layers, whereas increasing the charge density at the surface will favour thicker adsorbed layers. The latter enables the production of thicker and rougher films, while the former results in thinner and smoother films (Fu *et al.*, 2005; Yoo *et al.*, 1998).

The addition of salts to the adsorbate solutions has been studied since ions can interfere with the electrostatic balance of molecules and the internal PEM structure. PAH, PSS and poly(diallyldimethylammonium) (PDADMA) have been used to demonstrate that the addition of salt increases the thickness of the layers due to a screening mechanism of charge–charge repulsions (Chen and McCarthy, 1997; Schlenoff *et al.*, 1998). Basically, the electrostatic repulsion among groups with the same charge is reduced by the interaction with salt ions present in solution and allows for a more stable build-up and interaction between layers. The adsorption process can be increased with increasing concentrations of salt, but since this mechanism is somewhat similar to that of pH, high concentrations of salt may render the materials uncharged, which may in turn induce desorption or 'stripping' rather than adsorption (Hoogeveen *et al.*, 1996; Linford *et al.*, 1998).

Temperature also affects the build-up mechanism of multilayer films. However, while most studies on temperature focus on its effect on stability and functionality after construction, few study temperature as a build-up variable. A study by Salomäki, Vinokurov and Kankare (Salomäki *et al.*, 2005) used films made of PSS and PDADMA to verify that temperature has a remarkable effect on the apparent mass of the resulting films. In aqueous solutions containing sodium bromide, the mass of adsorbed material at 55°C was approximately 5 times higher than a construction performed at a temperature of 25°C, along with a tendency for the mass to increase exponentially. In the same report, similar results were found for PSS and PAH films. The conclusion was that a higher temperature during deposition increases the mobility of the adsorbed molecules, resulting in deeper layer interpenetration through the PEMs and a greater likelihood of finding more energetically optimal binding sites at the deposition site.

Polyelectrolyte multilayers made using synthetic polymers have been extended to applications other than simple interface tuning. For instance, a dye may be used as film ingredient or it may be loaded after film assembly (Becker *et al.*, 2010). The latter takes advantage of the swelling properties of multilayer films, in which a dye can be loaded – and reversibly unloaded – into the film by immersion in a saturated solution. Ideally, a dye may be replaced by a therapeutic molecule or growth factor to act as a drug delivery device upon implantation or for use in cell culture. The versatility of LbL is not limited to polymers. Other extended functionalities include the use of inorganic particles, such as bioactive nanoparticles (BNPs), to develop surfaces that could mineralize into calcium phosphate crystals under the presence of a simulated body fluid (Couto *et al.*, 2009). Carbon nanotubes have also been used as LbL ingredients along with polyelectrolytes for substrates with improved mechanical properties (Komatsu, 2012; Liang *et al.*, 2003; Mamedov *et al.*, 2002). Work with clay platelets (Elzbieciak *et al.*, 2010) and graphene (Kong *et al.*, 2009) has also taken advantage of the LbL methodology, although polymers offer more mobility to the internal structure of the film. In summary, LbL can be used as a means to coat surfaces and control the interface in almost any fashion desired, due to the various mechanisms that influence its thickness, roughness and loading capabilities. Such surface property control makes LbL a powerful technique to develop tunable cutting-edge devices whose surfaces influence the fate of cells, important for a range of applications in conventional medicine and tissue engineering.

4.2.2   Natural-based polymeric materials

Natural materials are intricate structures which have arisen from hundreds of millions of years of evolution. Compared to current technology, natural structures are more multifaceted than man-made materials, forming complex arrays and hierarchical structures with multiple functionalities (Ashby and Jones, 2006; Langer and Tirrell, 2004; Meyers *et al.*, 2008). The advantage of using natural-based polymers for biomedical and tissue engineering applications lies in their similar with biological macromolecules, thus avoiding triggering chronic or immunological reactions and toxicity (Mano *et al.*, 2007). Natural polymers are generally more similar to the extracellular matrix (ECM) than synthetic materials and are therefore appropriate multifunctional ingredients for constructing multilayer films and coatings which better interface with biological environments. The range of potential LbL ingredients spans the gamut from synthetic to natural materials, including polysaccharides, polypeptides, proteins, nucleic acids, DNA, and even viruses, allowing highly biomimetic surfaces to be engineered (Boudou *et al.*, 2010; Chen and McCarthy, 1997; Gribova *et al.*, 2011; Lyklema and Deschênes, 2011; Tang *et al.*, 2006).

One of the natural polymers most exploited to construct multilayer films is hyaluronic acid (HA). Hyaluronic acid, also known as hyaluronan or hyaluronate, is an important glycosaminoglycan component of the connective tissue, synovial fluid and the vitreous humour of the eye (Drury and Mooney, 2003). It is a linear polysaccharide composed of numerous repelling anionic groups capable of binding cations and water molecules. Solutions of HA exhibit clear viscoelastic properties that make them excellent biological absorbers and lubricants. HA may also be obtained through a process of microbial fermentation, avoiding the risk of animal-derived pathogens (Sutherland, 1998).

Several authors have focused on the use of HA as a polyanion (Chua et al., 2008; Johansson et al., 2005; Picart et al., 2001, 2005b). In particular, HA has been combined with poly(L-lysine) (PLL) (Garza et al., 2004; Khademhosseini et al., 2004; Zhang et al., 2005). PLL is a linear polypeptide containing an amino group and has been widely used in biomaterial applications to enhance cell adhesion (Mazia et al., 1975; Richert et al., 2002; Veerabadran et al., 2007). A thorough investigation of the PLL/HA system provided a better understanding of the exponential growth mechanism of layer thickness, as opposed to the linear growth mechanism introduced by Decher (Elbert et al., 1999; Picart et al., 2002; Porcel et al., 2007). For this system, the exponential growth mechanism relies on the diffusion of at least one of the polymers 'in-and-out' of the molecular chains during each 'bilayer' deposition step. In these studies, it was PLL that diffused through the film.

Polysaccharides of marine origin form another important class of natural materials. Chitosan (CHI) is a polysaccharide exhibiting cationic properties and solubility under acidic conditions. Its precursor, chitin, is the second most important natural polymer in the world and can be extracted from marine crustaceans including shrimp and crabs. CHI possesses several characteristics favourable for promoting dermal regeneration and accelerated wound healing (Rinaudo, 2006). Alginate (ALG) is another polysaccharide of marine origin, namely from brown algae. Its structure contains carboxyl groups, which makes it negatively charged (Percival and McDowell, 1990). Both of these polysaccharides possess adhesive, non-toxic, bacteriostatic, fungistatic, haemostatic, and anti-microbial properties, making them ideal for tissue engineering applications. Because of these unique features, CHI and ALG have been frequently used in biomedical applications including burn dressings, drug delivery, and implants (Berger et al., 2004; Boateng et al., 2008; George and Abraham, 2006). They have been used as LbL ingredients, both together or in conjugation with other materials. PLL/ALG and CHI/HA systems have been shown to possess an internal structure typically produced by the exponential growth mechanism (Elbert et al., 1999; Picart et al., 2001). Martins et al. (2010) demonstrated the loading capability of CHI/ALG films with human serum albumin (HSA) which depended on the availability of amino groups to bind the negatively charged protein. Crosslinking with glutaraldehyde reduced the amino group availability and HSA adsorption compared to uncrosslinked films.

The molecular weight of LbL ingredients is an important property influencing the build-up mechanism. However, few studies using natural polymers have systematically investigated the influence of polymer molecular weight on the physical properties of polyelectrolyte multilayers, since it is difficult to obtain monodisperse polymers from natural sources. As such, studies related to molecular weight have mostly focused on the use of well-characterized synthetic polymers (Ariga *et al.*, 1997; Stockton and Rubner, 1997; Sui *et al.*, 2003). Nevertheless, using CHI and HA films, Kujawa *et al.* (2005) found that the thickness increased when the molecular weight of these polysaccharides was higher. They demonstrated that the higher the molecular weight the thicker the film: a longer polymer tends to adsorb in a 'loopier' way, while a smaller molecule adsorbs in an extended conformation. Therefore, with longer polymers, there are more anchoring sites available for the next layer to adsorb and potentially higher interpenetrating capabilities than with smaller molecules.

The mechanical properties of multilayer films can affect cell adhesion and differentiation. Increasing the Young's modulus (i.e. stiffness) of multilayer films improves cell adhesion (Richert *et al.*, 2004; Schneider *et al.*, 2006, 2007). This behaviour makes LbL substrates suitable as tunable platforms to control cell adhesion. Moreover, LbL substrates with varying support stiffness could potentially induce stem cell differentiation. Multilayers can also be used as reservoirs for therapeutic molecules (discussed below) and can incorporate naturally occurring smart biomaterials. By fabricating a film containing stimuli-responsive ingredients, interface properties could be changed on demand.

All films constructed by ESA from weak polyelectrolytes – synthetic or natural – are inherently pH-responsive. Properties like Young's modulus, roughness, and wettability may be tailored by varying the film assembly conditions, due to the dissociation of the weak acidic and alkaline functional groups on the chains. Wood *et al.* (2005) demonstrated the pH controlled degradation of a film, from which therapeutic molecules were released to the bulk medium. Ionic strength may also be used as a stimuli-response mechanism for films made of weak, and, unlike pH, strong polyelectrolytes (Chen and McCarthy, 1997; Hoogeveen *et al.*, 1996; Linford *et al.*, 1998; Schlenoff *et al.*, 1998). A screening effect caused by the interaction between free counterions and charged polymer groups leads to a reduction of the electrostatic repulsion between charged groups of the same sign. Such mechanisms are potentially useful for drug delivery, because they could provide sustained and controlled drug release with greater effectiveness, lower toxicity and improved patient convenience over conventional formulations. However, the sharp variations of pH and ionic strength associated with these mechanisms may be harmful for cells. On the other hand, cells are less sensitive to temperature variations, at least for temperatures near 37°C. Therefore, temperature responsive materials could be a more favourable choice for use with cells. One of the most popular temperature responsive polymers is poly(N-isopropylacrylamide) (PNIPAAm). PNIPAAm has a critical solution temperature (LCST) around 32°C enclosing both hydrophobic and hydrophilic domains (Schild, 1992). Below the LCST, the polymer is soluble in aqueous solution, with an expanded or hydrated coil conformation; above it, the polymer becomes insoluble, preferring a folded globular structure. PNIPAAm is a synthetic material, but combinations with natural polymers are possible. CHI, a weak polycation, has been grafted with PNIPAAm to construct films with dual responsiveness (Martins *et al.*, 2011). Elastin-like recombinamers (ELRs) are another class of temperature-responsive materials; a class of the elastin-like polymers (ELPs) family, they are genetically engineered peptide-based macromolecules obtained through the fermentation of modified bacterial strains (Rodríguez-Cabello *et al.*, 2009). ELRs are rooted in the peptide sequence of natural elastin present in tissues that require elasticity as part of their function, such as skin, tendons and muscles. In an aqueous environment they are known to self-assemble and exhibit a transition temperature (T_t): below T_t, the free polymer chains adopt random coil conformations; above it, they fold into organized structures known as β-spirals. Although ELRs respond primarily to temperature, the effects of other stimuli, such as pH, ionic strength, and concentration also affect the transition phenomenon. Their genetically-engineered origin also allows motifs to be integrated that have biological relevance, such as cell adhesion (Girotti *et al.*, 2004) and mineralization enhancers (Prieto *et al.*, 2011). In LbL, they have already been suggested as suitable ingredients for constructing films with multiple stimuli-responsive properties (Barbosa *et al.*, 2009; Costa *et al.*, 2011b; Golonka *et al.*, 2011). They are an elegant example of how nature inspires engineers to develop new multifunctional materials for tissue engineering applications.

4.3   Scalability to three dimensions

LbL was first introduced as a surface engineering technique for planar surfaces. However, this technology can be used to coat surfaces on substrates or templates with a variety of shapes and sizes beyond planar. As a result, devices such as scaffolds that could only be fabricated through conventional methods may now be engineered to exhibit new tunable properties according to the desired application.

4.3.1   Membranes

Most polyelectrolyte multilayers may be assembled with up to dozens of layers to a thickness that may reach tens of micrometres. However, PEMs are assembled and adhere to underlying substrates and cannot be detached without damaging the films. Detached films, usually called 'free-standings', are potentially useful for a range of nanostructured membrane based devices for a range of sensing, detection, and drug delivery applications (Choi *et al.*, 2005; Jiang and Tsukruk, 2006).

A typical approach to fabricate free-standings involves a sacrificial layer or substrate that is dissolved after build-up (Lee *et al.*, 2009; Ott *et al.*, 2010). Cellulose acetate, for instance, can be used as a substrate and dissolved with acetone after build-up (Mamedov and Kotov, 2000). A potential disadvantage of such an approach is that the films may undergo both chemical and physical changes during the post-processing steps to separate the film from the underlying substrate and interfere with the internal structure of the multilayers. More recently, other strategies have been introduced that avoid using chemicals to remove the template. Hydrophobic substrates have been used to assemble free-standing PEMs, exploiting a peeling mechanism of the film from the substrate. Larkin *et al.* (2010) used Teflon and polypropylene as planar templates from which CHI and HA films could be easily peeled without damage. Another method consists of assembling films virtually 'on air without templates'. Mallwitz and Laschewsky (2005) prepared ultrathin freestanding arrays of LbL films in the holes of electron microscopy grids (100×100 µm). The precise mechanism was not clear. Apparently, the pores were so small that the multilayer ingredients were suspended across the orifices instead of coating their walls, so that thin polymeric lamellae could remain trapped in the pores. However, the authors did not develop a strategy to extract the free-standings from the grid.

4.3.2   Capsules

LbL has been proposed as a technology to produce capsules as tunable drug carriers for drug delivery applications and cell differentiation (De Cock *et al.*, 2010; Peyratout and Dähne, 2004). Typically, the template for a hollow or loaded capsule is a nano- or micro-particle. Similar to some free-standings, the template particle is sacrificial. After assembling the multilayers, the core is removed with a solvent. The result is a hollow multilayer shell which may encapsulate molecules of interest such as DNA and proteins (Caruso *et al.*, 1998; Liang *et al.*, 2003). The sacrificial cores were organic microparticles (typically 1–10 µm in diameter) such as polystyrene or melamine formaldehyde, which were then dissolved by organic solvents or strong acids (Donath *et al.*, 1998). The drawback of such an approach is that it employs aggressive solvents that could damage the remaining shell and destroy the functionality and purpose of the structure. Also, organic residues from the aggressive solvents could affect biological environments, making the approach less appealing for tissue engineering. Such drawbacks led to the introduction of other templates that could be dissolved under less aggressive conditions, such as mesoporous silica (with diameters of 500 nm to 5 μm, dissolved in hydrogen fluoride) and porous calcium carbonate (with diameters of 3–5 μm, dissolved in ethylenediamine tetraacetic acid, EDTA) (Mauser *et al.*, 2006; Wang and Caruso, 2006).

The concept of permeability after build-up is of great interest for drug delivery and tissue engineering applications. A carrier must be able to retain and preserve the drug and release it when necessary in a controlled fashion. The most obvious way to control such permeability is to vary the number of layers assembled around a template: more layers result in higher resistance to molecular diffusion (She *et al.*, 2010). In order to fill the interior of a multilayer capsule with a drug, both pre- and post-loading encapsulation mechanisms have been exploited (De Geest *et al.*, 2007). In pre-loading, the sacrificial core may already contain the molecule of interest that will remain upon dissolution, or crystals of the molecule of interest that can be used as a template and liquefied after assembly, without removal. In the post-loading approach, prefabricated hollow capsules can be loaded by altering the permeability of the capsule shell (Köhler and Sukhorukov, 2007; Sukhorukov *et al.*, 2001). Several mechanisms based on the effect of external variables have been exploited to reversibly change the permeability of a multilayer shell. Multilayer capsules are inherently responsive to pH and ionic strength due to their interference with the electrostatic balance between layers. The permeability of such responsive capsules was studied by Shutava *et al.* (2005) for capsules based on tannic acid and PAH. The capsules seemed impermeable to fluorescein isothiocyanate (FITC)-dextrans at neutral pH, but became permeable at both low and high pH values. Ibarz *et al.* (2001) reported that the permeability of hollow PSS/PAH polyelectrolyte capsules sharply improved upon increasing the concentration of salts. Temperature has also been suggested as a mechanism for post-loading. It is believed that a rearrangement occurs with increasing temperature due to a tendency of shells to reduce the shell/water interface (Köhler *et al.*, 2006). However it has only been achieved for temperatures in the range 50–70°C, and is thus not suitable for biological applications (Bédard *et al.*, 2009; Glinel *et al.*, 2003; Huang and Chang, 2009). For drug release, the same mechanisms used to post-encapsulate a molecule of interest could also be used to modulate its release in a controlled fashion.

4.3.3   Tubes

Hollow tubes of micron or submicron diameters and high aspect ratios are interesting for various applications. They possess large specific surface areas, making them appealing for catalysis, sensing, or as tissue scaffolding (Murugan and Ramakrishna, 2006; Sanchez-Castillo *et al.*, 2004; Wang *et al.*, 2004). Hollow tubes could be used for encapsulation purposes and find application in fields like drug delivery and as transport channels in microfluidics (Choi *et al.*, 2009). In biology, tubes of submicron dimensions, such as protein nanotubes and microtubules are commonly found and play an essential role providing mechanical reinforcement in cells and small blood vessels (Komatsu, 2012).

Tube-like structures may be fabricated in a similar manner to other shapes. Templates (cylindrical in this case) are coated and subsequently dissolved under conditions that do not destroy the coating, thus only tube-like objects remain. To date, this morphology has not been extensively used and only a few reports exist of multilayer nanostructured hollow tubes compatible with tissue engineering applications. Tube-like structures have previously been fabricated by assembling polyelectrolytes onto micrometric glass fibres that were then dissolved by hydrogen fluoride (Mueller *et al.*, 2007). Another strategy to fabricate tube-like structures is to coat porous membranes (the templates) with polyelectrolytes, which in turn adsorb to the wall of the pores, after which the membranes are dissolved (Liang *et al.*, 2003). He *et al.* (2008) pursued an interesting approach by assembling tubes of PSS and PAH using the porous membrane template method and transforming them into capsules by varying temperature. This structural transition from tubular to vesicular or vesicle-like geometry is often considered for drug delivery, gene delivery, self-assembled nanoreactors, or biochemical sensors (Vriezema *et al.*, 2005).

In the future, multilayer tubular structures may be of significant interest in tissue engineering. For instance, nanostructured artificial blood vessels could be developed as an alternative to the conventional autologous grafting treatment by the incorporation of collagen and elastin to mimic the intima, media and adventitia layers (Koens *et al.*, 2010; Patel *et al.*, 2006).

4.3.4   Three-dimensional scaffolds

LbL is not limited to the simple geometries shown thus far. The technique may be applied to more complex geometries by using more intricate structures as templates. Porous agglomerates of particles, for instance, can be coated and, as for previous structures, the template can be chelated, leaving only the multilayer assembly to serve as a support for cell culture. Miranda *et al.* (2011) and Sher *et al.* (2010) have developed nanostructured multilayer structures based on agglomerates of particles of chitosan (not leachable) and paraffin (leachable), respectively. Their work, which will be referred to again in the next section, demonstrates that more complex and functional structures may be made using LbL.

4.4   Application of nanostructured multilayer devices in tissue engineering

The numerous possibilities for adjusting the chemical, physical, and mechanical properties of multilayer films have garnered attention from tissue engineers hoping to develop systems to control protein and cell adhesion and release therapeutic molecules and growth/differentiation factors. In addition, researchers wish to tune the surface properties of such microsystems by adding specific ligands or stimuli-responsive biomaterials.

4.4.1   Controlling cell adhesion and proliferation

When a biomedical device comes in contact with a cell-populated biological environment, adhesion is the first event to occur. Adhesion will be followed by proliferation and, in the case of stem cell cultures, differentiation.

We have already mentioned that changing the mechanical properties of a multilayer film may affect cellular adhesion. Schneider *et al.* (2006) studied the cell adhesion of HCS-2/8 human chondrosarcoma cells on PLL/HA films with various degrees of crosslinking. The Young's modulus of the film could be varied over two orders of magnitude simply by varying the concentration of a crosslinker, carbodiimide, from 3 to 400 kPa for the lowest and highest concentrations, respectively. Increasing the film stiffness also increased the roughness and number of attached cells, demonstrating that a simple modification to the mechanical properties of multilayer films can affect their response in a biological environment. Phelps *et al.* (2011) also applied this concept to selectively crosslink the outer region of a PLL and poly(glutamic acid) (PGA) film, resulting in a rigid outer skin to promote cell attachment, while leaving the film interior (and any embedded bioactive species) unaffected. Crosslinked CHI and HA films have also been studied *in vivo* by implantation in a rat oral cavity (Etienne *et al.*, 2005b). Crosslinked films demonstrated up to 60% higher resistance to degradation compared to uncrosslinked samples.

Multilayer films may be engineered with anti-fouling properties to control cellular and bacterial adhesion by enhanced resistance to serum proteins.

A straightforward way to build 'stealth' substrates of any morphology is to use derivatives of poly(ethylene glycol) (PEG), a highly hydrated polymer effective in rendering surfaces non-adsorbent to proteins (Ochs *et al.*, 2008; Wagner *et al.*, 2004; Wattendorf *et al.*, 2008). Another strategy is to add antimicrobial agents. Etienne *et al.* have added defensin (Etienne *et al.*, 2004) and chromofungin (Etienne *et al.*, 2005a) to the film constitution and reported growth inhibitions of up to 98% for *Escherichia coli* with defensin and 65% for *Candida albicans* with chromofungin. Rendering a surface non-fouling can serve as immuno-'camouflage' to prevent the immune rejection of implanted biomaterials or to enhance the efficiency of injected drug delivery vehicles.

The application of LbL to more complex structures is highly important for tissue engineering applications such as cell culture platforms, where hierarchical organization better resembles native tissues. Besides bidimensional, spherical, and tubular shapes, more complex geometries such as agglomerated shapes can be fabricated and used as scaffolds for cell culture. LbL may be used to agglomerate beads (Miranda *et al.*, 2011) or to coat packet-free leachable templates (Sher *et al.*, 2010) leading in both cases to porous structures that may be populated by cells. Recently, Ye *et al.* (2012) coated a decellularized porcine aortic valve with CHI and heparin showing anti-thrombotic properties. The device was shown to be haemocompatible by reducing platelet adhesion while allowing the proliferation of endothelial progenitor cells. The work of Sher *et al.* (2010) introduced an interconnected three-dimensional porous scaffold using a new drop-wise perfusion technique over a three-dimensional (3D) template formed by free-form mouldable paraffin wax spheres held together only by the LbL coating (Fig. 4.1). The paraffin was dissolved by dichloromethane, leaving only the assembled layers as a scaffold. SaOs-2 human osteoblast-like cells were seeded and then attached and proliferated. The authors did not characterize the system for its mechanical properties, however. It is possible that the resulting structure is not adequate for applications involving high loads, such as bone regeneration, but may be more appropriate for regenerating soft tissues, such as cartilage or nerves.

4.4.2   Controlling cellular metabolism: encapsulation and differentiation

Multilayer films may be of use for cell encapsulation since the films can coat various cell types and can even be used to build multilayered cell architectures. Developing a device that provides immune protection via micro-encapsulation as well as enhanced control of both transport and surface

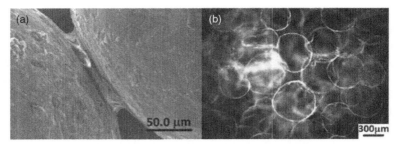

4.1 (a) Individual connections between two paraffin wax particles.
(b) Fluorescence images showing the multilayered 3D assembly
after leaching of the core material. (*Source*: Reprinted (adapted) with
permission from Sher *et al.* (2010). © 2010 Wiley-VCH.)

physicochemical properties is an important and ambitious step in the development of an effective encapsulating barrier. One approach to cell encapsulation involves coated liquefied capsules. For example, Costa *et al.* (2011a)
have used liquefied alginate beads coated with CHI and ALG to immobilize
SaOs-2 human osteoblast-like cells. Another cell encapsulation approach
exploits cell surface charge to assemble polyelectrolytes. Krol *et al.* (2006)
pursued this surface charge approach with pancreatic islet cells to assemble
a multilayer shell that effectively encapsulated them for the treatment of
diabetes. By incubating Langerhans cells with alternating solutions of poly-
cationic PAH or PDADMA and polyanionic PSS, the authors built a barrier
that covered and protected the cells. The shell was permeable and the cells
were immunologically protected and retained their insulin secretion capability (Fig. 4.2). The coating, however, consisted of synthetic materials which
could be cytotoxic in the long term. Veerabadran *et al.* (2007) followed a
similar approach to encapsulate individual mouse mesenchymal stem cells
(MSCs) using PLL and HA. The cells were encapsulated within shells with
thicknesses of approximately 6–9 nm. Cell morphology and viability were
maintained for up to 1 week.

Encapsulation has tremendous potential for the maintenance and differentiation of stem cells. LbL allows the encapsulating shells to be tuned to
achieve optimal permeability. Nutrients, waste, and chemical differentiation
factors could be exchanged between the cell and its external environment
and the differentiation process triggered on demand. Other approaches
have been introduced to drive the differentiation of cells by incorporating specific factors in a multilayer formulation. Dierich *et al.* (2007) presented a formulation with poly(lactic-co-glycolic acid) (PLGA), PLL and
poly(L-lysine succinylated) (PLLs), along with bone morphogenetic proteins (BMPs) and tissue growth factors (TGFs). Using this system, they

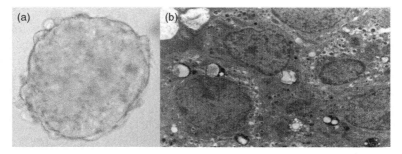

4.2 (a) Light microscopy image of trypan blue-stained islets after encapsulation with three pairs of polyelectrolyte layers (PDADMA/ PSS). (b) Electron micrographs of PAH/PSS/PAH-coated islets. (*Source*: Reprinted (adapted) with permission from Krol *et al.* (2006). © 2006 American Chemical Society.)

were able to drive the differentiation of embryoid bodies into cartilage and bone. Another interesting approach is to embed differentiation factors into spherical shapes, which have higher surface areas and improved accessibility compared to planar substrates. Facca *et al.* (2010) created capsules templated in silica particles (1 μm) coated with a similar formulation. The incubation with embryonic stem (ES) cells led to both *in vitro* and *in vivo* bone formation. It is unclear if the delivery of BMP was due to the degradation of the capsules or to their internalization by the cells. The latter option is a possibility: Rivera-Gil *et al.* (2009) have recently demonstrated the internalization of multilayer capsules by embryonic NIH/3T3 fibroblasts.

These encapsulation methods have proven to be novel approaches to shield cells from certain external factors and to controllably differentiate stem cells either by taking advantage of the permeable nature of multilayer films or by targeting the intracellular environment directly. Therefore, LbL structures have potential use for tissue engineering applications where cell transplantation or damaged tissue replacement is necessary.

4.4.3   Biomimetic and 'intelligent' substrates

Systems used for biomedical and tissue engineering applications must be biocompatible and not trigger toxic, inflammatory, or rejection responses. The structures and properties found in natural materials have inspired the development of multilayer systems with higher degrees of functionalization. One approach, discussed in Section 4.2.2, is to incorporate natural materials directly into multilayer systems. For instance, nanostructured films containing CHI retain some of CHI's anti-microbial properties (Joshi *et al.*, 2011). Another approach consists of integrating ingredient molecules, such as bioactive ligands, that can drive specific phenomena mimicking their native function. Picart *et al.* (2005a) incorporated arginine-glycine-(aspartic acid) (RGD), a cell adhesion promoter, with the substrate using PLL and RGD-grafted-PGA in the terminating layer. The proliferation of primary human osteoblast cells was enhanced on films containing RGD compared to films without. RGD-enhanced proliferation was observed not only for simple adsorbed layers, but also for crosslinked films. To mimic the microenvironments of bone marrow and the thymus, Lee and Kotov (2009) incorporated delta-like 1 Notch ligands (DL-1) at the surface of PEM films. The Notch ligands stimulated haematopoietic stem cells (HSCs) to differentiate into T-cell lymphocytes and progressively develop their own ECM.

Adding stimuli-responsive materials to multilayer formulations offers additional functionalization, by controlling surface properties and the conformational rearrangement of molecular chains. Stimuli-repsonsive properties have been engineered into several types of systems, such as grafted surfaces and hydrogels (Mano, 2008), and are beginning to be used in multilayer systems for controlling biological events such as cell attachment/detachment. As mentioned before, pH and ionic strength may be useful stimuli response mechanisms for drug delivery applications (refer to Section 4.2.2), but changes of these variables beyond physiological conditions may be too aggressive for cell cultures. Temperature changes, on the other hand, are usually better tolerated by cells for temperatures up to 37°C. Costa *et al.* used an ELR containing RGD to generate bilayer coatings (Costa *et al.*, 2009) and multiple stimuli-responsive multilayers (Costa *et al.*, 2011b) with enhanced cell adhesion. By varying the pH, ionic strength, and temperature of CHI/ELR films, it was possible to switch the wettability of the substrates from moderately hydrophobic (contact angle around 70°) to superhydrophilic (contact angle of 0°) (Fig. 4.3). However, the variation was observed for values not suitable for biological studies: pH 11; ionic strength 0.15 M; and temperature 50°C. Nevertheless, at physiological conditions, the cellular adhesion of an SaOs-2 osteoblast-like cell line was higher than in films ending in CHI or a scrambled non-functional RGD sequence. Martins *et al.* (2011) described a multilayer system of ALG and PNIPAAm-grafted CHI to reproduce the cell-sheet technology. SaOs-2 human osteoblast-like cells seeded onto coated substrates ending in the modified CHI were allowed to achieve confluence at 37°C. Upon decreasing the temperature to 4°C, the peeling effect enabled cells to be detached from the surface as a thin layer. Zahn *et al.* (2012) demonstrated a similar approach by using the ion-induced

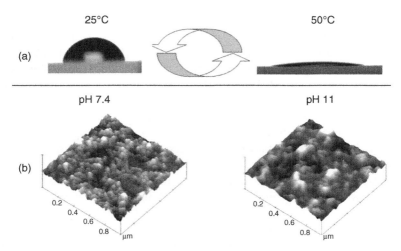

4.3 Physical changes induced in ELR modified surfaces. (a) Wettability variations with different temperatures. (*Source*: Reprinted (adapted) with permission from Costa *et al.* (2009). © 2009 Wiley-VCH.) (b) Topography at distinct pH values. (*Source*: Reprinted (adapted) with permission from Costa *et al.* (2011b). © 2011 Wiley-VCH.)

detachment of myoblast layers by the eroding effect of ferrocyanide on PLL and HA multilayers.

Incorporating biofunctional ingredients and bioinspired concepts into multilayer systems has enhanced the functionality of such systems and is producing promising results for tissue engineering applications.

4.4.4 Multilayers in controlled drug delivery

Current advances in biomedicine have been driven by the need to develop systems that can effectively encapsulate, protect, and deliver active agents with specific kinetics (Wohl and Engbersen, 2012). At the same time, distinct classes of biomolecules are widely available and related to the treatment of several diseases or the development of new biomedical devices. Ideally, a biomolecule carrier should act as a compartment with tunable drug permeability. The multilayered capsule systems described above are ideal for such drug delivery applications, including the associated methods to load and control the release of therapeutic molecules and growth/differentiation factors. The adequate loading efficiency and controlled release of the molecules of interest such as growth factors, are necessary to ensure the survival of cells in tissue engineering applications. Erel *et al.* (2012) fabricated hydrogen-bonded multilayers of micelles of poly[2-(*N*-morpholino)ethyl methacrylate-*block*-2-(diisopropylamino)ethyl methacrylate] (PMEMA-*b*-PDPA), a dicationic block copolymer, encapsulating pyrene onto a surface with tannic acid. The films were dual responsive and released pyrene upon increasing either temperature or pH. Crouzier *et al.* (2009) developed crosslinked films of PLL and HA as reservoirs loaded with a recombinant human BMP-2 (rhBMP-2) to control the differentiation of myoblasts into fibroblasts. They demonstrated that the amount of loaded rhBMP-2 could be regulated by film thickness and assembly conditions. Also, the rhBMP-2 loaded in the films retained its bioactivity and was protected for three consecutive cultures.

Another area targeted for multilayer drug delivery systems is cancer treatment, where the uncontrolled systemic administration of chemotherapeutics has several drawbacks to patients. Zhao *et al.* (2007) developed CHI/ALG microcapsules loaded with doxorubicin, an effective anticancer drug, and demonstrated that the apoptosis of HepG2 tumour cells could be induced *in vitro* by incubating with loaded microcapsules. A suspension of microcapsules was also injected directly to tumours induced in mice, inducing greater tumour inhibition compared to free drug administration. De Geest *et al.* (2012) demonstrated that polymeric multilayer capsules (PMLCs) increase antigen delivery toward antigen-presenting cells *in vivo*, enforcing antigen presentation and stimulating T-cell proliferation. Poon *et al.* (2011) recently addressed the stability and biodistribution of electrostatic self-assembled nanoparticles for systemic administration and off-body tracing by imaging techniques. They provided evidence of the importance of a larger number of layers and the role of the terminating layer in terms of functional retention. A formulation of PLL and dextran films topped with a layer of HA was assembled onto gold nanoparticles or quantum dots. In mice, these coated particles had a blood elimination half-life of 9 h and low liver accumulation. These reports represent a step forward for the use of LbL systems in *in vivo* applications such as cancer treatment, and augment the mounting evidence that such systems can be designed to have minimal or no negative effects on living organisms.

4.4.5   Biomineralized multilayers

Developing systems with LbL technology to facilitate bone regeneration or to improve the mechanical properties of substrates is an active area of research. Such research is adding versatility to biomineralization processes by developing controlled nanostructured assemblies that tune the kinetics of calcium phosphate, apatite formation, and ultimately bone regeneration (Kokubo *et al.*, 2003; Rezwan *et al.*, 2006; Suchanek and Yoshimura, 1998).

Bioactive coatings on orthopedic implants may create an environment compatible with osteogenesis and provide bone-friendly interfaces for forming natural bonding junctions between the implant and host bone (Boccaccini and Blaker, 2005). Couto, Alves and Mano (2009) developed hybrid films of CHI and bioactive glass that mimicked the brick-and-mortar layered structure of nacre. They showed that the incorporation of BNPs rendered the substrates prone to mineralization upon immersion in SBF solution. The mineralization was also more pronounced with an increasing number of layers. Fukui and Fujimoto (2011) described a hybrid nanocapsule templated in phospholipids that encapsulated phosphate ions (Fig. 4.4). The phospholipids were coated using CHI, dextran or DNA and then incubated in a solution containing calcium ions. They demonstrated that

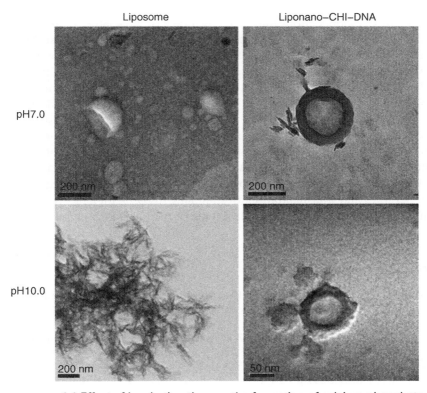

4.4 Effect of incubation time on the formation of calcium phosphate crystals using liposome and liponano–CHI–DNA templates. Each reaction was carried out at pH 7.0 or pH 10.0 for 18 h, using $CaCl_2$ or Na_2HPO_4/NaH_2PO_4 as ion species, respectively. (*Source*: Reprinted with permission from Fukui and Fujimoto (2011). © 2011 American Chemical Society.)

the chemical surface change could influence the counter-ion diffusion and therefore the mineralization rate.

4.5   Conclusion

For the past 20 years, the use of LbL as a processing technique has evolved from traditional surface modification on two-dimensional (2D) planar surfaces to three-dimensional (3D) environments. The procedure can even be performed using mild conditions and in a cell-friendly environment. The most recent results show that LbL may be used to encapsulate cells, control cell adhesion and metabolism, tune the release of encapsulated molecules, engineer substrates with smart responsiveness, and enhance mineralization. We believe that the technology will follow two trends in the near future. The first trend will involve using existing nanostructured multilayer devices in more *in vivo* applications that take advantage of the spatial organization and bioactivity, adjustable stiffness and chemistry, or adjustable stiffness and bioactivity, of multilayer devices. The other trend will involve the development of more intricate structures with hierarchical complexity embedded in devices with multiple functionalities, such as controlling the diffusion of gases and nutrients to cultured cells. Higher levels of functionalization should enable the development of systems exhibiting intelligent responses to external stimuli. For example, drug release systems could respond to an external stimulus by releasing encapsulated drugs and therefore influencing cellular metabolism on demand. The inclusion of specific ligands or metallic particles could enable the development of targeted delivery strategies for cell therapies and transplantation. In summary, research on nanostructured multilayer devices will likely result in innovative low-cost technologies to meet critical biomedical needs.

4.6   References

Ariga, K., Hill, J. P. and Ji, Q. (2008). Biomaterials and biofunctionality in layered macromolecular assemblies. *Macromolecular Bioscience*, **8**, 981–990. (DOI: 10.1002/mabi.200800102).

Ariga, K., Lvov, Y. and Kunitake, T. (1997). Assembling alternate dye–polyion molecular films by electrostatic layer-by-layer adsorption. *Journal of the American Chemical Society*, **119**, 2224–2231. (DOI: 10.1021/ja963442c).

Ashby, M. F. and Jones, D. R. H. (2006). *Engineering Materials 2: An Introduction to Microstructures, Processing and Design*, Oxford, United Kingdom, Elsevier.

Barbosa, J. S., Costa, R. R., Testera, A. M., Alonso, M., Rodríguez-Cabello, J. C. and Mano, J. F. (2009). Multi-layered films containing a biomimetic stimuli-responsive recombinant protein. *Nanoscale Research Letters*, **4**, 1247–1253. (DOI: 10.1007/s11671-009-9388-5).

Becker, A. L., Johnston, A. P. R. and Caruso, F. (2010). Layer-by-layer-assembled capsules and films for therapeutic delivery. *Small*, **6**, 1836–1852. (DOI: 10.1002/smll.201000379).

Bédard, M. F., Munoz-Javier, A., Mueller, R., Del Pino, P., Fery, A., Parak, W. J., Skirtach, A. G. and Sukhorukov, G. B. (2009). On the mechanical stability of polymeric microcontainers functionalized with nanoparticles. *Soft Matter*, **5**, 148–155. (DOI: 10.1039/b812553h).

Berger, J., Reist, M., Mayer, J. M., Felt, O., Peppas, N. A. and Gurny, R. (2004). Structure and interactions in covalently and ionically crosslinked chitosan hydrogels for biomedical applications. *European Journal of Pharmaceutics and Biopharmaceutics*, **57**, 19–34. (DOI: 10.1016/s0939–6411(03)00161–9).

Bertrand, P., Jonas, A., Laschewsky, A. and Legras, R. (2000). Ultrathin polymer coatings by complexation of polyelectrolytes at interfaces: Suitable materials, structure and properties. *Macromolecular Rapid Communications*, **21**, 319–348. (DOI: 10.1002/(sici)1521–3927(20000401)21:7<319::aid-marc319>3.0.co;2–7).

Boateng, J. S., Matthews, K. H., Stevens, H. N. E. and Eccleston, G. M. (2008). Wound healing dressings and drug delivery systems: A review. *Journal of Pharmaceutical Sciences*, **97**, 2892–2923. (DOI: 10.1002/jps.21210).

Boccaccini, A. R. and Blaker, J. J. (2005). Bioactive composite materials for tissue engineering scaffolds. *Expert Review of Medical Devices*, **2**, 303–317. (DOI: 10.1586/17434440.2.3.303).

Boudou, T., Crouzier, T., Ren, K., Blin, G. and Picart, C. (2010). Multiple functionalities of polyelectrolyte multilayer films: New biomedical applications. *Advanced Materials*, **22**, 441–467. (DOI: 10.1002/adma.200901327).

Caruso, F., Caruso, R. A. and Möhwald, H. (1998). Nanoengineering of inorganic and hybrid hollow spheres by colloidal templating. *Science*, **282**, 1111–1114. (DOI: 10.1126/science.282.5391.1111).

Chen, W. and Mccarthy, T. J. (1997). Layer-by-layer deposition: A tool for polymer surface modification. *Macromolecules*, **30**, 78–86. (DOI: 10.1021/ma961096d).

Choi, H. J., Brooks, E. and Montemagno, C. D. (2005). Synthesis and characterization of nanoscale biomimetic polymer vesicles and polymer membranes for bioelectronic applications. *Nanotechnology*, **16**, S143-S149. (DOI: 10.1088/0957–4484/16/5/002).

Choi, S-W., Zhang, Y. and Xia, Y. (2009). Fabrication of microbeads with a controllable hollow interior and porous wall using a capillary fluidic device. *Advanced Functional Materials*, **19**, 2943–2949. (DOI: 10.1002/adfm.200900763).

Chua, P.-H., Neoh, K.-G., Kang, E.-T. and Wang, W. (2008). Surface functionalization of titanium with hyaluronic acid/chitosan polyelectrolyte multilayers and RGD for promoting osteoblast functions and inhibiting bacterial adhesion. *Biomaterials*, **29**, 1412–1421. (DOI: 10.1016/j.biomaterials.2007.12.019).

Clark, S. L. and Hammond, P. T. (2000). The role of secondary interactions in selective electrostatic multilayer deposition. *Langmuir*, **16**, 10206–10214. (DOI: 10.1021/la000418a).

Costa, N. L., Sher, P. and Mano, J. F. (2011a). Liquefied capsules coated with multilayered polyelectrolyte films for cell immobilization. *Advanced Engineering Materials*, **13**, B218-B224. (DOI: 10.1002/adem.201080138).

Costa, R. R., Custódio, C. A., Arias, F. J., Rodríguez-Cabello, J. C. and Mano, J. F. (2011b). Layer-by-layer assembly of chitosan and recombinant biopolymers into biomimetic coatings with multiple stimuli-responsive properties. *Small*, **7**, 2640–2649. (DOI: 10.1002/smll.201100875).

Costa, R. R., Custódio, C. A., Testera, A. M., Arias, F. J., Rodríguez-Cabello, J. C., Alves, N. M. and Mano, J. F. (2009). Stimuli-responsive thin coatings using elastin-like polymers for biomedical applications. *Advanced Functional Materials*, **19**, 3210–3218. (DOI: 10.1002/adfm.200900568).

Couto, D. S., Alves, N., M. and Mano, J., F. (2009). Nanostructured multilayer coatings combining chitosan with bioactive glass nanoparticles. *Journal of Nanoscience and Nanotechnology*, **9**, 1741–1748. (DOI: 10.1166/jnn.2009.389).

Crouzier, T., Ren, K., Nicolas, C., Roy, C. and Picart, C. (2009). Layer-by-layer films as a biomimetic reservoir for rhBMP-2 delivery: Controlled differentiation of myoblasts to osteoblasts. *Small*, **5**, 598–608. (DOI: 10.1002/smll.200800804).

De Geest, B. G., Sanders, N. N., Sukhorukov, G. B., Demeester, J. and De Smedt, S. C. (2007). Release mechanisms for polyelectrolyte capsules. *Chemical Society Reviews*, **36**, 636–649. (DOI: 10.1039/b600460c).

De Geest, B. G., Willart, M. A., Hammad, H., Lambrecht, B. N., Pollard, C., Bogaert, P., De Filette, M., Saelens, X., Vervaet, C., Remon, J. P., Grooten, J. and De Koker, S. (2012). Polymeric multilayer capsule-mediated vaccination induces protective immunity against cancer and viral infection. *ACS Nano*, **6**, 2136–2149. (DOI: 10.1021/nn205099c).

De Cock, L. J., De Koker, S., De Geest, B. G., Grooten, J., Vervaet, C., Remon, J. P., Sukhorukov, G. B. and Antipina, M. N. (2010). Polymeric multilayer capsules in drug delivery. *Angewandte Chemie International Edition*, **49**, 6954–6973. (DOI: 10.1002/anie.200906266).

Decher, G., Hong, J. D. and Schmitt, J. (1992). Buildup of ultrathin multilayer films by a self-assembly process: III. Consecutively alternating adsorption of anionic and cationic polyelectrolytes on charged surfaces. *Thin Solid Films*, **210–211**, 831–835. (DOI: 10.1016/0040-6090(92)90417-a).

Dierich, A., Le Guen, E., Messaddeq, N., Stoltz, J-F., Netter, P., Schaaf, P., Voegel, J-C. and Benkirane-Jessel, N. (2007). Bone formation mediated by synergy-acting growth factors embedded in a polyelectrolyte multilayer film. *Advanced Materials*, **19**, 693–697. (DOI: 10.1002/adma.200601271).

Donath, E., Sukhorukov, G. B., Caruso, F., Davis, S. A. and Möhwald, H. (1998). Novel hollow polymer shells by colloid-templated assembly of polyelectrolytes. *Angewandte Chemie International Edition*, **37**, 2201–2205. (DOI: 10.1002/(sici)1521-3773(19980904)37:16<2201::aid-anie2201>3.0.co;2-e).

Drury, J. L. and Mooney, D. J. (2003). Hydrogels for tissue engineering: scaffold design variables and applications. *Biomaterials*, **24**, 4337–4351. (DOI: 10.1016/s0142-9612(03)00340-5).

Elbert, D. L., Herbert, C. B. and Hubbell, J. A. (1999). Thin polymer layers formed by polyelectrolyte multilayer techniques on biological surfaces. *Langmuir*, **15**, 5355–5362. (DOI: 10.1021/la9815749).

Elzbieciak, M., Wodka, D., Zapotoczny, S., Nowak, P. and Warszynski, P. (2010). Characteristics of model polyelectrolyte multilayer films containing laponite clay nanoparticles. *Langmuir*, **26**, 277–283. (DOI: 10.1021/la902077j).

Erel, I., Karahan, H. E., Tuncer, C., Bütün, V. and Demirel, A. L. (2012). Hydrogen-bonded multilayers of micelles of a dually responsive dicationic block copolymer. *Soft Matter*, **8**, 827–836. (DOI: 10.1039/c1sm06248d).

Etienne, O., Gasnier, C., Taddei, C., Voegel, J-C., Aunis, D., Schaaf, P., Metz-Boutigue, M-H., Bolcato-Bellemin, A-L. and Egles, C. (2005a). Antifungal coating by bio-functionalized polyelectrolyte multilayered films. *Biomaterials*, **26**, 6704–6712. (DOI: 10.1016/j.biomaterials.2005.04.068).

Etienne, O., Picart, C., Taddei, C., Haikel, Y., Dimarcq, J. L., Schaaf, P., Voegel, J. C., Ogier, J. A. and Egles, C. (2004). Multilayer polyelectrolyte films functionalized by insertion of defensin: A new approach to protection of implants from bacterial colonization. *Antimicrobial Agents and Chemotherapy*, **48**, 3662–3669. (DOI: 10.1128/aac.48.10.3662–3669.2004).

Etienne, O., Schneider, A., Taddei, C., Richert, L., Schaaf, P., Voegel, J. C., Egles, C. and Picart, C. (2005b). Degradability of polysaccharides multilayer films in the oral environment: an in vitro and in vivo study. *Biomacromolecules*, **6**, 726–733. (DOI: 10.1021/bm049425u).

Facca, S., Cortez, C., Mendoza-Palomares, C., Messadeq, N., Dierich, A., Johnston, A. P. R., Mainard, D., Voegel, J-C., Caruso, F. and Benkirane-Jessel, N. (2010). Active multilayered capsules for in vivo bone formation. *Proceedings of the National Academy of Sciences of the United States of America*, **107**, 3406–3411. (DOI: 10.1073/pnas.0908531107).

Falconnet, D., Csucs, G., Grandin, H. M. and Textor, M. (2006). Surface engineering approaches to micropattern surfaces for cell-based assays. *Biomaterials*, **27**, 3044–3063. (DOI: 10.1016/j.biomaterials.2005.12.024).

Fendler, J. H. (1996). Self-assembled nanostructured materials. *Chemistry of Materials*, **8**, 1616–1624. (DOI: 10.1021/cm960116n).

Fetters, L. J., Lohse, D. J., Richter, D., Witten, T. A. and Zirkel, A. (1994). Connection between polymer molecular-weight, density, chain dimensions, and melt viscoelastic properties. *Macromolecules*, **27**, 4639–4647. (DOI: 10.1021/ma00095a001).

Fu, J., Ji, J., Yuan, W. and Shen, J. (2005). Construction of anti-adhesive and anti-bacterial multilayer films via layer-by-layer assembly of heparin and chitosan. *Biomaterials*, **26**, 6684–6692. (DOI: 10.1016/j.biomaterials.2005.04.034).

Fukui, Y. and Fujimoto, K. (2011). Control in mineralization by the polysaccharide-coated liposome via the counter-diffusion of ions. *Chemistry of Materials*, **23**, 4701–4708. (DOI: 10.1021/cm201211n).

Garza, J. M., Schaaf, P., Muller, S., Ball, V., Stoltz, J.-F., Voegel, J.-C. and Lavalle, P. (2004). Multicompartment films made of alternate polyelectrolyte multilayers of exponential and linear growth. *Langmuir*, **20**, 7298–7302. (DOI: 10.1021/la049106o).

George, M. and Abraham, T. E. (2006). Polyionic hydrocolloids for the intestinal delivery of protein drugs: Alginate and chitosan – a review. *Journal of Controlled Release*, **114**, 1–14. (DOI: 10.1016/j.jconrel.2006.04.017).

Girotti, A., Reguera, J., Rodríguez-Cabello, J.C., Arias, F.J., Alonso, M. and Testera, A.M. (2004). Design and bioproduction of a recombinant multi(bio)functional elastin-like protein polymer containing cell adhesion sequences for tissue engineering purposes. *Journal of Materials Science: Materials in Medicine*, **15**, 479–484. (DOI: 10.1023/b:jmsm.0000021124.58688.7a).

Glinel, K., Sukhorukov, G. B., Möhwald, H., Khrenov, V. and Tauer, K. (2003). Thermosensitive hollow capsules based on thermoresponsive polyelectrolytes. *Macromolecular Chemistry and Physics*, **204**, 1784–1790. (DOI: 10.1002/macp.200350033).

Golonka, M., Bulwan, M., Nowakowska, M., Testera, A. M., Rodríguez-Cabello, J. C. and Zapotoczny, S. (2011). Thermoresponsive multilayer films based on ionic elastin-like recombinamers. *Soft Matter*, **7**, 9402–9409. (DOI: 10.1039/c1sm06276j).

Gribova, V., Auzely-Velty, R. and Picart, C. (2012). Polyelectrolyte multilayer assemblies on materials surfaces: From cell adhesion to tissue engineering. *Chemistry of Materials*, **24**, 854–869. (DOI: 10.1021/cm2032459).

Hammond, P. T. (2004). Form and function in multilayer assembly: New applications at the nanoscale. *Advanced Materials*, **16**, 1271–1293. (DOI: 10.1002/adma.200400760).

He, Q., Song, W., Möhwald, H. and Li, J. (2008). Hydrothermal-induced structure transformation of polyelectrolyte multilayers: From nanotubes to capsules. *Langmuir*, **24**, 5508–5513. (DOI: 10.1021/la703738m).

Hoogeveen, N. G., Cohen Stuart, M. A. and Fleer, G. J. (1996). Can charged (block co)polymers act as stabilisers and flocculants of oxides? *Colloids and Surfaces A: Physicochemical and Engineering Aspects*, **117**, 77–88. (DOI: 10.1016/0927–7757(96)03699–0).

Hook, A. L., Voelcker, N. H. and Thissen, H. (2009). Patterned and switchable surfaces for biomolecular manipulation. *Acta Biomaterialia*, **5**, 2350–2370. (DOI: 10.1016/j.actbio.2009.03.040).

Huang, C.-J. and Chang, F.-C. (2009). Using click chemistry to fabricate ultrathin thermoresponsive microcapsules through direct covalent layer-by-layer assembly. *Macromolecules*, **42**, 5155–5166. (DOI: 10.1021/ma900478n).

Hubbell, J. A. (1999). Bioactive biomaterials. *Current Opinion in Biotechnology*, **10**, 123–129. (DOI: 10.1016/s0958–1669(99)80021–4).

Ibarz, G., Dähne, L., Donath, E. and Möhwald, H. (2001). Smart micro- and nanocontainers for storage, transport, and release. *Advanced Materials*, **13**, 1324–1327. (DOI: 10.1002/1521–4095(200109)13:17<1324::aid-adma1324>3.0.co;2-l).

Izumrudov, V. A., Chaubet, F., Clairbois, A.-S. and Jozefonvicz, J. (1999). Interpolyelectrolyte reactions in solutions of functionalized dextrans with negatively charged groups along the chains. *Macromolecular Chemistry and Physics*, **200**, 1753–1763. (DOI: 10.1002/(sici)1521–3935(19990701)200:7<1753::aid-macp1753>3.0.co;2-n).

Jiang, C. and Tsukruk, V. V. (2006). Freestanding nanostructures via layer-by-layer assembly. *Advanced Materials*, **18**, 829–840. (DOI: 10.1002/adma.200502444).

Johansson, J. Å., Halthur, T., Herranen, M., Söderberg, L., Elofsson, U. and Hilborn, J. (2005). Build-up of collagen and hyaluronic acid polyelectrolyte multilayers. *Biomacromolecules*, **6**, 1353–1359. (DOI: 10.1021/bm0493741).

Joshi, M., Khanna, R., Shekhar, R. and Jha, K. (2011). Chitosan nanocoating on cotton textile substrate using layer-by-layer self-assembly technique. *Journal of Applied Polymer Science*, **119**, 2793–2799. (DOI: 10.1002/app.32867).

Khademhosseini, A., Suh, K. Y., Yang, J. M., Eng, G., Yeh, J., Levenberg, S. and Langer, R. (2004). Layer-by-layer deposition of hyaluronic acid and poly-L-lysine for patterned cell co-cultures. *Biomaterials*, **25**, 3583–3592. (DOI: 10.1016/j.biomaterials.2003.10.033).

Klitzing, R. V. (2006). Internal structure of polyelectrolyte multilayer assemblies. *Physical Chemistry Chemical Physics*, **8**, 5012–5033. (DOI: 10.1039/b607760a).

Koens, M. J. W., Faraj, K. A., Wismans, R. G., van der Vliet, J. A., Krasznai, A. G., Cuijpers, V. M. J. I., Jansen, J. A., Daamen, W. F. and van Kuppevelt, T. H. (2010).

Controlled fabrication of triple layered and molecularly defined collagen/elastin vascular grafts resembling the native blood vessel. *Acta Biomaterialia*, **6**, 4666–4674. (DOI: 10.1016/j.actbio.2010.06.038).

Köhler, K., Möhwald, H. and Sukhorukov, G. B. (2006). Thermal behavior of polyelectrolyte multilayer microcapsules: 2. Insight into molecular mechanisms for the PDADMAC/PSS system. *The Journal of Physical Chemistry B*, **110**, 24002–24010. (DOI: 10.1021/jp062907a).

Köhler, K. and Sukhorukov, G. B. (2007). Heat treatment of polyelectrolyte multilayer capsules: A versatile method for encapsulation. *Advanced Functional Materials*, **17**, 2053–2061. (DOI: 10.1002/adfm.200600593).

Kokubo, T., Kim, H-M. and Kawashita, M. (2003). Novel bioactive materials with different mechanical properties. *Biomaterials*, **24**, 2161–2175. (DOI: 10.1016/s0142–9612(03)00044–9).

Komatsu, T. (2012). Protein-based nanotubes for biomedical applications. *Nanoscale*, **4**, 1910–1918. (DOI: 10.1039/c1nr11224d).

Kong, B.-S., Geng, J. and Jung, H.-T. (2009). Layer-by-layer assembly of graphene and gold nanoparticles by vacuum filtration and spontaneous reduction of gold ions. *Chemical Communications*, 2174–2176. (DOI: 10.1039/b821920f).

Krol, S., del Guerra, S., Grupillo, M., Diaspro, A., Gliozzi, A. and Marchetti, P. (2006). Multilayer nanoencapsulation. New approach for immune protection of human pancreatic islets. *Nano Letters*, **6**, 1933–1939. (DOI: 10.1021/nl061049r).

Kujawa, P., Moraille, P., Sanchez, J., Badia, A. and Winnik, F. M. (2005). Effect of molecular weight on the exponential growth and morphology of hyaluronan/chitosan multilayers: A surface plasmon resonance spectroscopy and atomic force microscopy investigation. *Journal of the American Chemical Society*, **127**, 9224–9234. (DOI: 10.1021/ja044385n).

Langer, R. and Tirrell, D. A. (2004). Designing materials for biology and medicine. *Nature*, **428**, 487–492. (DOI: 10.1038/nature02388).

Larkin, A. L., Davis, R. M. and Rajagopalan, P. (2010). Biocompatible, detachable, and free-standing polyelectrolyte multilayer films. *Biomacromolecules*, **11**, 2788–2796. (DOI: 10.1021/bm100867h).

Lee, J. and Kotov, N. A. (2009). Notch ligand presenting acellular 3D microenvironments for ex vivo human hematopoietic stem-cell culture made by layer-by-layer assembly. *Small*, **5**, 1008–1013. (DOI: 10.1002/smll.200801242).

Lee, S., Lee, B., Kim, B. J., Park, J., Yoo, M., Bae, W. K., Char, K., Hawker, C. J., Bang, J. and Cho, J. (2009). Free-standing nanocomposite multilayers with various length scales, adjustable internal structures, and functionalities. *Journal of the American Chemical Society*, **131**, 2579–2587. (DOI: 10.1021/ja8064478).

Liang, Z., Susha, A. S., Yu, A. and Caruso, F. (2003). Nanotubes prepared by layer-by-layer coating of porous membrane templates. *Advanced Materials*, **15**, 1849–1853. (DOI: 10.1002/adma.200305580).

Linford, M. R., Auch, M. and Möhwald, H. (1998). Nonmonotonic effect of ionic strength on surface dye extraction during dye–polyelectrolyte multilayer formation. *Journal of the American Chemical Society*, **120**, 178–182. (DOI: 10.1021/ja972133z).

Luz, G. M. and Mano, J. F. (2009). Biomimetic design of materials and biomaterials inspired by the structure of nacre. *Philosophical Transactions of the Royal Society A: Mathematical, Physical and Engineering Sciences*, **367**, 1587–1605. (DOI: 10.1098/rsta.2009.0007).

Lyklema, J. and Deschênes, L. (2011). The first step in layer-by-layer deposition: Electrostatics and/or non-electrostatics? *Advances in Colloid and Interface Science*, **168**, 135–148. (DOI: 10.1016/j.cis.2011.03.008).

Mallwitz, F. and Laschewsky, A. (2005). Direct access to stable, freestanding polymer membranes by layer-by-layer assembly of polyelectrolytes. *Advanced Materials*, **17**, 1296–1299. (DOI: 10.1002/adma.200401123).

Mamedov, A. A. and Kotov, N. A. (2000). Free-standing layer-by-layer assembled films of magnetite nanoparticles. *Langmuir*, **16**, 5530–5533. (DOI: 10.1021/la000560b).

Mamedov, A. A., Kotov, N. A., Prato, M., Guldi, D. M., Wicksted, J. P. and Hirsch, A. (2002). Molecular design of strong single-wall carbon nanotube/polyelectrolyte multilayer composites. *Nature Materials*, **1**, 190–194. (DOI: 10.1038/nmat747).

Mano, J.F. (2008). Stimuli-responsive polymeric systems for biomedical applications. *Advanced Engineering Materials*, **10**, 515–527. (DOI: 10.1002/adem.200700355).

Mano, J. F., Silva, G. A., Azevedo, H. S., Malafaya, P. B., Sousa, R. A., Silva, S. S., Boesel, L. F., Oliveira, J. M., Santos, T. C., Marques, A. P., Neves, N. M. and Reis, R. L. (2007). Natural origin biodegradable systems in tissue engineering and regenerative medicine: Present status and some moving trends. *Journal of The Royal Society Interface*, **4**, 999–1030. (DOI: 10.1098/rsif.2007.0220).

Martins, G. V., Mano, J. F. and Alves, N. M. (2011). Dual responsive nanostructured surfaces for biomedical applications. *Langmuir*, **27**, 8415–8423. (DOI: 10.1021/la200832n).

Martins, G. V., Merino, E. G., Mano, J. F. and Alves, N. M. (2010). Crosslink effect and albumin adsorption onto chitosan/alginate multilayered systems: An in situ QCM-D study. *Macromolecular Bioscience*, **10**, 1444–1455. (DOI: 10.1002/mabi.201000193).

Mauser, T., Déjugnat, C., Möhwald, H. and Sukhorukov, G. B. (2006). Microcapsules made of weak polyelectrolytes: Templating and stimuli-responsive properties. *Langmuir*, **22**, 5888–5893. (DOI: 10.1021/la060088f).

Mayer, G. (2005). Rigid biological systems as models for synthetic composites. *Science*, **310**, 1144–1147. (DOI: 10.1126/science.1116994).

Mazia, D., Schatten, G. and Sale, W. (1975). Adhesion of cells to surfaces coated with polylysine. Applications to electron microscopy. *The Journal of Cell Biology*, **66**, 198–200. (DOI: 10.1083/jcb.66.1.198).

Meyers, M. A., Chen, P.-Y., Lin, A. Y.-M. and Seki, Y. (2008). Biological materials: Structure and mechanical properties. *Progress in Materials Science*, **53**, 1–206. (DOI: 10.1016/j.pmatsci.2007.05.002).

Miranda, E. S., Silva, T. H., Reis, R. L. and Mano, J. F. (2011). Nanostructured natural-based polyelectrolyte multilayers to agglomerate chitosan particles into scaffolds for tissue engineering. *Tissue Engineering Part A*, **17**, 2663–2674. (DOI: 10.1089/ten.tea.2010.0635).

Mueller, R., Daehne, L. and Fery, A. (2007). Hollow polyelectrolyte multilayer tubes: Mechanical properties and shape changes. *The Journal of Physical Chemistry B*, **111**, 8547–8553. (DOI: 10.1021/jp068762p).

Murugan, R. and Ramakrishna, S. (2006). Nano-featured scaffolds for tissue engineering: A review of spinning methodologies. *Tissue Engineering*, **12**, 435–447. (DOI: 10.1089/ten.2006.12.435).

Ochs, C. J., Such, G. K., Städler, B. and Caruso, F. (2008). Low-fouling, biofunctionalized, and biodegradable click capsules. *Biomacromolecules*, **9**, 3389–3396. (DOI: 10.1021/bm800794w).

Okano, T., Yamada, N., Sakai, H. and Sakurai, Y. (1993). A novel recovery system for cultured cells using plasma-treated polystyrene dishes grafted with poly(N-isopropylacrylamide). *Journal of Biomedical Materials Research*, **27**, 1243–1251. (DOI: 10.1002/jbm.820271005).

Ott, P., Trenkenschuh, K., Gensel, J., Fery, A. and Laschewsky, A. (2010). Free-standing membranes via covalent cross-linking of polyelectrolyte multilayers with complementary reactivity. *Langmuir*, **26**, 18182–18188. (DOI: 10.1021/la1035882).

Patel, A., Fine, B., Sandig, M. and Mequanint, K. (2006). Elastin biosynthesis: The missing link in tissue-engineered blood vessels. *Cardiovascular Research*, **71**, 40–49. (DOI: 10.1016/j.cardiores.2006.02.021).

Percival, E. and McDowell, R. H. (1990). Algal polysaccharides. In Dey, P. M. (ed.) *Methods in Plant Biochemistry. Carbohydrates*. London, United Kingdom: Academic Press.

Peyratout, C. S. and Dähne, L. (2004). Tailor-made polyelectrolyte microcapsules: From multilayers to smart containers. *Angewandte Chemie International Edition*, **43**, 3762–3783. (DOI: 10.1002/anie.200300568).

Phelps, J. A., Morisse, S., Hindié, M., Degat, M-C., Pauthe, E. and Van Tassel, P. R. (2011). Nanofilm biomaterials: Localized cross-linking to optimize mechanical rigidity and bioactivity. *Langmuir*, **27**, 1123–1130. (DOI: 10.1021/la104156c).

Picart, C., Elkaim, R., Richert, L., Audoin, T., Arntz, Y., Da Silva Cardoso, M., Schaaf, P., Voegel, J-C. and Frisch, B. (2005a). Primary cell adhesion on RGD-functionalized and covalently crosslinked thin polyelectrolyte multilayer films. *Advanced Functional Materials*, **15**, 83–94. (DOI: 10.1002/adfm.200400106).

Picart, C., Lavalle, P., Hubert, P., Cuisinier, F. J. G., Decher, G., Schaaf, P. and Voegel, J-C. (2001). Buildup mechanism for poly(l-lysine)/hyaluronic acid films onto a solid surface. *Langmuir*, **17**, 7414–7424. (DOI: 10.1021/la010848g).

Picart, C., Mutterer, J., Richert, L., Luo, Y., Prestwich, G. D., Schaaf, P., Voegel, J-C. and Lavalle, P. (2002). Molecular basis for the explanation of the exponential growth of polyelectrolyte multilayers. *Proceedings of the National Academy of Sciences of the United States of America*, **99**, 12531–12535. (DOI: 10.1073/pnas.202486099).

Picart, C., Schneider, A., Etienne, O., Mutterer, J., Schaaf, P., Egles, C., Jessel, N. and Voegel, J-C. (2005b). Controlled degradability of polysaccharide multilayer films in vitro and in vivo. *Advanced Functional Materials*, **15**, 1771–1780. (DOI: 10.1002/adfm.200400588).

Poon, Z., Lee, J. B., Morton, S. W. and Hammond, P. T. (2011). Controlling in vivo stability and biodistribution in electrostatically assembled nanoparticles for systemic delivery. *Nano Letters*, **11**, 2096–2103. (DOI: 10.1021/nl200636r).

Porcel, C., Lavalle, P., Decher, G., Senger, B., Voegel, J-C. and Schaaf, P. (2007). Influence of the polyelectrolyte molecular weight on exponentially growing multilayer films in the linear regime. *Langmuir*, **23**, 1898–1904. (DOI: 10.1021/la062728k).

Prakash, S., Karacor, M. B. and Banerjee, S. (2009). Surface modification in microsystems and nanosystems. *Surface Science Reports*, **64**, 233–254. (DOI: 10.1016/j.surfrep.2009.05.001).

Prieto, S., Shkilnyy, A., Rumplasch, C., Ribeiro, A., Arias, F. J., Rodríguez-Cabello, J. C. and Taubert, A. (2011). Biomimetic calcium phosphate mineralization with multifunctional elastin-like recombinamers. *Biomacromolecules*, **12**, 1480–1486. (DOI: 10.1021/bm200287c).

Rezwan, K., Chen, Q. Z., Blaker, J. J. and Boccaccini, A. R. (2006). Biodegradable and bioactive porous polymer/inorganic composite scaffolds for bone tissue engineering. *Biomaterials*, **27**, 3413–3431. (DOI: 10.1016/j.biomaterials.2006.01.039).

Richert, L., Lavalle, P., Payan, E., Shu, X. Z., Prestwich, G. D., Stoltz, J.-F., Schaaf, P., Voegel, J.-C. and Picart, C. (2004). Layer by layer buildup of polysaccharide films: Physical chemistry and cellular adhesion aspects. *Langmuir*, **20**, 448–458. (DOI: 10.1021/la035415n).

Richert, L., Lavalle, P., Vautier, D., Senger, B., Stoltz, J-F., Schaaf, P., Voegel, J-C. and Picart, C. (2002). Cell interactions with polyelectrolyte multilayer films. *Biomacromolecules*, **3**, 1170–1178. (DOI: 10.1021/bm0255490).

Rinaudo, M. (2006). Chitin and chitosan – properties and applications. *Progress in Polymer Science*, **31**, 603–632. (DOI: 10.1016/j.progpolymsci.2006.06.001).

Rivera-Gil, P., De Koker, S., De Geest, B. G. and Parak, W. J. (2009). Intracellular processing of proteins mediated by biodegradable polyelectrolyte capsules. *Nano Letters*, **9**, 4398–4402. (DOI: 10.1021/nl902697j).

Rodríguez-Cabello, J. C., Martín, L., Alonso, M., Arias, F. J. and Testera, A. M. (2009). 'Recombinamers' as advanced materials for the post-oil age. *Polymer*, **50**, 5159–5169. (DOI: 10.1016/j.polymer.2009.08.032).

Salomäki, M., Vinokurov, I. A. and Kankare, J. (2005). Effect of temperature on the buildup of polyelectrolyte multilayers. *Langmuir*, **21**, 11232–11240. (DOI: 10.1021/la051600k).

Sanchez-Castillo, M.A., Couto, C., Kim, W.B. and Dumesic, J.A. (2004). Gold-nanotube membranes for the oxidation of CO at gas–water interfaces. *Angewandte Chemie International Edition*, **43**, 1140–1142. (DOI: 10.1002/anie.200353238).

Schild, H. G. (1992). Poly(N-isopropylacrylamide): Experiment, theory and application. *Progress in Polymer Science*, **17**, 163–249. (DOI: 10.1016/0079–67 00(92)90023-r).

Schlenoff, J. B. and Dubas, S. T. (2001). Mechanism of polyelectrolyte multilayer growth: Charge overcompensation and distribution. *Macromolecules*, **34**, 592–598. (DOI: 10.1021/ma0003093).

Schlenoff, J. B., Ly, H. and Li, M. (1998). Charge and mass balance in polyelectrolyte multilayers. *Journal of the American Chemical Society*, **120**, 7626–7634. (DOI: 10.1021/ja980350+).

Schneider, A., Francius, G., Obeid, R., Schwinté, P., Hemmerlé, J., Frisch, B., Schaaf, P., Voegel, J.-C., Senger, B. and Picart, C. (2006). Polyelectrolyte multilayers with a tunable Young's modulus: Influence of film stiffness on cell adhesion. *Langmuir*, **22**, 1193–1200. (DOI: 10.1021/la0521802).

Schneider, A., Vodouhê, C., Richert, L., Francius, G., Le Guen, E., Schaaf, P., Voegel, J. C., Frisch, B. and Picart, C. (2007). Multifunctional polyelectrolyte multilayer films: Combining mechanical resistance, biodegradability, and bioactivity. *Biomacromolecules*, **8**, 139–145. (DOI: 10.1021/bm060765k).

She, Z., Antipina, M. N., Li, J. and Sukhorukov, G. B. (2010). Mechanism of protein release from polyelectrolyte multilayer microcapsules. *Biomacromolecules*, **11**, 1241–1247. (DOI: 10.1021/bm901450r).

Sher, P., Custódio, C. A. and Mano, J. F. (2010). Layer-by-layer technique for producing porous nanostructured 3D constructs using moldable freeform assembly of spherical templates. *Small*, **6**, 2644–2648. (DOI: 10.1002/smll.201001066).

Shiratori, S. S. and Rubner, M. F. (2000). pH-dependent thickness behavior of sequentially adsorbed layers of weak polyelectrolytes. *Macromolecules*, **33**, 4213–4219. (DOI: 10.1021/ma991645q).

Shutava, T., Prouty, M., Kommireddy, D. and Lvov, Y. (2005). pH responsive decomposable layer-by-layer nanofilms and capsules on the basis of tannic acid. *Macromolecules*, **38**, 2850–2858. (DOI: 10.1021/ma047629x).

Stevens, M. M. and George, J. H. (2005). Exploring and engineering the cell surface interface. *Science*, **310**, 1135–1138. (DOI: 10.1126/science.1106587).

Stockton, W. B. and Rubner, M. F. (1997). Molecular-level processing of conjugated polymers. 4. Layer-by-layer manipulation of polyaniline via hydrogen-bonding interactions. *Macromolecules*, **30**, 2717–2725. (DOI: 10.1021/ma9700486).

Suchanek, W. and Yoshimura, M. (1998). Processing and properties of hydroxyapatite-based biomaterials for use as hard tissue replacement implants. *Journal of Materials Research*, **13**, 94–117. (DOI: 10.1557/jmr.1998.0015).

Sui, Z., Salloum, D. and Schlenoff, J. B. (2003). Effect of molecular weight on the construction of polyelectrolyte multilayers: Stripping versus sticking. *Langmuir*, **19**, 2491–2495. (DOI: 10.1021/la026531d).

Sukhorukov, G. B., Antipov, A. A., Voigt, A., Donath, E. and Möhwald, H. (2001). pH-controlled macromolecule encapsulation in and release from polyelectrolyte multilayer nanocapsules. *Macromolecular Rapid Communications*, **22**, 44–46. (DOI: 10.1002/1521–3927(20010101)22:1<44::aid-marc44>3.0.co;2-u).

Sutherland, I. W. (1998). Novel and established applications of microbial polysaccharides. *Trends in Biotechnology*, **16**, 41–46. (DOI: 10.1016/s0167–7799(97)01139–6).

Tang, Z., Wang, Y., Podsiadlo, P. and Kotov, N. A. (2006). Biomedical applications of layer-by-layer assembly: From biomimetics to tissue engineering. *Advanced Materials*, **18**, 3203–3224. (DOI: 10.1002/adma.200600113).

Tirrell, M., Kokkoli, E. and Biesalski, M. (2002). The role of surface science in bioengineered materials. *Surface Science*, **500**, 61–83. (DOI: 10.1016/S0039–6028(01)01548–5).

Veerabadran, N. G., Goli, P. L., Stewart-Clark, S. S., Lvov, Y. M. and Mills, D. K. (2007). Nanoencapsulation of stem cells within polyelectrolyte multilayer shells. *Macromolecular Bioscience*, **7**, 877–882. (DOI: 10.1002/mabi.200700061).

Vriezema, D. M., Aragonès, M. C., Elemans, J. A. A. W., Cornelissen, J. J. L. M., Rowan, A. E. and Nolte, R. J. M. (2005). Self-assembled nanoreactors. *Chemical Reviews*, **105**, 1445–1489. (DOI: 10.1021/cr0300688).

Wagner, V. E., Koberstein, J. T. and Bryers, J. D. (2004). Protein and bacterial fouling characteristics of peptide and antibody decorated surfaces of PEG-poly(acrylic acid) co-polymers. *Biomaterials*, **25**, 2247–2263. (DOI: 10.1016/j.biomaterials.2003.09.020).

Wang, X., Kim, Y.-G., Drew, C., Ku, B.-C., Kumar, J. and Samuelson, L. A. (2004). Electrostatic assembly of conjugated polymer thin layers on electrospun nanofibrous membranes for biosensors. *Nano Letters*, **4**, 331–334. (DOI: 10.1021/nl034885z).

Wang, Y. and Caruso, F. (2006). Nanoporous protein particles through templating mesoporous silica spheres. *Advanced Materials*, **18**, 795–800. (DOI: 10.1002/adma.200501901).

Wattendorf, U., Kreft, O., Textor, M., Sukhorukov, G. B. and Merkle, H. P. (2008). Stable stealth function for hollow polyelectrolyte microcapsules through a poly(ethylene glycol) grafted polyelectrolyte adlayer. *Biomacromolecules*, **9**, 100–108. (DOI: 10.1021/bm700857s).

Wohl, B. M. and Engbersen, J. F. J. (2012). Responsive layer-by-layer materials for drug delivery. *Journal of Controlled Release*, **158**, 2–14. (DOI: 10.1016/j.jconrel.2011.08.035).

Wood, K. C., Boedicker, J. Q., Lynn, D. M. and Hammond, P. T. (2005). Tunable drug release from hydrolytically degradable layer-by-layer thin films. *Langmuir*, **21**, 1603–1609. (DOI: 10.1021/la0476480).

Ye, X., Hu, X., Wang, H., Liu, J. and Zhao, Q. (2012). Polyelectrolyte multilayer film on decellularized porcine aortic valve can reduce the adhesion of blood cells without affecting the growth of human circulating progenitor cells. *Acta Biomaterialia*, **8**, 1057–1067. (DOI: 10.1016/j.actbio.2011.11.011).

Yoo, D., Shiratori, S. S. and Rubner, M. F. (1998). Controlling bilayer composition and surface wettability of sequentially adsorbed multilayers of weak polyelectrolytes. *Macromolecules*, **31**, 4309–4318. (DOI: 10.1021/ma9800360).

Zabet-Khosousi, A. and Dhirani, A-A. (2008). Charge transport in nanoparticle assemblies. *Chemical Reviews*, **108**, 4072–4124. (DOI: 10.1021/cr0680134).

Zahn, R., Thomasson, E., Guillaume-Gentil, O., Vörös, J. and Zambelli, T. (2012). Ion-induced cell sheet detachment from standard cell culture surfaces coated with polyelectrolytes. *Biomaterials*, **33**, 3421–3427. (DOI: 10.1016/j.biomaterials.2012.01.019).

Zhang, J., Senger, B., Vautier, D., Picart, C., Schaaf, P., Voegel, J.-C. and Lavalle, P. (2005). Natural polyelectrolyte films based on layer-by layer deposition of collagen and hyaluronic acid. *Biomaterials*, **26**, 3353–3361. (DOI: 10.1016/j.biomaterials.2004.08.019).

Zhao, Q., Han, B., Wang, Z., Gao, C., Peng, C. and Shen, J. (2007). Hollow chitosan-alginate multilayer microcapsules as drug delivery vehicle: Doxorubicin loading and in vitro and in vivo studies. *Nanomedicine: Nanotechnology, Biology and Medicine*, **3**, 63–74. (DOI: 10.1016/j.nano.2006.11.007).

5

Synthesis of carbon based nanomaterials for tissue engineering applications

D. A. STOUT and N. G. DURMUS, Brown University, USA
and T. J. WEBSTER, Northeastern University, USA

DOI: 10.1533/9780857097231.1.119

Abstract: The possibility of incorporating carbon based nanomaterials into living systems has opened the way for their potential applications in the emerging field of nanomedicine. Within this chapter, we will investigate and explain how carbon based nanomaterials are synthesized and provide a few tissue engineering examples as they pertain to the carbon based shape that was created (i.e., carbon nanotube and fibers, graphene and graphite, fullerenes, diamonds, and carbon-nanostructured materials). We put a special emphasis on tissue engineering based carbon nanomaterials as a scaffold or implantable device. Hybrids of carbon nanomaterials embedded into other materials will also be discussed.

Key words: carbon, synthesis, nanomaterials, biomaterials, nanotechnology, regenerative medicine, orthopedics, cardiovascular, proteins, bacteria.

5.1 Introduction

The main goal of tissue engineering is to replace diseased and/or damaged tissues with biological substitutes or biocompatible materials that can restore and maintain normal tissue or biological functions. There have been major advances in the areas of cell and organ transplantation, as well as advances in materials science and engineering which have aided in the continuing development of tissue engineering and regenerative medicine (Fong *et al.*, 2005; Silva, 2006; Harrison and Atala, 2007; Mironov *et al.*, 2008). But with all of these advancements, there still is a need for better materials and techniques to facilitate tissue growth. One field that can promote this advancement is nanotechnology, which focuses on gathering simple nanosized materials to form complex structures. In a sense, nanotechnology involves utilizing materials which possess at least one physical dimension between 1 and 100 nm to construct structures, devices, and systems that have novel properties (Silva, 2006) – in fact, many biological components, like DNA, involve some aspect of nano-dimensionality (Silva, 2006). With the knowledge of

119

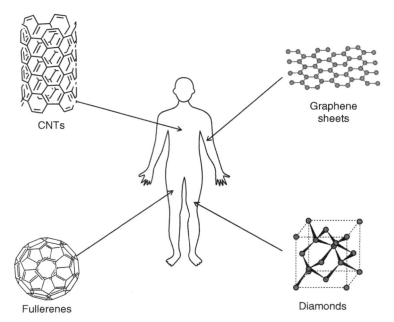

CNTs

Graphene
sheets

Fullerenes

Diamonds

5.1 The possibility of incorporating carbon based nanomaterials
into living systems has opened the way for their applications in
the emerging field of nanomedicine. A wide variety of different
nanomaterials based on allotropic forms of carbon, such as nanotubes,
graphene, fullerenes, and nanodiamonds, are currently being explored
towards different biomedical applications from orthopedic implants
(nanodiamonds) to strengthening existing carbon composites
(nanotubes) and fighting cancer (fullerenes).

nano-dimensionality in nature, it has logically given rise to interest in using
nanomaterials for tissue engineering with these materials having the potential for a significant impact in tissue engineering (see Fig. 5.1).

With the advent of the nanomaterial research boom, many synthesis
methods were devised to alter size, shape, chirality, and production yields,
alongside which came the occasional unintended discovery of new materials.
Another reason for intensive research on carbon based nanostructures and
nanomaterials such as fullerenes (Kroto *et al.*, 1985), single- or multi-walled
carbon nanotubes (SWCNTs and MWCNTs) (Iijima, 1991), or nanodiamonds (Talapatra *et al.*, 2005), is the high technological importance of these
systems owing to their unique mechanical and electronic properties, which,
again, can be tailored through their synthesis method (Krasheninnikov and
Banhart, 2007). When looking at carbon based nanomaterials, the variation in the hybridization type and hydrogen content, yields different structures (see Fig. 5.2). The synthesis in generating these materials lies within

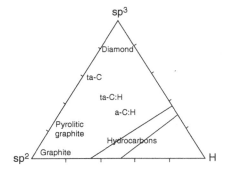

5.2 Schematic representation of the relative ternary composition of different carbon materials in thin films, showing variations in the degree of bond type and hydrogen content. The boundaries indicate the limits under which thin films can be formed as described in Jacog and Moller (1993).

exploiting these many bond types. Within this chapter we will investigate important carbon based nanomaterial synthesis (see Table 5.1) as it pertains to the carbon based shape that was created (carbon nanotube, graphene and graphite, fullerenes, diamonds, and carbon-nanostructured materials). We will place special emphasis on tissue engineering-based carbon nano-materials used as a scaffold or implantable device while omitting hybrids of carbon nanomaterials embedded into other materials. For each carbon nanomaterial shape, a few examples will be given of how these materials are used within the tissue engineering sphere.

5.2   Carbon nanotubes and fibers

Following the discovery of carbon nanotubes (CNTs) (Iijima and Ichihashi, 1993), carbon based nanotechnology has been rapidly developing as a platform technology for a variety of uses including biomedical applications. Among the various types of carbon based nanomaterials, CNTs have attracted increasing attention due to their mechanical, electrical, thermal, optical, and structural properties. CNTs are well-ordered, hollow nanostructures consisting of carbon atoms bonded to each other via sp^2 bonds which are stronger than sp and sp^3 bonds and are the key factors rendering excellent mechanical strength and high electrical and thermal conductivity to CNTs. Abstractly, CNTs can be imagined to form when a graphene sheet is rolled into a cylinder. One graphene sheet rolled up will form a SWCNT, while more than one concentric graphene sheet creates a MWCNT. Looking at their dimensions, SWCNTs typically have diameters ranging from 0.5 to 1.5 nm and a length ranging from 100 nm up to several micrometers. MWCNTs

Table 5.1 Summary of synthesis methods to produce carbon based nanomaterials for tissue engineering applications

Method	Nano shape	Advantages	Disadvantages
Aqueous dispersion	Graphene	Simple, low production costs	Low control over product dimensions, low yield
Carbide-derived carbon	CNTs	Ability to align CNTs, no catalyst required	Very small pore size distribution
CVD	CNTs, fullerenes, NCDs, nanowalls, graphene	Low set-up cost, high product yield, high purity, easy to scale-up, control over dimension and shapes	Many variables, hard to control
Electric arc discharge	CNTs, fullerenes, NCDs, graphene	Simple, high productive yield, high purity	High production costs, low control over product dimensions, toxic by-products
Graphite sonication	Graphene	Simple, low production costs, eco-friendly, no catalyst required	Low control over product dimensions
Ion beam deposition	NCDs, fullerenes, CNTs	High control over dimension size, no catalyst required	Low yield
Laser vaporization	CNTs, fullerenes, NCDs	High purity, low production costs	Low control over product dimensions
Liquid–liquid interface precipitation	Fullerenes	No physical template or catalyst required, simple	Low yield
Metal–carbon growth melts	Graphene	High purity, low production costs, no catalyst required	Low yield
Organic	Fullerenes	Simple (usually one step)	Very low yield
Plasma	CNTs, fullerenes, NCDs	Mass production, very versatile, simple	Low productive yield
Solar production	CNTs	High purity, potential to dope with elements (e.g. tin), eco-friendly	Low yield

Abbreviations: CNT – carbon nanotube; CVD – chemical vapor deposition; NCD – nanocrystalline diamond.

have larger diameters (which can be more than 100 nm) due to multilayer structures.

Due to the superb mechanical, electrical and surface properties of CNTs, they are the ideal candidate for a wide range of applications such as structural materials (Qian *et al.*, 2000), sensors (Wang, 2005), field emission displays (de Heer *et al.*, 1995), hydrogen storage materials (Chen *et al.*, 2001), tips for scanning probe microscopy (Snow *et al.*, 2002), and nanometer-sized semiconductor devices (Menon and Srivastava, 1997). For example, metallic SWCNTs can carry currents over 50 fold greater than those of normal metals (Yao *et al.*, 2000), but more importantly, by having a high aspect ratio (i.e., the length to diameter ratio) and high surface areas with many dangling bonds on the side walls, CNTs have the potential for biomedical applications ranging from neural regeneration (Cellot *et al.*, 2009), cardiomyocyte growth (Stout *et al.*, 2011) and muscle regeneration (Baughman *et al.*, 2002), to orthopedic (Christenson *et al.*, 2007) and induced pluripotent stem cell applications (Takahashi and Yamanaka, 2006).

In recent years, CNTs have become the backbone of many tissue engineering composite materials that include biodegradable polymers, metals, plastics, and natural materials (i.e., proteins and peptides). Though these advances have shed further insight into CNT–tissue interactions and propelled CNT–composite tissue engineering into public view, this section will solely focus on the synthesis of the CNTs themselves and provide tissue engineering examples where CNTs are the main focus of the investigation.

5.2.1   Synthesis methods

Electric arc discharge method

One of the easiest and optimal ways of creating nanotubes is using a technique involving the passage of a direct current through two high-purity graphite electrodes separated by 1–2 mm, usually in a helium atmosphere. When an arc is formed, a deposit forms on the cathode (negative electrode) while the anode (positive electrode) is consumed (Ebbesen and Ajayan, 1992). The deposit exhibits a cigar-like structure in which the inner core contains CNTs which are nested with polyhedral graphene particles (Ebbesen and Ajayan, 1992) (see Fig. 5.3).

While changing these basic parameters, one can optimize and alter CNT parameters for specific biomaterial applications. For example, to create SWCNTs, Iijima and Ichihashi (1993) were the first to arc iron-graphite electrodes in a methane-argon atmosphere to create tube diameters ranging from 0.75 to 13 nm. By involving iron-cobalt-nickel-graphite electrode mixtures in a pure helium atmosphere, Bethune *et al.* (1993) created SWCNTs that exhibited an average diameter of 1.2 nm.

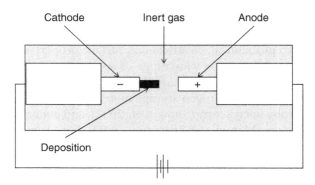

Cathode   Inert gas   Anode

−   +

Deposition

5.3 Schematic of the electric arc method.

Today, SWCNTs can be produced using the electric arc discharge method with many different metals such as gadolinium (Subramoney *et al.*, 1993), cobalt-platinum (Lambert *et al.*, 1994), cobalt-ruthenium (Lambert *et al.*, 1994), cobalt (Ajayan *et al.*, 1993), nickel-yttrium (Journet *et al.*, 1997), rhodium-platinum (Saito *et al.*, 1998), and cobalt-nickel-iron-cerium (Huang *et al.*, 2001), with nickel-yttrium having the highest yield (< up to 90%) with an average diameter of ~1.4 nm. This nickel-yttrium mixture is now used worldwide for the production of SWCNTs in high yield (Akiladevi and Basak, 2011; Li *et al.*, 2011).

Even with this simplicity and high yield, the arc method usually involves high-purity graphite electrodes, metal powders (only for producing SWCNTs), and high-purity helium and argon gases; thus, the costs associated with the production of SWCNTs and MWCNTs are high. Although the crystallinity of the material is also high, there is still very little to no control over dimensions (length and diameter) of the tubes. Also, the by-products formed, such as polyhedral graphite particles (in the case of MWCNTs), encapsulated metal particles (for SWCNTs), and amorphous carbon, can have toxicological affects (Pulskamp *et al.*, 2007; Hull *et al.*, 2009).

Laser vaporization method

The use of a yttrium aluminum garnet type laser to vaporize a metal material to create CNTs is a technique known as laser vaporization (see Fig. 5.4). If a laser targets a pure graphite target inside a furnace at 1200°C, in an argon atmosphere, a host of MWCNTs can be generated (Guo *et al.*, 1995). The laser beam has sufficient energy density to convert the graphite into amorphous carbon, which, within a flowing buffer gas such as argon, leads to SWCNT growth downstream of the target. Therefore, nanotube growth is created when the graphite carbon atoms attach at the edges of the adjacent

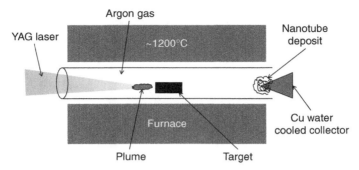

5.4 Schematic of an experimental set-up for the basic neodymium-doped yttrium aluminum garnet (Nd:YAG) laser evaporation system. (*Source*: Adapted from Guo *et al.* (1995).)

growing graphene tubules in a lip–lip interaction, which results in a multi-layered tube (Guo *et al.*, 1995).

SWCNTs have been shown to be a little more difficult to produce and with mixed outcomes. As compared to MWCNT laser vaporization, SWCNT laser vaporization requires the addition of metal particles as catalysts to the graphite targets to hinder the agglomerative attachment of carbon molecules around the same tube (Thess *et al.*, 1996; Qin *et al.*, 1997). Thess *et al.* (1996) were the first to obtain SWCNT bundles using graphite-cobalt-nickel targets with diameters around 14 Å packed in a crystalline form. But when others tried to produce the same results, groups found that the tubules showed narrower diameters (~1–2 Å) and fewer than 20% of the tubes possessed an armchair chirality (Qin *et al.*, 1997), thus supporting the reproducibility problems of the laser vaporization method.

Other groups enhanced this method to be able to produce SWCNTs in the absence of an oven using a carbon dioxide (CO_2) laser focused on a graphite metal target (Maser *et al.*, 1998; Muñoz *et al.*, 2000), where argon and nitrogen were determined to be the best atmosphere to generate SWCNT bundles. It was also shown that the diameter of the tubes was dependent upon the laser power (Dillon *et al.*, 2000): as the laser pulse power increased, the diameter of the tubes narrowed. A more promising laser method CNT fabrication techniques revealed that ultrafast laser pulses could produce large amounts of SWCNTs (1.5 g/h) (Eklund *et al.*, 2002); using a porous target of graphite-cobalt-nickel(nitrate) it was possible to yield twice as much SWCNT material compared to the original standard metal-carbon target (Yudasaka *et al.*, 1997).

With these advantages, the laser technique has shown to be quite economically advantageous due to processes involving high-purity graphite rods, high laser power, and low production costs compared to other methods.

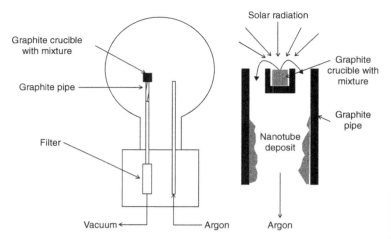

5.5 Schematic of the solar synthesis method as described in Laplaze *et al.* (1998).

Solar production method

Looking at alternative and 'green' chemistry methods, researchers have developed an alternative way to produce CNTs. It was first introduced by Laplaze and colleagues (1998), who were able to produce SWCNTs and MWCNTs when solar energy was focused on a carbon-metal target in an inert atmosphere. In brief, the reaction chamber consists of a Pyrex balloon flask located on top of a water-cooled brass support (see Fig. 5.5). In its center, a graphite crucible is placed vertically containing graphite-metal powders as a target material. The crucible is surrounded by a graphite pipe which is connected to a filter inside the water-cooled brass cylinder. The chamber is purged and a constant flow of a buffer gas is established passing through the graphite pipe on its way out. A parabolic mirror above the chamber focuses the collected sunlight on top of the target material. The evaporated material is drawn immediately through the graphite pipe, which acts as a thermal screen. On its walls, the soot material produced in the form of rubbery sheets can be collected. The diameter distribution of the SWCNTs lies in the range of 1.2–1.5 nm and the sample purity can be as good as one of the materials produced by the laser method where the nanotubes are accompanied by amorphous carbon and metal nanoparticles. The solar energy is able to vaporize graphite-metal targets due to the average incident solar flux of the experiments being close to 500 W/cm^2 that can reach a front temperature of 2800 K (Guillard *et al.*, 2001); this results in temperatures close to 3400 K that can increase the yield of nanotubes (Alvarez *et al.*, 1999, 2000).

Other groups have been able to scale up this process by increasing the solar flux range of 600 to above 900 W/cm^2, thus resulting in the generation of soot very rich in nanotubes (Terrones, 2003). Also, pyrolytic processes in a solar furnace have been shown to produce MWCNTs, particularly with the catalytic decomposition of CH_4 and C_4H_{10} in the presence of nickel-aluminum oxide or cobalt-magnesium oxide (Meier *et al.*, 1999).

Plasma method

In accordance with the growth of nanotube research and technology, researchers have looked for new ways to synthesize CNTs on a mass scale. Due to this material niche, the 3-phase AC plasma processing method was developed (see Fig. 5.6). The 3-phase AC plasma system, initially developed and optimized for the synthesis of novel grades of carbon black (Fulcheri *et al.*, 2002), and also continuous synthesis of fullerenes (Fulcheri *et al.*, 2000), was adapted to make carbon nanotubes and tube-like structures (Gruenberger *et al.*, 2004).

Adapted by Gruenberger *et al.* (2004), the plasma method works as: the plasma is generated by an arc discharge between three graphite electrodes placed in the top portion of the reactor. At the same time, an inert gas flows across the arc and the carbonaceous precursor and catalyst are injected into a high temperature region after the arc by a special powder injection system. Due to the high density of enthalpy obtained, the carbon and the metal catalysts are vaporized completely while passing through the graphite nozzle.

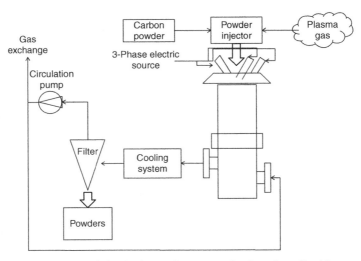

5.6 Schematic of the 3-phase plasma method as described by Gruenberger *et al.* (2004).

The quenching/sampling system collects the high temperature gas at a pre-determined position in the reactor and cools it rapidly. The gas is filtered and a part of it is re-injected into the reactor. The fraction that corresponds to the flow rate initially entering the reactor as plasma gas is exhausted through the exhaust nozzle. The set-up allows for the extraction of a gas volume superior to the initial flow of plasma gas and therefore disconnects the dependency of the two parameters, which leads to an additional degree of freedom in relation to process operation. Therefore, the carbon mass flow is no longer limited by a physical ablation rate (limiting step in the production rate of the classical nanotube processes), but is freely adjustable. Also, the process is operated at atmospheric pressure and the nanotube rich soot can be extracted continuously (Gruenberger et al., 2004).

This system can yield CNTs in the range of 10–50% in the as-produced soot with a production rate of 250 g/h, giving the freedom to make SWCNTs, carbon nanofibers, and carbon necklaces for research purposes (Fulcheri et al., 2000, 2002; Gruenberger et al., 2004).

Chemical vapor deposition method

The chemical vapor deposition method (CVD) is the most popular and widely used commercial method of producing carbon nanotubes due to its low set-up cost, high production yield, and ease of scale-up (Mukhopadhyay et al., 1998; Shigeo et al., 2002; Oncel and Yurum, 2006; Kumar and Ando, 2010). Compared to other synthesis methods, such as arc-discharge and laser-ablation, CVD is a more simple and economical technique. Moreover, it offers better control on growth, structural, and architectural parameters for CNT synthesis. CVD can harness hydrocarbons in any state (solid, liquid, or gas), and it allows CNT growth in a variety of forms, such as powder, thin or thick films, aligned or entangled, and straight or coiled nanotubes (Kumar and Ando, 2010).

Figure 5.7 shows the schematic of a typical CVD set up. The CVD process involves passing the vapor of the hydrocarbon precursor through a tubular reactor for 15–60 min. In the tubular reactor, a catalyst material is present at sufficiently high temperature (600–1200°C) to decompose the hydrocarbon. CNTs grown on the catalyst in the reactor are collected upon cooling the system to room temperature. In the case of a liquid hydrocarbon source (e.g., benzene and alcohol), the liquid is heated in a flask and an inert gas is purged through it, which in turn carries the hydrocarbon vapor into the reaction zone. If a solid hydrocarbon is to be used as the CNT precursor, it can be directly kept in the low-temperature zone of the reaction tube (Oncel and Yurum, 2006; Kumar and Ando, 2010).

Carbon precursors, catalyst materials for nanotube deposition, synthesis temperature, and reaction time are some of the important parameters which

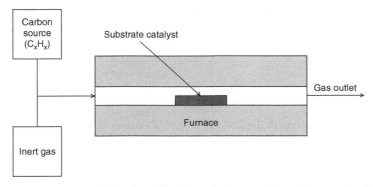

5.7 Schematic of the simplified chemical vapor deposition method.

affect the structure and quality of CNTs produced via the CVD method (Oncel and Yurum, 2006). The most commonly used CNT precursors are methane (Hernadi *et al.*, 1996), ethylene (Fan *et al.*, 1999), acetylene (Li *et al.*, 1996), benzene (Sen *et al.*, 1997), xylene (Wei *et al.*, 2002), and carbon monoxide (Nikolaev *et al.*, 1999; Kumar and Ando, 2010). The molecular structure of a hydrocarbon precursor has an important effect on the morphology of the CNTs grown via CVD (Kumar and Ando, 2010). For example, linear hydrocarbons (e.g., methane, ethylene, and acetylene) usually produce straight hollow CNTs, while cyclic hydrocarbons (e.g., benzene, xylene, and cyclohexane) produce relatively curved CNTs with the tube walls often bridged inside (Nerushev *et al.*, 2003; Morjan *et al.*, 2004; Kumar and Ando, 2010). In addition, metal nanoparticles are used to enable hydrocarbon decomposition during CNT synthesis (Kumar and Ando, 2010). Table 5.2 lists some of the catalysts used for CVD synthesis of CNTs. Carbon is highly soluble in these metals at high temperatures and it has a high diffusion rate (Kumar and Ando, 2010). In addition, temperature has an effect on the growth and structure of the CNTs grown via CVD. Low-temperature CVD reactions (600–900°C) usually yield MWCNTs, while high-temperature (900–1200°C) reactions favor SWCNT growth (Kumar and Ando, 2010).

Carbide-derived carbon method

Carbon produced by the extraction of metals from metal carbides (e.g., silicon carbide (SiC), titanium carbide (TiC), and aluminum carbide (Al_4C_3)) (Gogotsi *et al.*, 2003) is called carbide-derived carbon (CDC) (Equations [5.1] and [5.2]). Leaching in supercritical water, high-temperature treatment in halogens, vacuum decomposition, and other methods can be used to remove metals from carbides, producing carbon coatings or bulk and powdered carbon. The linear reaction kinetics, which were demonstrated for

Table 5.2 Examples of catalyst materials used in CNT with CVD synthesis

Catalyst	Reference
Cobalt (Co)	Sinha *et al.*, 2000
Cobalt and iron (Co/Fe)	Colomer *et al.*, 2000
Cobalt and vanadium (Co/V)	Mukhopadhyay *et al.*, 1999
Iron (Fe)	Singh *et al.*, 2003
Iron and cobalt (Fe/Co)	Couteau *et al.*, 2003
Iron and molybdenum (Fe/Mo)	Su *et al.*, 2000
Nickel (Ni)	Choi *et al.*, 2003

chlorination of both bulk-sintered polycrystalline (Ersoy *et al.*, 2001) and single crystalline SiC nanowhiskers (Cambaz *et al.*, 2006), allow transformation to any depth, until the complete particle or component is converted into carbon. This selective etching of carbides is considered to be a promising technique for the synthesis of various carbon structures (Baraton and Uvarova, 2001).

$$SiC + 2Cl_2(g) = SiCl_2(g) + C \qquad [5.1]$$

$$Ti_3SiC_2 + 8Cl_2(g) = SiCl_4(g) + 3TiCl_4(g) + 2C \qquad [5.2]$$

For CNTs, the thermal decomposition of SiC arose in 1997 when Kusunoki *et al.* (1997) discovered the growth of self-organized CNT films during the decomposition of SiC. In comparison to the more conventional CVD growth technique, CDC allows for the formation of highly aligned nanotubes with no catalyst required for CNT development (Kusunoki *et al.*, 1998, 1999). Moreover, selective preparation of CNTs of the zigzag type is possible which permits preparation of CNTs of the same chirality (Kusunoki *et al.*, 2002).

5.2.2   Tissue engineering examples

The development of nanotechnologies and synthesis of new nanomaterials have a great impact on existing macro- and micro-technologies and provide the ability to better mimic the native microenvironment for biomedical applications (Harrison and Atala, 2007; Dvir *et al.*, 2011). Due to their exceptional electrical, mechanical, and chemical properties, as well as their high aspect ratio, CNTs have been used in biomedical applications since 2004 (Harrison and Atala, 2007). For tissue engineering purposes, CNTs have been important materials for four main areas such as improved cell tracking and labeling, sensing cellular behavior and microenvironments, delivery of transfection agents and growth factors, and scaffolding and enhancing tissue matrices (Harrison and Atala, 2007). Table 5.3 summarizes the possibility of incorporating carbon based nanomaterials into living systems and has opened the way for the investigation of their potential applications in the emerging field of nanomedicine. In this review, we will focus on their use as structural support as scaffolds in tissue engineering applications.

The extracellular matrix (ECM) is a dynamic and highly-organized nanocomposite which defines the space the native tissue occupies. It is an essential component for tissue engineering as it regulates essential cellular functions such as cellular differentiation, proliferation, adhesion, and migration (Tsang *et al.*, 2010). Nanomaterials have been explored to create extracellular matrix (ECM)-like structures and compensate for some limitations that some popular synthetic polymers (such as polylactic acid (PLA) and poly(lactic-*co*-glycolic acid) (PLGA)) lack (Harrison and Atala, 2007; Dvir *et al.*, 2011). These limitations include weak mechanical properties, lack of electrical conductivity, the absence of adhesive and microenvironment-defining moieties, and the inability of cells to self-assemble in to three-dimensional (3D) tissues (Dvir *et al.*, 2011). In addition, CNTs can be easily functionalized for different biomedical applications.

CNTs might provide structural support due to their extraordinary mechanical strength and hold great promise as scaffolds for tissue engineering applications. In addition, the viscoelastic behavior of CNTs is similar to that observed in soft-tissue membranes (Suhr *et al.*, 2007; Dvir *et al.*, 2011). Many studies have shown that mechanical properties were significantly improved when CNTs were dispersed in conventional polymer solutions. For example, MWCNTs blended with a chitosan polymer showed significant improvement in mechanical properties compared with those of chitosan (Wang *et al.*, 2005). Chitosan is a natural biopolymer and has been widely used in many tissue engineering applications as it has biocompatibility, biodegradability, and multiple functional groups, as well as great solubility in the aqueous medium (Kumar, 2000; Sashiwa and Aiba, 2004; Wang *et al.*, 2005; Albanna *et al.*, 2012). On the other hand, chitosan has low to moderate mechanical strength which limits its use in load-bearing biomedical applications. To overcome this limitation, chitosan/MWCNTs nanocomposites were prepared by a simple solution-evaporation method. MWCNTs were homogeneously dispersed within the chitosan polymer matrix. Blending of only 0.8 wt% MWCNTs, the tensile modulus and strength of the chitosan/MWCNT nanocomposites were greatly improved by about 93% and 99%, respectively, when compared to pure chitosan (Wang *et al.*, 2005).

CNTs can also improve the mechanical properties of hydrogels, creating porous hydrogels with tunable mechanical properties. Highly cross-linked, 3D hydrogel microenvironments have high stiffnesses but they might limit cellular proliferation, migration, and morphogenesis (Patel *et al.*, 2005; Engler *et al.*, 2006; Shin *et al.*, 2012). To create 3D hydrogels with controlled

Table 5.3 Examples of cells types which have been successfully cultured on CNTs or CNT-incorporated nanocomposites

Cell type	Scaffold	Biological response	Reference
Smooth muscle cells	SWCNT – collagen	Cell viability in nanocomposites was consistently above 85% at both Day 3 and Day 7.	MacDonald et al., 2005
Human Neonatal Dermal Fibroblasts (HNDFb)	Carboxylated SWCNT – collagen – fibrin nanocomposite	Cell viability and morphology were similar in both collagen-fibrin composites and SWCNT-loaded nanocomposites.	Voge et al., 2008
Human embryonic stem cells (hESCs) (H9 cell line)	MWCNT – silk nanocomposite	Neuronal lineage differentiation of hESCs was observed.	Chen et al., 2012
Mouse embryonic neural stem cells (NSCs)	SWCNT – polyelectrolyte multilayer thin films	NSCs differentiated into neurons, astrocytes, and oligodendrocytes. Clear formation of neuritis was observed.	Jan and Kotov, 2007
Osteosarcoma cells (ROS 17/2.8 cells)	Nitric acid-treated SWCNT, functionalized SWCNT, MWCNT	Osteosarcoma cells were successfully cultured on SWCNT and MWCNT. Cell growth was controlled by chemical modifications on the CNTs.	Zanello et al., 2006
Mouse fibroblasts (cell line L929)	3D MWCNT-based networks	Fibroblasts were successfully cultured. Extensive growth, spreading, and adhesion of cells were observed.	Correa-Duarte et al., 2004
Chinese hamster ovary (CHO) cells	Aligned MWCNT films	CHO cells adhered to and grew on MWCNT films. They also aligned strongly with the axis of the bundles of the MWCNTs.	Abdullah et al., 2011
Osteoblast cells 11	Poly-D-lactide (PLA) – MWCNT nanocomposite	Aligned nanocomposites enhanced the extension and directed the outgrowth of osteoblast cells better than random composites.	Shao et al., 2011
Human umbilical vein endothelial cells (HUVECs)	13 Polyurethane – MWCNT nanocomposite	Aligned structure of MWCNTs functioned as extracellular signals and stimulated HUVEC cell growth, proliferation, and extracellular collagen secretion. In addition, nanocomposite preserved the anticoagulant function.	Meng et al., 2010

mechanical properties, Shin *et al.* (2012) coated CNTs with a thin layer of photo-crosslinkable gelatin methacrylate (GelMA), which is known to allow cell encapsulation and proliferation. The addition of CNTs significantly improved the mechanical properties of GelMA hydrogels without decreasing their porosity. The elastic modulus of the GelMA hydrogels increased from 15 to 60 kPa when 0.5 mg/mL of CNTs were coated. In addition, the CNT-GelMA hybrid hydrogels showed high toughness and strong tensile strength and they maintained their ability to elongate. NIH-3T3 cells – a standard fibroblast cell line – and human mesenchymal stem cells (hMSCs) were encapsulated in CNT-GelMA hybrid microgels with varying concentrations of CNTs from 0 to 0.5 mg/mL. Both cell types were spread and proliferated after encapsulation in CNT-GelMA hybrid microgels. 3T3 cells had greater than 90% cellular viability for 48 h for all CNT concentrations (Shin *et al.*, 2012). Apart from polymer and hydrogel enhancement, CNTs can also be used to reinforce ceramic matrices. For example, CNT/barium titanate ($BaTiO_3$) composites were fabricated by adding 1 wt% of CNTs. This increased the fracture toughness by about 240% compared to pure $BaTiO_3$ (Gao *et al.*, 2006). Moreover, CNTs were homogenously distributed into a hydroxyapatite (HA) coating using plasma spraying. Incorporating CNTs into a brittle bioceramic coating improved its fracture toughness by 56% (Balani *et al.*, 2007). Hence, CNTs can be used to enhance the mechanical properties of hydrogels, polymers, and ceramics to create better scaffolds to mimic native tissues.

It is important to note that CNTs are very attractive materials for tissue engineering applications due to their promising chemical, mechanical, and electrical properties. On the other hand, their extremely low solubility in organic and inorganic solvents as well as physiological fluids is the main limiting factor to integrating these unique materials for biological applications (Pastorin *et al.*, 2005; Kharisov *et al.*, 2009). Pristine CNTs are difficult to dissolve or disperse in biological systems due to their long-structured morphology, large molecular size, and aggregation within the solvent (Kharisov *et al.*, 2009). There are many approaches to increase the solubility of CNTs for biological applications. These functionalization methods can be grouped into two main procedures: (a) non-covalent functionalization and (b) covalent functionalization (Pastorin *et al.*, 2005). Non-covalent functionalization preserves the aromatic structures of CNTs. Therefore, their electrical properties are also preserved. Non-covalent functionalization methods usually use organic solvents, polymers, peptides, nucleic acids, sugars, and detergents (Pastorin *et al.*, 2005). Covalent functionalization usually includes esterification or amidation of oxidized CNTs, or attaching a functional group to the side walls of CNTs (Pastorin *et al.*, 2005). Moreover, CNTs can be modified by a natural or a bioactive species, such as starch (Star *et al.*, 2002) and single-stranded DNA (Zheng *et al.*, 2003). Table 5.4 summarizes examples

Table 5.4 Examples of non-covalent and covalent modification methods to increase the solubility of CNTs for biological applications

Functionalization	Type of CNT	Solubilization method	Solubility (mg/mL)	Reference
Non-covalent	SWCNT	Poly(m-phenylene -vinylene) ester	0.50 (in DMF or CHCl$_3$)	Star *et al.*, 2001
Non-covalent	SWCNT	Peptide	0.70 (in H$_2$O)	Dieckmann *et al.*, 2003
Non-covalent	SWCNT	Tween 20	0.05 (in H$_2$O)	Chen *et al.*, 2003
Non-covalent	SWCNT	SDS	< 0.10 (in H$_2$O)	Connell *et al.*, 2001
Non-covalent	SWCNT	Starch	0.50 (in H$_2$O)	Star *et al.*, 2002
Covalent	SWCNT/ MWCNT	Oxidation via HNO$_3$/H$_2$SO$_4$	1.77 (in H$_2$O)	Hamon *et al.*, 1999
Covalent	SWCNT/ MWCNT	1,3-dipolar cycloaddition	~20 (in H$_2$O)	Georgakilas *et al.*, 2002
Covalent	SWCNT/ MWCNT	Nitrenes	1.20 (in DMSO or TCE)	Holzinger *et al.*, 2003

Abbreviations: DMF – dimethylformamide; DMSO – dimethyl sulfoxide; SDS – sodium dodecyl sulfate; TCE – trichloroethylene; CHCl$_3$ – chloroform; H$_2$O – water; H$_2$SO$_4$ – sulfuric acid; HNO$_3$ – nitric acid.

of non-covalent and covalent modification methods to increase the solubility of CNTs and disperse them for biological applications.

5.3 Fullerenes (C$_{60}$)

Carbon 60 or 'Buckminsterfullerene' is a closed structure composed of 60 carbon atoms arranged in the form of a hollow ball. Twenty carbon hexagons are linked and bent into the shape of a sphere by the inclusion of twelve equally-spaced pentagonal 'holes'. The structure was 'discovered', or proved to exist, by Kroto *et al.* (1985) for which three of the authors received the Nobel Prize in Chemistry in 1996. The name was a tribute to Buckminster Fuller, whose geodesic domes it resembles. Coincidently, the structure was also recognized 5 years earlier by Sumio Iijima (1980), from an electron microscope image, where it formed the core of a 'bucky onion'. Also, a small percentage of fullerenes are formed in any process which involves burning of hydrocarbons (e.g., burning a candle). Since their discovery, fullerenes have since been found to occur in nature (Buseck *et al.*, 1992) and have been detected in outer space (Cami *et al.*, 2010).

It all started when Kroto and colleagues employed laser irradiation of graphite and produced stable and isolatable C$_{60}$ or buckminsterfullerene, a graphitic material in an icosahedron or soccer ball shape (Kroto *et al.*, 1985). The discovery was extended to a family of spheroidal graphitic materials including C_{70}, C_{60} nanowhiskers, multiwalled fullerenes, ultra-short nanotubes of lengths 20–80 nm (versus SWCNTs of micron length), and water-dispersed derivative of the C_{60} (usually known as C_{60}-neodymium nanoparticles). Fullerenes can be synthesized from previously described methods – CVD (Gruen *et al.*, 1994), laser (Dimitrijevic and Kamat, 1992), plasma (Oohara and Hatakeyama, 2003), and electric arc discharge (Stevenson *et al.*, 1999). Below are some special synthesis methods for fullerenes that have not yet been discussed.

5.3.1   Synthesis methods

Chemical vapor deposition with fluorination method

Unlike conventional CVD, as described within Section 5.2.1, this method uses fluorine in its system; every other aspect is the same. Fluorination allows the researcher to create different carbon based nanomaterials that use the interaction between carbon and fluorine; these can vary from covalent, through semi-ionic, to ionic, and van der Waals interactions, with bonding between carbon and hydrogen (conventional CVD is essentially limited to covalent bonds). This allows for new functionalities due to the ionization as it governs its reaction with fluorine and fluorides for the formation of ionic carbon intercalation compounds. Also, when the formation of a thermodynamically stable covalent C–F bond is not kinetically possible, a fluorine molecule can choose to form fluoride ions owing to the weak F–F bond and allow for large electron affinity.

One such example is the formation of $C_{60}F_{46}$. The $C_{60}F_{46}$ powder was made by the direct fluorination of fullerene C_{60} at 225°C under 101.32 kPa of F_2 for 2 days (Okino *et al.*, 1996). Subsequently, the single crystals and fullerenes can be grown by the sublimation of $C_{60}F_{46}$ powder.

Liquid–liquid interfacial precipitation method

By using the well-known liquid–liquid interfaces that act as nucleation sites of crystals (Sica *et al.*, 1996), researchers have been able to apply liquid–liquid interfacial precipitation (LLIP) to a system of toluene solution of C_{60}/isopropyl alcohol and have discovered that very fine needle-like crystalline precipitates of C_{60} are formed in the solution, known as C_{60}-whiskers. Therefore, the synthesis method has a great advantage since it requires no physical templates or catalysts. In short, a toluene solution of C_{60} powder with a concentration of 0.3% mass C_{60} can be poured into a glass bottle (9 mL) and then isopropyl alcohol is gently added into the bottle to form a liquid–liquid interface, where the upper phase is isopropyl alcohol and the lower phase is the toluene solution of C_{60}. The bottle then can be capped and sealed with aluminum foil and kept still at room temperature until the precipitates of needle-like crystals of C_{60} form (Miyazawa *et al.*, 2002). By altering the different types of solvents – one type has a lone electron pair and the other does not – the formation of hollow C_{60} fullerenes and solid C_{60} fullerenes can be achieved. Whether the hollow or solid form is synthesized strongly depends on whether or not the solvent possesses a lone electron pair, thus indicating that the type of solvent plays a critical role in the formation of the final structure of C_{60} fullerenes and derivatives (Qu *et al.*, 2011).

Organic synthesis method

In 2002, a group described an organic synthesis of the compound starting from simple organic compounds (Scott *et al.*, 2002). In short, a 12 step synthesis method from commercially available starting materials by rational chemical methods is needed. In the final step, a large polycyclic aromatic hydrocarbon consisting of 13 hexagons and three pentagons is submitted to flash vacuum pyrolysis at 1100°C and 1.3 Pa. The three carbon–chlorine bonds serve as free radical incubators and the ball is stitched up in an undoubtedly complex series of radical reactions. Even with this advancement, the chemical yield is low, ranging from 0.1% to 1% (Scott *et al.*, 2002). A similar exercise aimed to construct a C_{78} cage in 2008 (Amsharov and Jansen, 2008) by leaving out the halogens of the precursor. The method worked successfully but did not result in a sufficient yield, thus the organic synthesis of fullerenes remains a challenge for chemistry and is not readily used.

5.3.2   Tissue engineering examples

Since their arrival, the fullerene group has garnered a lot of attention due to their novel properties associated with their high surface area, low-dimensionality, and potential quantum confinement effect (Lieber and Wang, 2007), which enabled them to serve as one-dimensional (1D) building blocks in future magnetic and template applications (Guss *et al.*, 1994; Minato and Miyazawa, 2006). Within tissue engineering applications, the fullerene group can be used for support of other biomaterials for tissue engineering applications. Shi *et al.* (2007) created a poly(propylene fumarate) composite, incorporating dodecylated ultra-short nanotubes. The intended application was a scaffold for bone tissue, and the porosity of the material was shown to be tunable to favor cell tissue ingrowth. It was also shown that the composite had equivalent or better mechanical properties compared to the pure polymer and non-functionalized nanotube composite with a porosity to be controlled by particulate leaching using varied sodium chloride

(NaCl) content. The composite supported cell adhesion and proliferation of rat marrow stromal cells under a 7-day *in vitro* culture.

Another experiment showed that ultra-short, single-walled nanotubes – synthesized by cleavage of purified single-walled nanotubes via fluorination, followed by pyrolysis – can promote bone growth (Gu *et al.*, 2002). Also, ultra-short tubes, functionalized by an alkylation-based reduction before being incorporated into the polymer composite under thermal cross-linking, exhibited *in vivo* biocompatibility of the composite when investigated by implantation in rabbit femoral condyles and subcutaneous pockets (Sitharaman *et al.*, 2008). The functionalized nanotube composite showed greater bone tissue ingrowth, reduced inflammatory response, and similar degradation after 12 weeks when compared to the pure polymer.

Water-soluble C_{60} derivatives have been shown to be able to enhance neurite outgrowth in nerve growth factor-treated pheochromocytoma of the rat adrenal medulla (PC12) cells that were used as a model of nerve cells (Tsumoto *et al.*, 2010). Experiments on PC12 using this fullerene derivative showed that it has the potential to prevent oxidative stress-induced cell death without evident toxicity (Hu *et al.*, 2010).

5.4   Graphene

Graphene is usually known for a structure that is a one-atom-thick planar sheet of sp^2 bonded carbon atoms densely packed in a honeycomb crystal lattice and is the basic building block for graphitic materials of all other dimensionalities. It can be wrapped up into zero-dimensional (0D) fullerenes, rolled into 1D nanotube, or stacked into 3D graphite. For many years graphene was not thought to exist in the 'real' world and believed to be unstable with respect to the formation of curved structures such as soot, fullerenes, and nanotubes (McClure, 1956; Slonczewski and Weiss, 1958). Then Andre Geim and Konstantin Novoselov at the University of Manchester unexpectedly formed the special material by using the well-established drawing method with adhesive tape (Novoselov, 2004, 2005). For their success, they were awarded the Nobel Prize in Physics in 2010.

Today, graphene has been touted as the perfect material for all aspects of science. Due to its unusual electronic spectrum, graphene has led to the emergence of a new model of 'relativistic' condensed-matter physics, where quantum relativistic phenomena can now be mimicked and tested in table-top experiments. More generally, graphene represents a conceptually new class of materials that are only one atom thick, and, on this basis, offers new inroads into low-dimensional physics and material engineering that has continued to provide a fertile ground for applications such as tissue engineering. Graphene can be synthesized by previously described methods – CVD (Kim *et al.*, 2009), laser (Williams *et al.*, 2008), and electric arc discharge (Subrahmanyam *et al.*, 2009). Below are some special methods for fabricating graphene that have not been discussed.

5.4.1   Synthesis methods

Graphite sonication method

In one of the easiest methods, graphene can be synthesized through sonicating graphite. This consists of dispersing graphite in a proper liquid medium that is then sonicated (applying energy as sound waves – usually ultrasound – to the liquid medium) (Hernandez *et al.*, 2008). Next, the non-exfoliated graphite is separated from graphene by centrifugation. This process can obtain varied graphene concentration by changing the solvent liquid solution ranging from 0.01 mg/mL in N-methylpyrrolidone (Hernandez *et al.*, 2008) to 5.33 mg/mL with an ionic liquid (Nuvoli *et al.*, 2011).

Aqueous dispersion method via graphite oxide reduction and sonication

Demonstrated by Stankovich and co-workers (2006, 2007), the solution-based route involves chemical oxidation of graphite to hydrophilic graphite oxide, which can be readily exfoliated as individual graphene oxide sheets by their ultrasonication in water. Next, the graphene oxide, which is electrically insulating, can be converted back to conducting graphene by chemical reduction, like using hydrazine. Because of this method, the precipitates will create irreversible agglomerates owing to their hydrophobic nature, but Li *et al.* (2008) combated this by taking into account electrostatic repulsion, rather than just the hydrophilicity of graphene oxide as previously presumed. They were able to synthesize graphene in stable aqueous colloids through electrostatic stabilization, via addition of ammonia using a low-cost solution processing technique for the large-scale production of aqueous graphene dispersions without the need for polymeric or surfactant stabilizers (Li *et al.*, 2008).

Metal–carbon growth melts method

One of the simplest methods, metal–carbon growth melts (MCGM), works by dissolving carbon atoms inside a transition metal melt at a certain temperature (~1500°C), and then allowing the dissolved carbon to precipitate out at lower temperatures (~1000°C) as single layer graphene (Amini *et al.*, 2010) (see Fig. 5.8). In this process, the metal is first melted in contact with a carbon source. This results in dissolution and saturation of carbon atoms in the melt based on the binary phase diagram of metal–carbon (Amini *et al.*, 2010). Upon lowering the temperature, solubility of the carbon in the molten metal will decrease and the excess carbon will precipitate on top of the melt; the floating layer can be skimmed for removal.

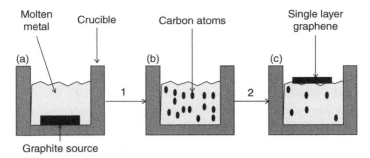

5.8 Schematic of the graphene growth from the metal–carbon growth melts method: (a) melting metal is in contact with graphite as carbon source, (b) dissolution of carbon inside the melt at a higher temperature than started, and (c) reducing the temperature for growth of graphene. (1) $T_a < T_b$. (2) $T_c < T_b$ and T_a. (*Source*: Adapted from Amini *et al.* (2010).)

5.4.2   Tissue engineering examples

Research on graphene was originally motivated by its peculiar electrical transport properties. However, owing to its exceptional thermal and mechanical properties and high electrical conductivity, it becomes of great interest in serving as new nanoscale building blocks to create unique biomaterials for tissue engineering applications (Chen *et al.*, 2008). For example, in the presence of an osteogenic medium, a graphene substrate helped remarkably by accelerating the differentiation of hMSCs at a rate comparable to differentiation under the influence of bone morphogenetic protein 2 (BMP-2) (Nayak *et al.*, 2011).

Fan and colleagues (2010) used graphene sheets, synthesized by a direct current arc-discharge method using ammonia (NH_3) as one of the buffer gases, dispersed in chitosan to prove that graphene/biomacromolecule composite materials could be made for tissue engineering applications. Chitosan was significantly reinforced upon the addition of a small amount of graphene sheets. It was shown that graphene/chitosan composites were biocompatible to normal subcutaneous aveolar and adipose tissue of a 100 day old male C3H/An mouse cell line (L929) and the absence of metallic impurities on the graphene sheets made them potential candidates as scaffolds for tissue engineering (Fan *et al.*, 2010).

5.5   Nanodiamond systems

In contrast to the materials above, diamond-like carbon has significant sp^3 bond content to which its metastable amorphous phase shares the hardness and chemical inertness of diamond (Robertson, 2002). Diamond-like carbon synthetically created – produced in a technological process as opposed to natural diamonds, which are created by geological processes – are also known as synthetic diamonds or cultivated diamonds. When formed, synthetic diamonds have similar properties to natural diamond; the hardness, thermal conductivity, and electron mobility, which depend totally on the details of the manufacturing processes, for example CVD.

Upon looking at diamond crystallinity, which may comprise one or many crystal structures, researchers noticed that polycrystalline diamond, consisting of numerous small grains, could be manufactured. Researchers were then able to alter the synthesis methods to create grain sizes ranging from nanometers to hundreds of micrometers, usually referred to as 'nanocrystalline' and 'microcrystalline' diamond, respectively (Denuault *et al.*, 2007). Within this chapter, we will primarily focus on nanocrystalline diamond (NCD). NCDs can be synthesized from previously described methods – CVD (Butler and Sumant, 2008), laser (Sun *et al.*, 2006), and plasma (Aharonovich *et al.*, 2008). Below are some special methods for tissue engineering NCDs that have not already been discussed.

5.5.1   Synthesis methods

Ion beam deposition method

Ion beam deposition (IBD) is a process of applying materials to a target through the application of a charged particle beam of ions to which other ions are then accelerated, focused, or deflected using high voltages or magnetic fields (see Fig. 5.9). From the technique point of view, the advantage of the IBD technique is the controllability of the energies and influences of the ions incident on the growing surface. The ion beam deposition technique also allows independent variation of the ion species, ion energy, and current density of the ion beam and the substrate temperature which convert to specific growth of controllable sizes of the diamonds. For example, Sun *et al.* (1999) used the IBD process in a vacuum chamber equipped with a Kaufman ion source to create NCDs with sizes ranging from 15 to 30 nm. A mixture of methane, hydrogen, and argon gases generated in the broad Kaufman ion source was used as the working gas and the ions. They were then extracted and accelerated to bombard the substrate perpendicularly. The energy of hydrocarbon and hydrogen ions bombarding the substrate varied from 80 to 200 eV where the substrate temperature was kept at 780°C (Sun *et al.*, 1999).

Microwave plasma chemical vapor deposition method

Diamonds grown by CVD are usually rough and have many defects which certain biomedical applications cannot tolerate. To overcome this

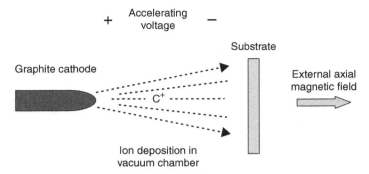

5.9 Schematic of the ion beam deposition system. (*Source*: Adapted from Aisenberg and Chabot (1971).)

obstacle, researchers used the microwave plasma chemical vapor deposition (MPCVD) method which utilizes plasma and microwave frequencies to enhance chemical reaction rates of the precursors (Kamo *et al.*, 1983) (see Fig. 5.10). Using this technique, Sharda *et al.* (2001), grew NCDs at 700°C which had a high optical absorption coefficient in the whole spectral region and a very high transmittance (~78%) in the near IR region, which is close to that of diamond, indicating that the NCD film has the potential for applications that need a higher purity of diamond.

5.5.2   Tissue engineering examples

NCD thin films exhibit a wide range of unique physicochemical and biological properties, such as mechanical hardness, chemical, and thermal resistance, as well as interesting optical and electrical properties. For applications in biology and bioelectronics, important properties have been shown to be non-cytotoxicity, relative ease of surface functionalization, and the ability to immobilize various biomolecules (e.g., nucleic acids, enzymes, and antibodies) on an NCD surface with strong covalent bonding (Hartl *et al.*, 2004; Christiaens *et al.*, 2006). Biocompatible NCD films have been shown to be a suitable coating for medical implants, particularly those designed for hard tissue surgery. Nanocrystalline diamond and diamond-like carbon have already been used for coating the heads and cups of artificial joint replacements, for example, metallic and polymeric prostheses of hip or temporomandibular joints (Lappalainen *et al.*, 2003; Papo *et al.*, 2004). For example, an *in vivo* study, using diamond layers deposited on metallic probes, implanted into rabbit femurs showed very high bonding strength to the metal base and also to the surrounding bone tissue, with no problems of corrosion (Specht *et al.*, 2004).

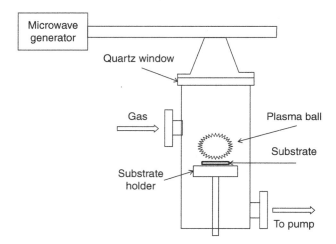

5.10 Schematic of the microwave plasma CVD system.

In addition, NCD films can be applied in the bone-anchoring stems of articular prostheses or other permanent bone implants in order to improve their interaction with the surrounding bone tissue. This beneficial action of NCD films can be anticipated not only due to their excellent mechanical and chemical resistance, but also due to their surface nanoroughness which has been shown to stimulate adhesion, growth, and maturation of osteoblasts, as well as differentiation of bone progenitor cells towards the osteoblastic phenotype (Price *et al.*, 2004; Dalby *et al.*, 2006; Yang *et al.*, 2009).

5.6   Carbon-nanostructured materials

Since the advent and development of multiple different synthesis techniques of CNTs, researchers have developed oddly different carbon based nanostructured materials due to altering or fine-tuning a known CNT method. For example, Ando *et al.* (1997) obtained petal-like graphite sheets as a by-product of a carbon nanotube synthesis through direct current arc-discharge evaporation. These odd carbon shapes have been categorized as carbon-nanostructured materials (CNMs) that encompass shapes which resemble flasks (Liu *et al.*, 2000, 2001; Kumar Rana *et al.*, 2003) to sheets of nanowalls (Wu *et al.*, 2002; Wang *et al.*, 2004; Zhu *et al.*, 2007). It only has recently been noted that these accidentally created shapes may have a niche within the nanobiomaterial world. Due to the irregularity of CNMs, instead of categorizing by synthesis method and then shape – as used above – we will compartmentalize CNMs by their shape and then describe the synthesis method behind them.

5.6.1   Carbon-nanostructured forms

Carbon nanowalls

Carbon nanowalls are created by peeling off the carbon from a natural graphite layer-by-layer and making each on a base. This is possible by joining the well-known fullerene structure (carbon arranged in spheres or tubes) and nanotube structure (microscopic carbon tubes) to form a family of carbon nanostructures through a microwave plasma enhanced chemical vapor deposition system (MPECVD) (Wu *et al.*, 2002). MPECVD differs from the CVD method explained above by the system being equipped with a 500 W microwave source and a transverse rectangular cavity to couple the microwave to a quartz tube for generating the plasma (Fan *et al.*, 1999, 2010; Ren *et al.*, 1999); this enhances chemical reaction rates of the precursors (Fan *et al.*, 2010) and has been widely used for diamond growth (Zhu *et al.*, 1991; Ye *et al.*, 2000) and CNT alignment (Qin *et al.*, 1998; Tsai *et al.*, 1999; Bower *et al.*, 2000). Using a sapphire substrate inside a quartz tube, with two parallel plate electrodes placed 2 cm apart in the longitudinal direction of the tube with gas mixtures of CH_4 and H_2, Wu *et al.* (2002) were able to create well-aligned carbon sheets (known as nanowalls) with a thickness in the range of several nanometers.

Zhu *et al.* (2007) were also able to create a carbon nanowall by inductively coupled radio frequency plasma enhanced chemical vapor deposition (RF PECVD) onto Si, SiO_2, Al_2O_3, Mo, Zr, Ti, Hf, Nb, W, Ta, Cu, and 304 stainless steel. Using a methane source for carbon with a volume concentration range of 5–100% in an H_2 atmosphere, RF (13.56 MHz) energy can be inductively coupled into the deposition chamber with a planar-coiled RF antenna (~20 cm in diameter) through a quartz window (Lieberman and Lichtenberg, 2005). The plasma density of this inductive plasma will get ~10 times greater than that in a capacitive mode at the same RF power input. During deposition, the RF power, total gas flow rate, and gas pressure are kept at 900 W, 10 sccm (standard cubic centimeters per minute) and ~12 Pa. What is so different from MPECVD is the use of a RF to generate energy which, in turn, will grow nanowalls without any catalyst or special substrate pretreatment (Wang *et al.*, 2004); the nanowalls are free from impurities (either catalyst and/or other allotropes of carbon), are atomically thin with the ability to alter their thickness (1 nm or less), and are free-standing (Zhu *et al.*, 2007).

Carbon nanoflasks

Experimentation with the catalytic formation of carbon nanotubes gave rise to yet another unique morphology resembling the shape of a flask (Liu *et al.*, 2000, 2001). The existence of these multi-shelled, flask-shaped carbon nanostructures, whose overall geometry differs from those of carbon nano-tubes or carbon nanoparticles, but is a hybrid of both, is known as a carbon nanoflask. From a geometric perspective, a nanoflask has an ovoid or a spherical bulbous base with a 'plus end' from which the nanotube grows, lengthening one-dimensionally; it was first discovered by Liu and colleagues (2000). What is more enticing about these different carbon-nanostructured materials is that, during the synthesis, the metal catalyst can be found encapsulated within the globular end of the nanoflask, which could give rise to future magnetic tissue engineering applications. Also, the unique structure of nanoflasks may lead to a range of different properties by enclosing various other metals or alloys inside the cavity, such as iron (Kumar Rana *et al.*, 2003).

Synthesis of carbon nanoflasks are prepared by a catalytic method, where the catalyst and the carbon sources are generated *in situ* by decomposing cobalt tricarbonyl nitrosyl ($Co(CO)_3NO$) at 900°C in a closed container (see Fig. 5.11) usually with magnesium (Mg) (Equations [5.3] and [5.4]) (Rana and Gedanken, 2002). In short, these molecules, on decomposition, act not only as a source of carbon, but also give rise to very small cobalt particles whose size can grow from several to hundreds of nanometers in the course of the $Co(CO)_3NO$ decomposition. These cobalt particles serve as nucleation centers for the carbon, which is formed by the reduction of the Co residue by the Mg atoms under the reaction conditions. When this process occurs, the growth proceeds by the addition of carbon as atoms, chains and rings (Harris and Tsang, 1998). The formation of the filled nanoflasks happens because the cobalt clusters coagulate very fast at the beginning. However, midway through the process, the coagulation slows down; the

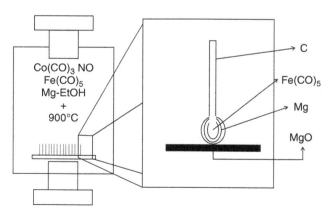

5.11 Schematic of the catalytic method to make nanoflasks. (*Source*: Adapted from Kumar Rana *et al.* (2003).)

carbon wraps most of the base and begins to creep up and climb, making a tube-like structure (Liu *et al.*, 2001).

$$Co(CO)_3NO(g) \rightarrow Co(s) + 3CO(g) + NO(g) \qquad [5.3]$$

$$CO(g) + Mg(l) \rightarrow MgO(s) + C(s) \qquad [5.4]$$

5.6.2   Tissue engineering examples

In tissue engineering some applications need specific locations where cell adherence is to be minimized. Using layers of carbon nanowalls, obtained by plasma assisted chemical vapor deposition following the injection of acetylene in radiofrequency argon plasma beam in the presence of hydrogen and ammonia, Stancu *et al.* (2010) were able to hinder the growth and proliferation of L929 fibroblasts in a specific location where nanowalls were formed. Carbon nanowalls can thus be used to mediate locational cell growth and migration (Stancu *et al.*, 2010).

Since most of these shapers are so new, most tissue engineering applications are still in the early stages. For example, since nanowall synthesis is catalyst-free and unconstrained by substrate surfaces, purification process(es) can be avoided; thus they may also have potential bioapplications as cell growth scaffolds, similar to those being demonstrated with carbon nanotubes (Chuang *et al.*, 2007). Carbon nanoflasks have the potential to hold cell-proliferating pharmaceuticals (Kumar Rana *et al.*, 2003) such that, when a scaffold is created, the medications can be excreted to promote tissue growth.

5.7   Conclusion

Following the discovery of fullerenes (Kroto *et al.*, 1985) and carbon nanotubes (Iijima, 1991), carbon based nanotechnology has developed rapidly as a platform technology for a variety of uses including biomedical applications which resulted in a new field of research known as nanomedicine. Carbon based nanomaterials for tissue engineering have the ability to be used on every aspect of promoting tissue and cellular growth – from cellular imaging, chemical and biological sensing, to bioactive agent delivery and matrix engineering. While new uses of carbon based nanomaterials for biomedical applications are being developed, concerns about cytotoxicity may be mitigated by chemical functionalization as well as better synthesis methods.

Carbon based nanomaterials may serve as an important component to imparting novel properties to the biomatrix for directing cell growth, resulting in a greater ability to mediate tissue engineering responses. Ultimately, the use of carbon based nanomaterials in combination with other biomaterials may be necessary to achieve the overall goals of tissue engineering, which will need further research in carbon synthesis methods. In conclusion, the use of carbon based nanomaterials for tissue engineering represents a challenging, but potentially rewarding, opportunity to develop the next generation of synthesis methods for future engineered biomaterials.

5.8   References

Abdullah, C. A., Asanithi, P., Brunner, E. W., Jurewicz, I., Bo, C., Azad, C. L., Ovalle-Robles, R., Fang, S., Lima, M. D., Lepro, X., Collins, S., Baughman, R. H., Sear, R. P. and Ab, D. (2011). Aligned, isotropic and patterned carbon nanotube substrates that control the growth and alignment of Chinese hamster ovary cells. *Nanotechnology*, **22**, 205102.

Aharonovich, I., Zhou, C., Stacey, A., Treussart, F., Roch, J.-F. and Prawer, S. (2008). Formation of color centers in nanodiamonds by plasma assisted diffusion of impurities from the growth substrate. *Applied Physics Letters*, **93**, 243112–243112-3.

Ajayan, P. M., Ichihashi, T. and Iijima, S. (1993). Distribution of pentagons and shapes in carbon nano-tubes and nano-particles. *Chemical Physics Letters*, **202**, 384–388.

Aisenberg, S. and Chabot, R. (1971). Ion-beam deposition of thin films of diamond-like carbon. *Journal of Applied Physics*, **42**, 2953–2958.

Akiladevi, D. and Basak, S. (2011). *Carbon Nanotubes (CNTs) Production, Characterisation and Its Applications*. International Journal of Advances in Pharmaceutical Sciences, Advanced Research Journals, Black Well HB.

Albanna, M. Z., Bou-Akl, T. H., Walters, H. L., 3rd and Matthew, H. W. (2012). Improving the mechanical properties of chitosan-based heart valve scaffolds using chitosan fibers. *Journal of the Mechanical Behavior of Biomedical Materials*, **5**, 171–180.

Alvarez, L., Guillard, T., Olalde, G., Rivoire, B., Robert, J. F., Bernier, P., Flamant, G. and Laplaze, D. (1999). Large scale solar production of fullerenes and carbon nanotubes. *Synthetic Metals*, **103**, 2476–2477.

Alvarez, L., Guillard, T., Sauvajol, J. L., Flamant, G. and Laplaze, D. (2000). Solar production of single-wall carbon nanotubes: Growth mechanisms studied by electron microscopy and Raman spectroscopy. *Applied Physics A: Materials Science and Processing*, **70**, 169–173.

Amini, S., Garay, J., Liu, G., Balandin, A. A. and Abbaschian, R. (2010). Growth of large-area graphene films from metal-carbon melts. *Journal of Applied Physics*, **108**, 094321-7.

Amsharov, K. Y. and Jansen, M. (2008). A C78 fullerene precursor: Toward the direct synthesis of higher fullerenes. *The Journal of Organic Chemistry*, **73**, 2931–2934.

Ando, Y., Zhao, X. and Ohkohchi, M. (1997). Production of petal-like graphite sheets by hydrogen arc discharge. *Carbon*, **35**, 153–158.

Balani, K., Anderson, R., Laha, T., Andara, M., Tercero, J., Crumpler, E. and Agarwal, A. (2007). Plasma-sprayed carbon nanotube reinforced hydroxyapatite coatings and their interaction with human osteoblasts in vitro. *Biomaterials*, **28**, 618–624.

Baraton, M. I. and Uvarova, I. (2001). *Functional Gradient Materials and Surface Layers Prepared by Fine Particles Technology*, Kluwer Academic Publishers.

Baughman, R. H., Zakhidov, A. A. and De Heer, W. A. (2002). Carbon nanotubes – the route toward applications. *Science*, **297**, 787–792.

Bethune, D. S., Klang, C. H., De Vries, M. S., Gorman, G., Savoy, R., Vazquez, J. and Beyers, R. (1993). Cobalt-catalysed growth of carbon nanotubes with single-atomic-layer walls. *Nature*, **363**, 605–607.

Bower, C., Zhu, W., Jin, S. H. and Zhou, O. (2000). Plasma-induced alignment of carbon nanotubes. *Applied Physics Letters*, **77**, 830–832.

Buseck, P. R., Tsipursky, S. J. and Hettich, R. (1992). Fullerenes from the geological environment. *Science*, **257**, 215–217.

Butler, J. E. and Sumant, A. V. (2008). The CVD of nanodiamond materials. *Chemical Vapor Deposition*, **14**, 145–160.

Cambaz, Z. G., Yushin, G. N., Gogotsi, Y., Vyshnyakova, K. L. and Pereselentseva, L. N. (2006). Formation of carbide-derived carbon on β-silicon carbide whiskers. *Journal of the American Ceramic Society*, **89**, 509–514.

Cami, J., Bernard-Salas, J., Peeters, E. and Malek, S. E. (2010). Detection of C60 and C70 in a Young planetary nebula. *Science*, **329**, 1180–1182.

Cellot, G., Cilia, E., Cipollone, S., Rancic, V., Sucapane, A., Giordani, S., Gambazzi, L., Markram, H., Grandolfo, M., Scaini, D., Gelain, F., Casalis, L., Prato, M., Giugliano, M. and Ballerini, L. (2009). Carbon nanotubes might improve neuronal performance by favouring electrical shortcuts. *Nature Nanotechnology*, **4**, 126–133.

Chen, C. S., Soni, S., Le, C., Biasca, M., Farr, E., Chen, E. Y. and Chin, W. C. (2012). Human stem cell neuronal differentiation on silk-carbon nanotube composite. *Nanoscale Research Letters*, **7**(1), 126.

Chen, H., Müller, M. B., Gilmore, K. J., Wallace, G. G. and Li, D. (2008). Mechanically strong, electrically conductive, and biocompatible graphene paper. *Advanced Materials*, **20**, 3557–3561.

Chen, R. J., Bangsaruntip, S., Drouvalakis, K. A., Kam, N. W., Shim, M., Li, Y., Kim, W., Utz, P. J. and Dai, H. (2003). Noncovalent functionalization of carbon nanotubes for highly specific electronic biosensors. *Proceedings of the National Academy of Sciences of the U S A*, **100**, 4984–4989.

Chen, Y., Shaw, D. T., Bai, X. D., Wang, E. G., Lund, C., Lu, W. M. and Chung, D. D. L. (2001). Hydrogen storage in aligned carbon nanotubes. *Applied Physics Letters*, **78**, 2128–2130.

Choi, G. S., Cho, Y. S., Son, K. H. and Kim, D. J. (2003). Mass production of carbon nanotubes using spin-coating of nanoparticles. *Microelectronic Engineering*, **66**, 77–82.

Christenson, E. M., Anseth, K. S., van den Beucken, J. J. J. P., Chan, C. K., Ercan, B., Jansen, J. A., Laurencin, C. T., Li, W. -J., Murugan, R., Nair, L. S., Ramakrishna, S., Tuan, R. S., Webster, T. J. and Mikos, A. G. (2007). Nanobiomaterial applications in orthopedics. *Journal of Orthopaedic Research*, **25**, 11–22.

Christiaens, P., Vermeeren, V., Wenmackers, S., Daenen, M., Haenen, K., Nesládek, M., Vandeven, M., Ameloot, M., Michiels, L. and Wagner, P. (2006). EDC-mediated DNA attachment to nanocrystalline CVD diamond films. *Biosensors and Bioelectronics*, **22**, 170–177.

Chuang, A. T. H., Robertson, J., Boskovic, B. O. and Koziol, K. K. K. (2007). Three-dimensional carbon nanowall structures. *Applied Physics Letters*, **90**, 123107–123113.

Colomer, J. F., Stephan, C., Lefrant, S., Tendeloo, G. V., Willems, I., Konya, Z., Fonseca, A., Laurent, C., Nagy, J. B. and Bianco, A. (2000). Large-scale synthesis of single-wall carbon nanotubes by catalytic chemical vapor deposition CCVD method. *Chemical Physics Letters*, **371**, 83–89.

Connell, M. J., Boul, P., Ericson, L. M., Huffman, C., Wang, Y., Haroz, E., Kuper, C., Tour, J., Ausman, K. D. and Smalley, R. E. (2001). Reversible water-solubilization of single-walled carbon nanotubes by polymer wrapping. *Chemical Physics Letters*, **342**, 265–271.

Correa-Duarte, M. A., Wagner, N., Rojas-Chapana, R., Morsczeck, C., Thie, M. and Giersig, M. (2004). Fabrication and biocompatibility of carbon nanotube-based 3D networks as scaffolds for cell seeding and growth. *Nano Letters*, 4, 2233–2236.

Couteau, E., Hernadi, K., Seo, J. W., Thien-Nga, L., Mik, C. S., Gaal, R. and Forr, L. (2003). CVD synthesis of high-purity multiwalled carbon nanotubes using $CaCO_3$ catalyst support for large-scale production. *Chemical Physics Letters*, **378**, 9–17.

Dalby, M. J., McCloy, D., Robertson, M., Agheli, H., Sutherland, D., Affrossman, S. and Oreffo, R. O. C. (2006). Osteoprogenitor response to semi-ordered and random nanotopographies. *Biomaterials*, **27**, 2980–2987.

De Heer, W. A., Châtelain, A. and Ugarte, D. (1995). A carbon nanotube field-emission electron source. *Science*, **270**, 1179–1180.

Denuault, G., Sosna, M. and Williams, K.-J. (2007). Classical experiments. In: Cynthia, G. Z. (ed.) *Handbook of Electrochemistry*. Amsterdam: Elsevier.

Dieckmann, G. R., Dalton, A. B., Johnson, P. A., Razal, J., Chen, J., Giordano, G. M., Muñoz, E., Musselman, I. H., Baughman, R. H. and Rk, D. (2003). Controlled assembly of carbon nanotubes by designed amphiphilic Peptide helices. *Journal of the American Chemical Society*, **125**, 1770–1777.

Dillon, A. C., Parilla, P. A., Alleman, J. L., Perkins, J. D. and Heben, M. J. (2000). Controlling single-wall nanotube diameters with variation in laser pulse power. *Chemical Physics Letters*, **316**, 13–18.

Dimitrijevic, N. M. and Kamat, P. V. (1992). Triplet excited state behavior of fullerenes: Pulse radiolysis and laser flash photolysis of fullerenes (C_{60} and C_{70}) in benzene. *The Journal of Physical Chemistry*, **96**, 4811–4814.

Dvir, T., Timko, B. P., Kohane, D. S. and Langer, R. (2011). Nanotechnological strategies for engineering complex tissues. *Nature Nanotechnology*, **6**, 13–22.

Ebbesen, T. W. and Ajayan, P. M. (1992). Large-scale synthesis of carbon nanotubes. *Nature*, **358**, 220–222.

Eklund, P. C., Pradhan, B. K., Kim, U. J., Xiong, Q., Fischer, J. E., Friedman, A. D., Holloway, B. C., Jordan, K. and Smith, M. W. (2002). Large-scale production of single-walled carbon nanotubes using ultrafast pulses from a free electron laser. *Nano Letters*, **2**, 561–566.

Engler, A. J., Sen, S., Sweeney, H. L. and Discher, D. E. (2006). Matrix elasticity directs stem cell lineage specification. *Cell*, **126**, 677–689.

Ersoy, D. A., McNallan, M. J. and Gogotsi, Y. (2001). Carbon coatings produced by high temperature chlorination of silicon carbide ceramics. *Materials Research Innovations*, **5**, 55–62.

Fan, H., Wang, L., Zhao, K., Li, N., Shi, Z., Ge, Z. and Jin, Z. (2010). Fabrication, mechanical properties, and biocompatibility of graphene-reinforced chitosan composites. *Biomacromolecules*, **11**, 2345–2351.

Fan, S., Chapline, M. G., Franklin, N. R., Tombler, T. W., Cassell, A. M. and Dai, H. (1999). Self-oriented regular arrays of carbon nanotubes and their field emission properties. *Science*, **283**, 512–514.

Fong, J., Wood, F. and Fowler, B. (2005). A silver coated dressing reduces the incidence of early burn wound cellulitis and associated costs of inpatient treatment: Comparative patient care audits. *Burns*, **31**, 562–567.

Fulcheri, L., Probst, N., Flamant, G., Fabry, F., Grivei, E. and Bourrat, X. (2002). Plasma processing: a step towards the production of new grades of carbon black. *Carbon*, **40**, 169–176.

Fulcheri, L., Schwob, Y., Fabry, F., Flamant, G., Chibante, L. F. P. and Laplaze, D. (2000). Fullerene production in a 3-phase AC plasma process. *Carbon*, **38**, 797–803.

Gao, L., Jiang, L. and Sun, J. (2006). Carbon nanotube-ceramic composites. *Journal of Electroceramics*, **17**, 51–55.

Georgakilas, V., Kordatos, K., Prato, M. and Guldi, D. M. (2002). Organic functionalization of carbon nanotubes. *Journal of the American Chemical Society*, **124**, 760–761.

Gogotsi, Y., Nikitin, A., Ye, H., Zhou, W., Fischer, J. E., Yi, B., Foley, H. C. and Barsoum, M. W. (2003). Nanoporous carbide-derived carbon with tunable pore size. *Nature Materials*, **2**, 591–594.

Gruen, D. M., Liu, S., Krauss, A. R., Luo, J. and Pan, X. (1994). Fullerenes as precursors for diamond film growth without hydrogen or oxygen additions. *Applied Physics Letters*, **64**, 1502–1504.

Gruenberger, T. M., Gonzalez-Aguilar, J., Fabry, F., Fulcheri, L., Grivei, E., Probst, N., Flamant, G., Okuno, H. and Charlier, J. C. (2004). Production of carbon nanotubes and other nanostructures via continuous 3-phase AC plasma processing. *Fullerenes, Nanotubes and Carbon Nanostructures*, **12**, 571–581.

Gu, Z., Peng, H., Hauge, R. H., Smalley, R. E. and Margrave, J. L. (2002). Cutting single-wall carbon nanotubes through fluorination. *Nano Letters*, **2**, 1009–1013.

Guillard, T., Flamant, G. and Laplaze, D. (2001). Heat, mass, and fluid flow in a solar reactor for fullerene synthesis. *Journal of Solar Energy Engineering*, **123**, 153–159.

Guo, T., Nikolaev, P., Thess, A., Colbert, D. T. and Smalley, R. E. (1995). Catalytic growth of single-walled nanotubes by laser vaporization. *Chemical Physics Letters*, **243**, 49–54.

Guss, W., Feldmann, J., Göbel, E. O., Taliani, C., Mohn, H., Müller, W., Häussler, P. and Ter Meer, H. U. (1994). Fluorescence from X traps in C60 single crystals. *Physical Review Letters*, **72**, 2644–2647.

Hamon, M. A., Chen, J., Hu, H., Chen, Y., Rao, A. M., Eklund, P. C. and Haddon, R. C. (1999). Dissolution of single-walled carbon nanotubes. *Advanced Materials*, **10**, 834–840.

Harris, P. J. F. and Tsang, S. C. (1998). A simple technique for the synthesis of filled carbon nanoparticles. *Chemical Physics Letters*, **293**, 53–58.

Harrison, B. S. and Atala, A. (2007). Carbon nanotube applications for tissue engineering. *Biomaterials*, **28**, 344–353.

Hartl, A., Schmich, E., Garrido, J. A., Hernando, J., Catharino, S. C. R., Walter, S., Feulner, P., Kromka, A., Steinmuller, D. and Stutzmann, M. (2004). Protein-modified nanocrystalline diamond thin films for biosensor applications. *Nature Materials*, **3**, 736–742.

Hernadi, K., Fonseca, A., Nagy, J. B., Bernaerts, D. and Lucas, A. (1996). Fe-catalyzed carbon nanotube formation. *Carbon*, **34**, 1249–1257.

Hernandez, Y., Nicolosi, V., Lotya, M., Blighe, F. M., Sun, Z., De, S., McGovern, I. T., Holland, B., Byrne, M., Gun'ko, Y. K., Boland, J. J., Niraj, P., Duesberg, G., Krishnamurthy, S., Goodhue, R., Hutchison, J., Scardaci, V., Ferrari, A. C. and Coleman, J. N. (2008). High-yield production of graphene by liquid-phase exfoliation of graphite. *Nature Nanotechnology*, **3**, 563–568.

Holzinger, M., Abraham, J., Whelan, P., Graupner, R., Ley, L., Hennrich, F., Kappes, M. and Hirsch, A. (2003). Functionalization of single-walled carbon nanotubes with (R-) oxycarbonyl nitrenes. *Journal of the American Chemical Society*, **125**, 8566–8580.

Hu, Z., Guan, W., Wang, W., Zhu, Z. and Wang, Y. (2010). Folacin C60 derivative exerts a protective activity against oxidative stress-induced apoptosis in rat pheochromocytoma cells. *Bioorganic and Medicinal Chemistry Letters*, **20**, 4159–4162.

Huang, H., Kajiura, H., Tsutsui, S., Hirano, Y., Miyakoshi, M., Yamada, A. and Ata, M. (2001). Large-scale rooted growth of aligned super bundles of single-walled carbon nanotubes using a directed arc plasma method. *Chemical Physics Letters*, **343**, 7–14.

Hull, M. S., Kennedy, A. J., Steevens, J. A., Bednar, A. J., Weiss, J. C. A. and Vikesland, P. J. (2009). Release of metal impurities from carbon nanomaterials influences aquatic toxicity. *Environmental Science and Technology*, **43**, 4169–4174.

Iijima, S. (1980). Direct observation of the tetrahedral bonding in graphitized carbon black by high resolution electron microscopy. *Journal of Crystal Growth*, **50**, 675–683.

Iijima, S. (1991). Helical microtubules of graphitic carbon. *Nature*, **354**, 56–58.

Iijima, S. and Ichihashi, T. (1993). Single-shell carbon nanotubes of 1-nm diameter. *Nature*, **363**, 603–605.

Jan, E. and Kotov, N. A. (2007). Successful differentiation of mouse neural stem cells on layer-by-layer assembled single-walled carbon nanotube composite. *Nano Letters*, **7**, 1123–1128.

Journet, C., Maser, W. K., Bernier, P., Loiseau, A., De La Chapelle, M. L., Lefrant, S., Deniard, P., Lee, R. and Fischer, J. E. (1997). Large-scale production of single-walled carbon nanotubes by the electric-arc technique. *Nature*, **388**, 756–758.

Kamo, M., Sato, Y., Matsumoto, S. and Setaka, N. (1983). Diamond synthesis from gas phase in microwave plasma. *Journal of Crystal Growth*, **62**, 642–644.

Kharisov, B. I., Kharissova, O. V., Gutierrez, H. L. and Mendez, U. O. (2009). Recent advances on the soluble carbon nanotubes. *Industrial and Engineering Chemistry Research*, **48**, 572–590.

Kim, K. S., Zhao, Y., Jang, H., Lee, S. Y., Kim, J. M., Kim, K. S., Ahn, J.-H., Kim, P., Choi, J.-Y. and Hong, B. H. (2009). Large-scale pattern growth of graphene films for stretchable transparent electrodes. *Nature*, **457**, 706–710.

Krasheninnikov, A. V. and Banhart, F. (2007). Engineering of nanostructured carbon materials with electron or ion beams. *Nature Materials*, **6**, 723–733.

Kroto, H. W., Heath, J. R., O'Brien, S. C., Curl, R. F. and Smalley, R. E. (1985). C60: Buckminsterfullerene. *Nature*, **318**, 162–163.

Kumar, M. and Ando, Y. (2010). Chemical vapor deposition of carbon nanotubes: A review on growth mechanism and mass production. *Journal of Nanoscience and Nanotechnology*, **10**, 3739–3758.

Kumar, M. N. V. R. (2000). A review of chitin and chitosan applications. *Reactive and Functional Polymers*, **46**, 1–27.

Kumar Rana, R., Brukental, I., Yeshurun, Y. and Gedanken, A. (2003). Synthesis and characterization of Fe3Co7 alloy encapsulated in carbon nanoflasks. *Journal of Materials Chemistry*, **13**, 467–472.

Kusunoki, M., Rokkaku, M. and Suzuki, T. (1997). Epitaxial carbon nanotube film self-organized by sublimation decomposition of silicon carbide. *Applied Physics Letters*, **71**, 2620–2622.

Kusunoki, M., Shibata, J., Rokkaku, M. and Hirayama, T. (1998). Aligned carbon nanotube film self-organized on a SiC wafer. *Japanese Journal of Applied Physics Part 2-Letters*, **37**, L605–L606.

Kusunoki, M., Suzuki, T., Honjo, C., Hirayama, T. and Shibata, N. (2002). Selective synthesis of zigzag-type aligned carbon nanotubes on SiC (0 0 0 1) wafers. *Chemical Physics Letters*, **366**, 458–462.

Kusunoki, M., Suzuki, T., Kaneko, K. and Ito, M. (1999). Formation of self-aligned carbon nanotube films by surface decomposition of silicon carbide. *Philosophical Magazine Letters*, **79**, 153–161.

Lambert, J. M., Ajayan, P. M., Bernier, P., Planeix, J. M., Brotons, V., Coq, B. and Castaing, J. (1994). Improving conditions towards isolating single-shell carbon nanotubes. *Chemical Physics Letters*, **226**, 364–371.

Laplaze, D., Bernier, P., Maser, W. K., Flamant, G., Guillard, T. and Loiseau, A. (1998). Carbon nanotubes: The solar approach. *Carbon*, **36**, 685–688.

Lappalainen, R., Selenius, M., Anttila, A., Konttinen, Y. T. and Santavirta, S. S. (2003). Reduction of wear in total hip replacement prostheses by amorphous diamond coatings. *Journal of Biomedical Materials Research Part B: Applied Biomaterials*, **66B**, 410–413.

Li, D., Muller, M. B., Gilje, S., Kaner, R. B. and Wallace, G. G. (2008). Processable aqueous dispersions of graphene nanosheets. *Nature Nanotechnology*, **3**, 101–105.

Li, N., Ma, Y., Wang, B., Huang, Y., Wu, Y., Yang, X. and Chen, Y. (2011). Synthesis of semiconducting SWCNTs by arc discharge and their enhancement of water splitting performance with TiO_2 photocatalyst. *Carbon*, **49**, 5132–5141.

Li, W. Z., Xie, S. S., Qian, L. X., Chang, B. H., Zou, B. S., Zhou, W. Y., Zhao, R. A. and Wang, G. (1996). Large-scale synthesis of aligned carbon nanotubes. *Science*, **274**, 1701–1703.

Lieber, C. M. and Wang, Z. L. (2007). Functional nanowires. *MRS Bulletin*, **32**, 99–108.

Lieberman, M. A. and Lichtenberg, A. J. (2005). *Principles of Plasma Discharges and Materials Processing*, Wiley-Interscience, Hoboken, NJ.

Liu, S., Boeshore, S., Fernandez, A., Sayagués, M. J., Fischer, J. E. and Gedanken, A. (2001). Study of cobalt-filled carbon nanoflasks. *The Journal of Physical Chemistry B*, **105**, 7606–7611.

Liu, S., Tang, X., Yin, L., Koltypin, Y. and Gedanken, A. (2000). Synthesis of carbon nanoflasks. *Journal of Materials Chemistry*, **10**, 1271–1272.

MacDonald, R. A., Laurenzi, B. F., Viswanathan, G., Ajayan, P. M. and Stegemann, J. P. (2005). Collagen-carbon nanotube composite materials as scaffolds in tissue engineering. *Journal of Biomedical Materials Research Part A*, **74**, 489–496.

Maser, W. K., Muñoz, E., Benito, A. M., Martínez, M. T., De La Fuente, G. F., Maniette, Y., Anglaret, E. and Sauvajol, J. L. (1998). Production of high-density single-walled nanotube material by a simple laser-ablation method. *Chemical Physics Letters*, **292**, 587–593.

Mcclure, J. W. (1956). Diamagnetism of graphite. *Physical Review*, **104**, 666–671.

Meier, A., Kirillov, V., A., Kuvshinov, G., G., Mogilnykh, Y., I., Weidenkaff, A. and Steinfeld, A. (1999). Production of catalytic filamentous carbon by solar thermal decomposition of hydrocarbons. *Journal de Physique IV France*, **09**, Pr3–393–Pr3–398.

Meng, J., Han, Z., Kong, H., Qi, X., Wang, C., Xie, S. and Xu, H. (2010). Electrospun aligned nanofibrous composite of MWCNT/polyurethane to enhance vascular endothelium cells proliferation and function. *Journal of Biomedical Materials Research A*, **95**, 312–320.

Menon, M. and Srivastava, D. (1997). Carbon nanotube 'T Junctions': Nanoscale metal-semiconductor-metal contact devices. *Physical Review Letters*, **79**, 4453–4456.

Minato, J.-I. and Miyazawa, K. I. (2006). C_{60} fullerene tubes as removable templates. *Journal of Materials Research*, **21**, 529–534.

Mironov, V., Kasyanov, V. and Markwald, R. R. (2008). Nanotechnology in vascular tissue engineering: From nanoscaffolding towards rapid vessel biofabrication. *Trends in Biotechnology*, **26**, 338–344.

Miyazawa, K., Kuwasaki, Y., Obayashi, A. and Kuwabara, M. (2002). C_{60} nanowhiskers formed by the liquid–liquid interfacial precipitation method. *Journal of Materials Research*, **17**, 83–88.

Morjan, R. E., Nerushev, O. A., Sveningsson, M., Rohmund, F., Falk, L. K. L. and Campbell, E. E. B. (2004). Growth of carbon nanotubes from C_{60}. *Applied Physics A: Materials Science and Processing*, **78**, 253–261.

Mukhopadhyay, K., Koshio, A., Tanaka, N. and Shinohara, H. (1998). A simple and novel way to synthesize aligned nanotube bundles at low temperature. *Japanese Journal of Applied Physics*, **37**, L1257–L1259.

Muñoz, E., Maser, W. K., Benito, A. M., Martínez, M. T., De La Fuente, G. F., Maniette, Y., Righi, A., Anglaret, E. and Sauvajol, J. L. (2000). Gas and pressure effects on the production of single-walled carbon nanotubes by laser ablation. *Carbon*, **38**, 1445–1451.

Nayak, T. R., Andersen, H., Makam, V. S., Khaw, C., Bae, S., Xu, X., Ee, P.-L. R., Ahn, J.-H., Hong, B. H., Pastorin, G. and Özyilmaz, B. (2011). Graphene for controlled and accelerated osteogenic differentiation of human mesenchymal stem cells. *ACS Nano*, **5**, 4670–4678.

Nerushev, O. A., Dittmar, S., Morjan, R. E., Rohmund, F. and Campbell, E. E. B. (2003). Particle size dependence and model for iron-catalyzed growth of carbon nanotubes by thermal chemical vapor deposition. *Journal of Applied Physics*, **93**, 4185–4190.

Nikolaev, P., Bronikowski, M. J., Bradley, R. K., Rohmund, F., Colbert, D. T., Smith, K. A. and Smalley, R. E. (1999). Gas-phase catalytic growth of single-walled carbon nanotubes from carbon monoxide. *Chemical Physics Letters*, **313**, 91–97.

Novoselov, K. S. (2004). Electric field effect in atomically thin carbon films. *Science*, **306**, 666–669.

Novoselov, K. S. (2005). Two-dimensional atomic crystals. *Proceedings of the National Academy of Sciences USA*, **102**, 10451–10453.

Nuvoli, D., Valentini, L., Alzari, V., Scognamillo, S., Bon, S. B., Piccinini, M., Illescas, J. and Mariani, A. (2011). High concentration few-layer graphene sheets obtained by liquid phase exfoliation of graphite in ionic liquid. *Journal of Materials Chemistry*, **21**, 3428–3431.

Okino, F., Kawasaki, S., Fukushima, Y., Kimura, M., Nakajima, T. and Touhara, H. (1996). Single crystal of C60F X and the x-ray structural analysis. *Fullerene Science and Technology*, **4**, 873–885.

Oncel, C. and Yurum, Y. (2006). Carbon nanotube synthesis via the catalytic CVD method: A review on the effect of reaction parameters. *Fullerenes, Nanotubes and Carbon Nanostructures*, **14**, 17–37.

Oohara, W. and Hatakeyama, R. (2003). Pair-ion plasma generation using fullerenes. *Physical Review Letters*, **91**, 205005.

Papo, M. J., Catledge, S. A., Vohra, Y. K. and Machado, C. (2004). Mechanical wear behavior of nanocrystalline and multilayer diamond coatings on temporomandibular joint implants. *Journal of Materials Science: Materials in Medicine*, **15**, 773–777.

Pastorin, G., Kostarelos, K., Prato, M. and Bianco, A. (2005). Functionalized carbon nanotubes: Towards the delivery of therapeutic molecules. *Journal of Biomedical Nanotechnology*, **1**, 1–10.

Patel, P. N., Gobin, A. S., West, J. L. and Patrick, C. W. Jr. (2005). Poly(ethylene glycol) hydrogel system supports preadipocyte viability, adhesion, and proliferation. *Tissue Engineering*, **11**, 1498–1505.

Price, R. L., Ellison, K., Haberstroh, K. M. and Webster, T. J. (2004). Nanometer surface roughness increases select osteoblast adhesion on carbon nanofiber compacts. *Journal of Biomedical Materials Research Part A*, **70A**, 129–138.

Pulskamp, K., Diabaté, S. and Krug, H. F. (2007). Carbon nanotubes show no sign of acute toxicity but induce intracellular reactive oxygen species in dependence on contaminants. *Toxicology Letters*, **168**, 58–74.

Qian, D., Dickey, E. C., Andrews, R. and Rantell, T. (2000). Load transfer and deformation mechanisms in carbon nanotube-polystyrene composites. *Applied Physics Letters*, **76**, 2868–2870.

Qin, L.-C., Iijima, S., Kataura, H., Maniwa, Y., Suzuki, S. and Achiba, Y. (1997). Helicity and packing of single-walled carbon nanotubes studied by electron nanodiffraction. *Chemical Physics Letters*, **268**, 101–106.

Qin, L. C., Zhou, D., Krauss, A. R. and Gruen, D. M. (1998). Growing carbon nanotubes by microwave plasma-enhanced chemical vapor deposition. *Applied Physics Letters*, **72**, 3437–3439.

Qu, Y., Liang, S., Zou, K., Li, S., Liu, L., Zhao, J. and Piao, G. (2011). Effect of solvent type on the formation of tubular fullerene nanofibers. *Materials Letters*, **65**, 562–564.

Rana, R. K. and Gedanken, A. (2002). Carbon nanoflask: A mechanistic elucidation of its formation. *The Journal of Physical Chemistry B*, **106**, 9769–9776.

Ren, Z. F., Huang, Z. P., Wang, D. Z., Wen, J. G., Xu, J. W., Wang, J. H., Calvet, L. E., Chen, J., Klemic, J. F. and Reed, M. A. (1999). Growth of a single freestanding multiwall carbon nanotube on each nanonickel dot. *Applied Physics Letters*, **75**, 1086–1088.

Robertson, J. (2002). Diamond-like amorphous carbon. *Materials Science and Engineering: R Reports*, **37**, 129–281.

Saito, Y., Tani, Y., Miyagawa, N., Mitsushima, K., Kasuya, A. and Nishina, Y. (1998). High yield of single-wall carbon nanotubes by arc discharge using Rh–Pt mixed catalysts. *Chemical Physics Letters*, **294**, 593–598.

Sashiwa, H. and Aiba, S. (2004). Chemically modified chitin and chitosan as biomaterials. *Progress in Polymer Science*, **24**, 887–908.

Scott, L. T., Boorum, M. M., McMahon, B. J., Hagen, S., Mack, J., Blank, J., Wegner, H. and De Meijere, A. (2002). A rational chemical synthesis of C60. *Science*, **295**, 1500–1503.

Sen, R., Govindaraj, A. and Rao, C. N. R. (1997). Carbon nanotubes by the metallocene route. *Chemical Physics Letters*, **267**, 276–280.

Shao, S., Zhou, S., Li, L., Li, J., Luo, C., Wang, J., Li, X. and Weng, J. (2011). Osteoblast function on electrically conductive electrospun PLA/MWCNTs nanofibers. *Biomaterials*, **32**, 2821–2833.

Sharda, T., Rahaman, M. M., Nukaya, Y., Soga, T., Jimbo, T. and Umeno, M. (2001). Structural and optical properties of diamond and nano-diamond films grown by microwave plasma chemical vapor deposition. *Diamond and Related Materials*, **10**, 561–567.

Shi, X., Sitharaman, B., Pham, Q. P., Liang, F., Wu, K., Edward Billups, W., Wilson, L. J. and Mikos, A. G. (2007). Fabrication of porous ultra-short single-walled carbon nanotube nanocomposite scaffolds for bone tissue engineering. *Biomaterials*, **28**, 4078–4090.

Shigeo, M., Ryosuke, K., Yuhei, M., Shohei, C. and Masamichi Kohnob (2002). Low-temperature synthesis of high-purity single-walled carbon nanotubes from alcohol. *Chemical Physics Letters*, **360**, 229–234.

Shin, S. R., Bae, H., Cha, J. M., Mun, J. Y., Chen, Y. C., Tekin, H., Shin, H., Farshchi, S., Dokmeci, M. R., Tang, S. and Khademhosseini, A. (2012). Carbon nanotube reinforced hybrid microgels as scaffold materials for cell encapsulation. *ACS Nano*, **6**, 362–372.

Sica, F., Adinolfi, S., Vitagliano, L., Zagari, A., Capasso, S. and Mazzarella, L. (1996). Cosolute effect on crystallization of two dinucleotide complexes of bovine seminal ribonuclease from concentrated salt solutions. *Journal of Crystal Growth*, **168**, 192–197.

Silva, G. A. (2006). Neuroscience nanotechnology: Progress, opportunities and challenges. *Nature Reviews Neuroscience*, **7**, 65–74.

Singh, C., Shaffer, M. S. P. and Windle, A. H. (2003). Production of controlled architectures of aligned carbon nanotubes by an injection chemical vapour deposition method. *Carbon*, **41**, 359–368.

Sinha, A. K., Hwang, D. W. and Hwang, L. P. (2000). A novel approach to bulk synthesis of carbon nanotubes filled with metal by a catalytic chemical vapor deposition method. *Chemical Physics Letters*, **332**, 455–460.

Sitharaman, B., Shi, X., Walboomers, X. F., Liao, H., Cuijpers, V., Wilson, L. J., Mikos, A. G. and Jansen, J. A. (2008). In vivo biocompatibility of ultra-short single-walled carbon nanotube/biodegradable polymer nanocomposites for bone tissue engineering. *Bone*, **43**, 362–370.

Slonczewski, J. C. and Weiss, P. R. (1958). Band structure of graphite. *Physical Review*, **109**, 272–279.

Snow, E. S., Campbell, P. M. and Novak, J. P. (2002). Single-wall carbon nanotube atomic force microscope probes. *Applied Physics Letters*, **80**, 2002–2004.

Specht, C. G., Williams, O. A., Jackman, R. B. and Schoepfer, R. (2004). Ordered growth of neurons on diamond. *Biomaterials*, **25**, 4073–4078.

Stancu, E. C., Ionita, M. D., Vizireanu, S., Stanciuc, A. M., Moldovan, L. and Dinescu, G. (2010). Wettability properties of carbon nanowalls layers deposited by a radiofrequency plasma beam discharge. *Materials Science and Engineering: B*, **169**, 119–122.

Stankovich, S., Dikin, D. A., Piner, R. D., Kohlhaas, K. A., Kleinhammes, A., Jia, Y., Wu, Y., Nguyen, S. T. and Ruoff, R. S. (2007). Synthesis of graphene-based nanosheets via chemical reduction of exfoliated graphite oxide. *Carbon*, **45**, 1558–1565.

Stankovich, S., Piner, R. D., Chen, X., Wu, N., Nguyen, S. T. and Ruoff, R. S. (2006). Stable aqueous dispersions of graphitic nanoplatelets via the reduction of exfoliated graphite oxide in the presence of poly(sodium 4-styrenesulfonate). *Journal of Materials Chemistry*, **16**, 155–158.

Star, A., Steuerman, D. W., Heath, J. R. and Stoddart, J. F. (2002). Starched carbon nanotubes. *Angewandte Chemie International Edition*, **41**, 2508–2512.

Star, A., Stoddart, J. F., Steuerman, D., Diehl, M., Boukai, A., Wong, E. W., Yang, X., Chung, S. W., Choi, H. and Heath, J. R. (2001). Preparation and properties of polymer-wrapped single-walled carbon nanotubes. *Angewandte Chemie International Edition*, **40**, 1721–1725.

Stevenson, S., Rice, G., Glass, T., Harich, K., Cromer, F., Jordan, M. R., Craft, J., Hadju, E., Bible, R., Olmstead, M. M., Maitra, K., Fisher, A. J., Balch, A. L. and Dorn, H. C. (1999). Small-bandgap endohedral metallofullerenes in high yield and purity. *Nature*, **401**, 55–57.

Stout, D. A., Basu, B. and Webster, T. J. (2011). Poly(lactic–co-glycolic acid): Carbon nanofiber composites for myocardial tissue engineering applications. *Acta Biomaterialia*, **7**, 3101–3112.

Su, M., Zheng, B. and Liu, J. (2000). A scalable CVD method for the synthesis of single-walled carbon nanotubes with high catalyst productivity. *Chemical Physics Letters*, **332**, 321–326.

Subrahmanyam, K. S., Panchakarla, L. S., Govindaraj, A. and Rao, C. N. R. (2009). Simple method of preparing graphene flakes by an arc-discharge method. *The Journal of Physical Chemistry C*, **113**, 4257–4259.

Subramoney, S., Ruoff, R. S., Lorents, D. C. and Malhotra, R. (1993). Radial single-layer nanotubes. *Nature*, **366**, 637.

Suhr, J., Victor, P., Ci, L., Sreekala, S., Zhang, X., Nalamasu, O. and Ajayan, P. M. (2007). Fatigue resistance of aligned carbon nanotube arrays under cyclic compression. *Nature Nanotechnology*, **2**, 417–421.

Sun, J., Hu, S.-L., Du, X.-W., Lei, Y.-W. and Jiang, L. (2006). Ultrafine diamond synthesized by long-pulse-width laser. *Applied Physics Letters*, **89**, 183115–3.

Sun, X. S., Wang, N., Zhang, W. J., Woo, H. K., Han, X. D., Bello, I., Lee, C. S. and Lee, S. T. (1999). Synthesis of nanocrystalline diamond by the direct ion beam deposition method. *Journal of Materials Research*, **14**, 3204–3207.

Takahashi, K. and Yamanaka, S. (2006). Induction of pluripotent stem cells from mouse embryonic and adult fibroblast cultures by defined factors. *Cell*, **126**, 663–676.

Talapatra, S., Ganesan, P. G., Kim, T., Vajtai, R., Huang, M., Shima, M., Ramanath, G., Srivastava, D., Deevi, S. C. and Ajayan, P. M. (2005). Irradiation-induced magnetism in carbon nanostructures. *Physical Review Letters*, **95**, 097201.

Terrones, M. (2003). Science and technology of the twenty-first century: Synthesis, properties and applications of carbon nanotubes. *Annual Review of Materials Research*, **33**, 419–501.

Thess, A., Lee, R., Nikolaev, P., Dai, H., Petit, P., Robert, J., Xu, C., Lee, Y. H., Kim, S. G., Rinzler, A. G., Colbert, D. T., Scuseria, G. E., Tománek, D., Fischer, J. E. and Smalley, R. E. (1996). Crystalline ropes of metallic carbon nanotubes. *Science*, **273**, 483–487.

Tsai, S. H., Chao, C. W., Lee, C. L. and Shih, H. C. (1999). Bias-enhanced nucleation and growth of the aligned carbon nanotubes with open ends under microwave plasma synthesis. *Applied Physics Letters*, **74**, 3462–3464.

Tsang, K. Y., Cheung, M. C., Chan, D. and Cheah, K. S. (2010). The developmental roles of the extracellular matrix: Beyond structure to regulation. *Cell Tissue Research*, **339**, 93–110.

Tsumoto, H., Kawahara, S., Fujisawa, Y., Suzuki, T., Nakagawa, H., Kohda, K. and Miyata, N. 2010. Syntheses of water-soluble 60 fullerene derivatives and their enhancing effect on neurite outgrowth in NGF-treated PC12 cells. *Bioorganic and Medicinal Chemistry Letters*, **20**, 1948–1952.

Voge, C. M., Kariolis, M., Macdonald, R. A. and Stegemann, J. P. (2008). Directional conductivity in SWCNT-collagen-fibrin composite biomaterials through strain-induced matrix alignment. *Journal of Biomedical Materials Research A*, **86**, 269–277.

Wang, J. (2005). Carbon-nanotube based electrochemical biosensors: A review. *Electroanalysis*, **17**, 7–14.

Wang, J., Zhu, M., Outlaw, R. A., Zhao, X., Manos, D. M. and Holloway, B. C. (2004). Synthesis of carbon nanosheets by inductively coupled radio-frequency plasma enhanced chemical vapor deposition. *Carbon*, **42**, 2867–2872.

Wang, S. F., Shen, L., Zhang, W. D. and Tong, Y. J. (2005). Preparation and mechanical properties of chitosan/carbon nanotubes composites. *Biomacromolecules*, **6**, 3067–3072.

Wei, B. Q., Vajtai, R., Jung, Y., Ward, J., Zhang, R., Ramanath, G. and Ajayan, P. M. (2002). Microfabrication technology: Organized assembly of carbon nanotubes. *Nature*, **416**, 495–496.

Williams, G., Seger, B. and Kamat, P. V. (2008). TiO_2-graphene nanocomposites. UV-assisted photocatalytic reduction of graphene oxide. *ACS Nano*, **2**, 1487–1491.

Wu, Y., Qiao, P., Chong, T. and Shen, Z. (2002). Carbon nanowalls grown by microwave plasma enhanced chemical vapor deposition. *Advanced Materials*, **14**, 64–67.

Yang, L., Sheldon, B. W. and Webster, T. J. (2009). The impact of diamond nanocrystallinity on osteoblast functions. *Biomaterials*, **30**, 3458–3465.

Yao, Z., Kane, C. L. and Dekker, C. (2000). High-field electrical transport in single-wall carbon nanotubes. *Physical Review Letters*, **84**, 2941–2944.

Ye, H., Sun, C. Q., Hing, P., Xie, H., Zhang, S. and Wei, J. (2000). Nucleation and growth dynamics of diamond films by microwave plasma-enhanced chemical vapor deposition (MPECVD). *Surface and Coatings Technology*, **123**, 129–133.

Yudasaka, M., Komatsu, T., Ichihashi, T. and Iijima, S. (1997). Single-wall carbon nanotube formation by laser ablation using double-targets of carbon and metal. *Chemical Physics Letters*, **278**, 102–106.

Zanello, L. P., Zhao, B., Hu, H. and Haddon, R. C. (2006). Bone cell proliferation on carbon nanotubes. *Nano Letters*, **6**, 562–567.

Zheng, M., Jagota, A., Strano, M. S., Santos, A. P., Barone, P., Chou, S. G., Diner, B. A., Dresselhaus, M. S., McLean, R. S., Onoa, G. B., Samsonidze, G. G., Semke, E. D., Usrey, M. and Walls, D. J. (2003). Structure-based carbon nanotube sorting by sequence-dependent DNA assembly. *Science*, **302**, 1545–1548.

Zhu, M., Wang, J., Outlaw, R. A., Hou, K., Manos, D. M. and Holloway, B. C. (2007). Synthesis of carbon nanosheets and carbon nanotubes by radio frequency plasma enhanced chemical vapor deposition. *Diamond and Related Materials*, **16**, 196–201.

Zhu, W., Stoner, B. R., Williams, B. E. and Glass, J. T. (1991). Growth and characterization of diamond films on nondiamond substrates for electronic applications. *Proceedings of the IEEE*, **79**, 621–646.

6

Fabrication of nanofibrous scaffolds for tissue engineering applications

H. CHEN, R. TRUCKENMÜLLER,
C. VAN BLITTERSWIJK and L. MORONI,
University of Twente, The Netherlands

DOI: 10.1533/9780857097231.1.158

Abstract: Nanofibrous scaffolds which mimic the structural features of a natural extracellular matrix (ECM) can be appealing scaffold candidates for tissue engineering as they provide similar physical cues to the native environment of the targeted tissue to regenerate. This chapter discusses different strategies to fabricate nanofibrous scaffolds for tissue engineering. We first describe three major methods for nanofibrous scaffold fabrication: molecular self-assembly, phase separation, and electrospinning. Then, approaches for surface modification of nanofibrous scaffolds including blending and coating, plasma treatment, wet chemical methods, and surface graft polymerization are presented. Finally, applications of nanofibrous scaffolds in tissue engineering are introduced.

Key words: electrospinning, self-assembly, phase separation, nanofibrous scaffolds, tissue engineering

6.1 Introduction

Tissue engineering aims to develop biological replacements through the integration of life science and engineering principles. In general, tissue engineering can be subdivided into three major approaches (Subbiah *et al.*, 2005): (1) injection or transplantation of isolated cells to a defect site or an injured tissue, (2) delivery of tissue-inducing biomolecules (e.g. growth factors, peptides, polysaccharides) to a targeted tissue, and (3) growth and differentiation of specific cell types in three-dimensional (3D) scaffolds. Among these approaches, scaffold-based tissue engineering has become the most popular, due to the potential of current scaffold fabrication technologies to incorporate chemical, physical, and biological stimuli at different scales that can direct cell activity.

The extracellular matrix (ECM) usually provides structural support to the cells in addition to performing various other important functions. Many extracellular proteins have a fibrous structure with a diameter at the

158

nanometer or sub-micrometer scales. For example, collagen, the most abundant ECM protein in our body, has a fibrous structure with a fiber diameter that ranges from 50 to 500 nm. For successful and functional engineering of tissues and organs in scaffold-based approaches, an artificial scaffold should mimic the structure and biological function of native ECM as much as possible both in terms of physical cues and chemical composition. Therefore, researchers have put much effort into developing nanofibrous scaffolds that mimic the fibrillar structure of native ECM. Currently, there are three basic techniques available for generating nanofibrous scaffolds: molecular self-assembly, phase separation, and electrospinning. These techniques allow nanofibrous scaffolds that mimic the ECM physical structure to be developed with a number of degradable and non-degradable synthetic polymers, but additional modifications are desirable to present biological cues, which exhibit a positive effect on cell adhesion, proliferation, and differentiation. Furthermore, recent studies have shown that submicron and nanoscale topographies can also modulate cell activity, thus acting as 'synthetic' biological cues (Watari *et al.*, 2011). Surface modifications of nanofibrous scaffolds include blending and coating, plasma treatment, wet chemical methods, and surface graft polymerization.

In this chapter, we describe nanofibrous scaffolds for tissue engineering. Processing technologies for achieving structural features resembling the extracellular matrix and approaches to improve cell-scaffold interactions will be introduced before discussing nanofibrous scaffold applications for tissue engineering.

6.2   Methods for nanofibrous scaffolds fabrication

Currently, there are three approaches available for the fabrication of nanofibers: (i) molecular self-assembly, (ii) phase separation and (iii) electrospinning. Although very different from each other, each method possesses its advantages and disadvantages (Table 6.1).

6.2.1   Molecular self-assembly

Molecular self-assembly has emerged as a useful approach for manufacturing scaffolds for tissue engineering (Goldberg *et al.*, 2007). Unlike electrospinning, self-assembly can be defined as a spontaneous process to form ordered and stable structures through a number of non-covalent interactions, such as hydrophilic, electrostatic, and van deer Waals interactions (Philp and Stoddart, 1996; Hartgerink *et al.*, 2001). The structural features of self-assembled materials can be tuned by controlling the kinetics, molecular chemistry, and assembly environment (e.g. pH, solvent, light, salt addition,

Table 6.1 Comparison of nanofiber processing techniques

Method	Level of productivity	Ease of processing	Repeatability	Advantages	Disadvantages
Self-assembly	Laboratory scale	Difficult	Yes	• Easy to get fiber diameter on lowest scale • Great control over 3D shape	• Low yield • Complex process • Little control over fiber dimension and orientation • Only short fiber can be obtained • Limited to a few polymers
Phase separation	Laboratory scale	Easy	Yes	• Tailorable mechanical properties • Great control over pore size and shape • Great control over 3D shape	• Low yield • Little control over fiber dimension and orientation • Difficult to control pore size and shape • Limited to a few polymers
Electrospinning	Laboratory/ industrial scales	Easy	Yes	• Well established and characterized technique • Long continuous fiber with diameter from micro-scale down to nano-scale • Control over fiber diameter and orientation • Tailorable mechanical properties • Plethora of polymers may be used	• Difficult to fabricate 3D shape • Difficult to control pore size and shape

and temperature). The key challenge in self-assembly is to design molecular building blocks that can undergo spontaneous organization into a well-defined pattern that mimic the structural features of biological systems (Smith *et al.*, 2008). Small building blocks, including small molecules, nucleic acids, and peptides, can self-assemble into nanofibrous structures. Among these building blocks, peptide-amphiphile (PA) units, which combine the functions of peptides with the characteristics of surfactants, have gained a lot of attention due to the versatility in their design for biological applications. The chemical structure of a representative PA molecule consists of four key structural features: a hydrophobic domain consisting of a long alkyl tail, a short peptide sequence capable of intermolecular hydrogen bonding, charged groups for enhanced solubility in water, and a bioactive epitope. Self-assembly of PA in water is governed by at least three major forces: hydrophobic interaction of alkyl tails, electrostatic repulsions between charged groups, and hydrogen bonding among peptide segments. The final structure of assembled PA reflects a balance of each force contribution (Cui *et al.*, 2010a).

The structural characteristics of nanofibers and nanofibrillar systems resulting from molecular self-assembly can be tailored by controlling processing parameters. Bundles of aligned PA nanofibers were obtained, for example, by self-assembling PA within parallel channels (Hung and Stupp, 2007). By introducing hydrophilic amino acids in peptide segments, specific sequences of PA could self-assemble into nanobelts with fairly monodisperse widths in the order of 150 nm and lengths of up to 0.1 mm; these can be used as cell culture systems and drug delivery carriers (Cui *et al.*, 2009). Nanofibrous scaffolds were also fabricated from self-assembly of β-sheet peptides containing phenylalanine for controlled release (Cui *et al.*, 2009; Zhao *et al.*, 2010). These studies proposed that the position of aromatic moieties is a significant determinant for supramolecular self-assembling conformations. Self-assembled chitin fibers with diameters of 3 nm and 10 nm could be also obtained from dissolving in hexafluoro-2-propanol (HFIP) and LiCl/N,N-dimethylacetamide (DMAC), respectively (Zhong *et al.*, 2010). These observations demonstrated that nanofiber assembly occurs at a critical concentration and fiber morphology can be controlled by solution concentration and solution evaporation rate.

Molecular self-assembly is a fairly new technique for developing nanofibrous scaffolds which have been studied for a variety of tissue engineering applications including nerve (Guo *et al.*, 2007), bone (Sargeant *et al.*, 2008), and cartilage (Shah *et al.*, 2010) regeneration. However, several technical hurdles still need to be addressed: first, it has not been demonstrated how to control the pore size and pore structure, which are important to allow for cell proliferation and migration. Secondly, their degradation profiles have not been systematically addressed (Smith *et al.*, 2009). Further, most of the self-assembled scaffolds are mechanically weak and do not effectively sustain and transfer mechanical loadings to the cells and surrounding tissues.

6.2.2   Thermally induced phase separation

Another interesting method used for manufacturing nanofibrous scaffold is phase separation, which is a thermodynamic process involving the separation of phases due to physical incompatibility. Specifically, a homogeneous polymer solution becomes thermodynamically unstable under certain temperature conditions, and will form a polymer-rich phase and a polymer-poor phase. When the solvent is removed, the polymer-rich phase will solidify to a 3D structure while the polymer-poor phase will become the void space. The process of developing nanofibrous scaffold from phase separation typically consists of five major steps: (i) raw material dissolution, (ii) gelation, (iii) solvent extraction, (iv) freezing, and (v) drying (Ramakrishna, 2005). The pore morphology of nanofibrous scaffolds can be tuned by varying processing parameters including polymer concentration, phase separation temperature, and solvent/non-solvent exchange (Zhang *et al.*, 2012). Furthermore, thermally induced phase separation can be combined with other processing techniques (e.g. particulate leaching and solid free-form fabrication) to generate scaffolds with complex porous structures and well-defined pore morphology (Holzwarth and Ma, 2011). For example, sodium chloride particles with diameters of 200–450 µm were mixed with a warm poly(lactic-co-glycolic acid)/tetrahydrofuran (PLGA/THF) solution and then cooled to a preset gelation temperature. The composite gels formed were extracted with cold ethanol and washed with distilled water to remove the solvent and leach the salt particles. The samples were freeze-dried, resulting in a nanofibrous scaffold with a macroporous structure left behind from the leached salt (Mao *et al.*, 2012). Nanofibrous poly(L-lactic acid) (PLLA) scaffolds have been developed by using phase separation to improve cell seeding, distribution, and mass transport (Woo *et al.*, 2003). When compared with solid pore-walled PLLA scaffolds, nanofibrous scaffolds allowed an approximately 2-fold increase in adhesion of osteoblast cells.

Thermally induced phase separation is a promising technique for developing nanofibrous scaffold with well-defined pore shape and pore size. Although this technique can be combined with other fabrication methods to control the final 3D structure, the drawbacks of this technique, including little control over fiber orientation and diameter, long fabrication time, and lack of mechanical properties, need to be addressed to further achieve a fine control of the resulting scaffolds at the macro-, micro-, and nano-scales.

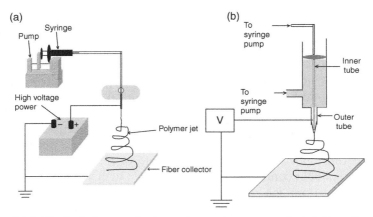

6.1 (a) Typical electrospinning set-up using a grounded static collector.
(b) Configuration of the coaxial electrospinning set-up used for
preparing core-shell structured nanofibers.

6.2.3   Electrospinning

Electrospinning is a versatile and well-established process capable of producing fibers with diameters down to the submicron or nanometer range
(Prabaharan *et al.*, 2011). A basic electrospinning setup (Fig. 6.1a) includes
a high voltage electric source with positive or negative polarity; a syringe
pump with capillaries or tubes to carry the solution (or melt) from a syringe
to the spinneret; and a conducting collector (Sill and Von Recum, 2008).
Under the electric field, the pulling force overcomes the surface tension of
the polymer solution (or melt) and creates a charged jet that travels through
the atmosphere allowing the solvent to evaporate, thus leading to the deposition of solid polymer fibers on the collector (Deitzel *et al.*, 2002; Sill and Von
Recum, 2008). The fiber formation and structure is affected by three general
types of variable: solution properties (concentration, viscosity, conductivity,
and surface tension), process factors (applied potential, collection distance,
emitting electrode polarity, and feed rate), and environmental parameters
(temperature, relative humidity, and velocity of the surrounding air in the
spinning chamber) (Lee *et al.*, 2003; Sill and Von Recum, 2008).

For many applications, it is necessary to fabricate a scaffold made of
aligned nanofibers, as the anisotropy in topography and structure can have a
great impact on mechanical properties and cell behavior. Currently, a number of collecting devices have been developed to attain aligned electrospun
fibers. These collecting devices can be divided into three main categories
according to the type of forces involved (Liu *et al.*, 2011): (1) mechanical
forces by using a rotating mandrel, (2) electrostatic forces by using parallel
electrodes, and (3) magnetic forces by using parallel permanent magnets.

Furthermore, electrospinning can be used for fabricating scaffold with complex architectures such as stacked arrays and tubular conduits. Stacked arrays of nanofiber scaffolds can be achieved by multilayering electrospinning. Aligned nanofibers were stacked into multilayered films with a controllable hierarchical porous structure by depositing fibers on an insulating substrate (e.g. quartz) onto which an array of electrodes have been patterned (Li *et al.*, 2004). Tubular conduits made of electrospun fibers can be fabricated by depositing fibers over a small diameter rod as a collector. These electrospun conduits are usually applied in vascular or neural tissue engineering since they mimic the structure of these tissues. For instance, electrospun fiber conduits were manufactured with a length of 12 cm and a thickness of 1 mm via electrospinning a mixture of collagen type I, elastin, and poly(D,L lactide-co-glycolide) on a rotating rod of diameter 4.75 mm (Stitzel *et al.*, 2006). Compliance tests demonstrated that the electrospun fiber conduit possesses a diameter change of 12~14% within the physiologic pressure range, a compliance behavior similar to that of a native vessel with a diameter change of 9%. Burst pressure testing results showed that the burst pressure for the electrospun fiber conduit was 1425 mmHg or nearly 12 times that of normal systolic pressure.

In order to tailor the structures of resultant fibers, the modification of electrospinning set-ups have been carried out not only on the fiber collector, but also on the spinneret. Coaxial spinnerets have been designed by many researchers for various aims in electrospinning (Fig. 6.1b). Using this technology, core-shell nanofibers can be achieved. By using a spinneret with double coaxial capillaries, two components can be fed through different coaxial capillary channels to generate composite nanofibers in the form of a core-sheath structure (He *et al.*, 2006; Zhang *et al.*, 2006; Venugopal *et al.*, 2008). Some difficult-to-process polymer solutions can be also electrospun to form an ultrafine core within the shell of spinnable polymers. Thus, when the polymer sheath is removed, the desired polymer nanofiber core is retained (Wang *et al.*, 2006). Using a similar concept, the core solution can be removed instead of the polymer sheath, resulting in hollow nanofibers. For example, an ethanol solution containing poly(vinyl pyrrolidone) (PVP) and tetraisopropyl titanate (Ti $(OiPr)_4$) was used as sheath solution while mineral oil was used as core solution. After simultaneous ejection through the inner and outer capillaries, hollow nanofibers were formed through removal of the inner core mineral oil phase (Li and Xia, 2004).

Shell-core and hollow nanofibers have shown their potential in tissue engineering and drug delivery due to their unique architecture and properties. Biologically active agents can be encapsulated inside a polymer shell to form a reservoir-type drug delivery device. Therefore, the polymer shell would offer a temporal protection for the bioactive agents and control their sustained release (He *et al.*, 2006). Tissue regeneration processes can be also facilitated via introduction of biomolecules (e.g. growth factors) into the core of shell-core nanofibrous scaffolds and, eventually, the release of multiple factors can be envisioned by adding different compounds in the shell and in the core fibers. Electrospun fibers with high surface to volume ratio and structures resembling ECM have shown great potential for applications in tissue engineering and drug delivery (Cui *et al.*, 2010b). Yet, a fine control over nanofiber population distribution is still lacking and is so far a limit in obtaining standardized scaffolds.

6.3   Surface modification of nanofibrous scaffolds

Surfaces play a vital role in biology and medicine, as most biological reactions take place at the interface between biological systems (Castner and Ratner, 2002). In tissue engineering, the chemical and physical characteristics of the biomaterial surface strongly impact on cell behavior, such as migration, attachment, and proliferation (Jiao and Cui, 2007). Although various degradable and non-degradable synthetic polymers have been used as tissue engineering scaffolding materials, a shortcoming of these materials is their lack of biological recognition (Smith *et al.*, 2008). Currently, several techniques have been developed to modify the scaffold surface including physical coating and blending, plasma treatment, graft polymerization, and wet chemical methods (Fig. 6.2).

6.3.1   Physical coating or blending

Perhaps the most straightforward and convenient means of modifying a polymer surface is blending functional molecules and active agents into the bulk polymer or just coating it on the polymer surface. Two or more materials are physically blended together to attain a new biocomposite with superior surface and/or mechanical properties. PLLA nanofibers with hydroxyapatite (HA) particles exposed on their surface were achieved by electrospinning different blended solutions (Sui *et al.*, 2007; Luong *et al.*, 2008). These composite fibers not only promoted osteoblast adhesion and growth, but also exhibited superior tensile properties as compared to the pure PLLA fibers, due to the internal ionic bonding between calcium ions in HA particles and ester groups in PLLA (Deng *et al.*, 2007). Jun *et al.* (2009) developed electrically conductive blends of poly(L-lactide-co-caprolactone) (PLCL) and polyaniline (PANi) submicron fibers and investigated their influence on the induction of myoblasts into myotubes. Incorporation of PANi into PLCL fibers significantly increased the conductivity. In addition, the tensile strength and elongation at break of PLCL fibers increased as the concentration of PANi in the fibers decreased, while the Young's modulus exhibited an

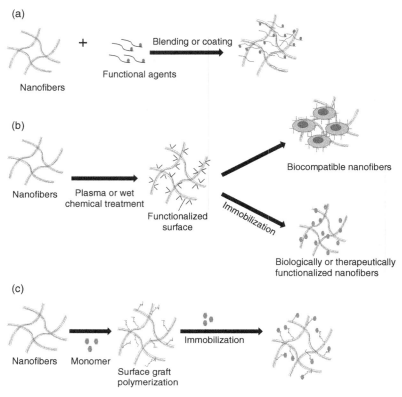

6.2 Surface modification techniques of nanofibrous scaffolds.
(a) Blending and coating; (b) plasma or wet chemical treatment;
(c) surface graft polymerization.

opposite trend. Taken together, these examples show that blending modifies not only the surface properties of substrate but also their bulk properties.

Coating is another physical approach of functionalizing nanofibrous scaffolds. Although this technique is limited in terms of controllability and maybe coatings prone to detachment from the scaffold, it remains one of the easiest and most convenient ways to functionalize the biomaterial surface. In general, a material containing the desired functional properties is used to coat nanofibrous scaffolds, aiming at improving their surface biocompatibility and enhancing cell–scaffold interactions. In recent years, coating of functional molecules onto polymers was significantly developed by combining new techniques including coaxial electrospinning and layer-by-layer assembly (LbL). Collagen-coated poly(ε-caprolactone) (PCL) nanofibers were developed by a coaxial electrospinning technique with collagen and PCL as inner and outer solutions, respectively, and cultured with human dermal fibroblasts (HDF) (Zhang *et al.*, 2005). The results demonstrated that collagen-coated PCL nanofibers from coaxial electrospinning showed higher efficiency in favoring cell proliferation compared to collagen-coated PCL nanofibers prepared by soaking the PCL matrix in a 10 mg/mL collagen solution. The mechanism behind LbL coatings is that positively and negatively charged macromolecules are alternatively introduced onto the substrate surface via strong electrostatic interaction (Ramakrishna, 2005). An environmentally benign surface modification process for poly(ethylene terephthalate) (PET) films was demonstrated by fabricating composite coatings through LbL assembly with cellulose and chitin nanofibrils (Qi *et al.*, 2012). The modified PET films exhibit high light transparency, flexibility, surface-hydrophilicity, and specific nanoporous structures.

6.3.2   Plasma-based surface modification

Plasma, also named the fourth state of matter, is a gaseous mixture of particles containing charged particles, excited neutral atoms and molecules, radicals, and UV photons (Denes and Manolache, 2004). Generally, plasma can be subdivided into two categories according to the temperature of particles: thermal plasma and non-thermal plasma. Non-thermal plasma, which is composed of low temperature particles has been commonly employed to modify the surface of heat-sensitive materials such as biodegradable polymers (Morent *et al.*, 2011). By using this technique, diverse functional groups can be effectively incorporated on the target surface of biodegradable polyesters to improve surface adhesion and wetting properties. For example, typical plasma treatment with different gaseous atmospheres such as oxygen, air, and ammonia can introduce carboxyl groups or amine groups to the target surface (Zhu *et al.*, 2005).

After introduction of specific functional groups on the surface of substrates by plasma treatment, synthetic and natural macromolecules could further immobilize monomers (e.g. extracellular matrix protein components) on their surface to enhance cellular adhesion and proliferation. This process is called plasma grafting. Nanofibrous scaffolds composed of PLLA and PLGA were fabricated by electrospinning (Park *et al.*, 2007b). Their surfaces were treated with oxygen plasma treatment and grafted with hydrophilic acrylic acid. Such surface-modified scaffolds were shown to improve fibroblast attachment and proliferation *in vitro*. Plasma polymerization, another plasma-based surface modification approach, is distinct from plasma grafting by coating the substrate instead of covalent binding species to a plasma-treated polymer surface (Barry *et al.*, 2006; Morent *et al.*, 2011). Under plasma conditions, gaseous or liquid monomers are converted into reactive fragments which can, in turn, recombine to form a polymer film which is deposited onto the substrate exposed to the plasma. To gain more insights into plasma polymerization processes, various –C:H:N thin films were deposited on glass under different NH_3/C_2H_4 gas ratios, power inputs (W), and gas flow rates (F) (Guimond et al., 2011). The results demonstrated that the film growth is determined by the ratio between the reaction parameters power inputs W and gas flow rates F, and the energy dissipated at its surface during the deposition given by ion flux times mean ion energy per deposition rate.

Plasma treatment, an effective and solvent-free technique, is commonly used for surface modification. The disadvantage of plasma treatment is that only localized surface areas can be treated (in depth from several hundred angstroms to 10 nm) without changing the bulk properties of the polymers (Jiao and Cui, 2007). Currently, most plasma-based surface modification is performed on two-dimensional (2D) polymer substrates, and in some cases on 3D porous scaffold.

6.3.3   Wet chemical methods

The cellular response to a biomaterial may be enhanced in synthetic polymer formulations by mimicking the surface roughness created by the associated nano-structured ECM components of natural tissue. PLGA films with nano-structured surface features were developed by treating them in a concentrated NaOH solution (Miller et al., 2004). The results from cell experiments indicated that NaOH-treated PLGA enhanced vascular smooth muscle cell adhesion and proliferation compared to conventional PLGA. This surface modification method is used not only on 2D film surfaces but also on 3D constructs. Chen et al. (2007) obtained PCL membranes with nanofibrous topology by coating the membrane surface with electrospun nanofibers. When the nanofiber-modified PCL was treated with 5M NaOH for 3h, a favorable 3T3 fibroblast cells morphology and strong cell attachment were observed on the modified surface, possibly due to the unique hydrophilic surface topography. Electrospun PLLA nanofibrous scaffolds were used as a matrix for mineralization of hydroxyapatite (Zhu et al., 2002). Carboxylic acid groups were introduced on the surface of PLLA scaffolds by hydrolysis in NaOH aqueous solution. Since calcium ions can bind to the carboxylate groups on the fiber surface, a significant improvement of the mineralization process was observed on modified PLLA electrospun scaffolds.

Although wet chemical methods have been widely applied to modify the surface wettability of biomaterials or to generate new functionalities, some major drawbacks should be noted. This modification technique can cause a partial degradation and scissions of the polymeric chains, resulting in loss of mechanical properties and a faster degradation process (Morent *et al.*, 2011). In addition, the use of hazardous organic solvents might affect cell viability.

6.3.4   Surface graft polymerization

Among the surface modification techniques, surface graft polymerization has emerged as a simple, effective, and versatile approach in the surface functionalization of polymers for a wide variety of applications (Ramakrishna, 2005). Grafting exhibits several advantages (Ramakrishna, 2005; Pettikiriarachchi *et al.*, 2012). Firstly, there is the ability to modify the polymer surface to possess desired properties through the choice of different monomers. Secondly, the ease and controllable introduction of graft chains with a high density and exact location of graft chains to the surfaces is possible without changing the bulk properties. Finally, a stable chemical bond is formed between the nanofiber surface and the preformed polymer, which offers endurance of the functional component, in contrast to physically coated polymer chains.

Surface graft polymerization is often initiated with plasma discharge, ultraviolet (UV) light, gamma rays, and electron beams (Yoo *et al.*, 2009; Pettikiriarachchi *et al.*, 2012). Chua *et al.* (2005) investigated how a functional poly(ε-caprolactone-co-ethyl ethylene phosphate) (PCLEEP) nanofibrous scaffold with surface-galactose ligand influenced behavior of rat hepatocytes. The modification of nanofibrous scaffolds was achieved by conjugating galactose ligands to a poly(acrylic acid) spacer UV-grafted onto the fiber surface. The functionalized nanofibrous scaffolds exhibited the unique property of promoting hepatocyte aggregates within the mesh and around the fibers. Furthermore, hepatocyte functions are maintained on these functional nanofiber substrates, similar to a galactosylated-film substrate configuration.

The grafting of extracellular protein like collagen, arginine-glycine-aspartic acid (RGD) peptides, and gelatin on nanofiber surfaces has become a popular method to develop biomimicking-tissue scaffolds. For example, the modification of electrospun polyethylene terephthalate (PET) nanofibers involved fiber treatment with formaldehyde to generate hydroxyl groups on the surface, followed by graft polymerization of methacrylic acid monomers initiated by Cerium (IV). Gelatin was then immobilized on the nanofibers by conjugation to the carboxylic acid moieties on their surface (Ma *et al.*, 2005). The grafting of gelatin enhanced the adhesion and spreading of cells on nanofibrous scaffolds. In another study, gelatin was grafted onto PLLA scaffolds via aminolysis of the PLLA fibers followed by glutaraldehyde coupling (Cui *et al.*, 2008).

6.4   Applications of nanofibrous scaffolds in tissue engineering

The characteristics of nanofibrous scaffolds that render them promising candidates for tissue engineering include high surface area and porosity, as well as the similarity of their fibrous structure to the physical features of natural ECM (Ma and Zhang, 1999; Park *et al.*, 2007a; Agarwal *et al.*, 2009; Pettikiriarachchi *et al.*, 2012). These features result in a more biologically compatible environment in which cells can grow and perform their regular functions. Therefore, nanofibrous scaffolds have been widely used as scaffolds for tissue engineering such as neural, cartilage, vascular, and bone tissue engineering (Table 6.2).

6.4.1   Nanofibrous scaffolds for neural tissue engineering

The nervous system coordinates the action of humans and transmits signals between different parts of the body. However, the nervous system has very limited capacity to repair itself after an injury. As a result, patients who have injures or traumas in the nervous system often suffer from the loss of sensory or motor function, and neuropathic pains (Venugopal and Ramakrishna, 2005; Prabaharan *et al.*, 2011). In order to facilitate the regrowth of nerves, many therapeutic approaches have been attempted. One of the most promising approaches is to adopt a neural tissue engineering strategy that employs a scaffold or conduit to facilitate nerve regeneration.

Yang *et al.* (2005) studied aligned and random PLLA electrospun nanofibrous scaffolds for the purpose of neural tissue engineering. Their results indicated that neural stem cell (NSCs) elongated and their neurites outgrew along the direction of the fiber orientation of the aligned nanofibers (Fig. 6.3). Furthermore, NSCs on aligned nanofibrous scaffolds showed a higher rate of differentiation than on microfibers. Thus, the aligned PLLA nanofibrous scaffolds showed potential for application in neural tissue engineering.

Arginine-alanine-aspartate (RAD)16-I are the most commonly used peptides in self-assembled peptide scaffolds for neural cell culture (Semino *et al.*, 2004; Silva *et al.*, 2004; Gelain *et al.*, 2006). Semino *et al.* (2004) developed RAD16-I self-assembled peptide scaffolds as a 3D culture system for cell encapsulation. Primary rat hippocampal cells were cultured on top of the self-assembled nanofibrous scaffold. Immunochemistry assays showed that glial cells and neurons increasingly migrated into the peptide scaffold to an approximate depth of 400–500 μm from the edge of the tissue in 1 week of cultures. Furthermore, mitotic activity of neural cells was maintained for 3 days after migration, which was attributed to the presence of the nanofibrous

Table 6.2 Nanofibrous scaffold for tissue engineering applications

Materials	*In vitro/in vivo* model	Cells	Potential application	Ref.
Self-assembled nanofiber				
Peptide amphiphile (PA) containing collagen (PA-collagen)	*In vitro*	Granule cells and Purkinje cells	Neural TE	Sur *et al.*, 2011
RADA16 (Ac-RADARADARADARADA-COHN2)	*In vitro*	Neural stem cells	Neural TE	Gelain *et al.*, 2006
RAD16-I peptide	Adult rats	Rat Schwann cells and embryonic NPCs	Neural TE	Guo *et al.*, 2007
PuraMatrix™	Mouse calvarial bone defect model		Bone TE	Misawa *et al.*, 2006
KLD-12 peptide (AcN-KLDLKLDLKLDL-CNH2)	*In vitro*	Bovine chondrocytes	Cartilage TE	Kisiday *et al.*, 2002
Peptide amphiphile (PA) combining with TGF	Omentum of Rats	Human mesenchymal stem cells	Cartilage TE	Shah *et al.*, 2010
Phase separation nanofiber				
Tyrosine-derived polycarbonate (TyrPC)	Rabbit calvarial critical-sized defect	–	Bone TE	Kim *et al.*, 2012
Chitosan/poly(ε-caprolactone) blend	*In vitro*	Bovine chondrocyte	Cartilage TE	Neves *et al.*, 2011
Poly-L-lactide	*In vitro*	Human aortic smooth muscle cells	Vascular TE	Hu *et al.*, 2010
Electrospinning nanofiber				
Poly(l-lactic acid)	*In vitro*	Neural stem cells	Neural TE	Yang *et al.*, 2005
Poly(ε-caprolactone)	*In vitro*	Human mesenchymal stem cells	Cartilage TE	Li *et al.*, 2005
Poly(L-lactic-co-ε-caprolactone) (75:25)	*In vitro*	Smooth muscle cell	Vascular TE	Mo and Weber, 2004
Gelatin-modified PET nanofibers	*In vitro*	Endothelial cells	Vascular TE	Ma *et al.*, 2005
Poly(ε-caprolactone)-collagen	Rabbit aorta-iliac bypass model	Vascular cells	Vascular TE	Tillman *et al.*, 2009
Poly(ε-caprolactone)	*In vitro*	Rabbit mesenchymal stem cells	Bone TE	Yoshimoto *et al.*, 2003
Poly(ε-caprolactone)	The omentum of rats	Rat mesenchymal stem cells	Bone TE	Shin *et al.*, 2004

TE refers to tissue engineering.

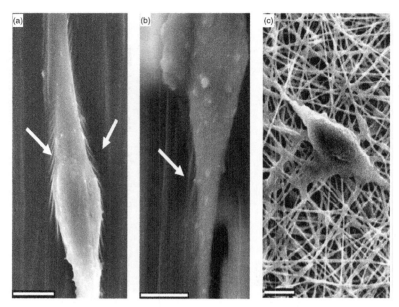

6.3 SEM micrographs of NSCs seeded on (a) nano-scale aligned fibers; (b) micro-scale aligned fibers; (c) nano-scale random fibers for 2 days showing the cell–matrix adhesion between the NSCs and the PLLA fibers. Arrows show some filament-like structures extend out from the NSC cell body and neurites and attach to the nanofibers. Bar = 5 mm (Yang *et al.* 2005).

scaffold environment resembling the native ECM. These results revealed that self-assembled peptide nanofibrous scaffolds are a potential substrate for supporting neural tissue regeneration. A hybrid nanofibrous matrix with homogeneous fiber diameter of 20–30 nm was designed by self-assembly of PA with the ability of presenting laminin epitopes (IKVAV or YIGSR) and collagen molecules (Sur *et al.*, 2011). Granule cells and Purkinje cells, two major neuronal subtypes of the cerebellar cortex, were cultured on the hybrid scaffold. The result showed the ability to modulate neuron survival and maturation by easy manipulation of epitope densities.

In addition to *in vitro* experiments, self-assembled nanofibrous scaffolds have been transplanted into animal models for the treatment of central nervous system injuries. Self-assembled RAD16-I peptide scaffold in combination with adult rat Schwann cells and embryonic neural progenitor cells (NPCs) were transplanted into adult rats (Guo *et al.*, 2007). The results indicated that the scaffolds integrated well with the host tissue with no obvious gap between the implants and the injured sites. A noteworthy observation from these experiments was that a large amount of host cells migrated into the scaffolds and extensive blood vessel formation was observed in the implants.

6.4.2   Nanofibrous scaffolds for cartilage tissue engineering

Cartilage, a predominantly avascular, alymphatic, and aneural tissue, is composed of chondroblasts embedded within a dense extracellular matrix (Chung and Burdick, 2008). Cartilage is classified into three types: elastic cartilage, hyaline cartilage, and fibrocartilage. Cartilage damage resulting from developmental abnormalities, trauma, aging related degeneration, and joint injury cause disability and extensive pain (Venugopal and Ramakrishna, 2005; Wang *et al.*, 2005). Due to its limited capacity to self-regenerate, cartilage is an ideal candidate for tissue engineering. In cartilage tissue engineering, chondrocytes and mesenchymal stem cells (MSCs) are commonly used for cartilage repair. Electrospun PCL nanofibrous scaffolds were seeded with human bone marrow-derived MSCs to investigate their ability to support *in vitro* MSC chondrogenesis (Li *et al.*, 2005). The results demonstrated that PCL nanofibrous scaffolds in the presence of transforming growth factor-β1 (TGF-β1) induced a differentiation of MSCs to chondrocytes comparable to pellet cultures. Although no inductive property of PCL nanofibrous scaffolds was observed, these meshes have better mechanical properties than cell pellets making them suitable for cartilage tissue engineering. Kisiday *et al.* (2002) investigated a self-assembling peptide hydrogel scaffold for cartilage regeneration. They combined the peptide KDK-12 with bovine chondrocytes and allowed them to self-assemble into a hydrogel. Their results indicated that chondrocytes proliferated well and maintained a chondrocytic phenotype in the hydrogel. Cells were able to produce cartilage-like ECM which was rich in type-II collagen and proteoglycans. Peptide amphiphiles (PAs) synthesized with a peptide binding sequence to TGF-β1 were designed to form nanofibers via self-assembly for cartilage regeneration (Shah *et al.*, 2010). The study demonstrated that these scaffolds support the survival and promote the chondrogenic differentiation of human mesenchymal stem cells. *In vivo* experiments showed that these scaffolds promoted regeneration of articular cartilage in a rabbit model with, or even in the absence of, exogenous growth factor (Plate VII, see color section between pages 264 and 265).

6.4.3   Nanofibrous scaffolds for vascular tissue engineering

Blood vessels perform a very important function of carrying and transporting blood to and from heart. In addition, they are also important places for the exchange of water and other chemicals between blood and the tissue (Ramakrishna, 2005; Cui *et al.*, 2010b). Mo and Weber (2004) developed an aligned electrospun nanofibrous scaffold from biodegradable poly(L-lactic acid-*co*-ε-caprolactone) (PLLA-CL) (75:25) with the goal of developing constructs for vascular tissue engineering. The fabricated nanometric fibers resembled the dimensions of natural ECM, possessed mechanical properties comparable to human coronary artery, and supported smooth muscle cell adhesion and proliferation well.

A major reason for graft failure is that the graft surface is only partially covered by endothelial cells following a degenerative state. To overcome this problem, PET nanofibrous scaffold were developed by electrospinning and their surfaces were modified by grafting gelatin. Their study indicated that gelatin-modified PET nanofibers were favorable for spreading and proliferation of endothelial cells and maintained cell phenotype (Ma *et al.*, 2005). Tillman *et al.* (2009) studied the *in vivo* stability of electrospun PCL-collagen scaffolds in vascular regeneration in a rabbit aortal-iliac bypass model. Their findings suggested that PCL-collagen scaffolds maintained enough mechanical strength for cell growth and other physiologic conditions.

6.4.4   Nanofibrous scaffolds for bone tissue engineering

Bone is a composite material made from an organic phase of collagen and glycoproteins, and an inorganic phase primarily consisting of hydroxyapatite (HA) (Rhoades and Pflanzer, 1996). The organic phase provides the resilient nature of bone and the inorganic minerals are responsible for bone hardness (Ramakrishna, 2005; Nisbet *et al.*, 2009). Bone loss may be caused by trauma, fractures, periodontitis, cancer, infectious disease, and osteoporosis (Kimakhe *et al.*, 1999). Presently, many approaches have been developed for bone regeneration, such as autografts, metal implants, and allografts. However, all of these methods have obvious drawbacks. For example, autografts present problems associated with a limited donor source as well as a secondary surgery site. Therefore, tissue engineering approaches are presently being investigated as a promising method for bone regeneration.

Yoshimoto *et al.* (2003) studied the potential of non-woven PCL scaffolds generated by electrospinning for bone tissue engineering. MSCs derived from bone marrow of neonatal rats were seeded on the nanofibers. The results demonstrated that MSCs migrated into the scaffolds and produced abundant ECM. Based on this study, Shin *et al.* (2004) implanted MSCs along with PCL nanofibrous scaffolds into the omentum of rats. Their study indicated ECM formation throughout the scaffolds along with the evidence of mineralization and type I collagen synthesis. When HA is incorporated into nanofibrous scaffolds it not only improves the mechanical properties, but also creates more biomimetic constructs. Either nanoparticles of HA, morphogenetic proteins (e.g. BMP-2) or both were incorporated into

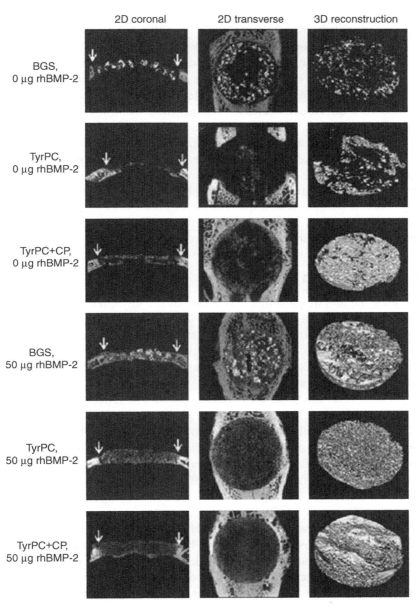

6.4 Representative microcomputed tomography images of bone regeneration in the rabbit model at 6 weeks post-implantation. White arrows in the first column identify the defect site. Remaining b-tricalcium phosphate fragments are identified in the 2D transverse images of the bone graft substitute (BGS) scaffolds (Kim *et al.*, 2012).

electrospun nanofibrous scaffolds of silk fibroin (Li *et al.*, 2006). Scaffolds were seeded with human MSCs for 31 days in osteogenic medium. HA nanoparticles were associated with improved bone formation. Furthermore, when silk fibroin scaffolds were combined with both nanoparticles of HA and BMP-2, they were associated with the highest observed mineralization.

Porous three-dimensional tyrosine-derived polycarbonate (TyrPC) scaffolds were developed for bone regeneration using a combination of salt-leaching and phase separation techniques (Kim *et al.*, 2012). The TyrPC scaffolds treated with or without recombinant human bone morphogenetic protein-2 (rhBMP-2) were implanted in a rabbit calvarial critical-sized defect model for 6 weeks. The *in vivo* studies showed that TyrPC scaffolds treated with rhBMP-2 or coated with calcium phosphate alone promoted bone regeneration at 6 weeks post-implantation (Fig. 6.4). Moreover, TyrPC polymeric scaffold, even without addition of any biological agents, induced similar bone regeneration to a commercially available bone graft substitute.

6.5   Conclusion

Mimicking the architecture of ECM is one of the key challenges of tissue engineering. Among all the methods used to generate artificial ECM, nanofibrous scaffolds demonstrated the most promising results. Nanofibrous scaffolds, irrespective of their method of synthesis, are characterized by high surface area and enhanced porosity, which are highly desired for tissue engineering and drug delivery applications.

At present, there are three dominant approaches to generate nanofibrous scaffolds: molecular self-assembly, phase separation, and electrospinning. Of these approaches, electrospinning is the most widely studied since it is a simple and versatile technique that can produce fibers with a diameter from the micrometer down to nanometer range. Furthermore, it allows one to tailor many aspects of the resulting scaffolds (Casper *et al.*, 2004; Moroni *et al.*, 2006; Liu *et al.*, 2011): (1) the fiber diameter can easily be controlled by varying the solution properties and the processing parameters, (2) the nanofibers can be collected with a rich variety of aligned/random structures using specially designed collectors, (3) they can be stacked and/or folded to form complex structures or architectures, (4) they can be obtained with a hollow or core-sheath structure by changing the configuration of the spinneret, and (5) nanofibers with porous surface structure can be achieved by altering the parameters such as solvent and humidity. The potential of nanofibrous scaffolds for the regeneration of various tissues is now under extensive investigation. In spite of the great achievements behind the design of nanofibrous scaffolds, there is still plenty of room for improvement.

The future research on nanofibrous architecture may be focused on the following aspects: (1) design the structure of nanofibrous architecture.

One strategy is to exploit new nanofabrication techniques. Recently, Badrossamay *et al.* (2010) developed a new technique for generation of continuous fibers by rotary jet-spinning. Using this technique, nanofiber structure can be fabricated into an aligned 3D structure or any arbitrary shape by varying the collector geometry. Another strategy is to integrate nanofiber into microfabricated 3D scaffolds to achieve more desirable scaffolds. For example, co-electrospun nanofluidic channels were integrated into stimuli-sensitive hydrogels to realize controlled release of biomolecules (Yang *et al.*, 2012). A third strategy is to fully explore current approaches of fabricating nanofibrous scaffolds. For instance, highly porous core-shell fiber networks were fabricated using an electrospinning system containing a water-immersed collector (Muhammad *et al.*, 2011). Finally, in combination with nanofabrication technologies, nanofibrous scaffold could be decorated with nanotopographic patterns, such as ridges and grooves, to better match the nanostructure of ECM. (2) Achieve a better control of ECM-mimicry. Nanofibrous scaffolds should be designed more and more as bioactive systems rather than just passive cell carriers. Integration of fabrication techniques with surface modification methods and the achievement of precise positioning of different bioactive cues will be a route to obtain nanofibrous scaffolds with closer properties to native ECM, and (3) Gain a better fundamental understanding of cell-scaffold interaction *in vitro*, followed by *in vivo* assessment. There are some important factors, such as mechanical properties and morphological optimization at the nano-, micro- and macroscale that need to be addressed at 'design stage' in tailoring nanofibrous scaffolds. All of these factors play an important role in governing cellular responses and host integration.

Although much joint effort by scientists from multiple disciplines is still needed for the development of nanofibrous scaffolds in different tissue engineering applications, it can be foreseen that nanofibrous scaffolds could approach 'off-the-shelf' surgically implantable constructs in the near future.

6.6   References

Agarwal, S., Greiner, A. and Wendorff, J. H. (2009). Electrospinning of manmade and biopolymer nanofibers – Progress in techniques, materials, and applications. *Advanced Functional Materials*, **19**, 2863–2879.

Badrossamay, M. R., McIlwee, H. A., Goss, J. A. and Parker, K. K. (2010). Nanofiber assembly by rotary jet-spinning. *Nano Letters*, **10**, 2257–2261.

Barry, J. J. A., Howard, D., Shakesheff, K. M., Howdle, S. M. and Alexander, M. R. (2006). Using a core–sheath distribution of surface chemistry through 3D tissue engineering scaffolds to control cell ingress. *Advanced Materials*, **18**, 1406–1410.

Casper, C. L., Stephens, J. S., Tassi, N. G., Chase, D. B. and Rabolt, J. F. (2004). Controlling surface morphology of electrospun polystyrene fibers: Effect of humidity and molecular weight in the electrospinning process. *Macromolecules*, **37**, 573–578.

Castner, D. G. and Ratner, B. D. (2002). Biomedical surface science: Foundations to frontiers. *Surface Science*, **500**, 28–60.

Chen, F., Lee, C. and Teoh, S. (2007). Nanofibrous modification on ultra-thin poly (e-caprolactone) membrane via electrospinning. *Materials Science and Engineering: C*, **27**, 325–332.

Chua, K. N., Lim, W. S., Zhang, P., Lu, H., Wen, J., Ramakrishna, S., Leong, K. W. and Mao, H. Q. (2005). Stable immobilization of rat hepatocyte spheroids on galactosylated nanofiber scaffold. *Biomaterials*, **26**, 2537–2547.

Chung, C. and Burdick, J. A. (2008). Engineering cartilage tissue. *Advanced Drug Delivery Reviews*, **60**, 243–262.

Cui, H., Muraoka, T., Cheetham, A. G. and Stupp, S. I. (2009). Self-assembly of giant peptide nanobelts. *Nano Letters*, **9**, 945–951.

Cui, H., Webber, M. J. and Stupp, S. I. (2010a). Self-assembly of peptide amphiphiles: From molecules to nanostructures to biomaterials. *Peptide Science*, **94**, 1–18.

Cui, W., Li, X., Chen, J., Zhou, S. and Weng, J. (2008). In situ growth kinetics of hydroxyapatite on electrospun poly (DL-lactide) fibers with gelatin grafted. *Crystal Growth and Design*, **8**, 4576–4582.

Cui, W., Zhou, Y. and Chang, J. (2010b). Electrospun nanofibrous materials for tissue engineering and drug delivery. *Science and Technology of Advanced Materials*, **11**, 014108.

Deitzel, J., Kosik, W., McKnight, S., Beck Tan, N., Desimone, J. and Crette, S. (2002). Electrospinning of polymer nanofibers with specific surface chemistry. *Polymer*, **43**, 1025–1029.

Denes, F. S. and Manolache, S. (2004). Macromolecular plasma-chemistry: An emerging field of polymer science. *Progress in Polymer Science*, **29**, 815–885.

Deng, X. L., Sui, G., Zhao, M. L., Chen, G. Q. and Yang, X. P. (2007). Poly (L-lactic acid)/hydroxyapatite hybrid nanofibrous scaffolds prepared by electrospinning. *Journal of Biomaterials Science, Polymer Edition*, **18**, 117–130.

Gelain, F., Bottai, D., Vescovi, A. and Zhang, S. (2006). Designer self-assembling peptide nanofiber scaffolds for adult mouse neural stem cell 3-dimensional cultures. *PLoS One*, **1**, e119.

Goldberg, M., Langer, R. and Jia, X. (2007). Nanostructured materials for applications in drug delivery and tissue engineering. *Journal of Biomaterials Science. Polymer Edition*, **18**, 241.

Guimond, S., Schütz, U., Hanselmann, B., Körner, E. and Hegemann, D. (2011). Influence of gas phase and surface reactions on plasma polymerization. *Surface and Coatings Technology*, **205**, S447–S450.

Guo, J., Su, H., Zeng, Y., Liang, Y. X., Wong, W. M., Ellis-Behnke, R. G., So, K. F. and Wu, W. (2007). Reknitting the injured spinal cord by self-assembling peptide nanofiber scaffold. *Nanomedicine: Nanotechnology, Biology and Medicine*, **3**, 311–321.

Hartgerink, J. D., Beniash, E. and Stupp, S. I. (2001). Self-assembly and mineralization of peptide-amphiphile nanofibers. *Science*, **294**, 1684–1688.

He, C. L., Huang, Z. M., Han, X. J., Liu, L., Zhang, H. S. and Chen, L. S. (2006). Coaxial electrospun poly (L-lactic acid) ultrafine fibers for sustained drug delivery. *Journal of Macromolecular Science Part B – Physics*, **45**, 515–524.

Holzwarth, J. M. and Ma, P. X. (2011). 3D nanofibrous scaffolds for tissue engineering. *Journal of Materials Chemistry*, **21**, 10243–10251.

Hu, J., Sun, X., Ma, H., Xie, C., Chen, Y. E. and Ma, P. X. (2010). Porous nanofibrous PLLA scaffolds for vascular tissue engineering. *Biomaterials*, **31**, 7971–7977.

Hung, A. M. and Stupp, S. I. (2007). Simultaneous self-assembly, orientation, and patterning of peptide-amphiphile nanofibers by soft lithography. *Nano Letters*, **7**, 1165–1171.

Jiao, Y. P. and Cui, F. Z. (2007). Surface modification of polyester biomaterials for tissue engineering. *Biomedical Materials*, **2**, R24.

Jun, I., Jeong, S. and Shin, H. (2009). The stimulation of myoblast differentiation by electrically conductive sub-micron fibers. *Biomaterials*, **30**, 2038–2047.

Kim, J., Magno, M. H. R., Waters, H., Doll, B. A., McBride, S., Alvarez, P., Darr, A., Vasanji, A., Kohn, J. and Hollinger, J. O. (2012). Bone regeneration in a rabbit critical-size calvarial model using tyrosine-derived polycarbonate scaffolds. *Tissue Engineering*, **18**, 1132–1139.

Kimakhe, S., Bohic, S., Larrose, C., Reynaud, A., Pilet, P., Giumelli, B., Heymann, D. and Daculsi, G. (1999). Biological activities of sustained polymyxin B release from calcium phosphate biomaterial prepared by dynamic compaction: An in vitro study. *Journal of Biomedical Materials Research*, **47**, 18–27.

Kisiday, J., Jin, M., Kurz, B., Hung, H., Semino, C., Zhang, S. and Grodzinsky, A. (2002). Self-assembling peptide hydrogel fosters chondrocyte extracellular matrix production and cell division: Implications for cartilage tissue repair. *Proceedings of the National Academy of Sciences*, **99**, 9996.

Lee, K., Kim, H., Khil, M., Ra, Y. and Lee, D. (2003). Characterization of nano-structured poly (ε-caprolactone) nonwoven mats via electrospinning. *Polymer*, **44**, 1287–1294.

Li, C., Vepari, C., Jin, H. J., Kim, H. J. and Kaplan, D. L. (2006). Electrospun silk-BMP-2 scaffolds for bone tissue engineering. *Biomaterials*, **27**, 3115–3124.

Li, D., Wang, Y. and Xia, Y. (2004). Electrospinning nanofibers as uniaxially aligned arrays and layer-by-layer stacked films. *Advanced Materials*, **16**, 361–366.

Li, d. and Xia, Y. (2004). Direct fabrication of composite and ceramic hollow nanofibers by electrospinning. *Nano Letters*, **4**, 933–938.

Li, W. J., Tuli, R., Okafor, C., Derfoul, A., Danielson, K. G., Hall, D. J. and Tuan, R. S. (2005). A three-dimensional nanofibrous scaffold for cartilage tissue engineering using human mesenchymal stem cells. *Biomaterials*, **26**, 599–609.

Liu, W., Thomopoulos, S. and Xia, Y. (2012). Electrospun nanofibers for regenerative medicine. *Advanced Healthcare Materials*, **1**(1), 10–25.

Luong, N. D., Moon, I. S., Lee, D. S., Lee, Y. K. and Nam, J. D. (2008). Surface modification of poly (l-lactide) electrospun fibers with nanocrystal hydroxyapatite for engineered scaffold applications. *Materials Science and Engineering: C*, **28**, 1242–1249.

Ma, P. X. and Zhang, R. (1999). Synthetic nano-scale fibrous extracellular matrix. *Journal of Biomedical Materials Research*, **46**, 60–72.

Ma, Z., Kotaki, M., Yong, T., He, W. and Ramakrishna, S. (2005). Surface engineering of electrospun polyethylene terephthalate (PET) nanofibers towards development of a new material for blood vessel engineering. *Biomaterials*, **26**, 2527–2536.

Mao, J., Duan, S., Song, A., Cai, Q., Deng, X. and Yang, X. (2012). Macroporous and nanofibrous poly (Lactide-co-Glycolide) (50/50) scaffolds via phase separation combined with particle-leaching. *Materials Science and Engineering: C*, **32**, 1407–1414.

Miller, D. C., Thapa, A., Haberstroh, K. M. and Webster, T. J. (2004). Endothelial and vascular smooth muscle cell function on poly (lactic-co-glycolic acid) with nano-structured surface features. *Biomaterials*, **25**, 53–61.

Misawa, H., Kobayashi, N., Soto-Gutierrez, A., Chen, Y., Yoshida, A., Rivas-Carrillo, J. D., Navarro-Alvarez, N., Tanaka, K., Miki, A., Takei, J., Ueda, T., Tanaka, M., Endo, H., Tanaka, N. and Ozaki, T. (2006). PuraMatrix™ facilitates bone regeneration in bone defects of calvaria in mice. *Cell Transplantation*, **15**, 903–910.

Mo, X. and Weber, H. J. (2004). Electrospinning P (LLA-CL) nanofiber: A tubular scaffold fabrication with circumferential alignment. *Macromolecules. Symposia*, **21**, 413–416.

Morent, R., de Geyter, N., Desmet, T., Dubruel, P. and Leys, C. (2011). Plasma surface modification of biodegradable polymers: A review. *Plasma Processes and Polymers*, **8**, 171–190.

Moroni, L., Licht, R., de Boer, J., de Wijn, J. R. and van Blitterswijk, C. A. (2006). Fiber diameter and texture of electrospun PEOT/PBT scaffolds influence human mesenchymal stem cell proliferation and morphology, and the release of incorporated compounds. *Biomaterials*, **27**, 4911–4922.

Muhammad, G., Lee, J. M., Kim, J., Lim, D. W., Lee, E. K. and Chung, B. G. (2011). Highly porous core-shell polymeric fiber network. *Langmuir*, **27**, 10993–10999.

Neves, S. C., Moreira Teixeira, L. S., Moroni, L., Reis, R. L., van Blitterswijk, C. A., Alves, N. M., Karperien, M. and Mano, J. F. (2011). Chitosan/Poly(ε-caprolactone) blend scaffolds for cartilage repair. *Biomaterials*, **32**, 1068–1079.

Nisbet, D., Forsythe, J., Shen, W., Finkelstein, D. and Horne, M. (2009). Review paper: A review of the cellular response on electrospun nanofibers for tissue engineering. *Journal of Biomaterials Applications*, **24**, 7–29.

Park, H., Cannizzaro, C., Vunjak-Novakovic, G., Langer, R., Vacanti, C. A. and Farokhzad, O. C. (2007a). Nanofabrication and microfabrication of functional materials for tissue engineering. *Tissue Engineering*, **13**, 1867–1877.

Park, K., Ju, Y. M., Son, J. S., Ahn, K. D. and Han, D. K. (2007b). Surface modification of biodegradable electrospun nanofiber scaffolds and their interaction with fibroblasts. *Journal of Biomaterials Science, Polymer Edition*, **18**, 369–382.

Pettikiriarachchi, J. T. S. , Parish, C. L., Nisbet, D. R. and Forsythe, J. S. (2012). Architectural and surface modification of nanofibrous scaffolds for tissue engineering, Available from: http://onlinelibrary.wiley.com/doi/10.1002/9783527610419.ntls0258/full (Accessed 19 April 2013).

Philp, D. and Stoddart, J. F. (1996). Self-assembly in natural and unnatural systems. *Angewandte Chemie International Edition in English*, **35**, 1154–1196.

Prabaharan, M., Jayakumar, R. and Nair, S. (2011). Electrospun nanofibrous scaffolds-current status and prospects in drug delivery. *Advances in Polymer Science*, **246**, 241–262.

Qi, Z. D., Saito, T., Fan, Y. and Isogai, A. (2012). Multifunctional coating films by layer-by-layer deposition of cellulose and chitin nanofibrils. *Biomacromolecules*, **13**, 553–558.

Ramakrishna, S. (2005). *An Introduction to Electrospinning and Nanofibers*, Singapore, World Scientific Pub Co Inc.

Rhoades, R. and Pflanzer, R. G. (1996). *Human Physiology*, USA Saunders College Pub.

Sargeant, T. D., Guler, M. O., Oppenheimer, S. M., Mata, A., Satcher, R. L., Dunand, D. C. and Stupp, S. I. (2008). Hybrid bone implants: Self-assembly of peptide amphiphile nanofibers within porous titanium. *Biomaterials*, **29**, 161–171.

Semino, C. E., Kasahara, J., Hayashi, Y. and Zhang, S. (2004). Entrapment of migrating hippocampal neural cells in three-dimensional peptide nanofiber scaffold. *Tissue Engineering*, **10**, 643–655.

Shah, R. N., Shah, N. A., Lim, M. M. D. R., Hsieh, C., Nuber, G. and Stupp, S. I. (2010). Supramolecular design of self-assembling nanofibers for cartilage regeneration. *Proceedings of the National Academy of Sciences*, **107**, 3293–3298.

Shin, M., Yoshimoto, H. and Vacanti, J. P. (2004). In vivo bone tissue engineering using mesenchymal stem cells on a novel electrospun nanofibrous scaffold. *Tissue Engineering*, **10**, 33–41.

Sill, T. J. and von Recum, H. A. (2008). Electrospinning: Applications in drug delivery and tissue engineering. *Biomaterials*, **29**, 1989–(2006.

Silva, G. A., Czeisler, C., Niece, K. L., Beniash, E., Harrington, D. A., Kessler, J. A. and Stupp, S. I. (2004). Selective differentiation of neural progenitor cells by high-epitope density nanofibers. *Science*, **303**, 1352–1355.

Smith, I., Liu, X., Smith, L. and Ma, P. (2009). Nanostructured polymer scaffolds for tissue engineering and regenerative medicine. *Wiley Interdisciplinary Reviews: Nanomedicine and Nanobiotechnology*, **1**, 226–236.

Smith, L. A., Liu, X. and Ma, P. X. (2008). Tissue engineering with nano-fibrous scaffolds. *Soft Matter*, **4**, 2144–2149.

Stitzel, J., Liu, J., Lee, S. J., Komura, M., Berry, J., Soker, S., Lim, G., van Dyke, M., Czerw, R. and Yoo, J. J. (2006). Controlled fabrication of a biological vascular substitute. *Biomaterials*, **27**, 1088–1094.

Subbiah, T., Bhat, G., Tock, R., Parameswaran, S. and Ramkumar, S. (2005). Electrospinning of nanofibers. *Journal of Applied Polymer Science*, **96**, 557–569.

Sui, G., Yang, X., Mei, F., Hu, X., Chen, G., Deng, X. and Ryu, S. (2007). Poly-L-lactic acid/hydroxyapatite hybrid membrane for bone tissue regeneration. *Journal of Biomedical Materials Research Part A*, **82**, 445–454.

Sur, S., Pashuck, E. T., Guler, M. O., Ito, M., Stupp, S. I. and Launey, T. (2011). A hybrid nanofiber matrix to control the survival and maturation of brain neurons. *Biomaterials*, **33**, 545–555.

Tillman, B. W., Yazdani, S. K., Lee, S. J., Geary, R. L., Atala, A. and Yoo, J. J. (2009). The in vivo stability of electrospun polycaprolactone-collagen scaffolds in vascular reconstruction. *Biomaterials*, **30**, 583–588.

Venugopal, J., Low, S., Choon, A. T. and Ramakrishna, S. (2008). Interaction of cells and nanofiber scaffolds in tissue engineering. *Journal of Biomedical Materials Research Part B: Applied Biomaterials*, **84**, 34–48.

Venugopal, J. and Ramakrishna, S. (2005). Applications of polymer nanofibers in biomedicine and biotechnology. *Applied Biochemistry and Biotechnology*, **125**, 147–157.

Wang, M., Jing, N., Su, C. B., Kameoka, J., Chou, C. K., Hung, M. C. and Chang, K. A. (2006). Electrospinning of silica nanochannels for single molecule detection. *Applied Physics Letters*, **88**, 033106.

Wang, Y., Kim, U. J., Blasioli, D. J., Kim, H. J. and Kaplan, D. L. (2005). In vitro cartilage tissue engineering with 3D porous aqueous-derived silk scaffolds and mesenchymal stem cells. *Biomaterials*, **26**, 7082–7094.

Watari, S., Hayashi, K., Wood, J. A., Russell, P., Nealey, P. F., Murphy, C. J. and Genetos, D. C. (2011). Modulation of osteogenic differentiation in hMSCs cells by submicron topographically-patterned ridges and grooves. *Biomaterials*, **33**, 128–136.

Woo, K. M., Chen, V. J. and Ma, P. X. (2003). Nano-fibrous scaffolding architecture selectively enhances protein adsorption contributing to cell attachment. *Journal of Biomedical Materials Research Part A*, **67**, 531–537.

Yang, F., Murugan, R., Wang, S. and Ramakrishna, S. (2005). Electrospinning of nano/micro scale poly (L-lactic acid) aligned fibers and their potential in neural tissue engineering. *Biomaterials*, **26**, 2603–2610.

Yang, H., Hong, W. and Dong, L. (2012). A controlled biochemical release device with embedded nanofluidic channels. *Applied Physics Letters*, **100**, 153510–153510–4.

Yoo, H. S., Kim, T. G. and Park, T. G. (2009). Surface-functionalized electrospun nanofibers for tissue engineering and drug delivery. *Advanced Drug Delivery Reviews*, **61**, 1033–1042.

Yoshimoto, H., Shin, Y., Terai, H. and Vacanti, J. (2003). A biodegradable nanofiber scaffold by electrospinning and its potential for bone tissue engineering. *Biomaterials*, **24**, 2077–2082.

Zhang, Y., Venugopal, J., Huang, Z. M., Lim, C. and Ramakrishna, S. (2005). Characterization of the surface biocompatibility of the electrospun PCL-collagen nanofibers using fibroblasts. *Biomacromolecules*, **6**, 2583–2589.

Zhang, Y., Wang, X., Feng, Y., Li, J., Lim, C. and Ramakrishna, S. (2006). Coaxial electrospinning of (fluorescein isothiocyanate-conjugated bovine serum albumin)-encapsulated poly (ε-caprolactone) nanofibers for sustained release. *Biomacromolecules*, **7**, 1049–1057.

Zhang, Z., Hu, J. and Ma, P. X. (2012). Nanofiber-based delivery of bioactive agents and stem cells to bone sites. *Advanced Drug Delivery Reviews*, **64**, 1129–1141.

Zhao, Y., Tanaka, M., Kinoshita, T., Higuchi, M. and Tan, T. (2010). Nanofibrous scaffold from self-assembly of [beta]-sheet peptides containing phenylalanine for controlled release. *Journal of Controlled Release*, **142**, 354–360.

Zhong, C., Cooper, A., Kapetanovic, A., Fang, Z., Zhang, M. and Rolandi, M. (2010). A facile bottom-up route to self-assembled biogenic chitin nanofibers. *Soft Matter*, **6**, 5298–5301.

Zhu, X., Chian, K. S., Chan-Park, M. B. E. and Lee, S. T. (2005). Effect of argon-plasma treatment on proliferation of human-skin–derived fibroblast on chitosan membrane in vitro. *Journal of Biomedical Materials Research Part A*, **73**, 264–274.

Zhu, Y., Gao, C., Liu, X. and Shen, J. (2002). Surface modification of polycaprolactone membrane via aminolysis and biomacromolecule immobilization for promoting cytocompatibility of human endothelial cells. *Biomacromolecules*, **3**, 1312–1319.

7

Fabrication of nanomaterials for growth factor delivery in tissue engineering

R. R. SEHGAL and R. BANERJEE, Indian Institute of
Technology, Bombay, India

DOI: 10.1533/9780857097231.1.183

Abstract: Advancement in recombinant technology and nanotechnology
have led to exciting opportunities to engineer different growth
factor-delivering nanomaterials with applications in tissue engineering.
Many conventional materials have been explored as scaffolds for growth
factor delivery but tuning pharmacokinetics of growth factor release
remains a challenge. Using nanostructured scaffolds with growth factor
delivery can mimic the physiological environment and also provide
sustained therapeutic doses of proteins with tunable release profiles.
This chapter focuses on different types of nanostructures and their
applications in tissue engineering, and highlights the nanomaterials used
for drug and gene delivery in tissue engineering.

Key words: tissue engineering, growth factor, controlled delivery,
nanostructures.

7.1 Introduction

Tissue engineering involves fabricating porous biocompatible scaffolds with
synthetic or natural biomaterials combined with biological triggers with
or without pre-seeded cells which help in cellular adhesion, proliferation,
and differentiation (Luo *et al.,* 2007). Conventional approaches to making three-dimensional (3D) porous and biodegradable scaffolds provide
sufficient vascularization for growing cells but do not provide the regular
physiochemical triggers for cellular differentiation as present in the natural
physiological tissue environment. In the natural environment, cells interact
with the nanofibrous extracellular matrix (ECM) which plays a role in cellular growth and differentiation. Release of multiple growth factors (GFs),
which are the signaling molecules, plays an important role in different processes like regulation of cell proliferation, differentiation, migration, and cell
adhesion (Goldberg *et al.,* 2007; Shi *et al.,* 2010). Released GFs bind directly
to transmembrane cell receptors of the target cell and finally perform their
action at gene transcription level through a cascade of processes (Lee *et al.,*

183

2009). Hence, recent advances in tissue engineering have targeted designing nanostructured scaffolds which incorporate growth factors loaded within nanoparticles or nanofibers to provide sustained release.

Initially, different strategies like direct entrapment in scaffold and chemical conjugation on scaffold surface have been used providing localized growth factor delivery but controlling the pharmacokinetic profile of growth factor and maintaining its bioactivity are big challenges (Lee *et al.*, 2011). Growth factor delivery through microcarriers inside scaffolds has been extensively studied, but encapsulation of growth factors inside nanocarriers not only provides natural nano-topography of ECM but also maintains its therapeutic level for longer durations (Shi *et al.*, 2010). Different strategies have been applied for encapsulating growth factors inside nanostructures. These nanostructures can be formed by scaling large materials down to nano-scale by electrospinning and nanopatterning or can be formed by self-assembly of molecules (Engel *et al.*, 2008; Zhang and Webster, 2009). Pharmacokinetics of growth factor release can be changed by tuning size, composition, and spatio-temporal position of nanostructures within scaffolds (Shi *et al.*, 2010). The present chapter focuses on different strategies for growth factor delivery from tissue engineered scaffolds with major emphasis on growth factor-delivering nanoengineered scaffolds and their applications for tissue engineering.

7.2   Strategies for controlled growth factor delivery in tissue engineering

Although different growth factors have been directly introduced by injection, they produce only minimal effects even at high doses due to short half-lives of the growth factors in plasma, for example, platelet derived growth factor (PDGF) has a half-life of less than 2 min in plasma (Bowen-Pope *et al.*, 1984). Different carriers given below have been used for increasing half-life of growth factors.

7.2.1   Encapsulation inside matrices

Different kinds of polymers like chitosan, gelatin, fibrin glue, alginate hydrogels, polylactic acid (PLA), poly(lactide-co-glycolate) (PLGA), and polyethylene glycol (PEG) have been used for direct encapsulation of growth factors within polymer matrices which protect growth factors from enzymatic degradation and allow sustained release (Sheridan *et al.*, 2000; Obara *et al.*, 2003; Yamamoto *et al.*, 2003; Simmons *et al.*, 2004; Taylor *et al.*, 2004; Piantino *et al.*, 2006). The release profile of growth factors depends on the interaction with the matrix material, molecular weight, water holding capacity, and degradation profile of scaffold (Sheridan *et al.*, 2000; Yamamoto *et al.*, 2003), for example, acid treated gelatin having a positive charge has been found to bind strongly with negatively charged growth factor (Yamamoto *et al.*, 2001). In one study gelatin with different cross-linking and water holding capacities was used for BMP-2 delivery. Increasing cross-linking of gelatin reduced both water holding capacity and degradation of polymer but prolonged the growth factor delivery (Yamamoto *et al.*, 2003). Contradictory to that, gamma-irradiated fast degrading alginate hydrogels encapsulating dual growth factor bone morphogenetic protein-2 (BMP-2) and transforming growth factor (TGF β3) showed more osteogenesis compared to non-irradiated alginate by providing space for growing bone tissue (Simmons *et al.*, 2004). Dependence of growth factor release on scaffold structural properties made it a limiting approach for tuning growth factor delivery.

7.2.2   Immobilization approaches

As discussed above, growth factor delivery through polymer matrices depends on scaffolding material properties and leads to changes in scaffold physical properties on optimizing growth factor delivery. Another option is that the growth factors may be immobilized by adsorption, ionic interactions or covalent linkages (Lee *et al.*, 2003; Wang *et al.*, 2008a; Zeng *et al.*, 2010). Affinity based immobilization with biological molecules like heparin and chondroitin sulfate have shown high binding efficiency for growth factors, which helps in cell adhesion and improves the bioactivity of growth factors (Lee *et al.*, 2003; Steffens *et al.*, 2004; Davis *et al.*, 2011). Another promising approach for improving bioavailability of growth factor is covalent immobilization. Although covalent bond formation increases the bioavailability of the protein and provides sustained delivery through hydrolysis of covalent bond, the number of growth factor binding groups on the scaffold and blockage of some groups in receptor binding sites results in drawbacks with this method (Gomez and Schmidt, 2007; Masters, 2011). Many approaches have proved that protein bioactivity is not affected after covalent immobilization (Mann *et al.*, 2001; Gomez and Schmidt, 2007; Zeng *et al.*, 2010); however, it does depend on the type of bond and group included in the bond formation (Chiu *et al.*, 2001; Masters *et al.*, 2011).

7.2.3   Encapsulation inside microstructures

Natural polymers like chitosan and gelatin, and synthetics like PLA, PLGA, and polycaprolactone (PCL) have been explored for growth factor delivery from microstructures (Holland *et al.*, 2003; Basmanav *et al.*, 2008; Huang *et al.*, 2008; Park *et al.*, 2009c; Kokai *et al.*, 2010). Natural polymers like gelatin and chitosan mimic the extracellular matrix; for example, TGF-loaded chitosan microspheres inside chitosan scaffold showed sustained delivery for 5 days, helping in cartilage regeneration (Kim *et al.*, 2003). However, one must avoid the use of harsh chemicals in microsphere fabrication as they may reduce bioactivity of the growth factor (Wu *et al.*, 2011; Park *et al.*, 2009c). Similarly, gelatin microspheres have been used for encapsulating growth factors like TGF and fibroblast growth factor (FGF) (Holland *et al.*, 2003; Huang *et al.*, 2008). To avoid harsh organic solvents, researchers have used direct growth factor immobilization by ionic interaction or affinity based heparin/growth factor immobilization on prefabricated microspheres. Huang *et al.* (2008) formed FGF/heparin loaded gelatin microsphere-cell constructs which showed synergistic effects on wound healing in athymic mice.

Synthetic polymers have been explored extensively because of easy availability, adequate mechanical strength, and adjustable growth factor delivery by changing polymeric composition, molecular weight, and degradation profile (Dhandayuthapani *et al.*, 2011; Lee *et al.*, 2011). Microspheres of PLGA, a polyester of lactic acid and glycolic acid, have been extensively explored for growth factor encapsulation (Perets *et al.*, 2003; Park *et al.*, 2009b) because the release profile of growth factor can be changed by changing molecular weight and PLA/PGA ratio. The water/oil/water double emulsion method is mostly used for preparing growth factor-loaded microspheres in which growth factor, in combination with heparin or bovine serum albumin (BSA), is used as first water phase protecting growth factor from organic solvent. For example, FGF/heparin-encapsulating PLGA microspheres prepared by this method and entrapped inside alginate scaffolds maintained bioactivity of growth factor and showed enhanced angiogenesis compared to unloaded scaffolds in a mouse subcutaneous model. One major disadvantage, however, was that PLGA leads to highly acidic degradation end products (Perets *et al.*, 2003). Xu *et al.* (2003) formed nerve growth factor (NGF)-loaded polyphosphoester microspheres for nerve regeneration which gave less acidic end products as compared to PLGA (Xu *et al.*, 2003).

7.2.4   Encapsulation inside nanostructures

Advancement in nanotechnology has given opportunities for fabricating different nanostructures that can be used as controlled growth factor and drug delivery systems. These approaches are discussed below in detail.

7.3   Nanostructures for growth factor delivery in tissue engineering

Most tissue engineering approaches work on the basis that cells seeded or encapsulated on the three dimensional scaffold should behave structurally and functionally similar to native tissues. Most of the approaches discussed earlier provide structural similarity only at microscopic level and it has been proved that an ideal scaffold should have nanofeatures providing tissue specific microenvironment to maintain cellular behavior and function (Shi *et al.*, 2010). Rationale of using nanofeatured surfaces and nanoparticulated scaffolds for growth factor delivery is discussed below in detail.

7.3.1   Rationale of using nanostructures

The major advantage of using nanostructures for controlled delivery is that it not only provides controlled growth factor delivery but also mimics natural nanofeatures of the ECM. Nanofeatured ECM works as a template for cellular attachment and growth, provides signaling molecules in a spatio-temporal manner, regulating final cell phenotype, and also provides biological cues like heparin and collagen for efficient cell binding (Goldberg *et al.*, 2007; Shi *et al.*, 2010). Thus, for mimicking the *in vivo* ECM microenvironment, a scaffold should have nanocomponents, a nanotopographical surface, and should secrete signaling molecules. The second advantage of making nanostructured scaffold is improved material property after scaling down to nanostructures, such as higher surface roughness, higher surface energy, high surface area, and large area to volume ratio, which confers advantage over a microstructure and provides more focal adhesion points for secure binding of cells (Liu and Webster 2007; Zhang and Webster, 2009). Numerous top-down and bottom-up approaches for synthesis of nanostructures have been proposed (Zhang and Webster, 2009). In the following sections we will discuss briefly the approaches used for encapsulation of growth factors.

7.3.2   Single growth factor delivery systems

Two-dimensional (2D) nanotextured coatings

Nanotechnology can be used to design growth factor-loaded, nanotextured surfaces by different techniques such as immobilization of nanoparticles on implant surfaces, chemical treatment inducing formation of nanorough surfaces, layer-by-layer self-assembly. Titanium implants used for total bone replacement have good biocompatibility and mechanical strength but material inertness provides less osteointegration (Murugan and Ramakrishna, 2005). Various studies have coated titanium implants with growth factor loaded calcium phosphates using microthick coating to improve the osseointegration. Liu *et al.* (2005) coated titanium implants with BMP-2 loaded bioactive calcium phosphate (CP) particles which showed better ectopic bone formation in mice after 5 weeks of implantation as compared to controls without any BMP-2 and without coating (Liu *et al.*, 2005). In other approaches, growth factors were directly coated on titanium discs by development of nanofunctionalized surface. This was done by (1) plasma treatment and surface oxidation, which was further coupled with protein through an amino alkylsilane spacer molecule (Nanci *et al.*, 1998), and (2) plasma treatment introducing amine (-NH$_2$) groups on the implant surface through allylamine polymerization on titanium discs and further 1-ethyl-3-[3-dimethylaminopropyl] carbodiimide hydrochloride (EDC)-induced coupling of proteins with amine group (Puleo *et al.*, 2002). These nanorough surfaces maintained biological activity of coupled bone morphogenetic protein-4 (BMP-4) for 3 days. Recently Steinmuller-Nethla *et al.* (2006) introduced diamond nanoparticles for BMP-2 physicoadsorption on a titanium implant. Polycrystalline diamond nanoparticles were deposited with BMP-2 after nanoparticle oxygen treatment which introduced reactive oxygen and hydrogen groups providing high binding efficiency. Although they have good adsorption, further research should be undertaken exploring their biocompatibility and efficacy.

Layer-by-layer assembly is another promising method where growth factor delivery can be tuned by sandwiching growth factors between different oppositely charged nanolayers and also by changing the number of layers. For example, in one dual delivery approach, PCL/β-tricalcium phosphate scaffold was coated with poly (β-aminoester)-2 with further coating of poly (acrylic acid) (high negative charge) in case of recombinant BMP-2 and chondroitin sulfate (less negative) for vascular endothelial growth factor (VEGF). Alternate tetralayer coating of growth factors with oppositely charged polymer gave sustained delivery of BMP-2 for 2 weeks and VEGF for 7 days (Shah *et al.*, 2011).

3D nanostructured scaffolds

Nanostructures can themselves form 3D scaffolds like electrospun scaffolds or can be embedded inside a 3D scaffold. Different kinds of materials such as ceramics, proteins, synthetic or natural polymers, and their composites have been used for encapsulating growth factors inside nanostructures. These are discussed in the following sections.

7.4   Nanofibers

Nanofibers are long thread-like structures which have at least one dimension in nanometers. The major advantage of nanofibers is that they form long thread-like structures that can interact with each other to provide three-dimensional scaffolds with high surface area for cell adhesion. Different strategies for nanofiber formation in tissue engineering are mentioned below in detail.

7.4.1   Nanofibers formed by electrospinning

Electrospinning is the most common technique used for designing nano-fibrous structures. Simple set-up provides control over fiber morphology and diameter by changing different parameters like voltage, solvent, and polymer composition (Briggs and Arinzeh, 2011). Different methodologies like pre-blending of growth factors with polymer solution before electro-spinning, coaxial electrospinning forming core-shell electrospun fibers, and post-adsorption and post immobilization approaches have been explored for incorporating growth factors into nanofibers. These approaches are also presented schematically in Fig. 7.1.

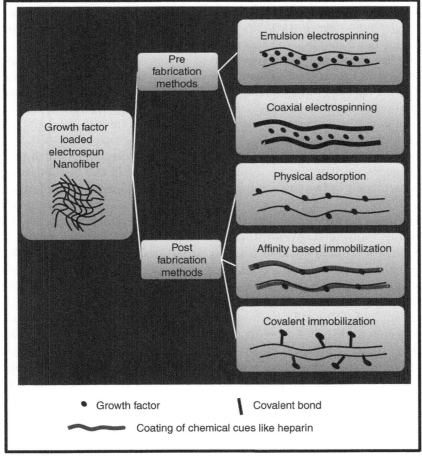

7.1 Different strategies for growth factor encapsulation inside nanofibers formed by electrospinning.

In the blending prefabrication method, growth factors are emulsified in polymer solution or composite solution before the electrospinning process. Encapsulation and the release profile of the growth factor depend on the type of polymer, polymer homogeneity, and its degradation profile, which can be easily optimized to obtain the desired release rate (Briggs and Arinzeh, 2011). Many growth factors like BMP-2, TGF, FGF, and PDGF have been successfully encapsulated inside different polymers and their composites to provide sustained delivery. Exposure of growth factor to harsh chemicals and techniques like sonication and stirring causes denaturation of protein. Hence, different techniques like addition of BSA, fetal calf serum, and cyclo-dextrin with growth factor solution (Sahoo et al., 2010b; Schofer et al., 2011b; Yang et al., 2011), and addition of hydrophilic components like collagen, PEG, and nanohydroxyapatite (nHA) have been explored to enhance the bioactivity of proteins (Lia et al., 2006; Fu et al., 2008; Schofer et al., 2011a). Fu et al. (2008) formed BMP-2/poly-L-lactide (PLLA)/nanohydroxyapatite scaffolds fabricated by a blending process showing sustained release over 2 months and better bone regeneration compared to unloaded PLLA/ nHA and PLLA scaffold. However, direct adsorption of the same amount of BMP-2 on PLLA/nHA nanofibrous scaffolds showed higher early bone formation compared to a BMP-2/PLLA/nHA blended scaffold, proving the role of initial high concentration of BMP-2 released from the nanofibrous scaffold surface (Fu et al., 2008).

Another prefabrication method is coaxial electrospinning which prevents exposure of growth factors to organic solvents. In this method, solutions are filled in two different diameter cylinders, one within the other, which are electrospun simultaneously to form core-shell nanofibrous structures. The final nanofiber structure depends upon flow rate of the solutions, polymer concentration, molecular weight, and miscibility of core-sheath solution after electrospinning (Briggs and Arinzeh, 2011). For example, in coaxial electrospinning to make PCL-r-fitcBSA (fluorescein isothiocyanate-conjugated bovine serum albumin)/PEG nanofibers, the diameter of nanofibers can be increased by changing the flow rate of the inner fitcBSA/PEG solution from 0.2 to 0.6 mL/h. This process allowed homogeneous distribution of the fitcBSA/PEG core solution, rather than making beads inside PCL fibers as in the blending process (Zhang et al., 2006). Encapsulation and release of growth factors from core-shell nanofibers depends upon flow rate of solutions, polymer-growth factor interaction, degradation and molecular weight of polymer, porosity, and swelling of polymeric shell. Liao et al. (2006) developed core-shell nanofibers of PDGF/PCL/PEG with growth factor inside the fiber core. A specific feed ratio of 3:1 (polymer:growth factor) gave maximum loading of growth factor by minimizing losses during electrospinning. The presence of PEG in the polymeric shell created pores after degradation that enhanced growth factor release, proving the roles of porosity and polymer degradation on release profile. All the growth factor was released within 40 days from PCL-PEG (PEG molecular weight = 3400 kD) compared to PCL alone which showed very little release (Liao *et al.*, 2006).

Although prefabrication methods gave good encapsulation and sustained release of growth factors, maintaining growth factor bioactivity and optimization of flow rate in coaxial electrospinning are big challenges (Ji *et al.*, 2010; Briggs and Arinzeh, 2011). In post fabrication methods, growth factor can be directly adsorbed or immobilized on nanofibrous scaffolds. Affinity based immobilization and covalent immobilization are more promising compared to direct adsorption because of diffusion-based burst release by direct adsorption. As heparin or heparin sulfate have good binding efficiency and protect growth factors from enzymatic degradation (Nissen *et al.*, 1999), nanofibers coated with them are promising. For example, in one study, collagen and gelatin nanofibrous scaffolds were coated with either biotinylated heparin or the perlecan domain I PlnDI segment of heparin sulfate glycosaminoglycan found in natural cartilage. These biotinylated-PlnDI-collagen and -gelatin membranes showed ten times higher binding of β-FGF2 compared to BSA- or heparin-coated gelatin and collagen nanofibers formed by affinity based immobilization without biotinylation (Casper *et al.*, 2007).

In the covalent immobilization method, growth factor forms direct chemical bonds with the material functional group. Most work in this field has been done on immobilization of growth factor on amine terminated PEG/PCL block copolymer (Choi *et al.*, 2008; Cho *et al.*, 2010). In this method, the carboxylic group of an aqueous protein forms a direct covalent bond with the amine group of PEG co-polymeric fibrous structure via EDC coupling reaction. These covalently conjugated growth factors show more sustained release compared to directly absorbed growth factor on fibrous mats (Cho *et al.*, 2010). One other covalent immobilization method involves coupling of growth factor with carboxy functionalized nanofibrous scaffolds. In a study by Lee *et al.* (2012), NGF was covalently immobilized on carboxy-pyrrole functionalized PLGA nanofibrous scaffolds for nerve tissue engineering. Pyrrole provides electrical conductivity that helps in neurite growth and extension. Although coupling of NGF reduces conductivity due to a decrease in surface functional groups, NGF-coated pyrrole-functionalized PLGA fibers showed more neurite extension and neurite outgrowth in PC 12 cells compared to control scaffolds without growth factors. This study suggested that the fiber surface properties also plays a role in tissue regeneration in addition to growth factors.

7.4.2   Nanofibers formed by phase separation method

Growth factor interactions with electrospun scaffolds can change material degradability, mechanical strength, and surface properties (Briggs and Arinzeh, 2011). Electrospinning also has the disadvantage of limited pore size which may cause poor cell infiltration (Hu and Ma, 2011). 3D macroporous nanofibrous scaffolds formed by phase separation and subsequent leaching have overcome these disadvantages because this method provides a 3D macroporous scaffold giving good cell filtration. Growth factor-loaded nanoparticles can also be immobilized on 3D macroporous, nanofibrous scaffolds instead of encapsulating growth factors directly in nanofibrous scaffold which eliminates chances of change in the physical properties of the scaffold while optimizing growth factor encapsulation. This process gives tunable growth factor delivery by changing particle properties like molecular weight, composition, growth factor to material ratio, and, finally, by varying the amount of particles on the scaffold surface. These nanofibrous scaffolds can be prepared by freezing polymeric solution at −20°C inside a sugar microsphere template, subsequent solvent exchange, removal, sugar leaching, and freeze drying. These nanofibrous macroporous scaffolds can be post-seeded with growth factor-loaded micro- and nanospheres for controlled delivery applications (Wei *et al.*, 2006, 2007; Jin *et al.*, 2008). In work done by Wei *et al.* (2006), PDGF was incorporated inside PLGA microspheres prepared by the double emulsion method, and PDGF-PLGA microspheres were immobilized evenly on PLLA nanofibrous macroporous scaffold. Different molecular weights of PLGA and different initial amounts of PDGF were used to show effects on release property, and it was found that low molecular weight (LMW) microspheres gave PDGF a higher burst release compared to high molecular weight (HMW) microspheres. Although, after the initial burst, LMW microspheres gave PDGF sustained release for 40 days, HMW microspheres started releasing PDGF at a higher rate after 35 days because of enhanced microsphere degradation.

7.4.3   Nanofibers formed by self-assembly

Self-assembly is the process whereby small molecules and macromolecules arrange themselves in a supramolecular design by different non-covalent interactions such as ionic bonds, hydrogen bonds, or hydrophobic interactions giving a nanofibrous structure. Injection of these molecules forms a 3D supramolecular, self-assembled nanofibrous gel network by ionic and pH triggers at physiological conditions. These gels have non-toxic breakdown products and do not require harsh environments for preparation making them suitable carriers for protein and cell delivery (Segers and Lee, 2007; Matson and Stupp, 2012).

Normal self-assembled peptides have alternative hydrophilic and hydrophobic parts with positively charged and negatively charged hydrophilic residues. Ionic interactions among these hydrophilic residues form β sheets, hairpin loop-like structures, micelles, and cylindrical structures in aqueous solution, based on peptide sequences and can be packaged into nanofibrous hydrogel networks by different triggers (Segers and Lee, 2007). For example, in one study, self-assembling peptide solution with PDGF was injected in a rat myocardial infarction model and showed sustained release of PDGF for 14 days compared to free growth factor released within 3 days (Hsieh *et al.*, 2006). Another method of enhancing growth factor affinity is tethering of biotinylated growth factors with biotinylated nanofibers using streptavidin as cross-linker. Using this approach biotinylated insulin-like growth factor (IGF) immobilized on nanofibers formed a gel after injection in the rat myocardium model and was used to treat infarction with sustained release of IGF for 28 days (Davis *et al.*, 2006). However, some other approaches have found a reduction in growth factor bioactivity after tethering via biotinylation (Kopesky *et al.*, 2011).

Another class of self-assembled fibers are the peptide amphiphilic (PA) molecules formed by fluorenylmethoxycarbonyl based chemistry where a large alkyl chain is attached to the N terminus of a peptide sequence. These amphiphilic molecules form self-assembled structures like micelles, nanotubes, nanospheres, or cylindrical structures in aqueous solution at a particular pH, in the presence of ions, or with the addition of growth factors (Matson and Stupp, 2012). The negative charge of the peptides at physiological pH causes electrorepulsion, but low pH and addition of cations or growth factors provides attractive forces for self-assembly with consequent hydrophobic interaction of alkyl chains (Hartgerink *et al.*, 2002). PA molecules have been used for delivering many growth factors like BMP-2, FGF2, and hepatocyte growth factor (HGF) (Hosseinkhani *et al.*, 2006a, 2006b, 2007). In a study by Hosseinkhani *et al.* (2006a), FGF2/PA solution was injected in a rat subcutaneous model for successful angiogenesis. *In vivo* sustained release for 20 days from degrading fibers (24 days) helped more in angiogenesis compared with free growth factor and PA solution only (Hosseinkhani *et al.*, 2006a). Furthermore, some authors have introduced epitope molecules into the chemical structure of PA fibers like a heparin binding peptide sequence at the time of peptide fabrication for efficient binding with growth factors and enhanced cell adhesion (Stendahl *et al.*, 2008).

Another novel approach is development of fusion proteins reducing enzymatic degradation of growth factors. In this approach, instead of the entire growth factor, cell receptor binding peptide sequences have been covalently attached to PA. After self-assembly, this surface exposed peptide does not lose bioactivity. Lee *et al.* (2009) formed a fusion protein of the osteopromotive domain (OPD) of BMP-2 with PA head group molecules and self-assembled the protein nanofibrous structure. Nanofibrous gel formed by these fusion proteins helped in osteogenesis and showed higher expression of Smad group proteins compared to PA molecules without the OPD peptide sequence (Lee *et al.*, 2009). Scaffolds have also been designed for cell triggered release of growth factors on demand. In a study by Sakiyama-Elbert *et al.* (2001), fusion protein of NGF was covalently immobilized on fibrin matrices for skin regeneration via transglutaminase plasmin cross-linker (TG-P-protein) that functions as substrate for plasmin. So, the final release of growth factor is actually triggered by cellular plasmin enzyme activity that works on fibrin degradation, releasing fusion protein, giving spatiotemporal release according to the requirement of growing cells (Sakiyama-Elbert *et al.*, 2001). Sustained delivery of growth factors from degradable natural molecules made them beneficial for tissue regeneration; however, the potential inflammatory and immune responses need to be evaluated *in vivo*.

7.5   Nanoparticles

Nanoparticles formed of natural and synthetic polymers, proteins, lipids, and ceramic particles have shown promising results for controlled growth factor delivery in different applications (Zhang and Uludağ, 2009), but polymers and proteins are mostly used in tissue engineering approaches from scaffolds. Common strategies used to encapsulate growth factor inside nanoparticles are presented schematically in Fig. 7.2.

7.5.1   Lipid based nanoparticles

The most common lipid based particles are liposomes, solid lipid particles, and lipid nanocapsules. Liposomes are nanostructures having an aqueous core surrounded by amphiphilic phospholipids arranged in bilayer configuration. These particles have advantages in terms of encapsulating hydrophobic drugs inside the lipid bilayer and hydrophilic moieties inside the aqueous core. It has been reported in many studies that liposomes can achieve very high encapsulation rates of hydrophobic drugs inside the lipid bilayer. Entrapment efficiency of hydrophobic drugs is basically affected by lipid-drug interactions. Entrapment of hydrophilic moieties, like growth factor and peptides, inside the aqueous core depends upon the capability of liposomes to entrap the hydrophilic drug-containing aqueous core inside the lipid bilayer (Chonn and Cullis, 1995). The second category of particle is solid lipid nanoparticles which are colloidal in nature and have a solid lipid core stabilized by a layer of emulsifying agent for particle stabilization. Although, these particles have many advantages such as biocompatible composition, and easy and fast preparation methods (easy to scale up for larger production) without any involvement of organic solvents retaining

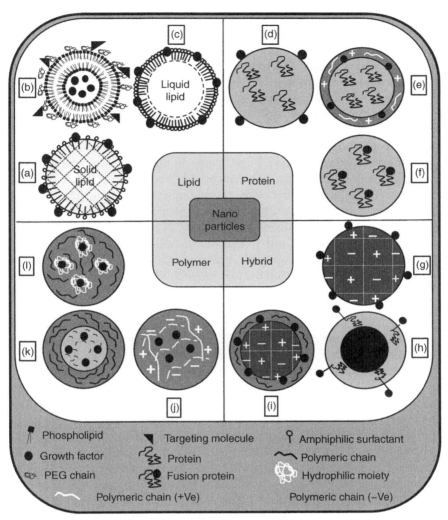

7.2 Schematic representation of different kinds of growth factor-loaded nanoparticles: (a) solid lipid nanoparticles; (b) liposome; (c) lipid nanocapsule; (d) protein based nanoparticle; (e) polymer-coated protein based nanoparticles; (f) nanoparticle of growth factor fusion protein; (g) ceramic nanoparticle; (h) protein coated magnetic nanoparticle; (i) polymer coated ceramic nanoparticles; (j) Ionically cross-linked polymeric nanoparticle; (k) polymeric nanocapsule; (l) polymeric nanosphere.

growth factor bioactivity, still they have issues of lower encapsulation efficiency because of their compact hydrophobic lipid core which can partition growth factor on its surface and lead to a fast release profile (Martins *et al.*, 2007).

Although lots of work has been done for delivering peptides and growth factor from lipid based particles through both oral and parenteral routes, no literature has reported a lipid nanoparticle based scaffolding system for tissue regeneration, and most of the work for growth factor delivery has been done on liposomes through a systemic route. The main advantage with lipid based particles is their biocompatibility by mimicking the cell membrane structure. Most of the phospholipids are surfactants and are primary constituents of the biological membrane. When given through the oral route, peptide transport through liposomes is basically dependent upon the mucoadhesiveness of liposomes, the internalization process, and the ability to enhance particle permeation though the gut wall (Martins *et al.*, 2007). Coating polymers like chitosan and polyethylene glycol (PEG) onto liposomes can enhance mucoadhesiveness and circulation in plasma, respectively. For example, in one study, PEG-coated dipalmitoylphosphatidylcholine (DPPC) and phosphatidylcholine (PC) liposomes were prepared to deliver recombinant epidermal growth factor for treatment of gastric ulcer. Although, epidermal growth factor (EGF) encapsulation efficiency was very low (5–15%), PEG-coated DPPC liposomes protected EGF through enzymatic degradation, enhanced its permeability and influx in $CaCO_3$ cells, and gave EGF higher bioavailability in rat plasma (through oral administration) compared to PEG-coated PC liposomes without EGF and with EGF in free solution (Li *et al.*, 2003).

As far as parenteral routes like intravenous injection are concerned, nanoparticles should have good cellular internalization properties, should escape the reticuloendothelial (RET) system, and should also be able to deliver drugs at a specific site. The internalization property of liposomes can be optimized by changing lipid composition and particle size, and coating with chemically modified natural compounds like PEG (formation of stealth liposomes) can help in avoiding RET systems (Martins *et al.*, 2007). Also, these coatings provide groups for further immobilization of targeting moieties like antibodies (Martin and Papahadjopoulos 1982; Scott *et al.*, 2009). In one study done by Li *et al.* (2008), hepatocyte growth factor was encapsulated inside sterically stabilized liposomes having arginine-glycine-aspartate (RGD) surface conjugating moiety. Cyclic RGD peptides (the targeting moiety for liver fibrosis) were combined with maleimide-[poly(ethylene glycol)]-1,2-dioleoyl-sn-glycero-3-phosphoethanolamine (MAL-PEG-DOPE) group of sterically stabilized liposome. These particles with particle size of 91 nm and encapsulation efficiency of 32% helped in treating liver cirrhosis through intraperitoneal injection and showed higher collagen fiber digestion, less collagen production, and higher apoptosis of α-smooth muscle actin (α-SMA)-positive cells compared to other controls (Li *et al.*, 2008). In a similar study, NGF was encapsulated inside liposomes (having composition of Soya PC, cholesterol (CHOL), and 1,2-distearoyl-sn-glycero-3-phosphoethanolamine (DSPE)-PEG). These liposomes were also surface conjugated with RMP-7 moiety to target the blood brain barrier (Xie *et al.*, 2005). In another different approach, BMP-2 was encapsulated inside liposomes in the presence of magnetic particles. This injectable system showed localized deposition of these magnetic liposomes and improved bone regeneration in a critical sized rat femoral defect model when a small magnet was implanted at the defect site (Matsuo *et al.*, 2003). Most work done with lipid particles has shown low encapsulation efficiency of proteins inside liposomes. Also, liposomes given by the systemic route have many disadvantages such as aggregation due to chemical and physical instability and enzymatic degradation of liposomes, finally exposing growth factors for fast release (Martins *et al.*, 2007; Zhang and Uludağ, 2009). To overcome this problem, some researchers coated lipid particles with a polymeric shell to prevent growth factor exposure to enzymes and also for controlled release of growth factor (Oh *et al.*, 2006; Haidar *et al.*, 2009).

7.5.2   Polymeric nanoparticles

Different kinds of polymeric particles like nanospheres and nanocapsules made of synthetic polymers like polyesters (PLGA, PLA, PGA, and poly(hydroxybutyrate-co-valerate) (PHBV)), poly(alkyl cyanoacrylate), polyanhydrides, and block copolymers of different categories have been explored extensively for encapsulating growth factors. These synthetic nanoparticles have potential advantages like tuned release based on molecular weight, polymeric concentration, addition of additives, changing composition, and easy processing. However, direct contact of protein with organic solvents, high temperature techniques for nanoparticle synthesis, and toxic degradation products are major challenges for preserving growth factor bioactivity (Zhang and Uludağ, 2009). PLGA has been extensively used because of its copolymeric nature and availability in different molecular weights. In one study, rhBMP-7 was dissolved in gelatin and BSA solution before emulsification for encapsulation inside PLGA nanospheres made by a double emulsion/solvent evaporation method (Wei *et al.*, 2007). These PLGA nanoparticles with various polymeric molecular weights were immobilized on PLLA nanofibrous scaffolds to optimize release profile. PLGA nanoparticles with LMW gave initial 48% burst release compared to 23% burst release with HMW nanoparticles and further release was sustained for 7–8 weeks with HMW nanoparticles showing molecular weight, and degradation profile play a pivotal role in tuning protein release profile.

Another synthetic polymer is poly(*N*-isopropylacrylamide) (PNIPAAM) which forms nanoparticles by free radical polymerization of N,N-methylene bisacrylamide and N-isopropylacrylamide monomer at 65°C. Hence, growth factor can be adsorbed only after synthesis in order to prevent denaturation of protein (Ertan *et al.*, 2013). In this study by Ertan *et al.* (2013), PLGA and PNIPAAM nanoparticles loaded with IGF and TGF, respectively, showed

85% encapsulation with PLGA nanoparticles because of encapsulation inside the nanosphere and only 15% encapsulation with PNIPAAM nanoparticles because of direct adsorption following nanoparticle fabrication.

Although synthetic proteins have many advantages discussed above, nanoparticles made from natural polymers like collagen, chitosan, and alginate are biocompatible and do not cause exposure to harmful chemicals (Zhang and Uludağ, 2009). Such polymers as chitosan and alginate form growth factor-encapsulating nanoparticles by charge interaction. Encapsulation and release profile of growth factor depends upon the intensity of ionic interactions between polymer and growth factor. For example, positively charged chitosan nanoparticles loaded with negatively charged hepatocyte growth factor (HGF) showed 85% HGF encapsulation because of strong ionic interactions between them and helped in hepatocytic differentiation of bone marrow mesenchymal stem cells with sustained release over periods of 31 days (Pulavendran, 2010). On the other hand, negatively charged alginate nanohydrogels encapsulating BMP-2, formed by reverse emulsification and evaporation, helped in osteoblastic differentiation of bone marrow stem cells (bMSC) but released protein within 7 days (67% encapsulation efficiency) with 40% burst release within 1 day, possibly due to less interaction between alginate and BMP-2 (Lim *et al.*, 2010). In another study by Binsalamah *et al.* (2011), alginate nanoparticles encapsulating placental growth factor (PLGF) was formed by cross-linking of chitosan solution for treating myocardial infarction. Although these particles showed less than 38% encapsulation of PLGF and release within 6 days, initial burst release was reduced to 20% (20% burst release within 24 h) and nanoparticles showed enhanced angiogenesis in a rat myocardial infarction model.

7.5.3   Protein and peptide nanoparticles

Proteins like collagen, albumin, fibrin, and elastin are potential candidates for growth factor delivery. Bovine serum albumin is an inexpensive and readily available protein that forms nanoparticles by coacervation but the use of glutaraldehyde as a cross-linker can cause toxicity (Segura *et al.*, 2005). In another approach to reduce burst release by direct adsorption of protein, albumin nanoparticles were coated with PLL and polyethyleneimine (PEI) aqueous solution to control BMP-2 growth factor release. These BMP-2 loaded particles showed 90% encapsulation efficiency and helped with *in vitro* bone regeneration (Wang *et al.*, 2008b; Zhang *et al.*, 2008).

Elastin is another protein which is the main constituent of dermal tissue and muscles. Elastin peptides have pentapeptide sequences [Val-Pro-Gly-X-Gly] where X can be any amino acid except proline and its aqueous solution forms nanoparticles by self-assembly above its inverse transition temperature (~30°C) by hydrophobic interactions (Bessa *et al.*, 2010). In a study by Koria *et al.* (2011), fusion protein of elastin-like peptides (ELP) with recombinant keratinocyte growth factor (KGF) was formed for chronic wound healing by coupling KGF at N terminus of ELP. The study showed advantages because the combined properties of KGF and ELP helped in epidermal and dermal growth, respectively (Koria *et al.*, 2011). Although bioactivity of these fused proteins was retained when these nanoparticles were embedded in fibrin matrices and used for wound healing in a mouse model, fused protein showed less keratinocyte growth and less granulation as compared to free KGF and ELP. So, new methodologies should be used in future to conserve the bioactivity of these proteins.

A new affinity based approach has been developed where different peptide sequences can be screened based on affinity with growth factors, and those peptides with the highest affinity can be used for growth factor encapsulation. In a study by Willerth *et al.* (2007), different peptide sequences taken from a phage library were screened for affinity with NGF by conjugated resin chromatograph. Fractions eluted at pH 4.5 showed highest levels of binding with NGF. Furthermore, this fraction showed more sustained release of NGF inside the fibrin matrix compared to other fractions (Willerth *et al.*, 2007).

Much research has been conducted into ionic nanoparticles of polymer and proteins which were formed by self-assembly of oppositely charged molecules. Most often heparin is used as a negatively charged counterpart because of growth factor binding sites present on its surface (Nissen *et al.*, 1999). Interaction of heparin/growth factor solution with a solution having positively charged groups like PLL, PEI, and chitosan forms polyionic, self-assembled nanoparticles with the potential advantages of high encapsulation efficiency, controlled growth factor release (20–30 days), and preservation of growth factor bioactivity (Park *et al.*, 2008, 2009a, 2009e, 2009f; Tan *et al.*, 2011). In a study by Park *et al.* (2008), TGF β-3-loaded heparin/PLL nanoparticles helped in chondrogenesis of bMSC cells and formed hyaline-like cartilage in animal models with controlled growth factor delivery for 30 days. Similarly, these FGF loaded heparin/PLL nanoparticles immobilized on PEI-coated PLGA microspheres gave better neurite extension of PC12 cells compared to microspheres without growth factor suggesting retention of bioactivity (Park *et al.*, 2009e).

7.5.4   Ceramic, magnetic and hybrid nanoparticles

Ceramic nanoparticles like nanocrystalline calcium sulfate have high surface area and positive charge making them beneficial for protein adsorption on their surface (Park *et al.*, 2011). In one strategy, aluminosilicate clay montmorillonite (MMT) was used to adsorb EGF between clay nanosheets to help in wound healing (EGF exchange with ions present between nanosheets). These silicate nanosheets showed toxicity at higher concentration; EGF releasing from lower MMT concentration helped in wound closure with human spontaneously transformed keratinocyte cell line (HaCaT cells) (Vaiana *et al.*, 2011). Different strategies have been explored where either protein-adsorbed nanoparticles form a nanocomposite with polymers or protein can be adsorbed on to the surface of nanoparticle-polymer nanocomposite to reduce chances of protein desorption from the nanoparticle surface. In one example of the first strategy, tunable delivery of BMP-2 was obtained by entrapping BMP-2 adsorbed nHA particles within PLGA microspheres of different molecular weight. These systems, using PLGA microspheres with 80 kD molecular weight, showed sustained delivery of BMP-2 for 7 weeks compared to collagen sponges for 3 weeks. Dissolution of basic calcium ions simultaneously with acidic degradation of PLGA reduced toxicity by balancing solution pH (Pitukmanorom *et al.*, 2008). In the second strategy, protein adsorbed on nanocomposite surface will exhibit a burst release profile but this strategy has the advantage of protecting the growth factor from harsh environments caused during nanocomposite fabrication (Duan and Wang, 2010). For example, in one approach, BMP-2 was adsorbed directly on silk fibroin/nHA nanocomposite scaffold (55% adsorption) and 7 days of protein release was observed with an initial burst of 48% within 1 day (Zhang *et al.*, 2011).

In a recent development, magnetic Fe_2O_3 nanoparticles conjugated with FGF were synthesized for differentiation of nasal olfactory mucosa cells. Growth factor was conjugated on the surface of gelatin-coated particles by Michael reaction. These FGF nanoparticles adsorbed on chitosan microcarriers showed higher migration and proliferation of olfactory cells inside hyaluronate based gels compared to an unloaded control, proving the role of FGF nanoparticles in skin regeneration (Skaat *et al.*, 2011).

7.6   Strategies for dual growth factor, drug and gene delivery

7.6.1   Dual growth factor delivery system

Dual delivery of two growth factors, growth factor along with drug, or growth factor with cells is more promising compared to single delivery systems as multiple cues can be provided at different phases of the tissue regeneration like natural ECM. For example, bone tissue secretes many different growth factors with distinct functionality like BMP, TGF, VEGF, PDGF, and angiopoietins at different stages of bone healing (Dimitriou *et al.*, 2005). Growth factor release profiles can be optimized by changing particle formulation and composition giving spatiotemporal release in a time dependent manner (Shi *et al.*, 2010). Schematically, all dual growth factor delivery approaches are shown in Fig. 7.3. Richardson *et al.* (2001) introduced for the first time

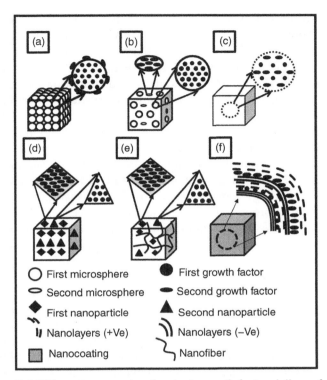

7.3 Different approaches for dual growth factor delivery from scaffolds: (a) one growth factor encapsulated inside microsphere and second immobilized on surface of microsphere; (b) different growth factors in different kinds of microspheres embedded inside porous matrix; (c) both growth factors embedded directly inside porous matrix; (d) different growth factors encapsulated inside different kinds of nanoparticles embedded inside porous matrix; (e) different growth factors encapsulated in different kinds of nanoparticles embedded inside nanofibrous matrix; (f) growth factors entrapped in oppositely charged layer-by-layer assembly of polymers coated on matrix surface.

VEGF and PDGF releasing dual growth factor delivery from PLGA microspheres. VEGF and PDGF are required in early and late phases of bone healing, respectively, hence the release profile of growth factors was optimized by encapsulating PDGF inside PLGA microspheres before fabrication and VEGF was adsorbed on PLGA microspheres after scaffold fabrication. This allowed a higher amount of VEGF to be released initially which helped endothelial cell recruitment and PDGF to be released at later phases for mature blood vessel formation (Richardson *et al.*, 2001).

Other than matrices and microstructures, recent advances in nanotechnology have provided different nanoparticles which can be modified for distinct release profiles. In a study by Ertan *et al.* (2013), PLGA and PNIPAAM nanoparticles were used to deliver IGF and TGF, respectively, for cartilage regeneration. First TGF delivery from PNIPAAM nanoparticles (TGF adsorbed on nanoparticle surface giving a burst release) helped in cell proliferation and then long-term delivery of IGF from PLGA nanoparticles (encapsulated inside the PLGA matrix giving a degradation-based release profile) caused cellular differentiation acting synergistically towards effective chondrogenesis (Ertan *et al.*, 2013).

The synergistic effect of a dual delivery system depends upon its release pattern – simultaneous or sequential. Sequential delivery of growth factors is more promising compared to simultaneous delivery. In one approach, BMP-2 and BMP-7 were encapsulated inside PLGA (fast degrading) and PHBV (slow degrading) nanocapsules and finally incorporated inside chitosan-PEG nanofibrous scaffolds. Sequential delivery of BMP-2 and BMP-7 (BMP-2 inside PLGA nanocapsule with fast 100% release and BMP-7 inside PHBV nanocapsule showed 50% sustained release for 21 days) showed synergistic effects in cellular differentiation compared to simultaneous release of both growth factors (Yilgor *et al.*, 2009).

Another more promising approach for increasing the efficacy of the growth factor delivery system is growth factor delivery with cells, mostly pluripotent stem cells inside the scaffold. Pluripotent stem cells can differentiate into any cell lineage based on tissue specific biochemical signaling. Scaffold properties such as hydrophilic environment, biocompatibility, degradability, and processing temperature are the key factors finalizing efficacy of the cell laden scaffold (Vo *et al.*, 2012). The most widely used approach is encapsulating cells inside a degradable, injectable hydrogel system which provides a biocompatible, porous environment, and which, upon degradation, can provide space for proliferating cells (Simmons *et al.*, 2004). Controlled delivery of growth factors helps in differentiation of encapsulated cells and further secretion of biochemical molecules from these cells helps in recruitment of host cells in animal study. Cardiomyocyte injection with growth factor (IGF)-loaded, self-assembled peptide hydrogels in a myocardial infarction model have shown the significance of co-delivered cells: growth factor-loaded peptide hydrogel did not show any improvement while the addition of cardiomyocyte to the hydrogel showed significant improvements (Davis *et al.*, 2006).

7.6.2   Nanostructures for drug delivery in tissue engineering

Tissue injury and invasive implantation cause inflammation and infection at the injury site, so controlled drug delivery from a nanostructured scaffold may provide therapeutic amounts of drug at the infectious site with reduction of side effects caused by systemic doses. Additionally, some drugs like statins and dexamethasone have bone inducing properties. These drugs may be encapsulated inside electrospun nanofibers and nanoparticles embedded in scaffolds for controlled drug delivery (Piskin *et al.*, 2008; Oliveira *et al.*, 2010). Encapsulation and the release profile of drugs inside nanostructures is affected by (1) material and drug mass ratio, (2) distribution profile of drug inside material, (3) interaction between drug and material (hydrophobic drug has more affinity for hydrophobic molecules and hydrophilic drug for hydrophilic ones), (4) degradation profile of material, and (5) exposed surface area of the scaffold (Kim *et al.*, 2004). Studies done by different authors for drug/nanoparticle formulation are listed in Table 7.1.

7.6.3 Nanostructures for gene delivery in tissue engineering

Although viral gene delivery provides high transduction and expression efficiency, there is a risk of immunogenic complications. On the other hand, non-viral approaches like lipofection and cationic nanoparticles, while less immunogenic, have low transfection efficiency. The most common commercially available non-viral system is lipid based Lipofectamine which has been shown to have a high toxicity problem in clinical trials; hence, a future demand is anticipated for non-viral transfection approaches using nanoparticles with decreased immunogenicity and toxicity compared to Lipofectamine (Partridge and Oreffo, 2004). In essence, nanoparticle size, uniformity, and scaffold degradability are the most important factors affecting gene transfection; toxicity, degradability, and bioactivity are other important factors affecting intracellular gene expression (Panyama and Labhasetwar, 2003). Different non-viral gene delivery approaches using different nanoparticles for tissue engineering application are summarized in Table 7.2.

7.7 Clinical prospective of nanostructures with growth factor delivery in tissue engineering

The clinical status of different nanostructured growth factor-releasing scaffolds in different tissue engineering application is discussed below with a summary in Table 7.3.

7.7.1 Bone and cartilage regeneration

The ideal scaffold for musculoskeleton regeneration should have an interconnected porous structure and should be biodegradable, biocompatible, mechanically strong, and release signaling molecules for successful regeneration (Hutmacher, 2000). BMP-2 adsorbed on absorbable collagen sponge

Table 7.1 Nanostructures with drug delivery in tissue engineering

Drug	Nanostructures/Scaffold	Key observation	References
Dexamethasone	PCL nanofibrous scaffold	*In vitro* controlled release for 15 days	Martins *et al.*, 2010
Dexamethasone	CMCht/PAMAM dendrimer nanoparticles with HA scaffold	*In vitro* bone regeneration with hMSC cells *In vitro* bone differentiation with rat bMSC	Oliveira *et al.*, 2009, 2010
Dexamethasone and BMP-2	PLLACL/collagen nanofibers	*In vivo* ectopic bone formation in rat *In vitro* controlled release for 22 days Tunable release by changing drug position from shell to core through coaxial nanofibers	Su *et al.*, 2012
Dexamethasone	PCL nanofibrous scaffold	*In vitro* bone regeneration with hMSC cells *In vivo* rat vascular grafting Formation of intima without hyperplasia	Nottelet *et al.*, 2007
Mefoxin	PLGA/PEG-b-PLA /PLA	*In vitro* release from 1 to 7 days *In vitro* anti-microbial activity	Kim *et al.*, 2004
Gentamicin	Chitosan nanoparticle and Quaternary Chitosan NP inside PMMA cement	*In vitro* anti-microbial activity *In vitro* biocompatibility with 3T3 fibroblast cells	Shi *et al.*, 2006
Simvastatin	PLGA nanofibrous scaffold	*In vitro* controlled release for 28 days *In vitro* bone regeneration with bMSC cells *In vivo* healing in rat ectopic model	Wadagaki *et al.*, 2011
Simvastatin	PCL nanofibrous scaffold	*In vivo* bone healing in rat cranial defect model	Piskin *et al.*, 2009
Simvastatin	LA oligomer grafted gelatin micelles inside gelatin hydrogel	*In vitro* controlled release for 20 days *In vitro* differentiation of MC3T3-E1 bone cells and *in vivo* healing in rabbit tooth extraction model	Tanigo *et al.*, 2010

CMCht/PAMAM, carboxymethylchitosan/poly(amidoamine); hMSC, human mesenchymal stem cells; NP, nanoparticle; PMMA, poly(methyl methacrylate).

Table 7.2 Nanostructures for gene delivery in tissue engineering

Gene	Nanoparticle/scaffold	In vitro release	Key features	References
p-TGF-β-1	Calcium phosphate NP in collagen/chitosan scaffold	15 days	In vitro biocompatible In vitro protein expression higher than lipofectamine (LFM 2000) In vitro chondrogenic differentiation of stem cells	Cao et al., 2010
β-Gal plasmid	PEI-DNA nanocomplex adsorbed on PLGA porous scaffold	4 days	Highest 60% transfection efficiency In vitro Gal expression DNA dose and release pattern affecting cell proliferation	Jang et al., 2006
p-FGF	PEI-DNA complex inside chitin/alginate nanofibrous scaffold	N/A	Higher transgenic expression compared to 2D system In vitro transgenic expression of FGF in HEK293 cells and dermal fibroblast cells	Lim et al., 2006
p-BMP-2	PEI-AC80 nanoparticles inside collagen/PGA scaffold	N/A	Enhanced biocompatibility by acetylating PEI Higher expression compared to unmodified PEI nanoparticles In vitro osteogenesis In vivo ectopic bone formation	Hosseinkhani et al., 2008
p-β-galacto sidase	Inside PLA–PEG–PLA triblock polymer with PLGA copolymer	20 days	Higher gene expression inside MC3T3 cells but less compared to LFM 2000	Luu et al., 2003
Condensed p-DNA	Inside PLA–PEG–PLA triblock polymer with PLGA copolymer	25% in 7 day	Higher expression in MC3T3 cells In vitro differentiation of cells DNA condensation used to enhance bioactivity	Liang et al., 2005
p-BMP-2	DNA/nHA complex inside collagen scaffold	N/A	12% transgenic efficiency, lower compared to 40% LFM 2000 Higher in vitro biocompatibility compared to LFM In vitro differentiation more compared to LFM 2000	Curtin et al., 2012
p-BMP-2	DNA/chitosan nanoparticles inside PLGA/nHA nanofibrous scaffold	60 days	DNA bioactivity maintained after release Higher BMP-2 expression and less toxicity with nanoparticle inside scaffold compared nanoparticle on nanofibrous surface	Nie and Wang, 2007
p-GL-3	Gold nanoparticles coated with DNA/chitosan/PEI	N/A	Highest internalization inside HeLa cells with complex nanoparticle Biocompatibility of particles less than LFM 2000 but complex particle comparable	Tencomnao et al., 2011

Table 7.3 Clinical status of growth factor loaded nanostructures in tissue engineering

Growth factor	Nanostructure and scaffold material	Clinical application	Animal models	References
BMP-2	PLLA NFS; PLGA/heparin NP/ fibrin matrix	Bone regeneration	Rat calvarial defect model	Chung et al., 2007; Schofer et al., 2011b
BMP-2	PLGA/nHA	Bone regeneration	Nude mouse tibia defect model	Fu et al., 2008
rhBMP-7	PLGA nanospheres immobilized on PLLA NFS	Bone regeneration	Rat subcutaneous model	Wei et al., 2007
BMP-2, bFGF	HCPN/fibrin matrix	Bone regeneration/ angiogenesis	Rat hind limb muscle	Jeon et al., 2006, 2008
BMP-2, HGF, bFGF	Peptide amphiphile (PA) NFS	Bone regeneration and angiogenesis	Rat and mice subcutaneous model	Hosseinkhani et al., 2006a, 2006b, 2007
TGF β-1 and TGF β-3	PEI/heparin NP/ poly(N-isopropyl acrylamideco- vinylimidazole) hydrogel; PLL/heparin NP/fibrin hydrogel	Cartilage regeneration	Nude mice subcutaneous injection	Park et al., 2009a, 2009d
VEGF	Chitosan/heparin NP in decellularized NFS	Angiogenesis	Mouse subcutaneous implantation	Tan et al., 2011
EGF	PCL-PEG-PCL NFS	Skin regeneration	Skin ulcers in diabetic mice	Choi et al., 2008
bFGF	Cyclodextrin/PELA core-sheath NFS		Diabetic rat wound healing model	Yang et al., 2011
KGF	KGF-ELP fusion NP inside fibrin matrix		Diabetic mice chronic wound healing model	Koria et al., 2011

Growth factor	System	Application	Model	Reference
GDNF	PCLEEP aligned NFS conduits	Peripheral nerve regeneration	Rat sciatic nerve gap healing	Chew et al., 2007
NGF	DOPE-PEG-RMP-7 liposome	Brain regeneration	Brain targeting by intravenous injection in rat	Xie et al., 2005
bFGF	Biotinylated bFGF conjugated to OX26-SA nanoparticles	Brain regeneration	Targeting by femoral injection in rats Cerebral Ischemia Model	Song et al., 2002
bFGF	Gelatin NP	Eyes photoreceptor regeneration	Targeting by intravitreal injection in RCS rats	Sakai et al., 2007
NGF	Polysorbate-80 coated PBCA NP	Brain regeneration	Systemic targeting in rat Parkinson model	Kurakhmaeva et al., 2008
PlGF	Chitosan/heparin NP	Myocardial regeneration	Rat myocardial infarction model	Binsalamah et al., 2011
PDGF-bb IGF-1	Self-assembled PA NFS IGF-1 immobilized on biotinylated PA NFS injected with cardiomyocytes or cardiac progenitor cells			Hsieh et al., 2006 Davis et al., 2006; Padin-Iruegas, 2009
SDF-1	Protease resistant s-SDF-1 fusion protein with self-assembling peptide			Segers et al., 2007

BFGF, basic fibroblast growth factor; GDNF, glial cell line derived growth factor; HCPN, heparin-conjugated PLGA nanospheres; NFS, nanofibrous scaffold; NGF, nerve growth factor; nHA, nanohydroxyapatite; PBCA, polybutylcyanoacrylate; PDGFbb, platelet derived growth factor; PELA, poly(ethylene glycol)-poly(dl-lactide); PlGF, placental growth factor; PLL, poly l lysine; PLLACL, poly(L-lactide-co-caprolactone); SDF, stromal cell derived factor-1.

and BMP-7 (OP-1) are clinically available formulations approved by the Food and Drug Administration (FDA) for interbody spinal fusion and tibial nonunions, respectively. But the very high doses of growth factors used in these formulations make it very costly (Friedlaender *et al.*, 2001; McKay *et al.*, 2007). Different materials such as alginate, fibrin, gelatin, and hydroxyapatite have been used for direct delivery of growth factors from polymer matrices, but trials in different animal models showed effectiveness only at high doses of growth factor due to poor control on release kinetics (Satoshi *et al.*, 2003; Yamamoto, 2003; Simmons *et al.*, 2004; Takashi *et al.*, 2005). Microspheres and nanospheres encapsulating growth factors inside scaffolds are potential candidates for clinical translation because growth factor release kinetics can be controlled by changing nanoparticle properties. There is much ambiguity in literature regarding the efficacy of growth factor-loaded, microparticle based scaffolds. Some of the dual delivery approaches using microspheres inside scaffolds have shown successful bone regeneration even at nanogram doses of growth factor (Riva *et al.*, 2010) while others reported effects only in early bone regeneration and failed to show effects at later stages in ectopic and orthotopic models (Patel *et al.*, 2008; Kempen *et al.*, 2009). So, new approaches using growth factor-loaded nanoparticulate scaffold may be a promising strategy for musculoskeleton healing because nanorough surfaces also play a role in cell adhesion and proliferation.

In a study by Wei *et al.* (2007), PLGA nanospheres loaded with BMP-7 (dose 5 µg per implant) and immobilized on PLLA 3D scaffolds showed long term controlled BMP-7 release for 50–70 days and showed higher ectopic bone formation compared to direct adsorption of BMP-7 on PLLA scaffolds after 3 weeks and 6 weeks of implantation. Although these PLGA nanospheres have potential for controlled delivery, toxicity created by organic solvent used in the nanoparticle fabrication may reduce growth factor bioactivity (Wei *et al.*, 2007). Hence, addition of biocompatible entities like heparin with growth factor have been explored to reduce toxicity concerns with PLGA. In a study by Jeon *et al.* (2008), BMP-2 loaded (4 µg per implant) heparinized PLGA nanospheres inside fibrin hydrogel showed sustained delivery of BMP-2 and healed bone with 80% defect closure within 4 weeks of implantation in a rat calvarial model. Injectable fibrin and poly(N-isopropylacrylamide-co-vinylimidazole) hydrogel with TGF-loaded heparin/PLL and heparin/PEI ionic nanoparticles have been used as minimally invasive delivery systems for cartilage regeneration (Park *et al.*, 2009a, 2009d). These injectable systems showed successful differentiation of hMSC encapsulated within them when implanted in a rat subcutaneous model.

Other than nanoparticles, BMP-2 loaded nanofibrous PLLA and peptide amphiphile scaffolds have been explored in different kinds of preclinical models for bone regeneration (Hosseinkhani *et al.*, 2007; Schofer *et al.*,

2011b). In a study, BMP-2 (174 ng per implant) encapsulating PLLA electrospun nanofibrous scaffolds showed successful bone formation after long term implantation in a rat calvarial defect model (Schofer *et al.*, 2011b). Results showed 30% bone area formation within 4 weeks and 50% after 12 weeks showing promise for nanofibrous structure in further research and efficacy studies in different animals.

7.7.2   Skin regeneration

Skin trauma, delayed wound healing in diabetic patients, and pressure ulcer are major indications where tissue engineered scaffolds are required providing suitable platforms for recruitment, proliferation, and differentiation of dermal and epidermal skin tissue (Martin, 1997; Zahedi *et al.*, 2010). Although different materials like collagen, fibrin, PLGA, PLLA, and PEG have been used for wound healing processes, encapsulation of growth factors inside scaffolds are particularly beneficial for early clot formation and keratinocyte differentiation for diabetic patients where wound healing is impaired (Zahedia *et al.*, 2010). Different kinds of natural polymeric hydrogel matrices showing topical EGF and FGF delivery like chitosan scaffolds (Obara *et al.*, 2003), collagen/PGA hybrid matrix (Nagato *et al.*, 2006), chondroitin sulfate-heparin ECM (Liu *et al.*, 2007) mimicking hydrogels have shown enhanced wound healing compared to unloaded scaffolds in normal and diabetic patient wound healing models. Although direct delivery from these scaffolds showed promising results, growth factor-loaded nanoparticles and microparticles can further increase growth factor half-life because of long-term tunable delivery. Gelatin microspheres have shown successful regeneration in different approaches like encapsulating FGF-loaded microspheres inside chitosan matrices for treating pressure ulcers in aged mice (Park *et al.*, 2009c), cell-FGF-polymeric constructs (fibroblasts grown on FGF releasing microspheres), and novel scaffolds where sweat gland cells-EGF-gelatin microspheres were designed and encapsulated in scaffolds formed by culturing keratinocytes on the surface of collagen/fibroblast matrices in organotype culture (Huang *et al.*, 2010). This last approach mimicking skin properties showed early wound closure in athymic mice and total healing within 6 weeks with expression of markers for sweat gland formation.

Nanostructured scaffolds with growth factor also have shown promising results in wound healing. Choi *et al.* (2008) formed an EGF-conjugating PCL/PEG electrospun membrane which showed more wound closure in a rat diabetic model compared to control after 7 days but did not show beneficial effects at later stages, probably because higher concentrations of EGF are required only at the initial stages of healing for recruitment and proliferation of keratinocytes (Choi *et al.*, 2008). For chronic wound closure where the whole skin is affected with impaired angiogenesis, scaffolds with pro-longed delivery of growth are required in order to recruit different kinds of cells and form mature capillaries for growing cells. Yang *et al.* (2011) formed FGF-encapsulating cyclodextrin/PEG-PLLA core-sheath electrospun scaf-folds which showed higher (around 80%) wound closure in diabetic mice with mature blood vessel formation within 2 weeks of implantation by con-trolled delivery of FGF for 4 weeks compared to other controls without growth factors. Using recent advances in genetic engineering, Koria *et al.* (2011) synthesized KGF-ELP fusion protein nanoparticle/fibrin hydrogel scaffolds introducing KGF and ELP within one particle which showed 36% coverage of skin tissue (epithelium formation) compared to 31% with free KGF in a fibrin matrix; this showed that bioactivity of KGF was not affected by fusion but gave less granulation compared to ELP particles alone inside the fibrin matrix. Hence further improvement to this approach is required for better therapeutic effects.

7.7.3   Angiogenesis

Angiogenesis is the process of new blood vessels forming from pre-existing ones and plays a substantial role in wound repair, tissue injury, and many different diseases like coronary heart disease and myocardial ischemic con-dition where cellular nourishment is impaired due to a loss of blood ves-sels. Different growth factors like angiopoietin, PDGF, VEGF, and FGF2 help in recruitment of endothelial cells from pre-existing blood vessels and form tube-like structures from proliferating cells with further recruitment of murine cells and smooth muscle cells (Murohara, 2003). Many approaches have been used for direct delivery of growth factors from tissue engineered scaffolds, for example, heparinized collagen matrix with immobilized VEGF showed better hemoglobin production compared to collagen scaffolds with-out heparin and growth factor after 14 days of implantation in a rat sub-cutaneous model (Steffens *et al.*, 2004). However, capillary formation and expression of different markers, which are crucial elements of angiogenesis, were not evaluated.

The efficacy of scaffold angiogenesis has been enhanced by incorporat-ing growth factor-loaded microspheres, nanoparticles, and nanofibers inside porous scaffolds which helps in maintaining therapeutic levels of growth factor for a longer duration. In most of the studies, PLGA microspheres with different molecular weights have been used providing sustained deliv-ery of VEGF, PDGF, and combinations thereof. Jin *et al.* (2008) immobilized PDGF-encapsulating PLGA microspheres on 3D PLLA nanofibrous scaf-folds and showed higher tissue invasion and more blood vessels (highest 75 blood vessels per mm^2) with higher amounts of PDGF and slow releasing PLGA-65kD microspheres compared to controls without PDGF, in a rat wound healing model. Similarly, Richardson *et al.* (2001) developed a tunable dual growth factor delivery system for VEGF and PDGF encapsulated within PLGA microspheres with different molecular weights and different method of growth factor incorporation. The dual delivery system showed synergistic effects with nanogram doses of growth factor and prolonged release for 31 days. This led to more mature and bigger blood vessels per square area compared to single growth delivered scaffolds in a rat subcutaneous model (Richardson *et al.*, 2001). On the other hand, dual delivery of VEGF/PDGF directly from heparinized polyurethane was fast (1–7 days) and helped in angiogenesis in a rat subcutaneous model after 2 months of implantation but did not make a significant difference in blood vessel number compared to the PDGF group after 2 months of implantation (Davies *et al.*, 2012).

In a recent study by Tan *et al.* (2011), chitosan/heparin nanoparticles chemically conjugated on decellularized jugular vein of buffalo were used for VEGF immobilization. Controlled release of chemically immobilized VEGF (40% in 30 days) on EDC-coupled nanoparticles/decellularized matrix showed prominent expression of different markers with mature blood vessel formation (140 capillaries/mm^2) compared to scaffolds without EDC coupling and without growth factor after 2 months of implantation in a rat subcutaneous model (Tan *et al.*, 2011). FGF is another growth factor helping in angiogenesis. It has been encapsulated inside PLGA microspheres embedded in an alginate hydrogel (Perets *et al.*, 2003) and in different nanoparticles and nanofibrous scaffolds, as discussed below. Jeona *et al.* (2006) physically immobilized FGF on heparin-conjugated PLGA nanospheres known as HCPN (heparin conjugated to amine functionalized PLGA particle via EDC coupling) with further embedding inside a fibrin hydrogel. This process helped in tuning the release kinetics by changing the amount of heparin, heparin molecular weight, and fibrinogen cross-linking and concentration. This FGF/nanosphere/fibrin hydrogel construct (25 μg FGF per implant) showed sustained release (60% release over period of 1 month) with promising results like higher numbers of capillaries (870 ± 105 capillaries/mm^2 from HCPN/fibrin gel as compared to 292 ± 89 capillaries /mm^2 in no treatment group) after 4 weeks implantation in mice ischemic limbs. Hosseinkhani *et al.* (2006a, 2006b) designed FGF- and HGF-encapsulating peptide amphiphile injectable, self-assembled, nanofibrous hydrogels with reduced doses of growth factor (Hosseinkhani *et al.*, 2006a, 2006b). FGF (10 μg per implant) showed sustained release from nanofibrous scaffold (*in vitro* and *in vivo* release for more than 31 days and 20 days and *in vivo* degradation for 30 days) leading to higher blood vessel formation (100 capillaries/mm^2) which was five times higher than control after 7 days of implantation in a rat subcutaneous model (Hosseinkhani *et al.*, 2006a). Although these nanoparticles and hydrogels are able to enhance the angiogenesis process by sustained growth factor release, further work in this area is required to reduce initial growth factor dose, and to prove its efficacy for prolonged period in animal models.

7.7.4   Nerve regeneration

Nerve regeneration is required for healing of injured neurons lost due to traumatic injury and for treatment of diseases like Parkinson's and Alzheimer's. Clinically available methods for peripheral nervous system (PNS) regeneration are suturing of two nerves for short injury and placement of autograft from other sites for long injury. Currently, the FDA has approved some material such as Integra Neurosciences Type 1 collagen tube and SaluMedica's SaluBridge Nerve Cuff™ for short nerve injury. No such method is beneficial for central nervous system (CNS) defects. Surgery can be done only for healing bone fragments to prevent secondary injury. Different kinds of materials are under investigation for nerve repair but simultaneous delivery of neurotropic factors like NGF, neurotropin-5 (NT), glial cell line derived growth factor (GDNF), ciliary neurotrophic factor (CNTF) can further enhance the efficacy of tissue engineered scaffolds for neuron regeneration (Schmidt and Leach, 2003). Designing scaffolds for nerve repair has many challenges like crossing the blood brain barrier without any inflammation in normal brain tissue, minimization of glial scar formation for spinal cord injury, and enhanced axonal extension of growing nerves from nerve conduit scaffold (where the conduit works as a bridge between proximal and distal ends of the injured nerve) with sustained delivery of signals inside these conduits (Willerth and Sakiyama-Elbert, 2007).

Different kinds of approaches like matrices, affinity based systems, microspheres, and nanoparticles have been used for growth factor delivery in PNS and CNS repair. Matrix based approaches helped in healing only short injuries of spinal cord. Growth factors neurotropin and GDNF releasing matrices like collagen type 1 (Houweling et al., 1998) and matrigels with Schwann cells (Iannotti et al., 2003), showed enhanced axonal regeneration in spinal cord injury compared to similar gels without growth factor delivery. CNTF-releasing photo-cross-linkable PEG hydrogels also showed improved neurite growth from an explanted retina (Burdick et al., 2006). Similarly, in one approach, IGF and PDGF dual delivery from collagen, laminin extracellular biomatrix, and 2% methyl cellulose gels were compared for axonal regeneration inside silicone tubing which acted as a nerve guidance channel after 8 weeks of implantation (Wells et al., 1997). Of the different gels, methylcellulose with dual delivery of growth factors showed the highest axonal growth proving that 2% methylcellulose conduits can be promising for nerve regeneration. Affinity based approaches gave further improvement in growth factor delivery. In one approach, physically adsorbed growth factors on heparin-immobilized fibrin-based hydrogels were used in different studies for short duration healing of small lesions in spinal cord injury (Taylor *et al.*, 2004) and long duration healing inside silicone tubing for long sciatic nerve regeneration (Lee *et al.*, 2003). Although these systems showed successful differences in short duration delivery healing spinal cord lesions, long duration delivery for 12 weeks did not show any significant differences between the different groups.

Limitation of these matrices for long duration therapy and chronic injury, where sustained delivery of growth factors for many months is required for proper integration of neurons, encapsulation of growth factors inside microspheres and nanospheres can be a promising strategy for prolonged delivery of growth factors. In one study, NGF-encapsulating polypropylene fumarate microspheres inside silicone tubing showed significant improvement in healing a 10 mm gap of rat sciatic nerve (Xu *et al.*, 2003). Other strategies are using growth factor-loaded microspheres inside nerve conduits. For example, prolonged release of GDNF over 50 days was seen from PLGA/PLLA double walled microspheres inside the inner surface of single lumen PCL conduit which helped in integrating a 1.5 cm long sciatic nerve gap with recruitment of Schwann cells (Kokai *et al.*, 2010).

Although different nanoparticles have been used for targeted NGF delivery to brain injury (Zhang and Uludağ, 2009), use of nanoparticulated scaffolds for PNS and CNS repair is under preclinical study. Different kinds of scaffolding systems with growth factor-loaded nanoparticles like polyionic nanoparticles (Park *et al.*, 2009a, 2009e, 2009f), rationally designed peptide molecules (Maxwell *et al.*, 2005), and fusion proteins (Sakiyama-Elbert *et al.*, 2001) are under preclinical investigation with different cell lines and dorsal root ganglia models. One clinical trial has been done on aligned poly-(caprolactone-coethyl-ethylene phosphate) (PCLEEP) nanofibrous conduit incorporating NGF for sciatic nerve regeneration (Chew *et al.*, 2007). These NGF-embedded aligned nanofibrous scaffolds rolled in the form of conduits showed synergistic effects forming myelinated axons in a rat sciatic nerve gap after 3 months of implantation compared to controls. Other than these scaffold based delivery systems, different nanoparticles have been used for treating diseases like Parkinson's, Alzheimer's, and also brain injury. Brain targeting of these pharmaceuticals depends upon particle efficiency at crossing the blood brain barrier (BBB) and final efficacy depends upon particle size, bioavailability, and side effects on normal tissue. Different strategies like coating with lipoprotein to penetrate the BBB and targeting nanoparticles by conjugating with BBB receptors have been used for enhancing efficacy of nanoparticles (Li and Duan, 2006).

In another study, polysorbate-80 surfactant (which causes enhanced BBB permeability) was coated on NGF-releasing polybutylcyanoacrylate nanoparticles as anti-Parkinson's therapy in C57B1/6 mice (Kurakhmaeva *et al.*, 2008). NGF delivering particles showed effective therapeutics persistent after 7 and 21 days of treatment. Xie *et al.* (2005), in targeting the BBB, designed sterically stabilized NGF-releasing liposomes by incorporating PEG and RMP-7 ligand. An *in vivo* study of this formulation through intravenous injection showed maximum retention of NGF liposomes in brain striatum, hippocampus, and cortex after 30 minutes of therapy compared to other controls like liposomes without RMP-7 and free NGF delivery without liposomes (Xie *et al.*, 2005). In a similar study for targeting brain ischemia, biotinylated FGF was immobilized to OX26-SA complex which was formed by conjugating OX2 (transferrin receptor antibody) with streptavidin (SA). This bFGF-immobilized OX26-SA complex crossed the BBB and showed an 80% reduction in infarct volume of ischemic brain in a rat model (Song *et al.*, 2002). Sakai *et al.* (2007) used bFGF-loaded gelatin nanoparticles for treating local photoreceptor degeneration in eyes. Intravenous injection of bFGF-gelatin particles showed retention of bFGF for 30 days and showed significant preservation of opsonin and a higher number of photoreceptors in the retina compared to other controls like injection of free bFGF, blank nanoaparticles and phosphate buffered saline (PBS) solution (Sakai *et al.*, 2007).

7.7.5   Other types of regeneration

Other than the above mentioned tissues, *in vivo* trials have been undertaken for regenerating different kinds of soft tissues like myocardium after myocardial infarction (Hsieh *et al.*, 2006), liver regeneration (Pulavendran *et al.*, 2010), and engrafting of pancreatic islets (Stendahl *et al.*, 2008) with the help of growth factor-delivering nanostructures. Direct injection of placental growth factor (PIGF)-loaded chitosan alginate nanoparticles was used in one approach for treating acute myocardial infarction and sustained delivery of PIGF over 120 hours gave significant reduction in scar area formation and improved functionality of myocardium (Binsalamah *et al.*, 2011). Mature heart tissue divides rarely and treating chronic myocardial infarction where infarcted area forms scar without any vasculature is a big challenge. Clinically, heart cells and other progenitor cells used to be injected at the site of heart injury but direct injection does not provide successful engrafting and is unable to maintain functionality of the transplanted cells (Curtis and Russell, 2009). In recent approaches, either growth factor alone or growth factors with different cells (like myocardiocytes or cardiac progenitor cells) have been injected in myocardial infarction models with self-assembled peptide molecules which, on injection, forms a hydrogel entrapping the delivery moieties (Davis *et al.*, 2006; Padin-Iruegas *et al.*, 2009). In work by Davis *et al.* (2006), a solution of peptide nanofibers tethered with biotinylated IGF molecules and entrapped with myocardiocytes significantly reduced apoptotic activity of cells and helped in treating myocardial infarction, while cell delivery using only nanofibers or with IGF-tethered nanofibers did not cause significant improvement (Davis *et al.*, 2006). Similarly, Padin-Iruegas *et al.* (2009) injected cardiac progenitor cells with IGF-tethered nanofibers in rat myocardial infarction and results showed synergistic effect with more differentiating myocytes and increased volume compared to only growth factor delivery and only cell delivery. In one recent novel approach, stromal cell derived factor-1 (SDF-1) which helps in stem cell recruitment but is easily degraded by enzymatic action was genetically modified to protect it from degradation (genetically modified form s-SDF-1) and subsequently a fusion protein of s-SDF-1 with nanofibers was formed for treating myocardial infarction in rats (Segers *et al.*, 2007). Intra-myocardial delivery of these fusion protein nanofibers recruited stem cells inside the infarcted site and showed enhanced angiogenesis.

Other than heart tissue regeneration, growth factor delivery from nanostructures has been used for engineering different soft tissues: VEGF- and FGF2-releasing heparin binding nanofibers (HBNF) for islet transplantation (Stendahl *et al.*, 2008), HGF-releasing chitosan nanoparticles for treating liver problems (Pulavendran *et al.*, 2010), and FGF-releasing PLGA electrospun fibers for ligament/tendon engineering (Sahoo *et al.*, 2010a).

7.8   Conclusion and future trends

Advances in nanotechnology and recombinant technology have enabled the design of different types of nanofibers and nanoparticles for controlled growth factor delivery, mostly made up of polymers and proteins. Their different properties of molecular weight, polymer ratio, polymer to protein ratio, and addition of additives can been changed for tunable growth factor delivery in spatiotemporal ways. These nanoparticulate growth factor-delivering scaffolds do not give a burst release like conventional scaffolding materials and also maintain bioactivity, which is the main limitation of conventional covalent immobilization methods.

Although nanoparticles have shown sustained delivery from days to months there are limitations to a universal procedure for *in vitro* release experiments because of variability in release media and volume; these should be standardized in order to compare different release patterns before final efficacy can be studied in animal models. Also, scaling up these nanoparticulate systems in a broader way is a challenge. The majority of work on nanoparticulate scaffolds has been done on single growth factor delivery; further exploration should be given to multicompartment systems delivering many growth factors in spatiotemporal ways at different phases of regeneration.

Growth factor-delivering nanostructures, either targeted directly or forming scaffolds, have shown enhanced tissue regeneration in different models compared to scaffolds without growth factors, but cost considerations will determine their suitability for translation. Most of the work using growth factor-loaded nanostructured scaffolds has been done in lower mammalian models; this should be further addressed in larger animals followed by clinical trials.

There are many areas like ligament, adipose tissue, pancreas, and liver regeneration where nanostructured delivery systems have shown wonderful results in preclinical trials and their clinical efficacy is the demand of the future. Recent advances in genetic engineering have opened ways for developing fusion proteins and affinity based nanoparticles which can be programmed according to cellular growth. In summary, different technologies should be combined to make cost effective scaffolds with potential advancement in future medicine.

7.9   References

Basmanav F B, Kose G T and Hasirci V (2008), 'Sequential growth factor delivery from complexed microspheres for bone tissue engineering', *Biomaterials*, **29**, 4195–4204.

Bessa P C, Machado R, Nürnberger S, Dopler D, Banerjee A, Cunha A M, Rodríguez-Cabello J C, Redl H, Griensven M V, Reis R L and Casal M (2010), 'Thermoresponsive self-assembled elastin-based nanoparticles for delivery of BMPs', *Journal of Controlled Release*, **142**, 312–318.

Binsalamah Z M, Paul A, Khan A A, Prakash S and Shum-Tim D (2011), 'Intramyocardial sustained delivery of placental growth factor using nanoparticles as a vehicle for delivery in the rat infarct model', *International Journal of Nanomedicine*, **6**, 2667–2678.

Bowen-Pope D F, Malpass T W, Foster D M and Ross R (1984), 'Platelet-derived growth factor *in vivo*: levels, activity, and rate of clearance', *Blood*, **64**, 458–469.

Briggs T and Arinzeh T L (2011), 'Growth factor delivery from electrospun materials', *Journal of Biomaterials and Tissue Engineering*, **1**, 129–138.

Burdick J A, Ward M, Liang E, Young M J and Langer R (2006), 'Stimulation of neurite outgrowth by neurotrophins delivered from degradable hydrogels', *Biomaterials*, **27**, 452–459.

Cao X, Deng W, Wei Y, Yang Y, Su W, Wei Y, Xu X and Yu J (2010), ' Incorporating pTGF-β1/calcium phosphate nanoparticles with fibronectin into 3-dimensional collagen/chitosan scaffolds: efficient, sustained gene delivery to stem cells for chondrogenic differentiation', *Europian Cells and Materials*, **23**, 81093.

Casper C L, Yang W, Farach-Carson M C and Rabolt J F (2007), 'Coating electrospun collagen and gelatin fibers with perlecan domain I for increased growth factor binding', *Biomacromolecules*, **8**, 1116–1123.

Chew S Y, Mi R, Hoke A and Leong K W (2007), 'Aligned protein–polymer composite fibers enhance nerve regeneration: A Potential Tissue-Engineering Platform', *Adv Funct Mater*, **17**(8), 1288–1296.

Chiu L L Y, Weisel R D, Li R K and Radisic M (2001), 'Defining conditions for covalent immobilization of angiogenic growth factors onto scaffolds for tissue engineering', *J Tissue Eng Regen Med*, **5**, 69–84.

Cho Y I, Choi J S, Jeong S Y and Yoo H S (2010), 'Nerve growth factor (NGF)-conjugated electrospun nanostructures with topographical cues for neuronal differentiation of mesenchymal stem cells', *Acta Biomaterialia*, **6**(12), 4725–4733.

Choi J S, Leong K W and Yoo H S (2008), '*In vivo* wound healing of diabetic ulcers using electrospun nanofibers immobilized with human epidermal growth factor', *Biomaterials*, **29**, 587–596.

Chonn A and Cullis P R (1995), 'Recent advances in liposomal drug-delivery systems', *Curr Opin Biotechnol*, **6**, 698–708.

Chung Y, Ahn K M, Jeon S H, Lee S Y, Lee J H and Tae G (2007), 'Enhanced bone regeneration with BMP-2 loaded functional nanoparticle–hydrogel complex', *J Control Release*, **121**, 91–99.

Curtin C M, Cunniffe G M, Lyons F G, Bessho K, Dickson G R, Duffy G P and Brien F J (2012), 'Innovative collagen nano-hydroxyapatite scaffolds offer a highly efficient non-viral gene delivery platform for stem cell-mediated bone formation', *Adv Mater*, **24**, 749–754.

Curtis M W and Russell B (2009), 'Cardiac tissue engineering', *J Cardiovasc Nurs*, **24**(2), 87–92.

Davis M E, Hsieh P C H, Takahashi T, Song Q, Zhang S, Kamm R D, Grodzinsky A J, Anversa P and Lee R T (2006), 'Local myocardial insulin-like growth factor 1 (IGF-1) delivery with biotinylated peptide nanofibers improves cell therapy for myocardial infarction', *PNAS*, **103**(21), 8155–8160.

Davies N H, Schmidt C, Bezuidenhout D and Zilla P (2012), 'Sustaining neovascularization of a scaffold through staged release of vascular endothelial growth factor-a and platelet-derived growth factor-bb', *Tissue Eng*, **18**(A), 1–2.

Dhandayuthapani B, Yoshida Y, Maekawa T D and Kumar D S (2011), 'Polymeric scaffolds in tissue engineering application: A review', *Int JPolymer Sci*, **2011**, Article ID 290602, 19 pages.

Dimitriou R, Tsiridis E and Giannoudis P V (2005), 'Current concepts of molecular aspects of bone healing', *Injury Int J Care Injured*, **36**, 1392–1404.

Duan B and Wang M (2010), 'Customized Ca-P/PHBV nanocomposite scaffolds for bone tissue engineering: Design, fabrication, surface modification and sustained release of growth factor', *J R Soc Interface*, **7**, 615–629.

Engel E, Michiardi A, Navarro M, Lacroix D and Planell J A (2008), 'Nanotechnology in regenerative medicine: the materials side', *Trends Biotechnol*, **126**(1), 39–47.

Ertan A B, Yılgor P, Bayyurt B, Çalıkoğlu A C, Kaspar C, Kök F N, Kose G T and Hasirci V (2013), 'Effect of double growth factor release on cartilage tissue engineering', *J Tissue Eng Regen MeD*, **7**, 149–160. doi: 10.1002/term.509.

Friedlaender G E, Perry C R, Cole J D, Cook S D, Cierny G, Muschler G F, Zych G A, Calhoun J H, Laforte A J and Yin S (2001), 'Osteogenic protein-1 (bone morphogenetic protein-7) in the treatment of tibial nonunions', *J Bone Joint Surg Am*, **83A**, 151–158.

Fu Y C, Nie H, Ho M L, Wang C K and Wang C H (2008), 'Optimized bone regeneration based on sustained release from three-dimensional fibrous PLGA/Hap composite scaffolds loaded with bmp-2', *Biotechnol Bioeng*, **99**, 996–1006.

Goldberg M, Langer R and Jia X (2007), 'Nanostructured materials for applications in drug delivery and tissue engineering', *J Biomater Sci Polym Ed*, **18**(3), 241–268.

Gomez N and Schmidt C E (2007), 'Nerve growth factor-immobilized polypyrrole: Bioactive electrically conducting polymer for enhanced neurite extension', *J Biomed Mater Res*, **81A**, 135–149.

Haidar Z S, Azari F, Hamdy R C and Tabrizian M (2009), 'Modulated release of OP-1 and enhanced preosteoblast differentiation using a core-shell nanoparticulate system', *J Biomed Mater Res*, **91A**, 919–928.

Hartgerink J D, Beniash E and Stupp S I (2002), 'Peptide-amphiphile nanofibers: A versatile scaffold for the preparation of self-assembling materials', *PNAS*, **99**(8), 5133–5138.

Holland T A, Tabata Y and Mikos A G (2003), '*In vitro* release of transforming growth factor-β1 from gelatin microparticles encapsulated in biodegradable, injectable oligo(poly(ethylene glycol) fumarate) hydrogels, *J Control Release*, **91**, 299–313.

Hosseinkhani H, Hosseinkhani M, Gabrielson N P, Pack D W, Khademhosseini A and Hisatoshi (2008), 'DNA nanoparticles encapsulated in 3D tissue-engineered scaffolds enhance osteogenic differentiation of mesenchymal stem cells', *J Biomed Mater Res*, **85A**, 47–60.

Hosseinkhania H, Hosseinkhani M, Khademhosseini A, Kobayashi H and Tabata Y (2006a), 'Enhanced angiogenesis through controlled release of basic fibroblast growth factor from peptide amphiphile for tissue regeneration', *Biomaterials*, **27**, 5836–5844.

Hosseinkhani H, Hosseinkhani M and Khademhosseini A (2006b), 'Tissue regeneration through self-assembled peptide amphiphile nanofibers', *Yakhteh Med J*, **8**(3), 204–209.

Hosseinkhani H, Hosseinkhani M, Khademhosseini A and Kobayashi H (2007), 'Bone regeneration through controlled release of bone morphogenetic protein-2 from 3-D tissue engineered nano-scaffold', *J Control Release*, **117**, 380–386.

Houweling D A, Lankhorst A J, Gispen W H, Bar P R and Joosten E A J(1998), 'Collagen containing neurotrophin-3 (nt-3) attracts regrowing injured corticospinal axons in the adult rat spinal cord and promotes partial functional recovery', *Exp Neurol*, **153**, 49–59.

Hsieh P C H, Davis M E, Gannon J, MacGillivray C and Lee R T (2006), 'Controlled delivery of PDGF-BB for myocardial protection using injectable self-assembling peptide nanofibers', *J Clin Invest*, **116**, 237–248.

Hu J and Ma P X (2011), 'Nano-fibrous tissue engineering scaffolds capable of growth factor delivery', *Pharm Res*, **28**, 1273–1281.

Huang S, Deng T, Wang Y, Deng Z, He L, Liu S, Yang J and Jin Y (2008), 'Multifunctional implantable particles for skin tissue regeneration: Preparation, characterization, *in vitro* and *in vivo* studies', *Acta Biomaterialia*, **4**, 1057–1066.

Huang S, Xu Y, Wu C, Sha D and Fu X (2010), '*In vitro* constitution and *in vivo* implantation of engineered skin constructs with sweat glands', *Biomaterials*, **31**, 5520–5525.

Hutmacher D W (2000), 'Scaffolds in tissue engineering bone and cartilage', *Biomaterials*, **21**, 2529–2543

Iannotti C, Li H, Yan P, Lu X, Wirthlin L and Xu X M (2003), 'Glial cell line-derived neurotrophic factor-enriched bridging transplants promote propriospinal axonal regeneration and enhance myelination after spinal cord injury', *Exp Neurol*, **183**, 379–393.

Jang J H, Bengali Z, Houchin T L and Shea L D (2006), 'Surface adsorption of DNA to tissue engineering scaffolds for efficient gene delivery', *J Biomed Mater Res*, **77A**, 50–58.

Jeon O, Song S J, Yang H S, Bhang S H, Kang S W, Sung M A, Lee S H and Kim B S (2008), 'Long-term delivery enhances *in vivo* osteogenic efficacy of bone morphogenetic protein-2 compared to short-term delivery', *Biochem Biophys Res Commun*, **369**, 774–780.

Jeona O, Kanga S W, Lima H W, Chung J H and Kim B S (2006), 'Long-term and zero-order release of basic fibroblast growth factor from heparin-conjugated poly(L-lactide-co-glycolide) nanospheres and fibrin gel', *Biomaterials*, **27**, 1598–1607.

Ji W, Yang F, Beucken J J P, Bian Z, Fan M, Chen Z and Jansen J A (2010), 'Fibrous scaffolds loaded with protein prepared by blend or coaxial electrospinning', *Acta Biomaterialia*, **6**, 4199–4207.

Jin Q, Wei G, Lin Z, Sugai J V, Lynch S E, Ma P X and Giannobile W V (2008), 'Nanofibrous scaffolds incorporating pdgf-bb microspheres induce chemokine expression and tissue neogenesis *in vivo*', *Plos One*, **3**(3), e1729.

Kempen D H R, Lu L, Heijink A, Hefferan T E, Creemers L B, Maran A, Yaszemski M J and Dhert W J A (2009), 'Effect of local sequential VEGF and BMP-2 delivery on ectopic and orthotopic bone regeneration', *Biomaterials*, **30**, 2816–2825.

Kim K, Luu Y K, Chang C, Fang D, Hsiao B S, Chu B and Hadjiargyrouc M (2004), 'Incorporation and controlled release of a hydrophilic antibiotic using poly(lactide-co-glycolide)-based electrospun nanofibrous scaffolds', *J Control Release*, **98**, 47–56.

Kim S E, Park J H, Cho Y W, Chung H, Jeong S Y, Lee E B and Kwon I C (2003), 'Porous chitosan scaffold containing microspheres loaded with transforming growth factor-b1: Implications for cartilage tissue engineering', *J Control Release*, **91**, 365–374.

Kokai L E, Ghaznavi A M and Marra C G (2010), 'Incorporation of double-walled microspheres into polymer nerve guides for the sustained delivery of glial cell line-derived neurotrophic factor', *Biomaterials*, **31**, 2313–2322.

Kopesky P W, Vanderploeg E J, Kisiday J D, Frisbie D D, Sandy J D and Grodzinsky A J (2011), 'Controlled delivery of transforming growth factor b1 by self-assembling peptide hydrogels induces chondrogenesis of bone marrow stromal cells and modulates smad2/3 signaling', *Tissue Eng*, **17**(A), 1–2.

Koria P, Yagia, H, Kitagawae Y, Megeeda Z, Nahmiasa Y, Sheridana R and Yarmusha M L (2011), 'Self-assembling elastin-like peptides growth factor chimeric nanoparticles for the treatment of chronic wounds', *PNAS*, **18**(3), 1034–1039.

Kurakhmaeva K B, Voronina T A, Kapica I G, Kreuter J, Nerobkova L N, Seredenin S B, Balabanian V Y and Alyautdin R N (2008), 'Antiparkinsonian effect of nerve growth factor adsorbed on polybutylcyanoacrylate nanoparticles coated with polysorbate-80', *Bull Exp Biol Med*, **145**(2), 259–262.

Lee A C, Yu V M, Lowe J B, Brenner M J, Hunter D A, Mackinnon S E and Sakiyama-Elberta S E (2003), 'Controlled release of nerve growth factor enhances sciatic nerve regeneration', *Exp Neurol*, **184**, 295–303.

Lee J Y, Bashur C A, Milroy C A, Forciniti L, Goldstein A S and Schmidt C E (2012), 'Nerve growth factor-immobilized electrically conducting fibrous scaffolds for potential use in neural engineering applications', *IEEE Trans Nanobioscience*, **11**(1), 15–21.

Lee J Y, Choo J E, Choi Y S, Suh J S, Lee S J, Chung C P and Park Y J (2009), 'Osteoblastic differentiation of human bone marrow stromal cells in self-assembled BMP-2 receptor-binding peptide-amphiphiles', *Biomaterials*, **30**, 3532–3541.

Lee K, Silva E A and Mooney D J (2011), 'Growth factor delivery-based tissue engineering: General approaches and a review of recent developments', *J R Soc Interface*, **8**, 153–170.

Li H and Duan X (2006), 'Nanoparticles for drug delivery to the central nervous system', *Nanoscience*, **11**(3), 207–209.

Li H, Song J, Park J and Han K (2003), 'Polyethylene glycol-coated liposomes for oral delivery of recombinant human epidermal growth factor', *Int J Pharm*, **258**, 11–19.

Li F, Sun J, Wang J, Du S, Lu W, Liu M, Chao Xie C and Shi J (2008), 'Effect of hepatocyte growth factor encapsulated in targeted liposomes on liver cirrhosis', *J Control Release*, **131**, 77–82.

Liang D, Luu Y K, Kim K, Hsiao B S, Hadjiargyrou M and Chu B (2005), 'In vitro non-viral gene delivery with nanofibrous scaffolds', *Nucleic Acids Res*, **33**(19) e170.

Liao I C, Chew S Y and Leong K W (2006), 'Aligned core shell nanofibers delivering bioactive proteins', *Nanomedicine*, **1**(4), 465–471.

Lim H J, Ghim H D and Choi J H (2010), 'Controlled release of BMP-2 from alginate nanohydrogels enhanced osteogenic differentiation of human bone marrow stromal cells', *Macromol Res*, **18**(8), 787–792.

Lim S H, Liao I C and Leong K W (2006), 'Nonviral gene delivery from nonwoven fibrous scaffolds fabricated by interfacial complexation of polyelectrolytes', *Mol Ther*, **13**(6), 1163–1172.

Liu H and Webster T J (2007), 'Nanomedicine for implants: A review of studies and necessary experimental tools', *Biomaterials*, **28**, 354–369.

Liu Y, Cai S, Shu X Z, Shelby J and Prestwich G D (2007), 'Release of basic fibroblast growth factor from a crosslinked glycosaminoglycan hydrogel promotes wound healing', *Wound Repair Regen*, **15**(2), 245–251.

Liu Y, Groot D and Hunziker E B (2005), 'BMP-2 liberated from biomimetic implant coatings induces and sustains direct ossification in an ectopic rat model', *Bone*, **36**, 745–757.

Luo Y, Engelmayr G, Auguste D T, Ferreira L S, Karp J M, Saigal R, Langer R (2007), 'Three-dimensional scaffolds'. In *Principles of Tissue Engineering*, Lanza R, Langer R, Vacanti J (eds.), Burlington: Academic Press, 3rd ed., Chapter 25, pp. 359–374.

Luu Y K, Kim K, Hsiao B S, Chua B and Hadjiargyrou M (2003), 'Development of a nanostructured DNA delivery scaffold via electrospinning of PLGA and PLA–PEG block copolymers', *J Control Release*, **89**, 341–353.

Mann B K, Schmedlen R H and West J L (2001), 'Tethered-TGF-b increases extracellular matrix production of vascular smooth muscle cells', *Biomaterials*, **22**, 439–444.

Martins A, Duarte A R, Faria S, Marques A P, Reis R L and Neves N M (2010), 'Osteogenic induction of hBMSCs by electrospun scaffolds with Dexamethasone release functionality', *Biomaterials*, **31**, 5875–5885.

Martin F J and Papahadjopoulos D (1982), 'Irreversible coupling of immunoglobulin fragments to preformed vesicles', *J Biol Chem*, **257**, 286–288.

Martin P (1997), 'Wound healing – aiming for perfect skin regeneration', *Science*, **276**, 75–81.

Martins S, Sarmento B, Ferreira D C and Souto E B (2007), 'Lipid-based colloidal carriers for peptide and protein delivery – liposomes versus lipid nanoparticles', *Int J Nanomed*, **2**(4), 595–607.

Masters K S (2011), 'Covalent growth factor immobilization strategies for tissue repair and regeneration', *Macromol Biosci*, **11**, 1149–1163.

Matson J B and Stupp S I (2012), 'Self-assembling peptide scaffolds for regenerative medicine', *Chem Commun*, **48**, 26–33.

Matsuo T, Sugita T, Kubo T, Yasunaga Y, Ochi M and Murakami T (2003), 'Injectable magnetic liposomes as a novel carrier of recombinant human BMP-2 for bone formation in a rat bone-defect model', *J Biomed Mater Res*, **66A**, 747–754.

Maxwell D J, Hicks B C, Parsons S, Sakiyama-Elbert S E (2005), 'Development of rationally designed affinity-based drug delivery systems', *Acta Biomaterialia*, **1**, 101–113.

McKay W F, Peckham S M and Badura J M (2007), 'Comprehensive clinical review of recombinant human bone morphogenetic protein-2 (INFUSE® Bone Graft),' *Int Orthop*, **31**, 729–734.

Murohara T (2003), 'Angiogenesis and vasculogenesis for therapeutic neovascularization', *Nagoya J Med Sci*, **66**, 1–7.

Murugan R and Ramakrishna S (2005), 'Development of nanocomposites for bone grafting', *Comp Sci Technol*, **65**, 2385–2406.

Nagato H, Umebayashi Y, Wako M, Tabata Y and Manabe M (2006), 'Collagen-poly glycolic acid hybrid matrix with basic fibroblast growth factor accelerated angiogenesis and granulation tissue formation in diabetic mice', *J Dermatol*, **33**(10), 670–675.

Nanci A, Wuest J D, Peru L, Brunet P, Sharma V, Zalzal S and McKee M D (1998), 'Chemical modification of titanium surfaces for covalent attachment of biological molecules', *J Biomed Mater Res*, **40**, 324–335.

Nie H and Wang C H (2007), 'Fabrication and characterization of PLGA/HAp composite scaffolds for delivery of BMP-2 plasmid DNA', *J Control Release*, **120**, 111–121.

Nissen N N, Shankar R, Gamelli R L, Singh A and Dipietro L A (1999), 'Heparin and heparan sulphate protect basic fibroblast growth factor from non-enzymic glycosylation', *Biochem J*, **338**, 637–642.

Nottelet B, Mandracchia D, Pektok E, Walpoth B, Gurny R and Möller M (2007), 'Electrospinning of drug loaded poly(ε-caprolactone) nanofibers: *in vivo* evaluation of novel degradable small-sized vascular grafts', *Euro Cells Mater*, **13**, 1473–2262.

Obara K, Ishiharab M, Ishizuka T, Fujita M, Ozekia Y, Maehara T, Saito Y, Yura H, Matsui T, Hattori H, Kikuchi M and Kurita A (2003), 'Photocrosslinkable chitosan hydrogel containing fibroblast growth factor-2 stimulates wound healing in healing-impaired db/db mice', *Biomaterials*, **24**, 3437–3444.

Oh K S, Han S K, Lee H S, Koo H M, Kim R S, Lee K E, Han S S, Cho S H and Yuk S H (2006), 'Core/shell nanoparticles with lecithin lipid cores for protein delivery', *Biomacromolecules*, **7**, 2362–2367.

Oliveira J M, Kotobuki N, Tadokoro M, Hirose M, Mano J F, Reis R L and Ohgushi H (2010), '*Ex vivo* culturing of stromal cells with dexamethasone-loaded carboxymethylchitosan/poly(amidoamine) dendrimer nanoparticles promotes ectopic bone formation', *Bone*, **46**, 1424–1435.

Oliveira J M, Sousa R A, Kotobuki N, Tadokoro M, Hirose M, Mano J F, Reis R L and Ohgushi H (2009), 'The osteogenic differentiation of rat bone marrow stromal cells cultured with dexamethasone-loaded carboxymethylchitosan/poly(amidoamine) dendrimer nanoparticles', *Biomaterials*, **30**, 804–813.

Padin-Iruegas M E, Misao Y, Davis M E, Segers V F M, Esposito G, Tokunou T, Urbanek K, Hosoda T, Rota M, Anversa P, Leri A, Lee R T and Kajstura J (2009), 'Cardiac progenitor cells and biotinylated igf-1 nanofibers improve endogenous and exogenous myocardial regeneration after infarction', *Circulation*, **120**(10), 876–887.

Panyama J and Labhasetwar V (2003), 'Biodegradable nanoparticles for drug and gene delivery to cells and tissue', *Adv Drug Deliver Rev*, **55**, 329–347.

Park C J, Clark S G, Lichtensteiger C A, Jamison R D and Johnson A J W (2009c), 'Accelerated wound closure of pressure ulcers in aged mice by chitosan scaffolds with and without bFGF', *Acta Biomaterialia*, **5**, 1926–1936.

Park J S, Nab K, Woo D G, Yang H N and Park K H (2009b), 'Determination of dual delivery for stem cell differentiation using Dexamethasone and TGF-b3 in/on polymeric microspheres', *Biomaterials*, **30**, 4796–4805.

Park J S, Yang H N, Woo D G, Chung H M and Park K H (2009d), '*In vitro* and *in vivo* chondrogenesis of rabbit bone marrow-derived stromal cells in fibrin matrix mixed with growth factor loaded in nanoparticles', *Tissue Eng*, **15**, 2163–2175.

Park J S, Park K, Woo D G, Yang H N, Chung H M and Park K H (2008), 'PLGA microsphere construct coated with tgf-β 3 loaded nanoparticles for neocartilage formation', *Biomacromolecules*, **9**, 2162–2169.

Park K H, Kim H, Moon S and Na K (2009f), 'Bone morphogenic protein-2 (BMP-2) loaded nanoparticles mixed with human mesenchymal stem cell in fibrin hydrogel for bone tissue engineering', *J Biosci Bioeng*, **108**(6), 530–537.

Park K H, Lee D H and Na K (2009a), 'Transplantation of poly(N-isopropylacrylamideco- vinylimidazole) hydrogel constructs composed of rabbit chondrocytes and growth factor-loaded nanoparticles for neocartilage formation', *Biotechnol Lett*, **31**, 337–346.

Park K H, Kim H and Na K (2009e), 'Neuronal differentiation of pc12 cells cultured on growth factor-loaded nanoparticles coated on plga microspheres', *J Microbiol Biotechnol*, **19**(11), 1490–1495.

Park Y B, Mohan K, Al-Sanousi A, Almaghrabi B, Genco R J, Swihart M T and Dziak R (2011), 'Synthesis and characterization of nanocrystalline calcium sulfate for use in osseous regeneration', *Biomed Mater*, **6**, 055007.

Partridge K A and Oreffo R O C (2004), 'Gene delivery in bone tissue engineering: progress and prospects using viral and nonviral strategies', *Tissue Eng*, **10**(1–2), 295–307.

Patel Z S, Young S, Tabata Y, Jansen J A, Wong M E K and Mikos A G (2008), 'Dual delivery of an angiogenic and an osteogenic growth factor for bone regeneration in a critical size defect model', *Bone*, **43**, 931–940.

Perets A, Baruch Y, Weisbuch F, Shoshany G, Neufeld G and Cohen S (2003), 'Enhancing the vascularization of three-dimensional porous alginate scaffolds by incorporating controlled release basic fibroblast growth factor microspheres', *J Biomed Mater Res*, **65A**, 489–497.

Piantino J, Burdick J A, Goldberg D, Langer R and Benowitz L I (2006), 'An injectable, biodegradable hydrogel for trophic factor delivery enhances axonal rewiring and improves performance after spinal cord injury', *Exp Neurol*, **201**, 359–367.

Piskin E, Isoglu A, Bölgen N, Vargel I, Griffiths S, Cavuşoğlu T, Korkusuz P, Güzel E and Cartmell S (2009), '*In vivo* performance of simvastatin-loaded electrospun spiral-wound polycaprolactone scaffolds in reconstruction of cranial bone defects in the rat model', *J Biomed Mater Res*, **90A**, 1137–1151.

Pitukmanorom P, Yong T H and Ying J Y (2008), 'Tunable release of proteins with polymer–inorganic nanocomposite microspheres', *Adv Mater*, **20**, 3504–3509.

Pulavendran S, Rajam M, Rose C and Mandal A B (2010), 'Hepatocyte growth factor incorporated chitosan nanoparticles differentiate murine bone marrow mesenchymal stem cell into hepatocytes *in vitro*', *IET Nanobiotechnol*, **4**(3), 51–60.

Puleo D A, Kissling R A and Sheu M S (2002), 'A technique to immobilize bioactive proteins, including bone morphogenetic protein-4 (BMP-4), on titanium alloy', *Biomaterials*, **23**, 2079–2087.

Richardson T P, Peters M C, Ennett A B and Mooney D J (2001), 'Polymeric system for dual growth factor delivery', *Nature*, **19**, 1029–1034.

Riva B D L, Sánchez E, Hernández A, Reyes R, Tamimi F, López-Cabarcos E, Delgado A and Évora C (2010), 'Local controlled release of VEGF and PDGF from a combined brushite–chitosan system enhances bone regeneration', *J Control Release*, **143**, 45–52.

Sahoo S, Ang L T, Goh J C H and Toh S L (2010a), 'Bioactive nanofibers for fibroblastic differentiation of mesenchymal precursor cells for ligament/tendon tissue engineering applications', *Differentiation*, **79**, 102–110.

Sahoo S, Ang L T, Goh J C H and Toh S L (2010b), 'Growth factor delivery through electrospun nanofibers in scaffolds for tissue engineering applications', *J Biomed Mater Res*, **93A**, 1539–1550.

Sakai T, Kuno N, Takamatsu F, Kimura E, Kohno H, Okano K and Kitahara K (2007), 'Prolonged protective effect of basic fibroblast growth factor–impregnated nanoparticles in royal college of surgeons rats', *Invest Ophthalmol Vis Sci*, **48**, 3381–3387.

Sakiyama-Elbert S E, Panitch A and Hubbell J A (2001), Development of growth factor fusion proteins for cell triggered drug delivery, *FASEB J*, **15**(7), 1300–1302.

Satoshi K, Ryuhei F, Shoji Y, Shinji F, Kazutoshi N, Koichiro T and Keiji M (2003), 'Bone regeneration by recombinant human bone morphogenetic protein-2 and a novel biodegradable carrier in a rabbit ulnar defect model', *Biomaterials*, **24**, 1643–1651.

Schmidt C E and Leach J B (2003), 'Neural tissue engineering: Strategies for repair and regeneration', *Annu Rev Biomed Eng*, **5**, 293–347.

Schofer M D, Veltum A, Theisen C, Chen F, Agarwal S, Fuchs-Winkelmann S and Paletta J R J (2011a), 'Functionalisation of PLLA nanofiber scaffolds using a possible cooperative effect between collagen type I and BMP-2: Impact on growth and osteogenic differentiation of human mesenchymal stem cell', *J Mater Sci Mater Med*, **22**, 1753–1762.

Schofer M D, Roessler P P, Schaefer J, Theisen C, Schlimme S, Heverhagen J T, Voelker M, Dersch R, Agarwal S, Fuchs S, Winkelmann and Paletta J R J

(2011b), 'Electrospun PLLA nanofiber scaffolds and their use in combination with BMP-2 for reconstruction of bone defects', *PLoS ONE*, **6**(9), e25462.

Scott R C, Rosano J M, Ivanov Z, Wang B, Chong P L, Issekutz A C, Crabbe D L and Kiani M F (2009), 'Targeting VEGF-encapsulated immunoliposomes to MI heart improves vascularity and cardiac function', *FASEB J*, **23**, 3361–3367.

Segers V F M and Lee R T (2007), 'Local delivery of proteins and the use of self-assembling peptides', *Drug Discov Today*, **12**, 13–14.

Segers V F M, Tokunou T, Higgins L J, MacGillivray C, Gannon J and Lee R T (2007), 'Local delivery of protease-resistant stromal cell derived factor-1 for stem cell recruitment after myocardial infarction', *Circulation*, **116**, 1683–1692.

Segura S, Espuelas S, Renedo M J and Irache J M (2005), 'Potential of albumin nanoparticles as carriers for interferon gamma', *Drug Dev Ind Pharm*, **31**, 271–280.

Shah N J, Macdonald M L, Beben Y M, Padera R F, Samuel R E and Hammond P T (2011), 'Tunable dual growth factor delivery from polyelectrolyte multilayer films', *Biomaterials*, **32**, 6183–6193.

Sheridan M H, Shea L D, Peters M C and Mooney D J (2000), 'Bioabsorbable polymer scaffolds for tissue engineering capable of sustained growth factor delivery', *J Control Release*, **64**, 91–102.

Shi J, Votruba A R, Farokhzad O C and Langer R (2010), 'Nanotechnology in drug delivery and tissue engineering: from discovery to applications', *Nano Lett*, **10**, 3223–3230.

Shi Z, Neoh K G, Kang E T and Wang W (2006), 'Antibacterial and mechanical properties of bone cement impregnated with chitosan nanoparticles', *Biomaterials*, **27**, 2440–2449.

Simmons C A, Alsberg E, Hsiong S, Kim W J and Mooney D J (2004), 'Dual growth factor delivery and controlled scaffold degradation enhance *in vivo* bone formation by transplanted bone marrow stromal cells', *Bone*, **35**, 562–569.

Skaat H, Ziv-Polat O, Shahar A and Margel S (2011), 'Enhancement of the growth and differentiation of nasal olfactory mucosa cells by the conjugation of growth factors to functional nanoparticles', *Bioconjugate Chem*, **22**(12), 2600–2610.

Song B W, Vinters H V, Wu D and Pardridge W M (2002), 'Enhanced neuroprotective effects of basic fibroblast growth factor in regional brain ischemia after conjugation to a blood-brain barrier delivery vector', *JPET*, **301**, 605–610.

Steffens G C M, Yao C, Prevel P, Markowicz M, Schenck P, Noah E M and Palluan (2004), 'Modulation of angiogenic potential of collagen matrices by covalent incorporation of heparin and loading with vascular endothelial growth factor', *Tissue Eng*, **10**, 9–10.

Steinmuller-Nethla D, Kloss F R, Najam-Ul-Haq M, Rainer M, Larsson K, Linsmeier C, Kohler G, Fehrer C, Lepperdinger G, Liu X, Memmel N, Bertel E, Huck C W, Gassner R and Bonn G (2006), 'Strong binding of bioactive BMP-2 to nanocrystalline diamond by physisorption', *Biomaterials*, **27**, 4547–4556.

Stendahl J C, Wang L J, Chow L W, Kaufman D B and Stupp S I (2008), 'Growth factor delivery from self-assembling nanofibers to facilitate islet transplantation', *Transplantation*, **86**(3), 478–481.

Su Y, Su Q, Liu W, Lim M, Venugopal J R, Moa X, Ramakrishna S, Al-Deyab S S and El-Newehy M (2012), 'Controlled release of bone morphogenetic protein 2 and dexamethasone loaded in core–shell PLLACL–collagen fibers for use in bone tissue engineering', *Acta Biomaterialia*, **8**, 763–771.

Takashi K, Akira M, Kunio T, Naoto S, Masataka N, Noriyuki T, Hajime O and Hideki Y (2005), 'Potentiation of the activity of bone morphogenetic protein-2 in bone regeneration by a PLA-PEG/hydroxyapatite composite', *Biomaterials*, **26**, 73–79.

Tan Q, Tang H, Hu J, Hu Y, Zhou X, Tao Y and Wu Z (2011), 'Controlled release of chitosan/heparin nanoparticle delivered VEGF enhances regeneration of decellularized tissue-engineered scaffolds', *Int J Nanomed*, **6**, 929–942.

Tanigo T, Takaoka R and Tabata Y (2010), 'Sustained release of water-insoluble simvastatin from biodegradable hydrogel augments bone regeneration', *J Control Release*, **143**, 201–206.

Taylor S J, McDonald III J W and Sakiyama-Elbert S E (2004), 'Controlled release of neurotrophin-3 from fibrin gels for spinal cord injury', *J Control Release*, **98**, 281–294.

Tencomnao T, Apijaraskul A, Rakkhithawatthan V, Chaleawlert-umpon S, Pimpa N, Sajomsang W and Saengkrit N (2011), 'Gold/cationic polymer nano-scaffolds mediated transfection for non-viral gene delivery system', *Carbohyd Polym*, **84**, 216–222.

Vaiana C A, Leonard M K, Drummy L F, Singh K M, Bubulya A, Vaia R A, Naik R R and Kadakia M P (2011), 'Epidermal growth factor: Layered silicate nanocomposites for tissue regeneration', *Biomacromolecules*, **12**, 3139–3146.

Vo T N, Kasper F K and Mikos A G (2012), 'Strategies for controlled delivery of growth factors and cells for bone regeneration', *Adv Drug Deliv Rev*, **64**, 1292–1309, http://dx.doi.org/10.1016/j.addr.2012.01.016.

Wadagaki R, Mizuno D, Yamawaki-Ogata A, Satake M, Kaneko H, Hagiwara S, Yamamoto N, Narita Y, Hibi H and Ueda M (2011), 'Osteogenic induction of bone marrow-derived stromal cells on simvastatin-releasing, biodegradable, nano- to microscale fiber scaffolds', *Ann Biomed Eng*, **39**(7), 1872–1881.

Wang A Y, Leong S, Liang Y C, Huang R C C, Chen C S and S Yu S M (2008a), 'Immobilization of growth factors on collagen scaffolds mediated by polyanionic collagen mimetic peptides and its effect on endothelial cell morphogenesis', *Biomacromolecules*, **9**, 2929–2936.

Wang G, Siggers K, Zhang S, Jiang H, Xu Z, Zernicke R F, Matyas J and Uludağ L (2008b), 'Preparation of bmp-2 containing bovine serum albumin (BSA) nanoparticles stabilized by polymer coating', *Pharm Res*, **25**(12), 2896–2909.

Wei G, Jin Q, Giannobile W V and Maa P X (2006), 'Nano-fibrous scaffold for controlled delivery of recombinant human PDGF-BB', *J Control Release*, **112**(1),103–110.

Wei G, Jin Q, Giannobile W V and Maa P X (2007), 'The enhancement of osteogenesis by nano-fibrous scaffolds incorporating rhBMP-7 nanospheres', *Biomaterials*, **28**, 2087–2096.

Wells M R, Kraus K, Batter D K, Blunt D G, Weremowitz J, Lynch S E, Antoniades H N and Hansson H A (1997), 'Gel matrix vehicles for growth factor application in nerve gap injuries repaired with tubes: A comparison of biomatrix, collagen, and methylcellulose', *Exp Neurol*, **146**, 395–402.

Willerth S M and Sakiyama-Elbert S E (2007), 'Approaches to neural tissue engineering using scaffolds for drug delivery', *Adv Drug Deliv Rev*, **59**(4–5), 325–338.

Willerth S M Johnson P J, Maxwell D J, Parsons S R, Doukas M E and Sakiyama-Elbert S E (2007), 'Rationally designed peptides for controlled release of nerve growth factor from fibrin matrices', *J Biomed Mater Res*, **80A**, 13–23.

Wu J, Liao C, Zhang J, Cheng W, Zhou N, Wang S and Wan Y (2011), 'Incorporation of protein-loaded microspheres into chitosan-polycaprolactone scaffolds for controlled release', *Carbohydr Polym*, **86**, 1048–1054.

Xie Y, Ye L, Zhang X, Cui W, Lou J, Nagai T and Hou X (2005), 'Transport of nerve growth factor encapsulated into liposomes across the blood–brain barrier: *In vitro* and *in vivo* studies', *J Control Release*, **105**, 106–119.

Xu X, Yee W C, Hwang P Y K, Yu H, Wan A C A, Gao S, Boon K L, Mao H Q, Leong K W and Wang S (2003), 'Peripheral nerve regeneration with sustained release of poly(phosphoester) microencapsulated nerve growth factor within nerve guide conduits', *Biomaterials*, **24**, 2405–2412.

Yamamoto M, Ikada Y and Tabata Y (2001), 'Controlled release of growth factors based on biodegradation of gelatin hydrogel', *J Biomater Sci*, **12**(1), 77–88.

Yamamoto M, Takahashi Y and Tabata Y (2003), 'Controlled release by biodegradable hydrogels enhances the ectopic bone formation of bone morphogenetic protein', *Biomaterials*, **24**, 4375–4383.

Yang Y, Xia T, Zhi W, Wei L, Weng J, Zhang C and Li X (2011), 'Promotion of skin regeneration in diabetic rats by electrospun core-sheath fibers loaded with basic fibroblast growth factor', *Biomaterials*, **32**, 4243–4254.

Yilgor P, Tuzlakoglu K, Reis R L, Hasirci N and Hasirci V (2009), 'Incorporation of a sequential BMP-2/BMP-7 delivery system into chitosan-based scaffolds for bone tissue engineering', *Biomaterials*, **30**(21), 3551–3559.

Zahedi P, Rezaeiana I, Ranaei-Siadatb S O, Jafaria S H and Supapho P (2010), 'A review on wound dressings with an emphasis on electrospun nanofibrous polymeric bandages', *Polym Adv Technol*, **21**, 77–95.

Zeng W, Yuan W, Li L, Mi J, Xu S, Wen C, Zhou Z, Xiong J, Sun J, Ying D, Yang M, Li X and Zhu C (2010), 'The promotion of endothelial progenitor cells recruitment by nerve growth factors in tissue-engineered blood vessels', *Biomaterials*, **31**, 1636–1645.

Zhang L and Webster T (2009), 'Nanotechnology and nanomaterials: Promises for improved tissue regeneration', *Nano Today*, **4**, 66–80.

Zhang S and Uludağ H (2009), 'Nanoparticulate systems for growth factor delivery', *Pharm Res*, **26**(7), 1561–1580.

Zhang S, Wang G, Lin X, Chatzinikolaidou M, Jennissen H P, Laub M and Uludag H (2008), 'Polyethylenimine-coated albumin nanoparticles for bmp-2 delivery', *Biotechnol Prog*, **24**, 945–956.

Zhang Y H, Zhu L J and Yao J M (2011), 'Studies on recombinant human bone morphogenetic protein 2 loaded nano-hydroxyapatite/silk fibroin scaffolds', *Adv Mater Res*, **175–176**, 253–257.

Zhang Y Z, Wang X, Feng Y, Li J, Lim C T and Ramakrishna S (2006), 'Coaxial electrospinning of (fluorescein isothiocyanate-conjugated bovine serum albumin)-encapsulated poly(e-caprolactone) nanofibers for sustained release', *Biomacromolecules*, **7**, 1049–1057.

Part II
Application of nanomaterials in soft tissue engineering

8

Nanomaterials for engineering vascularized tissues

D. F. COUTINHO, M. E. GOMES and
R. L. REIS, 3B's Research Group – Biomaterials,
Biodegradables and Biomimetics, University of Minho, Portugal

DOI: 10.1533/9780857097231.2.229

Abstract: The vascularization of tissues is the basis of their function. In cases of damage, living cells hold the potential to remodel their microenvironment and form new vessels. However, in cases of severe damage or disease, that inherent ability may be hampered. Although tissue engineering strategies are being developed, the ability to provide an effective engineered blood supply is hindered by the three-dimensionality of constructs. As a result, tissue engineers are now looking to new tools that are able to control at a nano-scale the spatial organization of the key biological elements for the formation of new vessels. The main aim of this chapter is to shed light on some of the recently proposed strategies to engineer vascularized tissues using biomaterials fabricated at a nano-scale.

Key words: tissue engineering, nanomaterials, angiogenic molecules, nano-structured substrates, vascularization.

8.1   Introduction

Tissue engineering offers the possibility of treating a number of diseases by engineering biomaterials that, combined with specific cells and growth factors (GFs), may help the native tissue to regenerate and regain functionality (Langer and Vacanti, 1993). Nevertheless, the inherent complexity of large and vascular tissues makes it difficult to develop a tissue engineering strategy that can effectively induce the native tissue regeneration. Therefore, the full clinical success of tissue engineering approaches is still limited to avascular or thin tissue types, such as cartilage or skin (Khademhosseini *et al.*, 2009). Of major importance is the lack of means to generate effective pathways for the homogenous distribution of oxygen and nutrients through the three-dimensional structure. As a result, there is a latent need in tissue engineering approaches to ensure the presence of a vascular-like network capable of diffusing through the whole construct and ensuring integration with the host vasculature (Carmeliet, 2005).

229

The advent of tissue engineering strategies enabling the development of biomaterials with integrated functional vasculature would allow not only assistance in the treatment of diseased tissue, restoring its functionality, but would also significantly contribute to restoring the circulation of ischaemic tissues. In order to achieve this, tissue engineers may rely on the intrinsic ability of cells to remodel their surrounding microenvironment in response to stimuli. Cellular microenvironments in living tissues consist of an extra-cellular matrix (ECM), soluble factors, and homeotypic and heterotypic cellular interactions (Milner and Bigas, 1999; Nguyen and D'Amore, 2001; Harlozinska, 2005; Sainson and Harris, 2008). When a defect occurs in a tissue, the supply of oxygen and nutrients is hampered and the neighbouring tissue becomes ischaemic. In response to the hypoxia, the cells surrounding the defect start producing soluble factors, such as vascular endothelial growth factor (VEGF), so that distant cells can remodel their microenvironment and direct the formation of new vessels towards that ischaemic site (Shweiki et al., 1992; Jewell et al., 2001). However, in cases of a critical size defect or with certain pathologies, cells might not be able to remodel the microenvironment, and the natural recovery might not occur (Carano and Filvaroff, 2003; Kanczler and Oreffo, 2008).

In an attempt to mimic the human vascular physiology, tissue engineers have been delivering GFs (Chen and Mooney, 2003; Rajagopalan et al., 2003; Patil et al., 2007), such as VEGF, to cells involved in the process of angiogenesis (Nehls and Drenckhahn, 1995; Keshaw et al., 2005; Correia et al., 2012), as endothelial cells (ECs), or biomaterials with structures similar to that of the ECM of tissues (Bettinger et al., 2008; Lesman et al., 2011), and even combinations of these (Musilli et al., 2012).

Advances in nanotechnologies in the last decades have given tissue engineers new tools to develop improved strategies for achieving vascularized tissues. Nanotechnology (Porter et al., 2008; Zhang and Webster, 2009), has been a particularly important tool for engineering vascularized tissues (Stephens-Altus and West, 2008) allowing: (1) development of new nano-sized biomaterials for improved delivery of GFs relevant for the formation of new vessels (Patil et al., 2007; Chan et al., 2010); (2) production of nano-patterned biomaterials that allow cell guiding towards a specific spatial arrangement (Mironov et al., 2008; Zhang and Webster, 2009); and (3) integration of these two approaches in instructive tissue engineering strategies (Moon et al., 2009; Barbick-Leslie et al., 2011). The ability to tune the spatial organization of GFs and cells at the nanometre-scale is no longer science fiction, and may be the missing approach for developing fully vascularized tissue engineered constructs.

In this chapter, we will discuss the biocomplexity of vascularized tissues and the inherent ability of tissues to adapt to a change in the microenvironment. We will further describe some of the current efforts for developing tissue engineering strategies aiming at overcoming vascularization hurdles, focusing on the advantage of using nanotechnologies to tightly control the properties of the engineered nano-scaled biomaterials, and thus improve tissue regeneration.

8.2   Biocomplexity of vascularized tissues

Blood vessels are responsible for nourishing nearly every organ in the body. They appear during embryogenesis to deliver to the body haematopoietic cells for immune protection, to supply oxygen and nutrients and to dispose of cellular metabolic waste (Noden, 1989; Hananhn, 1997). There are three types of blood vessels: the arteries (80–100 μm), which carry the blood from the heart to the different tissues, the capillaries (10–15 μm), which are responsible for the direct exchange of oxygen and nutrients from the blood to the cells, and the veins (80–100 μm), which carry the blood back to the heart (Kannan *et al.*, 2005).

The generic structure of blood vessels consists of three distinctive layers: the adventitia, the media and the intima (Vito and Dixon, 2003; Rhodin, 2011). The outer layer, the adventitia, consists primarily of collagen fibres and connective tissue, permitting the blood vessels to return to their initial shape after pumping the blood. The media layer is mainly composed of elastic fibres and surrounded by mural cells, which can be either pericytes (in medium-sized vessels) or smooth muscle cells (in large-sized vessels). The intima is the part that contacts the blood fluid and is constituted by a monolayer of endothelial cells and a basement membrane, which is deposited by cells and is rich in collagens and laminins. Although these three layers can be observed in larger vessels, capillaries only present the inner layer of endothelial cells. Arteries and veins are distinct, not only in their anatomical location and physiological function, due to the need to withstand different blood pressures, but also in the origin of the endothelial and smooth muscles that form them.

The formation of new vessels in a healthy scenario can, in a simplistic way, occur by: (1) vasculogenesis, which consists on the assembly of endothelial progenitors, or (2) angiogenesis, which is based on sprouting of new blood vessels from pre-existing ones (Carmeliet and Collen, 1997; Carmeliet, 2000, 2003).

The term vasculogenesis denotes the formation of new vessels when there are no pre-existing ones, as occurs during early vertebrate development. The *de novo* formation of blood vessels involves the differentiation of angioblasts from the mesodermal cells during embryonic development. These endothelial progenitor cells migrate and coalesce to create a primitive vascular network (primary vascular plexus). After blood vessel assembly, endothelial cells can differentiate into either arterial or venous endothelial cells in response to a combination of stimuli. The Notch-Gridlock signal-ling pathway is known to dictate the arterial/venous decision (Milner and Bigas, 1999; Holderfield and Hughes, 2008). For instance, the activation of the Notch pathway promotes the arterial fate, by increasing the expression of arterial marker ephrin-B2. On the other hand, a reduced expression of Gridlock gene leads to an increase in the expression of the venous marker EphB4, promoting venous differentiation (Zhong *et al.*, 2001; Holderfield and Hughes, 2008).

New vessels can subsequently be formed from these arterial and venous blood vessels by angiogenesis. Before forming new vessels, endothelial cells in blood vessels are retained in a quiescent state. The concentration of oxygen available to cells, dependent on the tissue type, is monitored by the hypoxia-inducible factor (HIF) (Jewell *et al.*, 2001). HIF is com-posed of an oxygen-sensitive subunit (HIF-1α, HIF-2α or HIF-3α) and an oxygen-insensitive and stable subunit (HIF-β) (Wang and Semenza, 1993; Lando *et al.*, 2002). Under normoxic conditions (basal oxygen concentra-tion), the prolyl hydroxylase domain (PDH) proteins hydroxylate the HIFs, targeting them for proteosomal degradation (Takeda *et al.*, 2007). However, under hypoxic conditions (low oxygen concentration), PDHs become inac-tive and enable HIFs to start the transcriptional activity of a wide range of genes involved in angiogenesis (such as VEGF, ANG-2, FGF or vari-ous other chemokines) and in cell-cycle regulation, in order to increase the concentration of pro-angiogenic cues and therefore the oxygen concentra-tion (Shweiki *et al.*, 1992; Wang and Semenza, 1993). Thus, in response to these pro-angiogenic cues, a complex cascade of events starts to take place. Pericytes begin to detach from the vessel wall through the action of ANG-2 and endothelial cells start to loosen their cell–cell junction contacts, by the action of vascular endothelial-cadherin molecules (Herber and Stainier, 2011). Activated matrix metalloproteinases start degrading the basement membrane, facilitating the migration of endothelial cells. A 'tip cell' is selected by VEGF-signalling to lead the migration towards the pro-angiogenic cue. While the activation of VEGF receptors on the tip cell stimulates their motile behaviour and the extension of filopodia, it also triggers the Notch signalling on adjacent endothelial cells that trail tip cells, inhibiting their tip cell fate and inducing a 'stalk cell' fate. Stalk cells are less motile cells but critical to maintaining connectivity with the parental vessel and respon-sible for the morphogenesis of the lumen, through which blood will flow (Herber and Stainier, 2011). Endothelial cells will continue migrating until tip cells contact with other vessels, leading to anastomosis. Once this contact is established, tip cell phenotype is repressed and tight endothelial cell junc-tions are re-established, forming a continuous lumen between the 'mother' vessel and the anastomosed vessel. Endothelial cells in the newly formed vessel resume their quiescent state and are subsequently stabilized by the establishment of mural cells (vascular smooth muscle cells and pericytes) on the outer wall of the vessel (Song and Bergers, 2005), recruited by factors such as platelet-derived growth factor-β (PDGF-β) and transforming growth factor-β1 (TGF-β1) (Carmeliet and Jain, 2011). Afterwards, protease inhibitors (tissue inhibitors of metalloproteinases, TIMPs) and plasminogen activator inhibitor-1 (PAI-1) stimulate the deposition of the basement membrane and the strengthening of cell–cell junctions, ensuring an optimal flow distribution in the newly formed vessel (Carmeliet and Jain, 2011).

The aforementioned molecular processes that regulate the formation of new vessels, as well as their maturation and remodelling, are open to manipulation, presenting an excellent opportunity for tissue engineers to improve vascularization.

8.3   Engineering nanomaterials to improve vascularization of tissues

The plurality of biological elements behind the phenomenon of microvessel formation indicates that the mimicry of tissue biocomplexity is a difficult but necessary task to accomplish, in order to achieve a successful clinical translation. Tissue engineers have been focusing on developing a synthetic microenvironment with the ability to manipulate endothelial cells to form new vessels, using biochemical and/or structural cues. The use of nano-scaled biomaterials to deliver GFs (Chen and Mooney, 2003) or guide cells allows not only protection of GFs and cells from the first immune response, but also tight control over the spatial organization of these entities, thus replicating more closely the organization of native tissues. Moreover, the use of GFs alone has already been proven not to be useful through the failure of clinical trials with GF therapy (Rajagopalan *et al.*, 2003). The following sections are devoted to understanding how tissue engineers are tackling the lack of strategies that could effectively mimic individually the biochemical and structural signals relevant in vascularization of tissues.

8.3.1   Delivery of angiogenic factors using engineered nanomaterials

During the formation of new blood vessels, cells act in response to a variety of chemical signals. Thus, in order to induce the formation of new blood vessels, investigators have tried to recapitulate the *in vivo* signalling process by delivering GFs to the ischaemic site (Chen and Mooney, 2003). For that, bolus injections with soluble GFs, delivered either locally or systemically, have been analyzed in human clinical trials (Rajagopalan *et al.*, 2003). However, these early clinical trials have failed to successfully replicate the *in vivo* cellular pathways. Several limitations of this approach have hampered its applicability, related both to the GF characteristics and the methodology used to deliver them. The short half-life of GFs in the human body leads to a reduced time-frame for effective use at the ischaemic site. High dosages of soluble GFs lead to poorly controlled delivery, often resulting in disorganized vascular growth, with transient cellular behaviours that resemble tumour immature vasculature.

Sustained and enduring stimulation with GFs is thus required for the formation of new blood vessels. Therefore, it is important to understand not only the concentration ranges of GFs but also the duration of the exposure of stimulation. Besides the dose response, the specific delivery site and the directionality of the angiogenic signalling are also of the utmost importance to achieve a successful outcome. An alternative strategy is to use polymeric-based systems to engulf the GFs. The encapsulation within biodegradable nanoparticles allows not only overcoming the short half-lives of GFs, but also controlling the release and dosage rate. The ability to tune the degradation rate of the polymeric system and to incorporate multiple stimuli-responsive polymers, has been opening new doors to a controlled and combinatorial delivery of angiogenic chemical stimuli. Silva and Mooney (2007) have encapsulated VEGF within alginate hydrogels for the sustained release of that growth factor in ischaemic hind limbs. It was found that the system sustains and localizes the delivery of VEGF and stimulates angiogenesis, returning limb perfusion to normal levels, and preventing limb necrosis. In a different study, recombinant human VEGF (rhVEGF) was encapsulated in poly(D,L-lactide-co-glycolide) (PLG) and assessed in a corneal implant model of angiogenesis (Cleland *et al.*, 2001). The free acid end groups of PLG appeared to retard the initial release of rhVEGF and the system showed it was able to generate local ocular neovascularization (Cleland *et al.*, 2001).

However, the delivery of a single GF does not replicate the complexity of chemical stimuli involved in the formation of new vessels *in vivo*. Therefore, in an attempt to mimic the complexity of the microenvironment during angiogenesis, the effect of a temporal-controlled delivery of GFs has also been investigated. For that, pH- or temperature-responsive hydrogels have been used to encapsulate different GFs, in order to take advantage of the local variations of pH and temperature at the ischaemic sites (Silva and Mooney, 2007; Garbern *et al.*, 2010; Zhou *et al.*, 2012). Hydrogels are polymeric networks that have the ability to absorb a large volume of water. In response to a variation in pH, temperature or ionic force of the solution, hydrogels can expand their network, releasing the bioentities entrapped within their matrix. Hydrogels present other remarkable properties that include their similarity with the natural tissue ECM, hydrophilicity, high permeability to oxygen and nutrients, and cellular biocompatibility. For this reason, they have become attractive biomaterials as drug and cell delivery agents. In other studies, hydrogels of poly(N-isopropylacrylamide-co-propylacrylic acid-co-butyl acrylate) (p[NIPAAm-co-PAA-co-BA]) were formed with basic fibroblast growth factor (bFGF) and injected into an infarcted rat myocardium. The ability to gel at acidic pH values, as found in the ischaemic myocardium, allowed the hydrogel to act as a GF supplier, and subsequently, once the oxygen levels were replaced, new vessels formed and the pH returned to physiological values, to promote polymer dissolution and elimination (Zhou *et al.*, 2012). This system showed increased microvessel density, improved blood flow and improved cardiovascular function after 28 days of treatment. In a different approach, the spatial-temporal control of GF delivery was achieved by designing 60 nm hybrid nanoparticles consisting of a lipid shell surrounding a polymer core, which was loaded with the drug (Plate VIII i, see colour section between pages 264 and 265). Target peptides for the vascular wall were immobilized on the surface of the nanoparticles, in order to achieve a spatial control over the delivery of the encapsulated drug. These nanoparticles were administrated *in vivo* in a carotid injury model, demonstrating the ability to localize the sites of injured vasculature and controlled drug release over 10–12 days *in vitro* (Chan *et al.*, 2010).

In an effort to mimic the directionality of the angiogenic signalling, gradient biomaterials have been developed, in which physical or chemical properties can be varied gradually and spatially (Aizawa *et al.*, 2010). As referred to in the previous section, the main driving force for EC migration is the presence of a gradient of VEGF molecules towards the ischaemic site. Therefore, researchers are developing gradient biomaterials that can not only give insights into the molecular mechanisms behind the cellular pathways, but also develop tissue-biomimetic gradient scaffolds. Chung and collaborators have developed a microfluidic platform in which capillary growth and EC migration could be evaluated (Plate VIII ii) (Chung *et al.*, 2009, 2010). They observed that ECs started to migrate towards the higher concentration of VEGF. In a more complex approach, Shin and collaborators developed a microfluidic platform where cells were attracted by a VEGF gradient and stabilized by an ANG-1 supplement, supporting *in vivo*-like three-dimensional (3D) capillary morphogenesis (Shin *et al.*, 2011). Nevertheless, despite their increasing complexity, these systems still do not take fully into account the multitude of biomolecules involved in angiogenesis and most importantly, do not provide clear evidence of the possibility of being integrated in tissue engineered constructs.

8.3.2   Engineering nanostructured substrates for endothelial cell guidance

As referred above, the ECM of the vascular microenvironment is composed of the basement membrane and a tightly cross-linked network of proteins and glycosaminoglycans (Kalluri, 2003). This nano-scaled fibrous mesh, although smaller than cells (which are micro-scaled), is useful in organizing cells spatially and in providing them with the correct chemical and physical cues to modulate their behaviour (Liliensiek *et al.*, 2009; Trkov *et al.*, 2010). In this context, vascular beds have been engineered to guide EC migration, adhesion, proliferation and differentiation. Nano/micro-fabrication technologies have been implemented in order to mimic the native vascular ECM at a nano- and micro-scale (Norman and Desai, 2006; Moon *et al.*, 2009; Barbick-Leslie *et al.*, 2011; Coutinho *et al.*, 2011). The key features considered essential for an engineered vascular bed are: (1) suitability of the material surface for cell attachment and migration; (2) a structure porosity that promotes vessel infiltration; and (3) the possibility of the material porosity allied to the degradation rate to allow growth of angiogenic sprouts.

Consequently, early studies were focused on developing structures with topographies that could enhance cell alignment and therefore promote angiogenesis. For example, bovine pulmonary artery smooth muscle cells (SMC) cultured onto nanopatterned (350 nm line width, 700 nm pitch, 350 nm depth) surfaces of poly(dimethylsiloxane) (PDMS) or poly(methyl methacrylate) (PMMA) showed a significantly improved alignment when compared to non-patterned surfaces (Yim *et al.*, 2005). In a different study, Bettinger *et al.* (2008) showed that ECs are able to respond to nano-scaled topographies. Patterns with a ridge/groove period of 1200 and 600 nm depth showed improved cell alignment and elongation, and reduced proliferation. Nanopatterned surfaces enhanced the formation of elongated clusters of endothelial progenitor cells (EPCs), in contrast to flat surfaces.

A number of studies have also used electrospun nanofibres to enhance EC alignment and therefore to promote the formation of new vessel-like structures (Xu *et al.*, 2004; Santos *et al.*, 2008, 2009). Electrospinning has been optimized for different polymers, allowing the creation of fibrous structures that resemble the native vascular bed. Huijun Lu and collaborators have developed nanofibres of synthetic poly-L-lactic acid (PLLA) by electrospinning (Lu *et al.*, 2009). The engineered fibres have been shown to promote and guide the proliferation of outgrowth endothelial cells (OECs) derived from the rabbit peripheral blood, indicating the possibility of using this synthetic fibre-mesh for vascularization strategies. In a more complex approach, Santos *et al.* (2008) developed a 3D scaffold which combined nano-networks electrospun onto microfibre meshes (Plate IX i). This ECM-inspired architecture allowed increasing not only in the adhesion surface area, but also of the interconnectivity in the construct. Micro- and macro-vascular ECs seeded onto the starch-based scaffold spanned between nanofibres, resulting in a complex organization of cells in the constructs. Their findings demonstrated that this ECM-like nano-network combined with the microfibre mesh provided the structural stability for the formation of capillary-like structures. In order to take advantage of the performance of the fibre-meshes, Santos and collaborators accelerated the establishment of a vascular bed by co-culturing human osteoblast cells (hOBs) with human dermal microvascular endothelial cells (HDMECs) onto the starch-based fibre-mesh (Santos *et al.*, 2009). Their study showed that the vascular-like structures evolved from a cord-like configuration to a more complex branched morphology with the presence of a lumen.

In an attempt to engineer vascular beds with higher similarity to the native tissue, hydrogels have been used to encapsulate ECs (Putra-Hanjaya *et al.*, 2012). As described in Section 8.3.1, hydrogels have proven to be suitable for cell encapsulation and delivery to the ischaemic site as an injectable system. However, these hydrogels might not have the desired mechanical stability and once implanted, their weak physical interactions can be loosened. As a result, synthetic and natural polymers have been chemically modified in order to allow production of hydrogels which are chemically cross-linkable by UV (Nichol *et al.*, 2010; Putra-Hanjaya *et al.*, 2012). A number of studies have used microfabrication technologies to develop patterned hydrogels using these chemically modified polymers (Aubin *et al.*, 2010; Putra-Hanjaya *et al.*, 2012). In a two-dimensional (2D) approach, Kawashima *et al.* (2010) have patterned human umbilical vein endothelial cells (HUVECs) into a non-adhesive substrate by an electrochemical-based bio-lithography technique. This patterned cell population was then transferred into cell-adhesive hydrogels by pouring a mixture of thrombin and fibrinogen over the patterns. Their results indicated that the formed fibrin hydrogel supports the formation of an aligned structure, highlighting the importance of the presence of the hydrogel microenvironment.

In a different study, methacrylated gelatin (GelMA) has been microfabricated into 2D and 3D structures. HUVECs encapsulated within GelMA microchannels showed elongation and reorganization, indicating the potential to create microvascularized biomimetic tissues (Nichol *et al.*, 2010). In fact, in a more complex study, Aubin and colleagues have studied the inherent potential of cells to self-organize into aligned structures when confined into the appropriate 3D microarchitecture (Aubin *et al.*, 2010). Their findings have demonstrated that cells which possess an intrinsic potential to organize into aligned tissues *in vivo* can also organize into aligned tissue constructs *in vitro*, elongating along the micropatterned GelMA hydrogels. Cells were encapsulated within hydrogel patterns with different widths, showing an increasing alignment with decreased width of the patterns. Rectangular micro-patterns, spaced by 200 μm, self-assembled into an aligned macro-construct through the convergence of contiguous micropatterns.

Considerable work has been devoted to the formation of lumens *in vitro*. Raghavan *et al.* (2010) have reported on the formation of lumen-like structures by culturing ECs within micro-scale channels, filled with collagen hydrogel. The 1 cm long tubes exhibited cell–cell junction characteristics of early stage capillary vessels. By varying the size and geometry of the micro-channelled template, lumens with different diameters and branched tubes were engineered, indicating the ability to develop geometrically defined vascular-like networks. Similarly, in a more recent study, Zheng *et al.* (2012) described the development of a platform for the study of angiogenesis. Microfluidic vascular networks with defined lumens were developed using type I collagen (Plate IX ii). In order to resemble the endothelium, HUVECs were seeded onto the walls of the microfluidic vascular networks and human brain vascular pericytes (HBVPCs) or human umbilical arterial smooth muscle cells (HUASMCs) were embedded in the collagen hydrogel. The developed platform not only allowed an effective mass transfer, but also supported the formation of sprouts into the collagen matrix.

8.4   Clinical progress

There are several challenges that face the application of angiogenic regenerative medicine therapies to a clinical scenario. These include the possibility, defended by many, of fuelling pre-existing tumour cells to proliferate, which has hampered the success of some clinical trials (Zhang *et al.*, 2000; Weis and Cheresh, 2005). In this context, it is still necessary to rigidly control the mechanisms that trigger the angiogenic switch, and tightly direct the action of the angiogenic cues delivered. Although there have been several clinical trials initiated following this unmet need (Rajagopalan *et al.*, 2003; Yoshito, 2006; Mitsos *et al.*, 2012), the majority have failed to clearly show evidence of improved therapy, thus not being approved for use. This failure occurs despite promising results in different animal models (Takeshita *et al.*, 1994), especially with strategies that rely on the use of pro-angiogenic GFs (Rajagopalan *et al.*, 2003). Therefore, it is clear that there is a great limitation in transferring the knowledge of pre-clinical models to the clinical setting.

8.5   Conclusion and future trends

In this chapter we have reviewed the recent developments regarding the use of nanotechnologies for improving the vascularization of tissues. Regardless of some hold-ups in earlier years related to the failure of several clinical trials with GF therapy, more recently, tissue engineers have been focusing on the development of new nano-scaled biomaterials that enable control of both spatial and temporal release of these GFs, providing added value over the earlier approaches. Nevertheless, a new generation of therapies is arising, which allows patterning these functionalized biomaterials and guiding endothelial cells in their process of forming new vessels. Further efforts to integrate these approaches can bring us closer to the ultimate goal of developing vascularized tissue engineering constructs that are able to anastomose with the host vasculature, leading to a clinically successful therapy.

In fact, the findings described in the previous sections give positive indications regarding the potential of nanotechnologies for the development of strategies to improve the vascularization of tissue engineered constructs. Although some of the technologies described are indeed capable of delivering GFs in a controlled fashion and of guiding endothelial cells, their integration in a fairly three-dimensional construct, able to aid in the regeneration of thick tissues, is not fully met. The major challenge now is thus to multiply the ability of controlling the formation of new vessels at a nano-scale, transposing it to a macro-sized construct for tissue engineering applications.

8.6   References

Aizawa, Y, Wylie, R and Shoichet, M (2010), 'Endothelial cell guidance in 3D patterned scaffolds', *Advanced Materials*, **22**, 4831–4835.

Aubin, H, Nichol, J W, Hutson, C B, Bae, H, Sieminski, A L, Cropek, D M, Akhyari, P and Khademhosseini, A (2010), 'Directed 3D cell alignment and elongation in microengineered hydrogels', *Biomaterials*, **31**, 6941–6951.

Barbick-Leslie, J E, Shen, C, Chen, C and West, J L (2011), 'Micron-scale spatially patterned, covalently immobilized vascular endothelial growth factor on hydrogels accelerates endothelial tubulogenesis and increases cellular angiogenic responses', *Tissue Engineering Part A*, **17**, 221–229.

Bettinger, C J, Zhang, Z, Gerecht, S, Borenstein, J T and Langer, R (2008), 'Enhancement of in vitro capillary tube formation by substrate nanotopography', *Advanced Materials*, **20**, 99–103.

Carano, R A D and Filvaroff, E H (2003), 'Angiogenesis and bone repair', *Drug Discovery Today*, **8**, 980–989.

Carmeliet, P (2000), 'Mechanisms of angiogenesis and arteriogenesis', *Nature Medicine*, **6**, 389–395.

Carmeliet, P (2003), 'Angiogenesis in health and disease', *Nature Medicine*, **9**, 653–660.

Carmeliet, P (2005), 'Angiogenesis in life, disease and medicine', *Nature*, **438**, 932–936.

Carmeliet, P and Collen, D (1997), 'Molecular analysis of blood vessel formation and disease', *American Journal of Physiology – Heart and Circulatory Physiology*, **273**, 2091–2104.

Carmeliet, P and Jain, R K (2011), 'Molecular mechanisms and clinical applications of angiogenesis', *Nature*, **473**, 298–307.

Chan, J M, Zhang, L, Tong, R, Ghosh, D, Gao, W, Liao, G, Yuet, K P, Gray, D, Rhee, J-W, Cheng, J, Golomb, G, Libby, P, Langer, R and Farokhzad, O C (2010), 'Spatiotemporal controlled delivery of nanoparticles to injured vasculature',

Proceedings of the National Academy of Sciences of the United States of America, **107**, 2213–2218.

Chen, R R and Mooney, D J (2003), 'Polymeric growth factor delivery strategies for tissue engineering', *Pharmaceutical Research*, **20**, 1103–1112.

Chung, S, Sudo, R, Mack, P J, Wan, C-R, Vickerman, V and Kamm, R D (2009), 'Cell migration into scaffolds under co-culture conditions in a microfluidic platform', *Lab on a Chip*, **9**, 269–275.

Chung, S, Sudo, R, Vickerman, V, Zervantonakis, I K and Kamm, R D (2010), 'Microfluidic platforms for studies of angiogenesis, cell migration, and cell–cell interactions', *Annals of Biomedical Engineering*, **38**, 1164–1177.

Cleland, J L, Duenas, E T, Park, A, Daugherty, A, Kahn, J, Kowalski, J and Cuthbertson, A (2001), 'Development of poly-(D,L-lactide–coglycolide) microsphere formulations containing recombinant human vascular endothelial growth factor to promote local angiogenesis', *Journal of Controlled Release*, **72**, 13–24.

Correia, C, Grayson, W, Park, M, Hutton, D, Zhou, B, Guo, X, Niklason, L, Sousa, R and Reis, R L (2012), 'In vitro model of vascularized bone: synergizing vascular development and osteogenesis', *PLoS ONE*, **6**, e28352.

Coutinho, D, Costa, P, Neves, N, Gomes, M E and Reis, R L (2011), 'Micro- and nanotechnology in tissue engineering'. In: Pallua, N. and Suscheck, C. (eds.) *Tissue Engineering*, 3–29.

Garbern, J C, Hoffman, A S and Stayton, P S (2010), 'Injectable pH- and temperature-responsive poly(N-isopropylacrylamide-co-propylacrylic acid) copolymers for delivery of angiogenic growth factors', *Biomacromolecules*, **11**, 1833–1839.

Hananhn, D (1997), 'Signaling vascular morphogenesis and maintenance', *Science*, **5322**, 48–50.

Harlozinska, A (2005), 'Progress in molecular mechanisms of tumor metastasis and angiogenesis', *Anticancer Research*, **25**, 3327–3333.

Herber, S P and Stainier, D Y R (2011), 'Molecular control of endothelial cell behaviour during blood vessel morphogenesis', *Nature Reviews*, **12**, 551–564.

Holderfield, M T and Hughes, C C W (2008), 'Crosstalk between vascular endothelial growth factor, notch and transforming growth factor-beta in vascular morphogenesis', *Circular Research*, **102**, 637–652.

Jewell, U R, Kvietikova, I, Scheid, A, Bauer, C, Wenger, R H and Gassmann, M (2001), 'Induction of HIF-1alpha in response to hypoxia is instantaneous', *The FASEB Journal*, **15**, 1312–1314.

Kalluri, R (2003), 'Basement membranes: Structure, assembly and role in tumor angiogenesis', *Nature Reviews*, **3**, 422–433.

Kanczler, J M and Oreffo, R O C (2008), 'Osteogenesis and angiogenesis: the potential for engineering bone', *European cells and materials*, **15**, 100–114.

Kannan, R Y, Salacinski, H J, Sales, K, Butler, P and Seifalian, A M (2005), 'The roles of tissue engineering and vascularisation in the development of micro-vascular networks: a review', *Biomaterials*, **26**, 1857–1875.

Kawashima, T, Yokoi, T, Kaji, H and Nishizawa, M (2010), 'Transfer of two-dimensional patterns of human umbilical vein endothelial cells into fibrin gels to facilitate vessel formation', *Chemical Communications*, **46**, 2070–2072.

Keshaw, H, Forbes, A and Day, R M (2005), 'Release of angiogenic growth factors from cells encapsulated in alginate beads with bioactive glass', *Biomaterials*, **26**, 4171–4179.

Khademhosseini, A, Vacanti, J P and Langer, R (2009), 'Progress in tissue engineering', *Scientific American*, **300**, 64–71.

Lando, D, Peet, D J, Whelan, D, Gorman, J J and Whitelaw, M (2002), 'Asparagine hydroxylation of the HIF transactivation domain: A hypoxic switch', *Science*, **295**, 858–861.

Langer, R and Vacanti, J (1993), 'Tissue engineering', *Science*, **260**, 920–926.

Lesman, A, Koffler, J, Atlas, R, Blinder, Y J, Kam, Z and Levenberg, S (2011), 'Engineering vessel-like networks within multicellular fibrin-based constructs', *Biomaterials*, **32**, 7856–7869.

Liliensiek, S J, Nealey, P and Murphy, C (2009), 'Characterization of endothelial basement membrane nanotopography in rhesus macaque as a guide for vessel tissue engineering', *Tissue Engineering Part A*, **15**, 2643–2651.

Lu, H, Feng, Z, Gu, Z and Liu, C (2009), 'Growth of outgrowth endothelial cells on aligned PLLA nanofibrous scaffolds', *Journal of Materials Science-Materials in Medicine*, **20**, 1937–1944.

Milner, L A and Bigas, A (1999), 'Notch as a mediator of cell fate determination in hematopoiesis: evidence and speculation', *Blood*, **93**, 2431–2448.

Mironov, V, Kasyanov, V and Markwald, R (2008), 'Nanotechnology in vascular tissue engineering: from nanoscaffolding towards rapid vessel biofabrication', *Trends in Biotechnology*, **26**, 338–344.

Mitsos, S, Katsanos, K, Koletsis, E, Kagadis, G C, Anastasiou, N, Diamantopoulos, A, Karnabatidis, D and Dougenis, D (2012), 'Therapeutic angiogenesis for myocardial ischemia revisited: basic biological concepts and focus on latest clinical trials', *Angiogenesis*, **15**, 1–22.

Moon, J J, Hahn, M S, Kim, I, Nsiah, B A and West, J L (2009), 'Micropatterning of poly(ethylene glycol) diacrylate hydrogels with biomolecules to regulate and guide endothelial morphogenesis', *Tissue Engineering Part A*, **15**, 579–585.

Musilli, C, Karam, J-P, Paccosi, S, Muscari, C, Mugelli, A, Montero-Menei, C N and Parenti, A (2012), 'Pharmacologically active microcarriers for endothelial progenitor cell support and survival', *European Journal of Pharmaceutics and Biopharmaceutics*, **81**, 609–616.

Nehls, V and Drenckhahn, D (1995), 'A microcarrier-based cocultivation system for the investigation of factors and cells involved in angiogenesis in three-dimensional fibrin matrices in vitro', *Histochemistry and cell biology*, **104**, 459–466.

Nguyen, L L and D'amore, P A (2001), 'Cellular interactions in vascular growth and differentiation', *International Review of Cytology*, **204**, 1–48.

Nichol, J W, Koshy, S T, Bae, H, Hwang, C M, Yamanlar, S and Khademhosseini, A (2010), 'Cell-laden microengineered gelatin methacrylate hydrogels', *Biomaterials*, **31**, 5536–5544.

Noden, D M (1989), 'Embryonic origins and assembly of blood vessels', *American Journal of Respiratory and Critical Care Medicine*, **140**, 1097–1103.

Norman, J J and Desai, T A (2006), 'Methods for fabrication of nanoscale topography for tissue engineering scaffolds', *Annals of Biomedical Engineering*, **34**, 89–101.

Patil, S D, Papadmitrakopoulos, F and Burgess, D J (2007), 'Concurrent delivery of dexamethasone and VEGF for localized inflammation control and angiogenesis', *Journal of Controlled Release*, **117**, 68–79.

Porter, A, Youtie, J, Shapira, P and Schoeneck, D J (2008), 'Refining search terms for nanotechnology', *Journal of Nanoparticle Research*, **10**, 715–728.

Putra-Hanjaya, D, Wong, K T, Hirotsu, K, Khetan, S, Burdick, J A and Gerecht, S (2012), 'Spatial control of cell-mediated degradation to regulate vasculogenesis and angiogenesis in hyaluronan hydrogels', *Biomaterials*, **33**, 6123–6131.

Raghavan, S, Nelson, C M, Baranski, J D, Lim, E and Chen, C S (2010), 'Geometrically controlled endothelial tubulogenesis in micropatterned gels', *Tissue Engineering Part A*, **16**, 2255–2263.

Rajagopalan, S, Mohler III, E R, Lederman, R J, Mendelsohn, F O, Saucedo, J F, Goldman, C K, Blebea, J, Macko, J, Kessler, P D, Rasmussen, H S and Annex, B H (2003), 'Regional angiogenesis with vascular endothelial growth factor in peripheral arterial disease', *Circulation*, **108**, 1933–1938.

Rhodin, J A (2011), 'Architecture of the vessel wall'. In: Ron Terjung (ed.) *Comprehensive Physiology*. John Wiley and Sons, 1–31.

Sainson, R C A and Harris, A L (2008), 'Regulation of angiogenesis by hemotypic and heterotypic notch signalling in endothelial cells and pericytes: from basic research to potential therapies', *Angiogenesis*, **11**, 45–51.

Santos, M I, Tuzlakoglu, K, Fuchs, S, Gomes, M E, Peters, K, Unger, R E, Piskin, E, Reis, R L and Kirkpatrick, C J (2008), 'Endothelial cell colonization and angiogenic potential of combined nano- and micro-fibrous scaffolds for bone tissue engineering', *Biomaterials*, **29**, 4306–4313.

Santos, M I, Unger, R E, Sousa, R A, Reis, R L and Kirkpatrick, C J (2009), 'Crosstalk between osteoblasts and endothelial cells co-cultured on a polycaprolactone-starch scaffold and the in vitro development of vascularization', *Biomaterials*, **30**, 4407–4415.

Shin, Y, Jeon, J S, Han, S, Jung, G-S, Shin, S, Lee, S-H, Sudo, R, Kamm, R D and Chung, S (2011), 'In vitro 3D collective sprouting angiogenesis under orchestrated ANG-1 and VEGF gradients', *Lab on a Chip*, **11**, 2175–2181.

Shweiki, D, Itin, A, Soffer, D and Keshet, E (1992), 'Vascular endothelial growth factor induced by hypoxia may mediate hypoxia-initiated angiogenesis', *Nature*, **359**, 843–845.

Silva, E A and Mooney, D J (2007), 'Spatiotemporal control of vascular endothelial growth factor delivery from injectable hydrogels enhances angiogenesis', *Journal of Thrombosis and Haemostasis*, **5**, 590–598.

Song, S and Bergers, G (2005), 'The role of pericytes in blood-vessel formation and maintenance', *Neuro-Oncology*, **7**, 452–464.

Stephens-Altus, J S and West, J L (2008), 'Nanotechnology for tissue engineering'. In Polak, J. M. (ed.) *Advances in Tissue Engineering*. London: Imperial College Press.

Takeda, K, Cowan, A and Fong, G-H (2007), 'Essential role for prolyl hydroxylase domain protein 2 in oxygen homeostasis of the adult vascular system', *Vascular Medicine*, **116**, 774–781.

Takeshita, S, Zheng, L, Brogi, E, Kearney, M, Pu, L Q, Bunting, S, Ferrara, N, Suymes, J F and Isner, J M (1994), 'Therapeutic angiogenesis. A single intra-arterial bolus of vascular endothelial growth factor augments revascularization in a rabbit ischemic hind limb mode', *Journal of Clinical Investigation*, **93**, 622–670.

Trkov, S, Eng, G, Liddo, R D, Parnigotto, P P and Nvunjak-Novakovic, G (2010), 'Micropatterned three-dimensional hydrogel system to study human endothelial-mesenchymal stem cell interactions', *Journal of Tissue Engineering and Regenerative Medicine*, **4**, 205–215.

Vito, R P and Dixon, S A (2003), 'Blood vessel consititutive models – 1995–2002', *Annual Review of Biomedical Engineering*, **5**, 413–439.

Wang, G L and Semenza, G L (1993), 'General involvement of hypoxia-inducible factor 1 in transcriptional response to hypoxia', *Proceedings of the National Academy of Sciences of the United States of America*, **90**, 4304–4308.

Weis, S M and Cheresh, D A (2005), 'Pathophysiological consequences of VEGF-induced vascular permeability', *Nature*, **437**, 497–504.

Xu, C Y, Inai, R, Kotaki, M and Ramakrishna, S (2004), 'Aligned biodegradable nanofibrous structure: a potential scaffold for blood vessel engineering', *Biomaterials*, **25**, 877–886.

Yim, E K F, Reano, R M, Pang, S W, Yee, A F, Chen, C S and Leong, K W (2005), 'Nanopattern-induced changes in morphology and motility of smooth muscle cells', *Biomaterials*, **26**, 5405–5413.

Yoshito, I (2006), 'Chapter 2 Animal and human trials of engineered tissues'. In: Yoshito, I. (ed.) *Interface Science and Technology.* Elsevier, San Diego, 91–233.

Zhang, L and Webster, T J (2009), 'Nanotechnology and nanomaterials: promises for improved tissue regeneration', *Nanotoday*, **4**, 66–80.

Zhang, Z G, Zhang, L, Jiang, Q, Zhang, R, Davies, K, Powers, C, Bruggen, N V and Chopp, M (2000), 'VEGF enhances angiogenesis and promotes blood-brain barrier leakage in the ischemic brain', *The Journal of Clinical Investigation*, **106**, 829–838.

Zheng, Y, Chen, J, Craven, M, Choi, N W, Totorica, S, Diaz-Santana, A, Kermani, P, Hempstead, B, Fischbach-Teschl, C, Lopez, J A and Stroock, A D (2012), 'In vitro microvessels for the study of angiogenesis and thrombosis', *Proceedings of the National Academy of Sciences of the United States of America*, **109**, 9342–9347.

Zhong, T P, Childs, S, Leu, J P and Fishman, M C (2001), 'Gridlock signalling pathway fashions the first embryonic artery', *Nature*, **414**, 216–220.

Zhou, J, Zhao, Y, Wang, J, Zhang, S, Liu, Z, Zhen, M, Liu, Y, Liu, P, Yin, Z and Wang, X (2012), 'Therapeutic angiogenesis using basic fibroblast growth factor in combination with a collagen matrix in chronic hindlimb ischemia', *The Scientific World Journal*, **2012**, 652794–652794.

9

Nanomaterials for cardiac tissue engineering

Y. SAPIR, Ben-Gurion University of the Negev, Israel,
B. POLYAK, Drexel University College of Medicine, USA and
S. COHEN, Ben-Gurion University of the Negev, Israel

DOI: 10.1533/9780857097231.2.244

Abstract: Cardiac tissue engineering (CTE) aims to create contractile tissues to replace damaged or missing myocardial tissues. Biomaterials/ scaffolds constitute a major component in various strategies of CTE, as stand-alone treatments or in combination with cells and/or bioactive molecules. This chapter describes the recent applications of nanomaterials and nanofabrication methods in CTE scaffolds to mimic the structural and architectural properties of the cardiac extracellular matrix (ECM) and to enable controllable and efficient single cell stimulation. Emphasis will be given to the implementation of magnetic field stimulation in CTE, attained by impregnating nano-magnetites into the CTE scaffold. Nanotechnology has already led to engineered cardiac tissues better suited for implantation.

Key words: alginate scaffolds, nanoparticles, nanostructures, nanofabrication, magneto-mechanical cell stimulation, tissue engineering, nanotechnology.

9.1   Introduction

Cardiac tissue engineering (CTE) aims to create contractile heart muscle tissues, able to replace missing (due to congenital heart defect) or dysfunctional parts of the heart (due to myocardial infarction), thus leading to cardiac repair. Currently, a leading CTE strategy is the *in vitro* engineering of a cardiac patch by seeding cardiomyocytes (CMs) as well as additional vascular cells in biodegradable scaffold, followed by applying inductive biological and physical (mechanical, electrical) cues to promote tissue maturation and function. The scaffolds play a critical role in CTE as they provide both the physical support and biological cues during tissue development. Initially, the materials used for scaffold fabrication were developed with an emphasis on their biocompatibility, mechanical properties, and degradability. However, as the field evolved, it became clear that the nanostructural features of a scaffold play a crucial role and can affect cell re-organization and tissue development.

244

This chapter describes the recent applications of nanomaterials and nanofabrication methods for reconstructing the optimal CTE scaffold. The nanomaterials can be structured as nanofibers, be integrated with specific recognition nanosites, or be bulk materials impregnated with nanoparticles. These features are incorporated into the CTE scaffolds to mimic the structure (chemical and physical) as well as the architecture of the cardiac extracellular matrix (ECM) and/or to enable controlled cell stimulation, with the aim of inducing the regeneration of a thick contractile cardiac tissue. Focus will be given herein to the detailed description of a composite alginate scaffold developed to enable magnetic stimulation of cells.

9.2   Heart muscle structure and diseases

9.2.1   Heart muscle structure

The human myocardium consists of 2–3 billion cardiomyocytes (CMs) (75% by volume, ~30% by number); the striated CMs are found only in the heart and are distinguishable from skeletal and smooth muscle cells. Apart from CMs, the heart tissue is mainly composed of fibroblasts (FBs, 60% by number) and endothelial cells (ECs). Unlike skeletal muscle cells, the CMs are controlled by the autonomic (involuntary) rather than the somatic (voluntary) nervous system. CMs can generate their own excitatory impulses, functioning as a biological pacemaker.

The myocardium has a unique structure, enabling it to synchronously contract (Fig. 9.1). It consists of cardiac muscle fibers, 50–100 μm long and with a diameter of ~14 μm. The ends of cardiac muscle fibers connect to neighboring fibers by intercalated discs containing desmosomes, which hold the fibers together, and gap junctions, which allow the muscle action potentials to conduct from one muscle fiber to its neighbors and to synchronously contract (Fig. 9.1a). Myofibrils, the contractile machinery in CMs, consist of repeating contractile units, the sarcomeres, which are composed of actin and myosin fibers and bounded by Z-lines of protein aggregates (Fig. 9.1b). During muscle contraction, the actin fibers move towards the inner space of the sarcomere by sliding along the fixed myosin fibers and the protein complexes enable the macroscopic movement associated with contractile activity (Fig. 9.1b) (Gregorio and Antin, 2000).

The myocyte and myocardial fiber orientation is held by the cardiac ECM – a highly organized nano/micro-fiber network of molecules, such as fibrillar type I and type III collagen, proteoglycans, and glycosaminoglycans that provide a structural integrity to adjoining myocytes and contribute to the overall left ventricular pump function (Spinale, 2007) (Fig. 9.1c). The myocardial ECM functions as a large reservoir of bioactive molecules that can directly influence its synthesis and degradation. Through cell–matrix

(a)

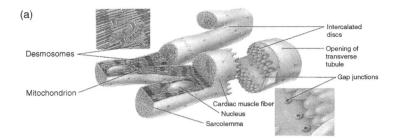

Intercalated
discs

Opening of
transverse
tubule

Gap junctions

Desmosomes

Mitochondrion

Cardiac muscle fiber
Nucleus
Sarcolemma

(b)

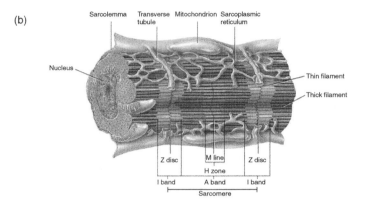

Sarcolemma   Transverse   Mitochondrion   Sarcoplasmic
               tubule   reticulum

Nucleus

Thin filament

Thick filament

Z disc   M line   Z disc
             H zone
I band   A band   I band
          Sarcomere

(c)

Protein fibres (collagen, elastin)

Adhesive proteins (fibronectin,
vitronectin, laminin)

Heparin/Heparan sulfate

Growth factors

Cell-ECM
interactions

9.1 Schematics of cardiac muscle structure. (a) overall organization of
the cardiac muscle fiber. Sarcolemma is plasma membrane of a muscle
cell. Sarcoplasmic reticulum (SR) is membranous sacs encircling each
myofibril. In a relaxed muscle fiber, the SR stores calcium ions. Release
of Ca^{2+} from the terminal cisterns of the SR triggers muscle contraction.
(b) Internal arrangement of the cardiac fiber showing basic sarcomere
structure. The assembly of contractile proteins into sarcomeres is a
complex process that requires coordinated synthesis of the constituent
proteins, the polymerization of actin and myosin (and many associated
proteins) into thin and thick filaments, respectively, and the association
of the two filament systems into highly organized sarcomeres. Newly
assembled sarcomeres consist of parallel arrays of ~1.0 μm-long thin
filaments that interdigitate with laterally aligned 1.6 μm-long thick

(Continued)

interactions, mechanical stimuli, such as stress or strain, are likely to be transduced from the myocardial ECM to the cardiac myocyte.

The complexity of the cardiac muscle structure presents great challenges in CTE. Clearly, the nanostructural features of the cardiac ECM and the nano-scale interactions of cells with ECM point to the potential benefits of implementing nanotechnology and nanoscience in CTE.

9.2.2   Myocardial infarction and congenital heart defect

Two major cardiac diseases may benefit from having cardiac patches available for transplantation: myocardial infarction (MI) and congenital heart defect.

MI is the most common manifestation of coronary heart disease (CHD), accounting for ~50% of all CHD cases in the US. MI usually results from occlusion in the main coronary arteries, thus causing a significant reduction in blood supply to the beating heart muscle (mainly of the left ventricle (LV)) and leading to ischemia and necrosis (Jessup and Brozena, 2003; Whelan et al., 2007; McMurray, 2010). Although CMs are the most vulnerable cell population, ischemia also kills vascular cells, fibroblasts, and nerves in the tissue. The sudden loss of a significant portion of myocardium leads to a decrease in heart contractility. To compensate for this loss, several structural changes occur, such as increase in LV volume, which leads to intensified stress on the ventricular wall. With time, formation of a noncontractile scar tissue together with thinning of the injured wall and dilation of the ventricular cavity, lead to congestive heart failure (CHF) (Jessup and Brozena, 2003; Whelan et al., 2007; McMurray, 2010). Over the last decade, major improvements have been made in the management of MI patients (Boersma et al.,

9.1 Continued.

filaments. Narrow, plate-shaped regions of dense protein material called Z discs separate one sarcomere from the next. Thus, a sarcomere extends from one Z disc to the next Z disc. The thick and thin filaments overlap one another to a greater or lesser extent, depending on whether the muscle is contracted, relaxed, or stretched. The pattern of their overlap, consisting of a variety of zones and bands (I, A and H), creates the striations that can be seen both in single myofibrils and in whole muscle fibers. The M line marks the middle of the sarcomere. (*Source*: (a, b) This material is reproduced with permission of John Wiley & Sons, Inc. from (Tortora, 2008).) (c) A schematic presentation of major cardiac ECM components, including fibrillar collagen, fibronectin, elastin, proteoglycans, and glycosaminoglycans. ECM also serves as a controlled reservoir of growth factors. The interaction of ECM with cells is mediated by integrins on the cell surface. Cadherins and gap junction proteins comprise the cell–cell interaction complex.

2003; Nabel and Braunwald, 2012). These include interventional therapies as well as drug regimens for prevention and long-term treatment. A major disadvantage of these therapies is their inability to regenerate a contractile tissue, at least partially, to compensate for the cardiac muscle loss after infarction.

The development of a cardiac patch is of utmost importance for correcting heart-related defects, which are the most common fatal birth defects (Kung *et al.*, 2008). Structural myocardial defects are diagnosed in approximately 1% of all newborns, with a risk of sudden cardiac death 25–100 times that of young patients in the general population (Silka *et al.*, 1998). In many cases, major heart defects are associated with other types of birth defects, cardiac or non-cardiac.

For these two diseases, the strategy of CTE has the potential to provide a solution, by engineering and transplantation of a contractile cardiac patch, thus empowering the heart function.

9.3   Cardiac tissue engineering (CTE)

The replacement of a large noncontractile scar after MI and the correction of congenital heart malformations, require the implantation of a fully-developed cardiac muscle patch to achieve cardiac repair. The patch should be thick and has to display the functional and morphological properties of the native cardiac muscle. Once integrated into the heart, the cardiac patch must develop a systolic force, withstand a diastolic load with appropriate compliance, and form an electrical and functional syncytium with the host myocardium.

Three main strategies for cardiac patch reconstruction have been developed over the last decade or so and they are based on seeding the cardiac cells in scaffolds (hydrogels, polymer layers, and pre-formed macroporous scaffolds) (Ruvinov *et al.*, 2008). The cell constructs are cultivated under medium perfusion for efficient nutrient and dissolved oxygen transport, while applying inductive biological and physical cues to generate a fully functional cardiac tissue. Since cells are 'the actual tissue engineers', the main goal of this strategy is to design the best inductive environment able to promote the development and maturation stages of the engineered tissue.

The extensive research in the last decade or so reveals that for the fabrication of thick functional cardiac patches, several challenges have yet to be resolved, such as patch pre-vascularization for rapid integration with the host, and the optimization of scaffold fabrication and stimulation patterns applied during patch cultivation for inducing alignment and synchronous contraction of the engineered cardiac tissue.

9.3.1   Pre-vascularization of cardiac patch

The engineering of thick functional cardiac patch demands its pre-vascularization to enable efficient mass transport during cultivation, induce the structural organization into a tissue, and accelerate the anastomosis of pre-formed vessels and host vessels upon implantation to promote patch integration and function (Levenberg et al., 2005).

While the role of endothelial capillaries in nutrient delivery to the functional tissue is clear, their attribution to the promotion of tissue regeneration *in vitro* is starting to be revealed (Caspi et al., 2007). *In vitro* co-culture experiments of endothelial cells (ECs) and cardiomyocytes (CMs) have demonstrated that EC signaling regulates cardiac growth and function *via* autocrinic and paracrinic pathways (Nishida et al., 1993). Such cross-talk has been monitored in cardiac tissue cultures by measuring parameters such as contractility (Ramaciotti et al., 1992), apoptosis (Kuramochi et al., 2004), and survival (Narmoneva et al., 2004). Reciprocally, CMs are believed to promote EC survival and assembly (Hsieh et al., 2006). Cardiofibroblasts also play a role in CM arrangement and tissue regeneration (Radisic et al., 2008; Shachar et al., 2011).

Patch pre-vascularization strategies include (1) using the body, for example, the blood vessel-enriched omentum, as a bioreactor to induce vessel penetration into the patch followed by its implantation onto the infarcted heart (Dvir et al., 2009), or (2) by seeding tri-cell cultures of CMs with ECs and stromal cells or embryonic fibroblasts (EmFbs) into polymeric scaffolds. Lesman et al. (2010) developed an engineered human cardiac tissue by co-culture of human embryonic stem cell (hESC)-derived CMs, EmFbs and human umbilical vein ECs (HUVECs). Transplantation of these cell constructs onto healthy rat hearts resulted in the formation of donor (human) as well as host (rat)-derived vasculature within the engrafted constructs, associated with functional integration of the human-derived vessels with host coronary vasculature (Lesman et al., 2010). Radisic et al. (2009) introduced a new strategy of seeding the tri-cell cultures sequentially into the scaffold. They showed that the sequential, but not simultaneous culture of ECs, FBs, and CMs resulted in elongated, beating cardiac organoids (Iyer et al., 2009). In a follow-up paper, the same group showed that the expression of Connexin 43 (Cx43) gap junction protein and construct contractile function are mainly mediated by vascular endothelial growth factor (VEGF) released from the non-myocytes during the preculture period (24 h) in the sequential seeded cultures (Iyer et al., 2012).

The pre-vascularization process of the cardiac patch could be enhanced by applying magneto-mechanical stimulation on ECs, as will be described later.

9.3.2   Stimulations in CTE

In engineering of the cardiac patch, it has been shown that apart from chemical stimulus (e.g., growth factors), additional physical cues are essential, such as electrical signaling, mechanical stimulation achieved upon medium perfusion, and stretching of the three-dimensional (3D) cellular constructs (Akhyari *et al.*, 2002; Zimmermann *et al.*, 2002; Radisic *et al.*, 2004a, 2004b). These signals promote the hypertrophy of neonatal CMs, increase contractile protein content, and encourage cell alignment into the bundles characteristic of a native heart tissue.

Mechanical stimulations can be implemented *via* bioreactors, which are developed to apply mechanical forces. Those can be applied *via* piston/compression systems such as substrate bending, hydrodynamic compression, and fluid shear (Carver and Heath, 1999; Roberts *et al.*, 2001; Guldberg, 2002; Dvir *et al.*, 2007). Zimmerman and colleagues were among the first groups to study the response of a cardiac cell construct to a mechanical stimulation. In their study, neonatal rat CMs were mixed with collagen type I and matrix factors and cast into circular molds, which were then subjected to a phasic mechanical stretch. This resulted in a ring-shaped engineered heart tissue (EHT) that displayed the important hallmarks of differentiated myocardium (Zimmermann *et al.*, 2002).

Other studies explored the influence of electrical stimulation on cardiac cells (McDonough and Glembotski, 1992; Inoue *et al.*, 2004; Kawahara *et al.*, 2006; Yamada *et al.*, 2007). Electrical signals mimicking those in the native heart (rectangular pulses, 2 ms, 5 V/cm, 1 Hz) were applied by Radisic and colleagues on neonatal rat CMs seeded onto collagen constructs. After 8 days of stimulation, the cardiac cell constructs already presented functional coupling between the cells, and an amplified contraction amplitude concomitant with a significant level of ultrastructural tissue maturation as compared to non-stimulated constructs (Radisic *et al.*, 2004a).

Our group developed a perfusion bioreactor combined with electrical stimulation (Barash *et al.*, 2010). These bioreactors provided a homogeneous interstitial fluid flow throughout the cardiac cell construct, thus attributing to the regeneration of a thick (>500 μm) cardiac tissue. The capability of medium perfusion to induce cardiac tissue formation was confirmed for the first time, from the cell signaling level up to the ultrastructural phenotype of the regenerated cardiac muscle tissue (Dvir *et al.*, 2007). When combined with electrical stimulation, the cardiac cell constructs revealed cell elongation and striation with enhanced expression level of the gap junction protein, Cx43, responsible for cell–cell coupling, by 4 days after cell seeding (Barash *et al.*, 2010).

Although these cell stimulation modes succeeded in improving the contractile features of thin cardiac patch, they were less effective in doing so in thicker patches.

9.4   Application of nanomaterials and nanofabrication methods in CTE

Tissue engineering has shown very promising results when working at both the macro- and micro-levels. Nevertheless, the idea of controlling the regenerating system at the single cell level is extremely fascinating. By constructing physiologically-relevant environments, delivering essential biomolecules to the cell itself, as well as by specifically stimulating the cell of interest, the tissue engineer could regulate the regeneration pattern with a high level of control. Nanotechnology also proposes various solutions for the existing challenges in CTE. For example, aligned scaffolds can be fabricated by unidirectional arrangement of nanofibers; such scaffolds would have the potential to induce the formation of anisotropic cardiac tissue. Similarly, nanopatterning of scaffold surfaces with recognition nanosites can do a similar job. In addition, the scaffolds can be impregnated with nanoparticles (NPs) to make them either conductive or responsive to magnetic fields, depending on the type of NPs used. These applications are described below in detail.

9.4.1   Nanofabrication of aligned surfaces and scaffolds

The initial attempts to induce cell alignment in cardiac patches consisted of applying physical signals during cultivation, such as mechanical stretch, medium perfusion, and electrical stimulation (Zimmermann *et al.*, 2002; Radisic *et al.*, 2004a; Dvir *et al.*, 2007; Barash *et al.*, 2010). Although such attempts were successful in promoting cell maturation and tissue contractility to some extent, they had limited success in inducing cell alignment (anisotropy) in thick cardiac patches. One possible explanation is that the scaffolds used in these studies had an internal isotropic structure and could not promote the development of anisotropic cell arrangement, even with construct stimulation.

Recent efforts to promote cardiac cell alignment in a physically-relevant 3D environment focused on the nanofabrication of anisotropic scaffolds, such as by electrospinning of nanofibers mimicking the nanofibrillar structure of the ECM network (Zhang *et al.*, 2011). The unidirectional arrangement of the nanofibers resulted in the formation of an anisotropic scaffold able to promote cell alignment. Such an aligned scaffold was produced from poly(methylglutarimide) (PMGI) by electrospinning (Orlova *et al.*, 2011). The nanofiber meshes were produced with prismatic cross sections (700–1000 nm in width and 300–500 nm in height); fiber alignment was controlled by a collector with oriented rectangular holes, and the fiber density was changed by varying the time of fiber deposition. Cardiac cells seeded in these anisotropic scaffolds proliferated, assumed an elongated morphology and were aligned as judged by the orientation of their α-actin filaments (Orlova *et al.*, 2011).

9.4.2 Impregnation of gold nanowires and nanoparticles for promoting scaffold conductivity

Nanostructures can be impregnated into tissue engineering scaffolds to enable more effective physical stimulations of the seeded cells and/or to influence the mechanical properties of the scaffold. For example, the impregnation of Au NPs into non-conductive scaffolds could contribute to a better propagation of the electrical signal, thus enabling the construction of thick contractile cardiac patches. Such an approach has recently been investigated by two groups.

One group fabricated an electrically-conductive hybrid hydrogel from thiol-2-hydroxyethyl methacrylate (thiol-HEMA)/HEMA impregnated with homogenously dispersed Au NPs that were synthesized using a polymer template gel (You *et al.*, 2011). According to the authors, the mechanical properties of the hybrid hydrogel matched those in cardiac muscle tissue and were 1–4 orders of magnitude better relative to matrices of conventional conductive polymers, such as polyaniline and polypyrrole. Neonatal rat CMs grown in these hybrid scaffolds exhibited a 2-fold increased expression of Cx43, with or without electrical stimulation compared to non-conductive HEMA scaffolds (You *et al.*, 2011).

Dvir *et al.* (2011) (Fig. 9.2) conferred electrical conductivity by creating a nanocomposite alginate scaffold impregnated with gold nanowires (with an average length of ~1 μm, to cross and bridge the electrically resistant scaffold pore walls) (Fig. 9.2c and 9.2d). Such composite alginate scaffold induced cell alignment with an improved electrical signal propagation. When electrically-stimulated, the engineered cardiac tissue contracted synchronously; calcium ion transient recordings were apparent at various sites in the nanocomposite scaffolds, while at pristine alginate scaffolds, they were revealed at the stimulation site only (Fig. 9.3). In addition, greater expression levels of contractile proteins (e.g., troponin I) and gap junction protein Cx43 were detected in the composite matrices (Dvir *et al.*, 2011).

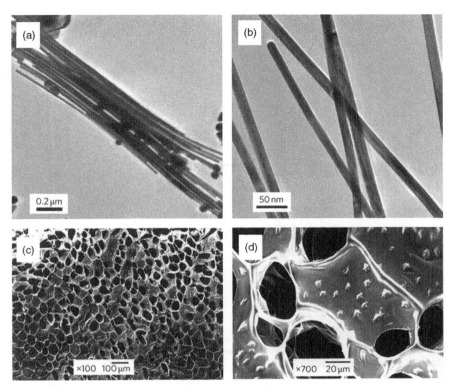

9.2 Impregnation of gold nanowires within macroporous alginate scaffolds. (a, b) Transmission electron microscope (TEM) images of a typical distribution of gold nanowires, which exhibited an average length of ~1 μm and average diameter of 30 nm. (c, d) Scanning electron microscope (SEM) images revealed that the nanowires (1 mg/mL) assembled in the pore walls of the porous alginate scaffold into star-shaped structures with a total length scale of 5 μm. The assembled wires were distributed homogeneously within the matrix at a distance of ~5 μm from one another. (*Source*: Figure reproduced with permission from Dvir *et al.* (2011).) Part (d) also appears in Plate II in the color section, see between pages 264 and 265.

9.4.3   Incorporation of biological signals into matrix

ECM nano-scale interactions play a critical role in affecting cell attachment, morphology, and function; thus major efforts have been invested in the intelligent design of novel materials bio-inspired by the ECM–cell interactions. Scaffolds fabricated from these bio-inspired materials provide not only the mechanical support for cells, but also specific inductive biological cues for cell migration, differentiation, and adhesion. One way to achieve such appropriate material surfaces is demonstrated by attaching ECM-derived

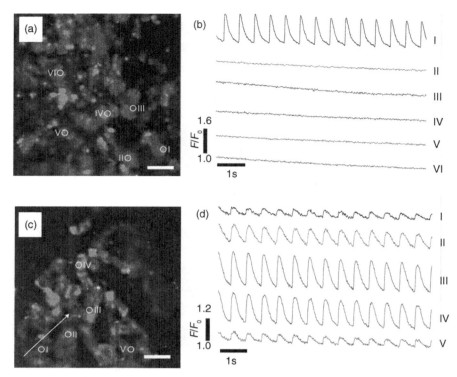

9.3 Calcium transient propagation within the engineered cardiac tissues in the nanowire impregnated alginate scaffold. Calcium transient was assessed at specified points (white circles) by monitoring calcium dye fluorescence. (a) Sites monitored in pristine scaffold, where site I is the stimulation point. (b) Calcium transients were only observed at the stimulation point in the pristine alginate scaffold. F/F0 refers to measured fluorescence normalized to background fluorescence. (c) Sites monitored in the nanowired alginate (Alg–NW) scaffold. The stimulation point was 2 mm diagonal to the lower left of point I (that is, off the figure). The white arrow represents the direction of propagation. (d) Calcium transients were observed at all points. (*Source*: Figure reproduced with permission from Dvir *et al.* (2011).) Parts (c, d) also appear in Plate II in the color section.

peptides. The most commonly used peptide is the sequence Arg-Gly-Asp (RGD), an adhesive moiety derived from fibronectin and laminin (Shin *et al.*, 2003). RGD peptide mediates cell adhesion and signaling *via* interactions with integrin receptors on the cell surface (Rosso *et al.*, 2004).

Our group has covalently-attached RGD peptide to alginate material in order to investigate its role within the 3D settings on engineering of a cardiac tissue *in vitro* (Tsur-Gang *et al.*, 2009; Shachar *et al.*, 2011). Since alginate is inert to cell interactions, it has been used as a 'blank canvas'

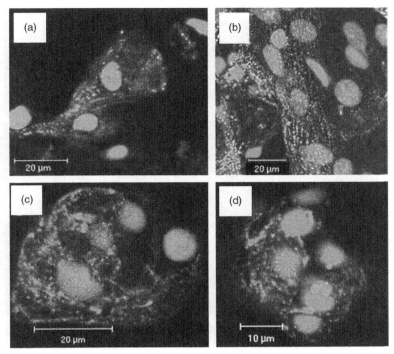

9.4 The effects of immobilized RGD peptide in macroporous alginate scaffolds on cardiac tissue engineering. Images from confocal microscope of immunostained section of cardiac cell constructs prepared and cultivated in RGD-immobilized alginate scaffolds (a, b) and in pristine alginate scaffolds (c, d) for 6 (a, c) and 12 (b, d) days. (*Source*: Figure reproduced with permission from Shachar *et al.* (2011).)

to investigate specific cell–matrix interactions by attaching RGD as well as other biological cues. The RGD peptide modification of alginate scaffold promoted cardiac tissue regeneration. A short time after seeding, the CMs were reorganized into myofibrils composed of multiple CMs in a typical myofiber bundle (Fig. 9.4). Cardio-fibroblasts supported the cardiac myofibers in a manner similar to that in the native cardiac tissue (although the overall cell mass was lower). Furthermore, the relative expression levels of contractile and cell–cell adhesion proteins were shown to be better maintained in the RGD-modified scaffolds as compared to the unmodified ones. Lastly, the signaling meditated through cell adhesion prevented cell apoptosis, further emphasizing the critical role of cell–matrix interactions in cell survival, organization, and tissue maturation (Shachar *et al.*, 2011).

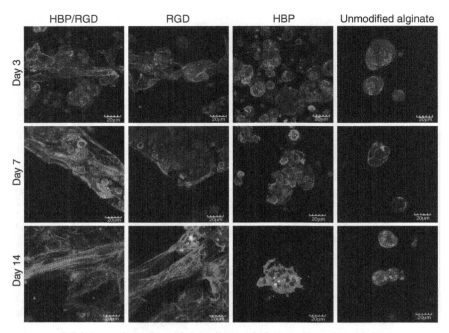

9.5 Integration of multiple cell–matrix interactions into alginate scaffolds promotes cardiac tissue organization. RGD and HBP peptides sequences were covalently attached to the matrix. Images from confocal microscope of immunostained section of cardiac cell constructs on days 3, 7, and 14 post seeding (Scale bar: 20 μm). (*Source*: Figure reproduced with permission from Sapir *et al.* (2011).)

Recognizing the multi-functional interactions in ECM, our group has recently shown that the co-integration of an additional molecular signal provided by heparin-binding peptides (HBPs) had promoted the formation of improved cardiac tissue *in vitro* (Sapir *et al.*, 2011). HBPs bind cell surfaces *via* the transmembrane proteoglycans syndecans (Cardin and Weintraub, 1989). We tested the cardiac tissue formed within macroporous alginate scaffolds bearing the two peptides, RGD and HBP, to those formed in a single peptide-attached or in unmodified alginate scaffolds (Sapir *et al.*, 2011). Confocal microscopy examination of 14-day cellular constructs revealed a characteristic striated fiber organization mainly in the HBP/RGD-attached cell constructs, while in the HBP-attached and pristine alginate cell constructs no such structures were observed (Fig. 9.5). In hematoxylin & eosin (H&E) stained histological sections, the HBP/RGD-attached cell constructs showed the formation of a consistent tissue with an anisotropic fiber arrangement, while in the RGD-attached scaffold no such arrangement was seen. Protein expression levels by western blot analyses for the contractile protein, sarcomeric α-actinin, and cell–cell

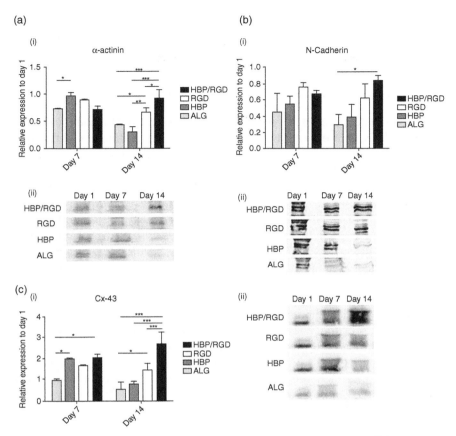

9.6 Integration of RGD and HBP peptides into alginate scaffolds promotes cardiac tissue features. Protein expression is presented as relative to day 1 for (a) α-actinin, (b) N-Cadherin, and (c) Cx-43. (i) Densitometry analysis of western blots, (ii) representative western blots. Asterisks denote significant difference (by 2-way ANOVA), $*p < 0.05$, $**p < 0.01$, and $***p < 0.005$ (Bonferroni's pos-hoc test used for comparison between the groups). (*Source*: Figure reproduced with permission from Sapir *et al.* (2011).)

adhesion molecule, N-cadherin, indicated that the expression level of these proteins was preserved in the HBP/RGD-attached scaffolds. For Cx43 gap junction protein, the expression level even increased with time, further supporting the notion of contractile muscle formation and tissue maturation (Fig. 9.6) (Sapir *et al.*, 2011). Together, the outcomes of these studies emphasize the importance of integration of multiple and complementary molecular signal sites into 3D matrices for the recreation of the cardiac inductive environment and successful formation of an engineered cardiac tissue *in vitro*.

Nanotechnology has significantly advanced the tissue engineering field by introducing various methods, such as nanopatterning of sequences peptides onto the surfaces of scaffolds in a precise and controlled manner.

9.4.4   Nanosensors and nanodevices in CTE

The capability of precisely monitoring the biophysical and biochemical activity of a cell/tissue is expected to increase our understanding of cell function in any unique 3D cell microenvironment, and to enable the follow-up of the tissue engineering process.

Over the past several decades, the study of electroactive cells, such as CMs, was carried out by recording and stimulating techniques, including voltage-sensitive dyes (Kleber and Rudy, 2004), glass micropipette, intracellular and patch-clamp electrodes (Klemic et al., 2002; Li et al., 2006), multi-electrode arrays (MEAs) (Halbach et al., 2003; Meyer et al., 2004; Heer et al., 2007), and planar field electron transistors (FETs) (Yeung et al., 2001; Ingebrandt et al., 2003). Most of these methods had inherent sensitivity limitations and could not monitor and record in physiologically-relevant 3D cell microenvironments.

A greater sensitivity than planar FETs was recently demonstrated by creating electrical interfaces from nanowire FETs (NWFETs) and carbon nanotubes (CNTs) protruding from the substrate plane, thus increasing NW/cell interfacial coupling (Patolsky et al., 2006; Cellot et al., 2009; Pui et al., 2009; Timko et al., 2009). Cohen-Karni et al. (2009) proposed a flexible approach to interface NWFETs with cells and demonstrated it for silicon NWFET arrays coupled to embryonic chicken CMs. They selected silicon nanowires due to their unique electronic properties and sizes (comparable with biological structures), as well as biocompatibility compared to CNTs. NWFET conductance signals recorded from CMs exhibited excellent signal-to-noise ratios and signal amplitudes that were tuned by varying device sensitivity through changes in water gate–voltage potential, Vg. For the first time, multiplexed recording of signals from NWFET arrays interfaced to CM monolayers enabled temporal shifts and signal propagation to be determined with good spatial and temporal resolution. Other groups have also described the coupling of electronics and tissues using flexible and/or stretchable planar devices that conform to natural tissue surfaces (Timko et al., 2009; Viventi et al., 2010; Kim et al., 2011a, 2011b; Viventi et al., 2011). Although these reports claim certain success in increasing sensitivity, the devices were still planar and could not address several key points relevant to 3D cell microenvironments. These parameters include being nano/micro-scale, three-dimensional, macroporous, and fit to mechanical properties of the desired body location.

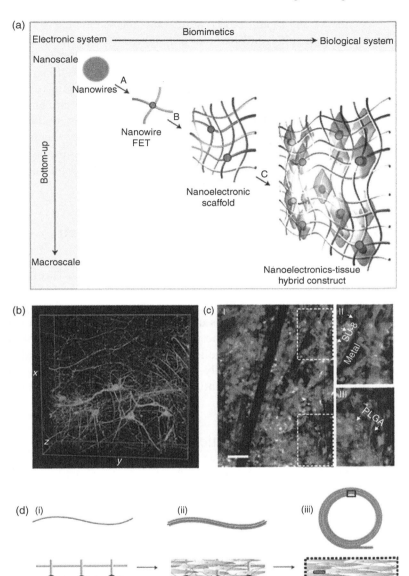

9.7 Hybrid macroporous nanoelectronic scaffolds. (a) Three biomimetic and bottom-up steps have been designed: step A, patterning, metallization, and epoxy passivation for single-nanowire FETs; step B, forming 3D nanowire FET matrices (nanoelectric scaffolds) by self- or manual-organization and hybridization with traditional ECMs; step C,

(Continued)

Recently, Tian *et al.* (2012) addressed this challenge by developing macroporous, flexible, and free-standing nanowire nanoelectronic scaffolds (nanoES), and their hybrids with synthetic or natural biomaterials, by using a stepwise fabrication method for incorporation. The resulting scaffold had an electrical sensory function as well as being appropriate for 3D culture of neurons, CMs, and smooth muscle cells (Fig. 9.7). The integrated sensory capability of the nanoES was demonstrated by real-time monitoring of the local electrical activity within 3D nanoES/CM constructs, recording the distinct pH changes inside and outside tubular vascular smooth muscle constructs and following the response of 3D nanoES-based neural and cardiac tissue models to drugs.

9.5   Case study: magneto-mechanical cell stimulation to promote CTE

The magnetic component of an electromagnetic field has recently received considerable attention as a potential means for cell stimulation. The magnetic field (MF) can penetrate deep into tissues, reaching a single cell and acting directly on its organelles unlike the electric field, which is shielded by the membrane potential. The mechanisms by which weak MFs initiate cascades leading to different cell processes are yet to be elucidated. MFs can be coupled to magnetically-responsive materials, making this strategy attractive for targeting and/or manipulation at a distance, on the nano-scale level.

9.5.1   Effects of magnetic fields on cells

In biological systems, the effects of magnetic field stimulation are usually attained using extremely low frequency electro-magnetic fields (working range 20–100 Hz). Ventura and colleagues induced murine embryonic stem

9.7 Continued.
incorporation of cells and growth of synthetic tissue through biological processes. (b) 3D reconstructed confocal images of rat hippocampal neurons after a 2-week culture in Matrigel on reticular nanoES. Dimensions x: 317 μm; y: 317 μm; z: 100 μm. (c) Confocal fluorescence micrographs of a synthetic cardiac patch. (II and III), zoomed-in view of the upper and lower dashed regions in I, showing metal interconnects, the SU-8 scaffold (arrows in II) and electrospun PLGA fibers (arrows in III). Scale bar, 40 μm. (d) Schematic of the synthesis of smooth muscle nanoES. The upper panels are side views, and the lower ones are either top views (I and II) or a zoom-in view (iii). (*Source:* Figure reproduced with permission from Tian *et al.* (2012).)

cell (ESC) differentiation to CMs by applying a MF of 50 Hz to the culture for up to 10 days. Cell stimulation triggered the expression of cardiac lineage-promoting genes, such as GATA-4 and Nkx-2.5, and the synthesis and secretion of dynorphin B, an endorphin playing a major role in cardiogenesis. MF effects ultimately led to a remarkable increase in the yield of ESC-derived CMs (Ventura et al., 2005).

Another study revealed that ECs respond to pulsed electro-magnetic fields generated between Helmholtz coils (15 Hz, 1Ga) and formed capillaries in the presence of heparin, endothelial cell growth factor, and a competent fibronectin matrix (Yen-Patton et al., 1988). Discrete stages of new vessel formation, similar to angiogenesis in vivo, were witnessed in EC culture monitored for up to 23 days in the presence of a MF. By contrast, in the control group not exposed to a pulsed magnetic field, no vessels were formed.

9.5.2   Magnetic induced cell and tissue patterning

Magneto-mechanical cell stimulation mediated via time-varying MFs and magnetic-responsive materials has been recently implemented in various tissue engineering strategies. Ferro- and ferri-magnetic materials are generally avoided in biomedical applications due to their tendency to magnetically interact and stick to each other, forming aggregates that can cause embolism within a blood stream. Thus, superparamagnetic NPs are typically chosen as a compromise between the desire to achieve strong magnetization and the need to avoid having permanently magnetized objects. Magnetite (Fe_3O_4) NPs are among those most widely used because they fulfill the requirements of high saturation magnetic moment (M_S ~90–98 emu/g, or ~450–500 emu/cm^3) and they have the lowest toxicity levels yet known in pre-clinical tests (Marchal et al., 1989; Weissleder et al., 1989).

Superparamagnetic NPs have been implemented in a recent technique named magnetic force-based tissue engineering (Mag-TE) (Ito et al., 2004). In this technique, magnetite cationic liposomes (MCLs) containing 10 nm magnetite NPs were loaded into cells. Then, by applying a magnetic field, the magnetically-labeled cells arranged themselves into a desired pattern on a steel substrate used as the cell culture surface. In this way, patterned lines of single cells and complex cell patterns (curved, parallel, or crossing patterns) were successfully fabricated. This technique was applied to a number of tissue engineering applications including preparation of skeletal muscles (Ito et al., 2009; Yamamoto et al., 2009), bone tissue (Ito et al., 2006; Shimizu et al., 2007a, 2007c), small-diameter vascular tissue (Ito et al., 2005b, 2006), retinal pigment epithelium (Ito et al., 2005a), and cardiac tissue (Shimizu et al., 2007b; Akiyama et al., 2010).

In CTE, the Mag-TE technique was applied to generate multi-layered CM sheets on ultra-low-attachment surface plates. Without MF stimulation, the CMs formed spheroids on the surfaces. In the presence of MF stimulation, the CM sheets had gap junctions in-between the layers, as judged by immunostaining to Cx43, and electrical connections were confirmed using multi-electrode extracellular potential mapping. The study demonstrated that the constructed cell sheet is not just 'cell aggregate' formed by magnetic attraction, but rather a magnetically shaped 'functional' cell cluster (Shimizu *et al.*, 2007b).

Using a combination of magnetic-induced tissue fabrication and ECM-based techniques, ring-shaped cardiac tissues were recently constructed (Shimizu *et al.*, 2007b; Akiyama *et al.*, 2010). Herein, a mixture of ECM precursor and CMs labeled with magnetite cationic liposomes (MCLs) containing 10 nm magnetite nanoparticles were added into a well containing a central polycarbonate cylinder. With the use of a magnet, the cells were attracted to the well bottom and formed a cell layer. During cultivation, the cell layer shrank towards the cylinder, leading to the formation of a ring-shaped tissue that possessed a multilayered cell structure and contractile properties. The cardiac tissue rings exhibited several advantages: (1) the CMs were densely packed in the engineered tissue and showed some unidirectional alignment; (2) the engineered tissues had sufficient strength for manipulation; and (3) they showed spontaneous contraction and were electrically excitable.

The Mag-TE technique was also used to investigate the therapeutic potential of mesenchymal stem cell (MSC) sheets in reparative angiogenesis. Human MSC sheets comprising 10–15 layered cells with 300 µm thickness were transplanted into a mouse hind limb ischemia model under conditions maintaining cell-to-cell connections. The MSC sheets, engrafted into the ischemic tissue, stimulated revascularization and inhibited host skeletal muscle cell apoptosis. This approach provides a novel modality in the field of regenerative medicine using cell sheets (Ishii *et al.*, 2011).

9.5.3 Magnetically induced release of bioactive molecules

In most tissue engineering approaches, multiple growth factors (GFs) are preloaded into the scaffold prior to cell cultivation and implantation to promote cell survival, angiogenesis, and integration into the host. Non-magnetic approaches for achieving controlled release of these GFs include their encapsulation within gelatin (Patel *et al.*, 2008) and microparticles (Perets *et al.*, 2003) or NPs (Yilgor *et al.*, 2009, 2010) that are further incorporated within the scaffold pores, or by coating the scaffold with a gelatin hydrogel loaded with growth factor (Kempen *et al.*, 2009). Controlled GF-release hydrogels were also constructed by the electrostatic affinity binding of a combination of heparin-binding proteins (via alginate-sulfate), resulting in a sequential factor release, at rates reflected by the specific factor equilibrium binding constants (Freeman *et al.*, 2008; Freeman and Cohen, 2009). Although the non-magnetic approaches for GF delivery often provide controlled release, the release profiles are pre-programmed prior to implantation. Magnetic approaches have the capability of modifying the release profile as needed *in vivo*.

Strategies to attain magnetically-induced GF delivery utilized the following mechanisms (Derfus *et al.*, 2007; Hu *et al.*, 2007; Liu *et al.*, 2009; Bock *et al.*, 2010; Brule *et al.*, 2011; Hoare *et al.*, 2011; Zhao *et al.*, 2011): (1) Large deformations in macroporous ferrogels caused enhanced medium flow via interconnected pores in the matrix, thus triggering GF release; (2) Encapsulation of magnetic nanoparticles (MNPs) and GFs in a thermoresponsive polymer device. Upon exposure to an alternating high frequency MF, the MNPs caused a local increase in temperature due to magnetic hysteresis loss and Brownian relaxation, leading to a polymer phase transition and consequent GF release; and (3) Exertion of mechanical force to a magnetically responsive scaffold through the application of an external time-varying or continuous field. Although none of these studies were relevant to CTE, it is clear that in future, magnetically induced GF release has high potential for CTE as it can offer combined advantages of GF release 'on demand' both *in vitro* and *in vivo*.

9.5.4 Magneto-mechanical stimulation of cells

The idea of combining the capability of a MF to intrude into a living substance, together with its ability to couple to a magnetically-responsive material within a substance, has initiated the promising novel approach of magneto-mechanostimulation. Such stimulation has a major advantage compared to other stimulation patterns, since the nano-magnetic actuation allows for 'action at a distance', thus enabling cell stimulation both *in vitro* and *in vivo*. The magnetic field can affect the cell itself together with actuation of a process within the target cell when coupled to magnetically responsive particles. In addition, stress parameters applied through MNPs can also be varied dynamically, simply by changing the frequency and strength of an applied field. Undoubtedly, the most advantageous feature of this system is its precision to target, manipulate, and activate individual ion channels or other targets within the cells by attaching targeting moieties on the MNPs. RGD-modified MNPs targeted to the cell receptors, as integrins, were able to deform cell cytoskeleton, thus causing activation of the mechanosensitive channels and cell response (Matthews *et al.*, 2006; Hughes *et al.*, 2007; Kanczler *et al.*, 2010).

The high specificity of integrin targeting initiated attempts to target the stress-responsive ion-channels, responsible for the cellular feedback, thus enabling actuation of the target only and not interrupting normal cell function. MNP modification with ion-channel-specific antibodies, enabled particle targeting and stimulation of the channels to open with appropriate cellular response (Hughes *et al.*, 2008). These examples proved the ability of a magnetic field to act remotely, while inducing cellular response in a highly controlled and accurate way.

Although the strategy of stress-responsive ion-channel stimulation has shown very promising results, the relatively rapid cell internalization of the attached MNPs limit their use to only short-term stimulation. This has led to extensive research to elucidate MNP internalization and the effect on internal cellular stimulation. Different cell types were shown to internalize different amounts of MNPs, depending on factors such as polymer hydrophobicity, particle size, and surface modification (for a review see Hughes *et al.*, 2005). Optimization studies are extensively performed as the potential of this field is expanding to applications such as cell transfections and internal manipulations within the cell of interest (Chorny *et al.*, 2007; McBain *et al.*, 2008; Pickard and Chari, 2010), or testing magnetic cell loading for use in targeted cell delivery (Polyak *et al.*, 2008; Chorny *et al.*, 2012; Macdonald *et al.*, 2012). Recently, MNP internalization was shown to be an active (requiring cytoskeleton reorganization) and a magnetic force-dependent process (Macdonald *et al.*, 2012).

Our group has recently developed magnetically-responsive alginate scaffolds and tested their ability to provide a means of physical stimulation to living cells seeded within the scaffold (Sapir *et al.*, 2012). The nanocomposite alginate scaffolds were impregnated with magnetic iron-oxide MNPs then seeded with bovine aortic ECs, and the cell construct was exposed to an alternating magnetic field. The MNP-impregnated scaffolds were found to be more elastic compared to pristine scaffolds, while incorporation of MNPs did not influence the macro-porosity structure of the scaffold (pore size and porosity) or their wetting extent by the culture medium (Fig. 9.8).

ECs cultivated in the magnetically stimulated constructs showed significantly elevated metabolic activity during the stimulation period, which was most likely related to cell migration and reorganization into loop-like structures. Immunostaining and confocal microscopy examination on day 14 revealed that the magnetically stimulated constructs, without supplementation of any angiogenic growth factors, contained vessel-like (loop) structures, while in the non-stimulated (control) scaffolds, the cells were mainly organized as sheets or aggregates (Plate X, see color section between pages 264 and 265).

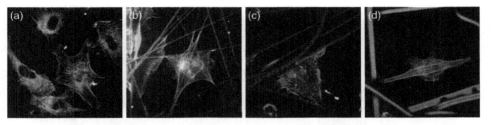

Plate I (Chapter 1) Cytoskeletal and focal adhesion of human aortic endothelial cells on electrospun nanofibers with various fiber diameters: (a) 270 nm, (b) 1 μm, (c) 2.39 μm, and (d) 4.45 μm (×600 magnification). Red: F-actin, green: vinculin. (*Source*: Reproduced with permission from Ju *et al.*, 2010, *Biomaterials*, 31, 4313–4321. © 2010 Elsevier.)

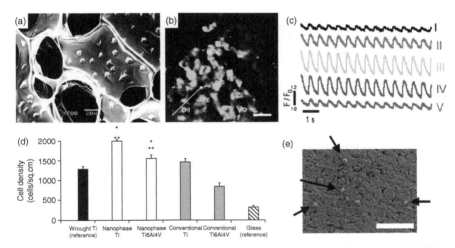

Plate II (Chapter 1) Gold nanowire-incorporating cardiac patch: (a) SEM image of nanowires assembled within the pore walls of the scaffold into star-shaped structures. (b, c) Calcium transient propagation at specific points (white circles) within the engineered tissues on alginate-nanowire scaffold. Green: calcium dye fluorescence; white arrow: the direction of propagation. (*Source* a-c): Reproduced with permission from Dvir *et al.*, 2011, *Nature Nanotechnology*, 6, 720–725. © 2011 Nature Publishing Group.) (d) Increased osteoblast adhesion on nanophase Ti and Ti6Al4V compacts. *$P < 0.01$ compared to respective conventional metal and **$P < 0.01$ compared to wrought Ti. (e) SEM image of osteoblasts on Ti compacts. (*Source* d-e): Reproduced from Webster and Ejiofor, 2004, *Biomaterials*, 25, 4731–4739. © 2004 Elsevier.)

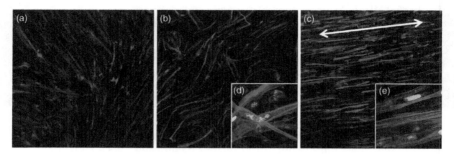

Plate III (Chapter 1) F-actin of human skeletal muscle cells (hSkMCs) on electrospun PCL/collagen nanofiber meshes: (a) culture dish, (b) randomly oriented, and (c) aligned electrospun nanofibrous scaffolds (×40 magnification). Laser confocal microscopy images of F-actin of hSkMCs seeded on (d) randomly oriented and (e) aligned scaffolds (×600 magnification). (*Source*: Reproduced with permission from Choi *et al.*, 2008, *Biomaterials*, 29, 2899–2906. © 2008 Elsevier.)

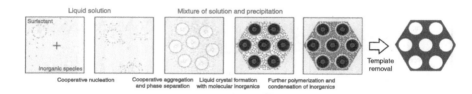

Plate IV (Chapter 3) Schematic of the cooperative self-assembly mechanism for mesoporous silica formation. (*Source*: Reproduced (adapted) with permission from Wan and Zhao, 2007, *Chemical Reviews*, 107, 2821–2860. © 2007 American Chemical Society.)

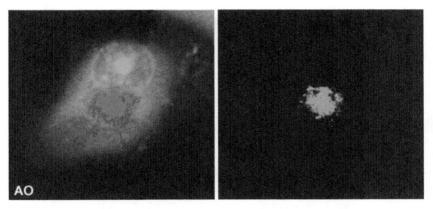

Plate V (Chapter 3) Uptake of the camptothecin-loaded mesoporous particles by pancreatic cancer cells; (left) stained with Acridine Orange; (right) fluorescence of the particles by themselves. (*Source*: Reproduced (adapted) with permission from Lu *et al.*, 2007, *Small*, 3, 1341–1346. © 2007 Wiley-VCH.)

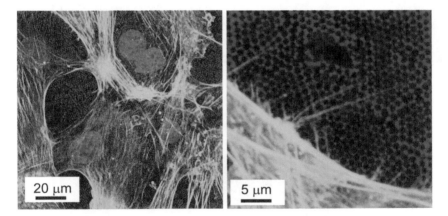

Plate VI (Chapter 3) Optical fluorescence of HMSCs on 1.3 μm pore SiO_2 inverse opal film, showing the actin filament networks (green) and the nuclei (red) (*Source*: Courtesy B. Hatton, M. Bucaro.)

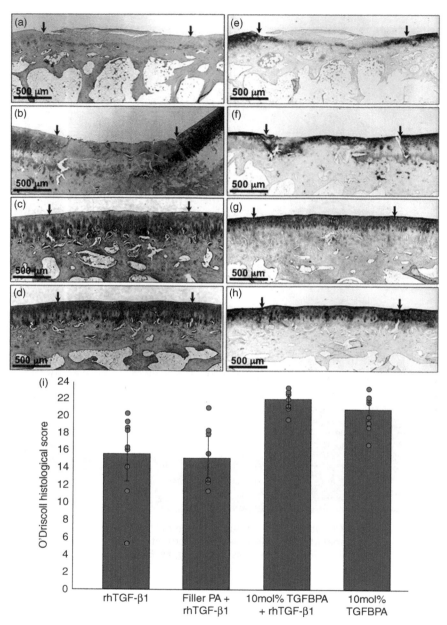

Plate VII (Chapter 6) (a–h) Histological evaluation of sample sections 12 weeks after treatment. Safranin-O staining for glycosaminoglycans (a–d) and type II collagen staining (e–h) in articular cartilage defects treated with (a, e) 100 ng/mL TGF-β1 (100TGF), (b, f) filler PA+TGF-β1, (c, g) 10%TGFBPA+TGF-β1, and (d, h) 10%TGFBPA alone 12 weeks post-op; (i) Histological scores of 12 week *in vivo* samples. Circles represent scores for individual specimens in each group (*n* = 8–10). The arrows show the boundaries of articular cartilage defects before treatment (Shah *et al.*, 2010).

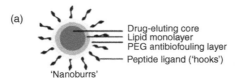

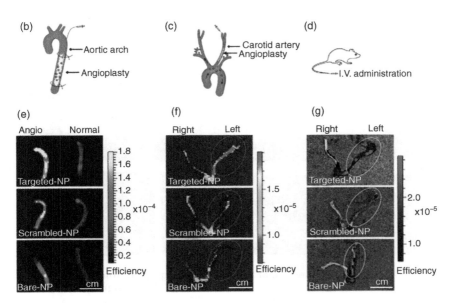

Plate VIII (Chapter 8) Engineered nanomaterials for the delivery of angiogenic factors. (i, a) Nanoparticles functionalized on the surface with target peptides for the vascular wall, containing a drug in its core. The system allowed spatial control of the delivery of the encapsulated drug, confirmed by: (b) *ex vivo* delivery in an abdominal aorta injury model, (c) *in vivo* IA delivery in a carotid injury mode, and (d) *in vivo* systemic delivery in a carotid angioplasty model. (e, f, g) Fluorescence images overlaid on photographs of: (e) balloon-injured aortas incubated with nanoparticles, compared with scrambled-peptide and non-targeted nanoparticles; (f, g) carotid arteries incubated with nanoparticles, compared with scrambled-peptide and non-targeted nanoparticles (Scale bar = 1 cm) (Chan *et al.*, 2010).

(*Continued*)

(ii) Gradients of GFs

(a)

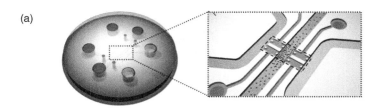

(b)

Day 1

Day 2

Day 3

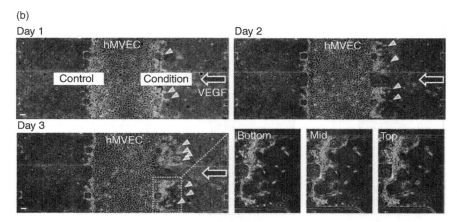

Plate VIII Continued. (ii, a) Schematics of a gradient microfluidic system that allows generation of a gradient of VEGF. (b) Sprouting of human microvascular endothelial cells (hMVEC), cultured in the centre, into the scaffold under the influence of a VEGF gradient. A biased angiogenic response on the condition side was observed, as confirmed by pictures taken at three focal planes (Chung *et al.*, 2010).

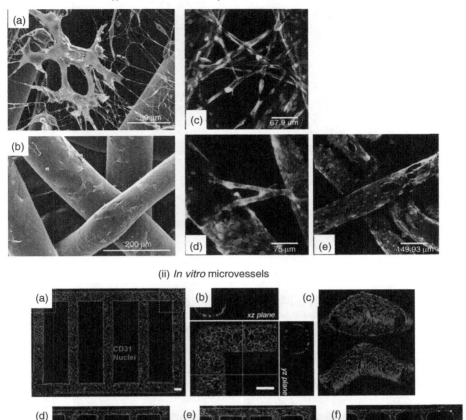

Plate IX (Chapter 8) Engineered nanomaterials for endothelial cell guidance.
(i) Nano/micro electrospun fibre mesh. (a, b) Scanning electron micrographs
of HUVEC cells seeded on (a) nano/microfibre-combined scaffold and (b)
microfibre scaffold after 3 days of culture. (c, d, e) Immunofluorescence (green)
micrographs of (c) vimentin and (d, e) PECAM-1 staining in HUVEC cells seeded
on (c, d) nano/microfibre-combined scaffold and (e) microfibre scaffold. Nuclei
were counterstained with Hoechst (blue fluorescence) (Santos *et al.*, 2008). (ii)
Engineered *in vitro* microvessels. (a) Z-stack projection of horizontal confocal
sections of micro-fluidic vessels with endothelial cells. Views of (b) the corner
and (c) branching sections (red, CD31; blue, nuclei; Scale bar = 100 μm). (d, e,
f) Sprouting angiogenesis and endothelia–pericyte interactions: (d) sprouting
into bulk collagen with pro-angiogenic stimuli (red, CD31; blue, nuclei); (e, f)
recruitment of pericyte-like cells from the collagen bulk with (e) endothelial
sprouting or (f) endothelial retraction (red, CD31; green, α-SMA; blue, nuclei)
(Zheng *et al.*, 2012).

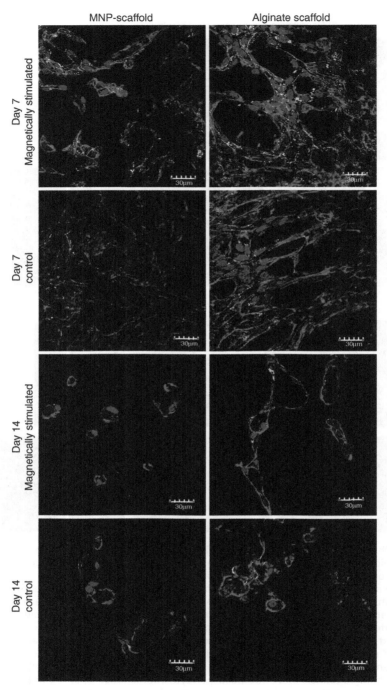

Plate X (Chapter 9) Endothelial cell organization in MNP-alginate and alginate constructs, on days 7 and 14 post cell seeding (bar: 30 μm). (*Source*: Figure reproduced with permission from Sapir *et al.* (2012).)

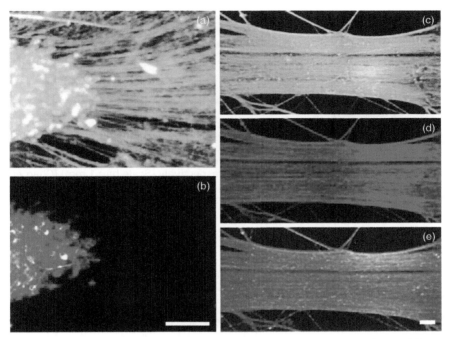

Plate XI (Chapter 10) Growth and orientation of nerve cell fibers: (a) axons stained with anti-tau-protein, (b) the nerve cell bodies stained with anti-MAP2 protein. Confocal microscope images of elongated axons: (c) tubulin, (d) phosphorylated neurofilament, (e) tau protein. (*Source:* Reprinted from Biazar *et al.* (2010), Copyright 2010, with permission from Dove Medical Press Ltd.)

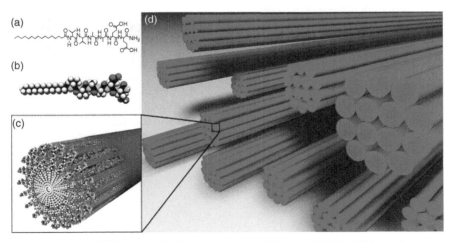

Plate XII (Chapter 10) Structure of peptide-amphiphile (PA) used to form mono-domain noodle gels. (a) Molecular model of the PA molecule, (b) nanofiber formation through self-assembly, (c) nanofiber bundles assembled in a longitudinal alignment after noodle formation. (*Source*: Reprinted from Angeloni *et al.* (2011), Copyright 2011, with permission from Elsevier.)

Plate XIII (Chapter 11) Silk hydrogels produce cartilaginous structures after 42 days of chondrocyte culture. (a) Construct gross appearance, (b) picrosirius red (left) and alcian blue (right) staining shows the presence of collagen and GAG, respectively (Scale bar: 1 mm) (Chao *et al.*, 2010).

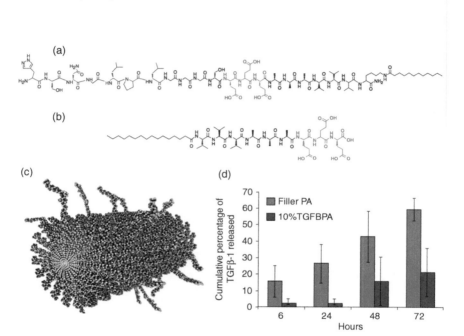

Plate XIV (Chapter 11) Design of PAs for articular cartilage regeneration. (a) Chemical structure of TGF-binding PA and (b) nonbioactive PA. (c) Illustration of co-assembly of TGF-binding and nonbioactive PA showing binding epitopes exposed on the surface of the nanofiber. (d) ELISA results show TGFβ-1 release for up to 72 h (Shah *et al.*, 2010).

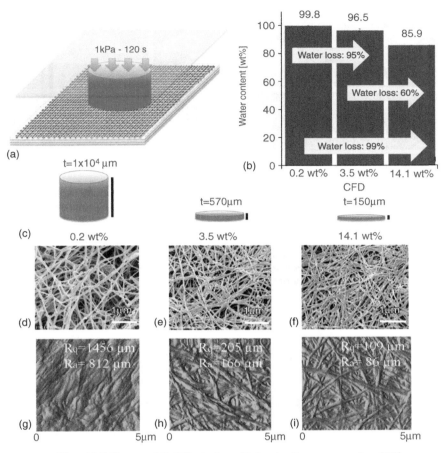

Plate XV (Chapter 14) Effect of multiple plastic compression (PC) on collagen gel fibrillar density and morphology. (a) Schematic diagram of the PC process applied on a collagen gel to generate sheets of controlled CFDs. (b) Bar chart representing the water content versus gel CFD (weight %), as made (0.2 wt%), post-single compression (3.5 wt%) and post-double compression (14.1 wt%). Arrows indicate the weight loss achieved through each compression stage. (c) Schematic diagram of the collagen gel obtained with different densities. (d, e, f) SEM micrographs of gels with increasing CFDs. (g, h, i) AFM profilometry micrographs of gels with increasing CFDs, with relative roughness values (Rq. RMS roughness, Ra. arithmetic roughness). For the SEM and the AFM micrographs, left, middle and right panels refer to CFDs of 0.2, 3.5 and 14.1 wt%, respectively (Zhao *et al.*, 2011). (*Source*: © 2011 The Royal Society of Chemistry.)

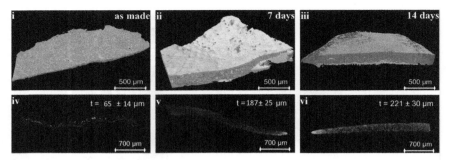

Plate XVI (Chapter 14) Morphological analysis of DC-µBG 40–60 scaffolds at different time points in SBF. MicroCT 3D reconstruction at days (i) 0, (ii) 7 and (iii) 14. Red indicates higher density regions. In the as made material they represent the presence of Bioglass®, while at days 7 and 14 in SBF, they represent zones of mineral formation. (iv, v, vi) microCT scan of DC-µBG 40–60, as made, and at days 7 and 14 in SBF, respectively. The lighter the grey, the higher the density. 2D analysis also permitted the evaluation of scaffold expansion through increase in the cross-sectional thickness (t). (*Source*: Copyright © 2011 The Royal Society of Chemistry.)

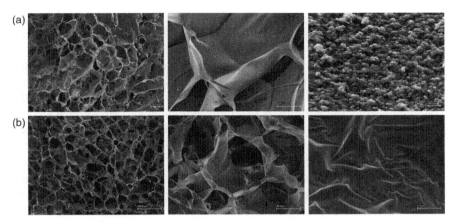

9.8 Scanning electron microscope (SEM) images of MNP-impregnated alginate scaffold. SEM of the (a) 1.2% (w/v) MNP-alginate and (b) non-magnetic alginate scaffolds. Note the presence of nanoparticles in MNP-alginate scaffold walls. (*Source*: Figure reproduced with permission from Sapir *et al.* (2012).)

The exact mechanisms acting on the cells within the nanocomposite scaffold under magnetic stimulation are still not clear, however some assumptions can be made. We showed that the impregnation process results in a high density of MNPs embedded within the scaffold wall, in close proximity, and interacting with each other. MNP interactions within the scaffold wall resemble the domain interactions existing in ferromagnetic materials (Frenkel and Doefman, 1930), leading to the phenomenon called magnetostriction. Magnetostriction causes the bulk materials to change their shape or dimensions during the process of magnetization, due to the domain interactions. Such an effect would lead to overall scaffold contraction and consequently to cell stimulation, migration, and organization into a tissue. This may explain the endothelial cell organization as observed in the magnetically stimulated alginate scaffold. Although this hypothesis appears to be plausible, it needs further investigation and confirmation.

9.6   Conclusion and future trends

The chapter presented the recent implementation of nanomaterials and nanofabrication techniques into the exciting area of CTE. The synthesis and incorporation of nanostructures into existing matrices, nano-actuation *via* highly controlled cell stimulation, and the precise presentation of inductive cues in close proximity to a target cell promise to bring us closer to the creation of a biomimetic environment with physiologically relevant properties that could promote the regeneration of thick contractile cardiac tissue. It is also clear that the nano-scale science will be the vision molding for future tissue engineers. All these are critically dependent on addressing a few challenges in this field.

First, the toxicity and side-effects of nanomaterials should be addressed. Recent studies pointed to possible toxicity and side-effects when using carbon nanotubes (Kunzmann *et al.*, 2011; Shen *et al.*, 2012). Although in tissue engineering, the nanofibers used are biological and ECM derived, still the biocompatibility of the materials needs to be thoroughly investigated before they can be safely applied in clinical trials.

Second, precise control over nanopatterning of topographic and biological cues has been mainly achievable in the two-dimensional (2D) level, leading to the successful creation of functional cell monolayers. Will it be possible to attain the same level of control in the 3D environment? Furthermore, will it be possible to attain the spatio-temporal distribution of multiple cues?

If such issues can be addressed, nanotechnology has the potential to solve many of the challenges still ahead in CTE. Pre-vascularization of the cardiac patch can be accelerated by preparing nanofibers releasing angiogenic factors (Montero *et al.*, 2012) or by impregnation of NPs releasing these factors (Mima *et al.*, 2012). In addition, we can continue and optimize the magneto-mechanical cell stimulation strategy to accelerate EC organization into vessels.

In summary, we envision the future of CTE as offering efficient solutions of 'smart' inductive scaffolds closely mimicking cardiac ECM, able to precisely control the tissue regeneration course by presenting inductive cues based on cellular demand, as well as by applying stimulation patterns in a selective and controlled way. These time-controlled processes could be driven by using nano-biosensors, able to sense and guide the release of growth factors and trigger cell stimulation within the complex scaffold. Finally, the engineering process may also be efficiently enabled *in vivo,* if needed, based on the cell stimulation patterns controlled from a distance (as in the case of magnetic field stimulation).

9.7   Acknowledgements

Yulia Sapir gratefully acknowledges the generous fellowship from the late Mr Daniel Falkner and his daughter Ms Ann Berger and thanks the Azrieli Foundation for awarding the Azrieli Fellowship supporting her PhD program. The study was partially supported by The Israel Science Foundation (grant # 1368/08) (SC), Israel Ministry of Science (SC), The Louis and Bessie Stein Family foundation through the Drexel University College of Medicine (BP, YS, SC), American Associates of the Ben-Gurion University of the Negev, Israel (YS), the European Union FWP7 (INELPY) (SC) and the National Heart, Lung and Blood Institute (Award Number 5R01HL107771) (BP).

The content is solely the responsibility of the authors and does not necessarily represent the official views of the National Heart, Lung and Blood Institute or the National Institutes of Health.

Professor Cohen holds the Claire and Harold Oshry Professor Chair in Biotechnology.

9.8   References

Akhyari, P., Fedak, P. W., Weisel, R. D., Lee, T. Y., Verma, S., Mickle, D. A. and Li, R. K. (2002). Mechanical stretch regimen enhances the formation of bioengineered autologous cardiac muscle grafts. *Circulation*, **106**, I137–42.

Akiyama, H., Ito, A., Sato, M., Kawabe, Y. and Kamihira, M. (2010). Construction of cardiac tissue rings using a magnetic tissue fabrication technique. *Int J Mol Sci*, **11**, 2910–20.

Barash, Y., Dvir, T., Tandeitnik, P., Ruvinov, E., Guterman, H. and Cohen, S. (2010). Electric field stimulation integrated into perfusion bioreactor for cardiac tissue engineering. *Tissue Eng Part C Methods*, **16**, 1417–26.

Bock, N., Riminucci, A., Dionigi, C., Russo, A., Tampieri, A., Landi, E., Goranov, V. A., Marcacci, M. and Dediu, V. (2010). A novel route in bone tissue engineering: Magnetic biomimetic scaffolds. *Acta Biomater*, **6**, 786–96.

Boersma, E., Mercado, N., Poldermans, D., Gardien, M., Vos, J. and Simoons, M. L. (2003). Acute myocardial infarction. *Lancet*, **361**, 847–58.

Brule, S., Levy, M., Wilhelm, C., Letourneur, D., Gazeau, F., Menager, C. and Le Visage, C. (2011). Doxorubicin release triggered by alginate embedded magnetic nanoheaters: A combined therapy. *Advanced Materials*, **23**, 787–790.

Cardin, A. D. and Weintraub, H. J. (1989). Molecular modeling of protein-glycosaminoglycan interactions. *Arteriosclerosis*, **9**, 21–32.

Carver, S. E. and Heath, C. A. (1999). Semi-continuous perfusion system for delivering intermittent physiological pressure to regenerating cartilage. *Tissue Eng*, **5**, 1–11.

Caspi, O., Lesman, A., Basevitch, Y., Gepstein, A., Arbel, G., Habib, I. H., Gepstein, L. and Levenberg, S. (2007). Tissue engineering of vascularized cardiac muscle from human embryonic stem cells. *Circ Res*, **100**, 263–72.

Cellot, G., Cilia, E., Cipollone, S., Rancic, V., Sucapane, A., Giordani, S., Gambazzi, L., Markram, H., Grandolfo, M., Scaini, D., Gelain, F., Casalis, L., Prato, M., Giugliano, M. and Ballerini, L. 2009. Carbon nanotubes might improve neuronal performance by favouring electrical shortcuts. *Nat Nanotechnol*, **4**, 126–33.

Chorny, M., Alferiev, I. S., Fishbein, I., Tengood, J. E., Folchman-Wagner, Z., Forbes, S. P. and Levy, R. J. (2012). Formulation and in vitro characterization of composite biodegradable magnetic nanoparticles for magnetically guided cell delivery. *Pharm Res*, **29**, 1232–41.

Chorny, M., Polyak, B., Alferiev, I. S., Walsh, K., Friedman, G. and Levy, R. J. (2007). Magnetically driven plasmid DNA delivery with biodegradable polymeric nanoparticles. *FASEB J*, **21**, 2510–9.

Cohen-Karni, T., Timko, B. P., Weiss, L. E. and Lieber, C. M. (2009). Flexible electrical recording from cells using nanowire transistor arrays. *Proc Natl Acad Sci U S A*, **106**, 7309–13.

Derfus, A. M., Von Maltzahn, G., Harris, T. J., Duza, T., Vecchio, K. S., Ruoslahti, E. and Bhatia, S. N. (2007). Remotely triggered release from magnetic nanoparticles. *Adv Mater*, **19**, 3932–3936.

Dvir,T.,Kedem,A.,Ruvinov,E.,Levy,O.,Freeman,I.,Landa,N.,Holbova,R.,Feinberg, M. S., Dror, S., Etzion, Y., Leor, J. and Cohen, S. (2009). Prevascularization of cardiac patch on the omentum improves its therapeutic outcome. *Proc Natl Acad Sci U S A*, **106**, 14990–5.

Dvir, T., Levy, O., Shachar, M., Granot, Y. and Cohen, S. (2007). Activation of the ERK1/2 Cascade via pulsatile interstitial fluid flow promotes cardiac tissue assembly. *Tissue Eng*, **13**, 2185–93.

Dvir, T., Timko, B. P., Brigham, M. D., Naik, S. R., Karajanagi, S. S., Levy, O., Jin, H., Parker, K. K., Langer, R. and Kohane, D. S. (2011). Nanowired three-dimensional cardiac patches. *Nat Nanotechnol*, **6**, 720–5.

Freeman, I. and Cohen, S. (2009). The influence of the sequential delivery of angiogenic factors from affinity-binding alginate scaffolds on vascularization. *Biomaterials*, **30**, 2122–31.

Freeman, I., Kedem, A. and Cohen, S. (2008). The effect of sulfation of alginate hydrogels on the specific binding and controlled release of heparin-binding proteins. *Biomaterials*, **29**, 3260–8.

Frenkel, J. and Doefman, J. (1930). Spontaneous and induced magnetisation in ferromagnetic bodies. *Nature*, **126**, 274–5.

Gregorio, C. C. and Antin, P. B. (2000). To the heart of myofibril assembly. *Trends Cell Biol*, **10**, 355–62.

Guldberg, R. E. (2002). Consideration of mechanical factors. *Ann N Y Acad Sci*, **961**, 312–4.

Halbach, M., Egert, U., Hescheler, J. and Banach, K. (2003). Estimation of action potential changes from field potential recordings in multicellular mouse cardiac myocyte cultures. *Cell Physiol Biochem*, **13**, 271–84.

Heer, F., Hafizovic, S., Ugniwenko, T., Frey, U., Franks, W., Perriard, E., Perriard, J. C., Blau, A., Ziegler, C. and Hierlemann, A. (2007). Single-chip microelectronic system to interface with living cells. *Biosens Bioelectron*, **22**, 2546–53.

Hoare, T., Timko, B. P., Santamaria, J., Goya, G. F., Irusta, S., Lau, S., Stefanescu, C. F., Lin, D. B., Langer, R. and Kohane, D. S. (2011). Magnetically triggered nanocomposite membranes: A versatile platform for triggered drug release. *Nano Lett*, **11**, 1395–400.

Hsieh, P. C., Davis, M. E., Lisowski, L. K. and Lee, R. T. (2006). Endothelial-cardiomyocyte interactions in cardiac development and repair. *Annu Rev Physiol*, **68**, 51–66.

Hu, S. H., Liu, T. Y., Tsai, C. H. and Chen, S. Y. (2007). Preparation and characterization of magnetic ferroscaffolds for tissue engineering. *J Magn Magn Mater*, **310**, 2871–3.

Hughes, S., Dobson, J. and El Haj, A. J. (2007). Magnetic targeting of mechanosensors in bone cells for tissue engineering applications. *J Biomech*, **40**, Supplement 1, S96–104.

Hughes, S., El Haj, A. J. and Dobson, J. (2005). Magnetic micro- and nanoparticle mediated activation of mechanosensitive ion channels. *Med Eng Phys*, **27**, 754–62.

Hughes, S., McBain, S., Dobson, J. and El Haj, A. J. (2008). Selective activation of mechanosensitive ion channels using magnetic particles. *J R Soc Interface*, **5**, 855–63.

Ingebrandt, S., Yeung, C. K., Staab, W., Zetterer, T. and Offenhausser, A. (2003). Backside contacted field effect transistor array for extracellular signal recording. *Biosens Bioelectron*, **18**, 429–35.

Inoue, N., Ohkusa, T., Nao, T., Lee, J. K., Matsumoto, T., Hisamatsu, Y., Satoh, T., Yano, M., Yasui, K., Kodama, I. and Matsuzaki, M. (2004). Rapid electrical stimulation of contraction modulates gap junction protein in neonatal rat cultured cardiomyocytes: involvement of mitogen-activated protein kinases and effects of angiotensin II-receptor antagonist. *J Am Coll Cardiol*, **44**, 914–22.

Ishii, M., Shibata, R., Numaguchi, Y., Kito, T., Suzuki, H., Shimizu, K., Ito, A., Honda, H. and Murohara, T. (2011). Enhanced angiogenesis by transplantation of mesenchymal stem cell sheet created by a novel magnetic tissue engineering method. *Arterioscler Thromb Vasc Biol*, **31**, 2210–5.

Ito, A., Akiyama, H., Yamamoto, Y., Kawabe, Y. and Kamihira, M. (2009). Skeletal muscle tissue engineering using functional magnetite nanoparticles. 2009 International Symposium on Micro-NanoMechatronics and Human Science (MHS), 9–11 Nov. 2009, 2009 Piscataway, NJ, USA. IEEE, 379–82.

Ito, A., Hibino, E., Kobayashi, C., Terasaki, H., Kagami, H., Ueda, M., Kobayashi, T. and Honda, H. (2005a). Construction and delivery of tissue-engineered human retinal pigment epithelial cell sheets, using magnetite nanoparticles and magnetic force. *Tissue Eng*, **11**, 489–96.

Ito, A., Ino, K., Hayashida, M., Kobayashi, T., Matsunuma, H., Kagami, H., Ueda, M. and Honda, H. (2005b). Novel methodology for fabrication of tissue-engineered tubular constructs using magnetite nanoparticles and magnetic force. *Tissue Eng*, **11**, 1553–61.

Ito, A., Ino, K., Shimizu, K., Honda, H. and Kamihira, M. Fabrication of 3D tissue-like structure using magnetite nanoparticles and magnetic force. 2006 IEEE International Symposium on Micro-Nano Mechanical and Human Science, MHS, November 5, 2006–November 8, 2006, 2006 Nagoya, Japan. Inst. of Elec. and Elec. Eng. Computer Society, IEEE Robotics and Automation Society; Nagoya University; City of Nagoya; Chubu Science and Technology Center; Nagoya Urban Industries Promotion Corporation.

Ito, A., Takizawa, Y., Honda, H., Hata, K. I., Kagami, H., Ueda, M. and Kobayashi, T. (2004). Tissue engineering using magnetite nanoparticles and magnetic force: Heterotypic layers of cocultured hepatocytes and endothelial cells. *Tissue Eng*, **10**, 833–40.

Iyer, R. K., Chiu, L. L. and Radisic, M. (2009). Microfabricated poly(ethylene glycol) templates enable rapid screening of triculture conditions for cardiac tissue engineering. *J Biomed Mater Res A*, **89**, 616–31.

Iyer, R. K., Odedra, D., Chiu, L. L., Vunjak-Novakovic, G. and Radisic, M. (2012). VEGF secretion by non-myocytes modulates connexin-43 levels in cardiac organoids. *Tissue Eng Part A*, **18**(17–18), 1771–83.

Jessup, M. and Brozena, S. (2003). Heart failure. *N Engl J Med*, **348**, 2007–18.

Kanczler, J. M., Sura, H. S., Magnay, J., Green, D., Oreffo, R. O., Dobson, J. P. and El Haj, A. J. (2010). Controlled differentiation of human bone marrow stromal cells using magnetic nanoparticle technology. *Tissue Eng Part A*, **16**, 3241–50.

Kawahara, Y., Yamaoka, K., Iwata, M., Fujimura, M., Kajiume, T., Magaki, T., Takeda, M., Ide, T., Kataoka, K., Asashima, M. and Yuge, L. (2006). Novel electrical stimulation sets the cultured myoblast contractile function to 'on'. *Pathobiology*, **73**, 288–94.

Kempen, D. H. R., Lu, L. C., Heijink, A., Hefferan, T. E., Creemers, L. B., Maran, A., Yaszemski, M. J. and Dhert, W. J. A. (2009). Effect of local sequential VEGF and BMP-2 delivery on ectopic and orthotopic bone regeneration. *Biomaterials*, **30**, 2816–25.

Kim, D. H., Lu, N., Ghaffari, R., Kim, Y. S., Lee, S. P., Xu, L., Wu, J., Kim, R. H., Song, J., Liu, Z., Viventi, J., De Graff, B., Elolampi, B., Mansour, M., Slepian, M. J., Hwang, S., Moss, J. D., Won, S. M., Huang, Y., Litt, B. and Rogers, J. A. (2011a). Materials for multifunctional balloon catheters with capabilities in cardiac electrophysiological mapping and ablation therapy. *Nat Mater*, **10**, 316–23.

Kim, D. H., Lu, N., Ma, R., Kim, Y. S., Kim, R. H., Wang, S., Wu, J., Won, S. M., Tao, H., Islam, A., Yu, K. J., Kim, T. I., Chowdhury, R., Ying, M., Xu, L., Li, M., Chung, H. J., Keum, H., McCormick, M., Liu, P., Zhang, Y. W., Omenetto, F. G., Huang, Y., Coleman, T. and Rogers, J. A. (2011b). Epidermal electronics. *Science*, **333**, 838–43.

Kleber, A. G. and Rudy, Y. (2004). Basic mechanisms of cardiac impulse propagation and associated arrhythmias. *Physiol Rev*, **84**, 431–88.

Klemic, K. G., Klemic, J. F., Reed, M. A. and Sigworth, F. J. (2002). Micromolded PDMS planar electrode allows patch clamp electrical recordings from cells. *Biosens Bioelectron*, **17**, 597–604.

Kung, H. C., Hoyert, D. L., Xu, J. and Murphy, S. L. (2008). Deaths: Final data for 2005. *Natl Vital Stat Rep*, **56**, 1–120.

Kunzmann, A., Andersson, B., Thurnherr, T., Krug, H., Scheynius, A. and Fadeel, B. (2011). Toxicology of engineered nanomaterials: Focus on biocompatibility, biodistribution and biodegradation. *Biochim Biophys Acta*, **1810**, 361–73.

Kuramochi, Y., Cote, G. M., Guo, X., Lebrasseur, N. K., Cui, L., Liao, R. and Sawyer, D. B. (2004). Cardiac endothelial cells regulate reactive oxygen species-induced cardiomyocyte apoptosis through neuregulin-1beta/erbB4 signaling. *J Biol Chem*, **279**, 51141–7.

Lesman, A., Habib, M., Caspi, O., Gepstein, A., Arbel, G., Levenberg, S. and Gepstein, L. (2010). Transplantation of a tissue-engineered human vascularized cardiac muscle. *Tissue Eng Part A*, **16**, 115–25.

Levenberg, S., Rouwkema, J., Macdonald, M., Garfein, E. S., Kohane, D. S., Darland, D. C., Marini, R., van Blitterswijk, C. A., Mulligan, R. C., D'Amore, P. A. and Langer, R. (2005). Engineering vascularized skeletal muscle tissue. *Nat Biotechnol*, **23**, 879–84.

Li, X., Klemic, K. G., Reed, M. A. and Sigworth, F. J. (2006). Microfluidic system for planar patch clamp electrode arrays. *Nano Lett*, **6**, 815–9.

Liu, T. Y., Hu, S. H., Liu, D. M., Chen, S. Y. and Chen, I. W. (2009). Biomedical nano-particle carriers with combined thermal and magnetic responses. *Nano Today*, **4**, 52–65.

MacDonald, C., Barbee, K. and Polyak, B. (2012). Force dependent internalization of magnetic nanoparticles results in highly loaded endothelial cells for use as potential therapy delivery vectors. *Pharm Res*, **29**(5), 1270–81.

Marchal, G., Van Hecke, P., Demaerel, P., Decrop, E., Kennis, C., Baert, A. L. and van der Schueren, E. (1989). Detection of liver metastases with superparamagnetic iron oxide in 15 patients: results of MR imaging at 1.5 T. *AJR Am J Roentgenol*, **152**, 771–5.

Matthews, B. D., Overby, D. R., Mannix, R. and Ingber, D. E. (2006). Cellular adaptation to mechanical stress: Role of integrins, Rho, cytoskeletal tension and mechanosensitive ion channels. *J Cell Sci*, **119**, 508–18.

McBain, S. C., Griesenbach, U., Xenariou, S., Keramane, A., Batich, C. D., Alton, E. W. and Dobson, J. (2008). Magnetic nanoparticles as gene delivery agents: Enhanced transfection in the presence of oscillating magnet arrays. *Nanotechnology*, **19**, 405102.

McDonough, P. M. and Glembotski, C. C. (1992). Induction of atrial natriuretic factor and myosin light chain-2 gene expression in cultured ventricular myocytes by electrical stimulation of contraction. *J Biol Chem*, **267**, 11665–8.

McMurray, J. J. (2010). Clinical practice. Systolic heart failure. *N Engl J Med*, **362**, 228–38.

Meyer, T., Boven, K. H., Gunther, E. and Fejtl, M. (2004). Micro-electrode arrays in cardiac safety pharmacology: A novel tool to study QT interval prolongation. *Drug Saf*, **27**, 763–72.

Mima, Y., Fukumoto, S., Koyama, H., Okada, M., Tanaka, S., Shoji, T., Emoto, M., Furuzono, T., Nishizawa, Y. and Inaba, M. (2012). Enhancement of cell-based therapeutic angiogenesis using a novel type of injectable scaffolds of hydroxyapatite-polymer nanocomposite microspheres. *PLoS One*, **7**, e35199.

Montero, R. B., Vial, X., Nguyen, D. T., Farhand, S., Reardon, M., Pham, S. M., Tsechpenakis, G. and Andreopoulos, F. M. (2012). bFGF-containing electro-spun gelatin scaffolds with controlled nano-architectural features for directed angiogenesis. *Acta Biomaterialia*, **8**, 1778–1791.

Nabel, E. G. and Braunwald, E. (2012). A tale of coronary artery disease and myocardial infarction. *N Engl J Med*, **366**, 54–63.

Narmoneva, D. A., Vukmirovic, R., Davis, M. E., Kamm, R. D. and Lee, R. T. (2004). Endothelial cells promote cardiac myocyte survival and spatial reorganization: Implications for cardiac regeneration. *Circulation*, **110**, 962–8.

Nishida, M., Springhorn, J. P., Kelly, R. A. and Smith, T. W. (1993). Cell-cell signaling between adult rat ventricular myocytes and cardiac microvascular endothelial cells in heterotypic primary culture. *J Clin Invest*, **91**, 1934–41.

Orlova, Y., Magome, N., Liu, L., Chen, Y. and Agladze, K. (2011). Electrospun nanofibers as a tool for architecture control in engineered cardiac tissue. *Biomaterials*, **32**, 5615–24.

Patel, Z. S., Young, S., Tabata, Y., Jansen, J. A., Wong, M. E. K. and Mikos, A. G. (2008). Dual delivery of an angiogenic and an osteogenic growth factor for bone regeneration in a critical size defect model. *Bone*, **43**, 931–940.

Patolsky, F., Timko, B. P., Yu, G., Fang, Y., Greytak, A. B., Zheng, G. and Lieber, C. M. (2006). Detection, stimulation, and inhibition of neuronal signals with high-density nanowire transistor arrays. *Science*, **313**, 1100–4.

Perets, A., Baruch, Y., Weisbuch, F., Shoshany, G., Neufeld, G. and Cohen, S. (2003). Enhancing the vascularization of three-dimensional porous alginate scaffolds by incorporating controlled release basic fibroblast growth factor microspheres. *J Biomed Mater Res A*, **65**, 489–97.

Pickard, M. and Chari, D. (2010). Enhancement of magnetic nanoparticle-mediated gene transfer to astrocytes by 'magnetofection': Effects of static and oscillating fields. *Nanomedicine (Lond)*, **5**, 217–32.

Polyak, B., Fishbein, I., Chorny, M., Alferiev, I., Williams, D., Yellen, B., Friedman, G. and Levy, R. J. (2008). High field gradient targeting of magnetic nanoparticle-loaded endothelial cells to the surfaces of steel stents. *Proc Natl Acad Sci U S A*, **105**, 698–703.

Pui, T. S., Agarwal, A., Ye, F., Balasubramanian, N. and Chen, P. (2009). CMOS-Compatible nanowire sensor arrays for detection of cellular bioelectricity. *Small*, **5**, 208–12.

Radisic, M., Park, H., Martens, T. P., Salazar-Lazaro, J. E., Geng, W., Wang, Y., Langer, R., Freed, L. E. and Vunjak-Novakovic, G. (2008). Pre-treatment of synthetic elastomeric scaffolds by cardiac fibroblasts improves engineered heart tissue. *J Biomed Mater Res A*, **86**, 713–24.

Radisic, M., Park, H., Shing, H., Consi, T., Schoen, F. J., Langer, R., Freed, L. E. and Vunjak-Novakovic, G. (2004a). Functional assembly of engineered myocardium by electrical stimulation of cardiac myocytes cultured on scaffolds. *Proc Natl Acad Sci U S A*, **101**, 18129–34.

Radisic, M., Yang, L., Boublik, J., Cohen, R. J., Langer, R., Freed, L. E. and Vunjak-Novakovic, G. (2004b). Medium perfusion enables engineering of compact and contractile cardiac tissue. *Am J Physiol Heart Circ Physiol*, **286**, H507–16.

Ramaciotti, C., Sharkey, A., McClellan, G. and Winegrad, S. (1992). Endothelial cells regulate cardiac contractility. *Proc Natl Acad Sci U S A*, **89**, 4033–6.

Roberts, S. R., Knight, M. M., Lee, D. A. and Bader, D. L. (2001). Mechanical compression influences intracellular Ca2+ signaling in chondrocytes seeded in agarose constructs. *J Appl Physiol*, **90**, 1385–91.

Rosso, F., Giordano, A., Barbarisi, M. and Barbarisi, A. (2004). From cell-ECM interactions to tissue engineering. *J Cell Physiol*, **199**, 174–80.

Ruvinov, E., Dvir, T., Leor, J. and Cohen, S. (2008). Myocardial repair: from salvage to tissue reconstruction. *Expert Rev Cardiovasc Ther*, **6**, 669–86.

Sapir, Y., Cohen, S., Friedman, G. and Polyak, B. (2012). The promotion of in vitro vessel-like organization of endothelial cells in magnetically responsive alginate scaffolds. *Biomaterials*, **33**, 4100–9.

Sapir, Y., Kryukov, O. and Cohen, S. (2011). Integration of multiple cell-matrix interactions into alginate scaffolds for promoting cardiac tissue regeneration. *Biomaterials*, **32**, 1838–1847.

Shachar, M., Tsur-Gang, O., Dvir, T., Leor, J. and Cohen, S. (2011). The effect of immobilized RGD peptide in alginate scaffolds on cardiac tissue engineering. *Acta Biomater*, **7**, 152–62.

Shen, W., Cai, K., Yang, Z., Yan, Y., Yang, W. and Liu, P. (2012). Improved endothelialization of NiTi alloy by VEGF functionalized nanocoating. *Colloid Surf B Biointerf*, **94**, 347–53.

Shimizu, K., Ito, A. and Honda, H. (2007a). Mag-seeding of rat bone marrow stromal cells into porous hydroxyapatite scaffolds for bone tissue engineering. *J Biosci Bioeng*, **104**, 171–7.

Shimizu, K., Ito, A., Lee, J. K., Yoshida, T., Miwa, K., Ishiguro, H., Numaguchi, Y., Murohara, T., Kodama, I. and Honda, H. (2007b). Construction of multi-layered cardiomyocyte sheets using magnetite nanoparticles and magnetic force. *Biotechnol Bioeng*, **96**, 803–9.

Shimizu, K., Ito, A., Yoshida, T., Yamada, Y., Ueda, M. and Honda, H. (2007c). Bone tissue engineering with human mesenchymal stem cell sheets constructed using magnetite nanoparticles and magnetic force. *J Biomed Mater Res – Part B Appl Biomater*, **82**, 471–80.

Shin, H., Jo, S. and Mikos, A. G. (2003). Biomimetic materials for tissue engineering. *Biomaterials*, **24**, 4353–64.

Silka, M. J., Hardy, B. G., Menashe, V. D. and Morris, C. D. (1998). A population-based prospective evaluation of risk of sudden cardiac death after operation for common congenital heart defects. *J Am Coll Cardiol*, **32**, 245–51.

Spinale, F. G. (2007). Myocardial matrix remodeling and the matrix metalloproteinases: Influence on cardiac form and function. *Physiol Rev*, **87**, 1285–342.

Tian, B., Liu, J., Dvir, T., Jin, L., Tsui, J. H., Qing, Q., Suo, Z., Langer, R., Kohane, D. S. and Lieber, C. M. (2012). Macroporous nanowire nanoelectronic scaffolds for synthetic tissues. *Nat Mater*, **11**, 986–94.

Timko, B. P., Cohen-Karni, T., Yu, G., Qing, Q., Tian, B. and Lieber, C. M. (2009). Electrical recording from hearts with flexible nanowire device arrays. *Nano Lett*, **9**, 914–8.

Tortora, G. J., Derrickson, B. H. (2008). *Principles of Anatomy and Physiology*, John Wiley & Sons.

Tsur-Gang, O., Ruvinov, E., Landa, N., Holbova, R., Feinberg, M. S., Leor, J. and Cohen, S. (2009). The effects of peptide-based modification of alginate on left ventricular remodeling and function after myocardial infarction. *Biomaterials*, **30**, 189–95.

Ventura, C., Maioli, M., Asara, Y., Santoni, D., Mesirca, P., Remondini, D. and Bersani, F. (2005). Turning on stem cell cardiogenesis with extremely low frequency magnetic fields. *FASEB J*, **19**, 155–7.

Viventi, J., Kim, D. H., Moss, J. D., Kim, Y. S., Blanco, J. A., Annetta, N., Hicks, A., Xiao, J., Huang, Y., Callans, D. J., Rogers, J. A. and Litt, B. (2010). A conformal, bio-interfaced class of silicon electronics for mapping cardiac electrophysiology. *Sci Transl Med*, **2**, 24ra22.

Viventi, J., Kim, D. H., Vigeland, L., Frechette, E. S., Blanco, J. A., Kim, Y. S., Avrin, A. E., Tiruvadi, V. R., Hwang, S. W., Vanleer, A. C., Wulsin, D. F., Davis, K., Gelber, C. E., Palmer, L., van der Spiegel, J., Wu, J., Xiao, J., Huang, Y., Contreras, D., Rogers, J. A. and Litt, B. (2011). Flexible, foldable, actively multiplexed, high-density electrode array for mapping brain activity in vivo. *Nat Neurosci*, **14**, 1599–605.

Weissleder, R., Stark, D. D., Engelstad, B. L., Bacon, B. R., Compton, C. C., White, D. L., Jacobs, P. and Lewis, J. (1989). Superparamagnetic iron oxide: Pharmacokinetics and toxicity. *AJR Am J Roentgenol*, **152**, 167–73.

Whelan, R. S., Mani, K. and Kitsis, R. N. (2007). Nipping at cardiac remodeling. *J Clin Invest*, **117**, 2751–3.

Yamada, M., Tanemura, K., Okada, S., Iwanami, A., Nakamura, M., Mizuno, H., Ozawa, M., Ohyama-Goto, R., Kitamura, N., Kawano, M., Tan-Takeuchi, K., Ohtsuka, C., Miyawaki, A., Takashima, A., Ogawa, M., Toyama, Y., Okano, H. and Kondo, T. (2007). Electrical stimulation modulates fate determination of differentiating embryonic stem cells. *Stem Cells*, **25**, 562–70.

Yamamoto, Y., Ito, A., Kato, M., Kawabe, Y., Shimizu, K., Fujita, H., Nagamori, E. and Kamihira, M. (2009). Preparation of artificial skeletal muscle tissues by a magnetic force-based tissue engineering technique. *J Biosci Bioeng*, **108**, 538–43.

Yen-Patton, G. P., Patton, W. F., Beer, D. M. and Jacobson, B. S. (1988). Endothelial cell response to pulsed electromagnetic fields: Stimulation of growth rate and angiogenesis in vitro. *J Cell Physiol*, **134**, 37–46.

Yeung, C. K., Ingebrandt, S., Krause, M., Offenhausser, A. and Knoll, W. (2001). Validation of the use of field effect transistors for extracellular signal recording in pharmacological bioassays. *J Pharmacol Toxicol Methods*, **45**, 207–14.

Yilgor, P., Hasirci, N. and Hasirci, V. (2010). Sequential BMP-2/BMP-7 delivery from polyester nanocapsules. *J Biomed Mater Res A*, **93**, 528–36.

Yilgor, P., Tuzlakoglu, K., Reis, R. L., Hasirci, N. and Hasirci, V. (2009). Incorporation of a sequential BMP-2/BMP-7 delivery system into chitosan-based scaffolds for bone tissue engineering. *Biomaterials*, **30**, 3551–9.

You, J. O., Rafat, M., Ye, G. J. and Auguste, D. T. (2011). Nanoengineering the heart: Conductive scaffolds enhance connexin 43 expression. *Nano Lett*, **11**, 3643–8.

Zhang, B., Xiao, Y., Hsieh, A., Thavandiran, N. and Radisic, M. (2011). Micro- and nanotechnology in cardiovascular tissue engineering. *Nanotechnology*, **22**, 494003.

Zhao, X. H., Kim, J., Cezar, C. A., Huebsch, N., Lee, K., Bouhadir, K. and Mooney, D. J. (2011). Active scaffolds for on-demand drug and cell delivery. *Proc Natl Acad Sci U S A*, **108**, 67–72.

Zimmermann, W. H., Schneiderbanger, K., Schubert, P., Didie, M., Munzel, F., Heubach, J. F., Kostin, S., Neuhuber, W. L. and Eschenhagen, T. (2002). Tissue engineering of a differentiated cardiac muscle construct. *Circ Res*, **90**, 223–30.

10
Nanomaterials for neural tissue engineering

M. E. MARTI, A. D. SHARMA, D. S. SAKAGUCHI
and S. K. MALLAPRAGADA, Iowa State University, USA

DOI: 10.1533/9780857097231.2.275

Abstract: This chapter reviews the use of strategies combining nanotechnology with other guidance cues (e.g., biomaterials, support cells, chemical factors) in neural tissue engineering. Nano-structure fabrication techniques and various nanomaterials in neural tissue engineering are presented.

Key words: nanomaterials, neural, regeneration, tissue engineering, electrospinning, biomaterials, self-assembly, conduits, degradable.

10.1   Introduction to neural tissue engineering

The major components of the nervous system are the central nervous system (CNS) and the peripheral nervous system (PNS). The CNS consists of the brain and spinal cord, whereas the PNS is comprised of motor and sensory neurons that transfer impulses from and to the CNS, respectively, and enable communication between the CNS and other parts of the body. Following injury, damage to the nervous system may appear as demyelination of the nerves, nerve degeneration, scar tissue formation, or a communication gap between neurons and/or support cells. As nerve degeneration progresses at the distal stump, this is likely to result in serious, long-term neurological deficits unless appropriate interventions are successfully implemented (Zhang et al., 2005a; Yucel et al., 2010). Unlike the CNS, the PNS has considerable endogenous regenerative capacity following injury, in large part due to its glial cell components, the Schwann cells (SCs) (Horner and Gage, 2000). The CNS does have glial support cells, the astrocytes and oligodendrocytes. The oligodendrocytes, like the peripheral SCs, form the insulating myelin sheath around the axons. However, unlike SCs, the oligodendrocytes hinder regeneration by the production of inhibitory molecules.

Currently, several techniques are used to repair nerve lesions. Coaptation is one of the most common surgical procedures that use sutures to repair damaged nerves. The parts to be repaired are forcefully connected to each other by means of direct sutures. This technique is used for closing wounds,

275

suturing lesions in severed nerves (Haastert *et al.*, 2010), or treating bone fractures (Weinstein and Ralphs, 2004).

Other forms of therapy involve transplantation or grafting procedures. An example is an allograft wherein cells, tissues, and organs are transplanted from donors of the same species. To prevent tissue rejection, immunosuppression (IS) is required (Karabekmez *et al.*, 2009; Grinyo *et al.*, 2010), which may cause infections and tumor formation in the recipient (Xiong *et al.*, 2006; Han *et al.*, 2012). Xenograft is the transplantation of tissue from a donor of a different species to the recipient. This procedure raises ethical concerns (Silva, 2006). In addition, the age between donor and host tissues must be considered, as tissues from different species age at different rates. The gold standard in transplantation therapy is autografting, wherein the patient's own tissue is transplanted from one part of the body to another. It is the best approach for repairing nerve lesions, especially for short gaps (Siemionow and Brzezicki, 2009; Zheng and Cui, 2012). This procedure avoids ethical issues, reduces the likelihood of rejection, and eliminates the need for IS. Drawbacks of this therapy include the need for multiple surgeries, and donor site morbidity; these limit its widespread clinical use, especially for longer nerve gaps. However, recent advances in neural tissue engineering (NTE) show promise for neural regeneration.

Neural tissue engineering is an interdisciplinary field wherein medicine, engineering, and life sciences work together to discover and develop alternatives for nerve regeneration (Langer and Vacanti, 1993). Various guidance cues are applied with engineering and clinical methods to repair nerve deficits and promote neural regeneration. These cues include use of support- and stem cells, trophic and growth factors, and biological and synthetic biomaterials. They are implemented on natural or artificial platforms called scaffolds. The scaffold design is crucial in neural regeneration since these guidance cues can mimic the microenvironment of the extracellular matrix (ECM) and promote attachment, proliferation, growth, migration, differentiation, and survival of the cells, and offer encouraging outcomes and useful alternatives.

Entubulized scaffolds are tubular devices with enhanced surface and structural properties that are used to enclose severed nerves and bridge nerve gaps; they offer interesting possibilities for therapeutic strategies (Dillon *et al.*, 1998). These artificial nerve guidance conduits (NGCs) are made of biological or synthetic materials and consist of various types of fibers with tubes, channels, or lumens of varying size. They can be combined with other guidance cues which can be physical (e.g., patterned substrates, nano-scaffolds), chemical (e.g., neurotrophic or other growth factors), and cellular (e.g., SCs, stem cells, etc.) (Miller *et al.*, 2001; Recknor *et al.*, 2006). In combination with NGCs, their use has resulted in enhanced and oriented

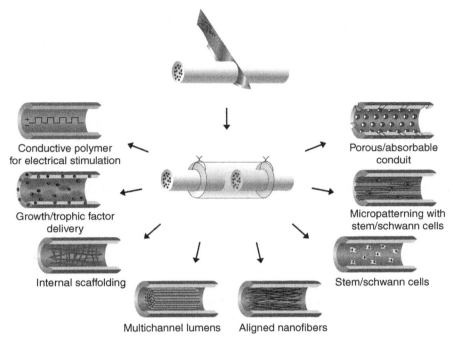

Conductive polymer
for electrical stimulation

Growth/trophic factor
delivery

Internal scaffolding

Multichannel lumens   Aligned nanofibers

Porous/absorbable
conduit

Micropatterning with
stem/schwann cells

Stem/schwann cells

10.1 Schematic diagram demonstrating the properties of an ideal nerve guidance conduit. Various guidance cues can be incorporated into NGCs to mimic the ECM. (*Source*: Reprinted from Marti *et al.* (2012), Copyright 2012, with permission from IOS Press.)

neural regeneration as well as restoration of function (Fig. 10.1). Studies have shown that use of NGCs with other guidance cues can be a favorable alternative to autologous nerve grafts (Gu *et al.*, 2011; Sedaghati *et al.*, 2011).

An ideal NGC must fulfill a number of requirements prior to its use for facilitated neural repair. Essential properties of an ideal NGC include:

1. Biocompatibility
2. Biodegradability
3. Low-toxicity
4. Infection resistance
5. Permeability
6. Porosity
7. Mechanical properties
8. Conductivity
9. Orientation.

Biocompatibility plays a critical role in designing a nerve conduit. The scaffold material should be non-toxic, and not cause inflammation or swelling of tissues/cells at the transplant site. Biodegradable conduits reduce the number of surgeries needed to repair neural lesions. Conduits must maintain structural integrity before complete regeneration is achieved. The porosity and permeability of the NGC are crucial for maintaining mechanical strength and for the transfer of nutrients and metabolites.

The NGCs can be a platform for guidance cues, for example, embedded growth factors, to facilitate neural regeneration. The controlled and continuous release of such factors during the lifetime of the conduit is necessary to avoid critical problems such as inflammation, swelling, and deterioration of surrounding tissues. Electric and magnetic cues may also stimulate enhanced regeneration. The conductivity of the NGC can be increased by using appropriate scaffold material or application of electrical cues; however, problems such as non-biodegradability of conductive polymers must be avoided.

Nanotechnology is the use of engineered materials on a nanometer (nm) scale. The sizes of these materials are generally between 1 and ~100 nm, and they are compatible with biological systems including regenerative neural tissue therapy. Nano-engineered structures, especially nanotubes (NTs) and nanofibers (NFs) imitate the structure of the ECM and can mimic the environment of the nervous system. Recent and remarkable improvements in nanotechnology offer promising alternatives for neural regeneration studies. Nanotechnology can be used to engineer scaffolds with oriented NFs, NTs, and other materials that are functionalized to serve as cell-binding domains or for the release of trophic or growth factors. These scaffolds also facilitate nutrient and oxygen diffusion, and provide topographical cues. Recent advances in nanotechnology, biomaterials, and tissue engineering have improved the effectiveness of NGCs and enabled their use with a control at the nano-scale to obtain enhanced and oriented neuroregeneration (Zhang and Webster, 2009; Biazar *et al.*, 2011) (Plate XI, see colour section between pages 264 and 265).

In the following few sections, the use of nanotechnology combined with other guidance cues (e.g., biomaterials, support cells, chemical factors) in neural tissue engineering strategies is reviewed. Nano-structure fabrication techniques and nanomaterials in NTE are presented. Applications and recent advances in research are also discussed.

10.2   Nano-scaffold design techniques

The structure and geometry of the nano-structural scaffolds are very important in promoting enhanced cell activity for neural regeneration. Thus, graft technology to produce these materials is critical to obtain optimum mechanical strength and surface properties for axon regeneration. Nano-structures made of several biomaterials have been shown to support axonal outgrowth. Several methods have been used to fabricate nano-structural scaffolds such as electrospinning, self-assembly, and phase separation (Cunha *et al.*, 2011).

10.2.1 Electrospinning

Electrospinning is a technique for producing NFs in micro-/nano-scale from natural and synthetic polymers or polymer composites. A solution of the starting material in an appropriate solvent is charged using a spinneret and a high voltage supply. This is done at a polymer concentration above the critical entanglement concentration, below which NFs are not produced. Under the influence of an electrical field, the surface tension of the polymer is overcome at the tip of the spinneret and the charged polymer jet is directed at a target, resulting in formation of a Taylor cone. Nanofibers emerging from the spinneret are then collected in a parallel orientation using a rotating collector (Fig. 10.2). The latter can be a drum or a disk as well as a plate. The use of a stationary collector may result in randomly oriented NFs.

By electrospinning, fibers of nano- to micrometer dimensions can be fabricated. Solution properties (i.e. concentration, conductivity, viscosity, and elasticity) and process parameters (i.e. voltage, distance from target, needle

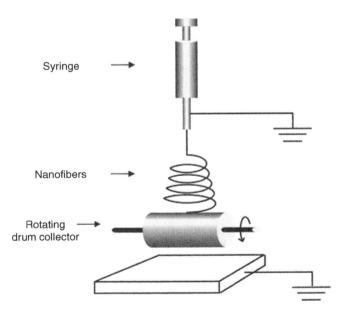

Syringe

Nanofibers

Rotating
drum collector

10.2 This schematic diagram illustrates the experimental set-up for the production of aligned nanofibers using electrospinning. (*Source*: Reprinted from Wang *et al.* (2011a), Copyright 2011, with permission from Elsevier.)

identity, and temperature) can be altered to fabricate various types of fibers for different applications. The mechanical properties of the NFs (e.g., thickness, composition, surface area, and porosity) are adjusted by controlling the solution and process parameters to optimize the alignment, stability, and morphology (Cunha *et al.*, 2011; Leach *et al.*, 2011b). For different applications, the collection schemes can be varied as single ground, rotating single ground, horizontal ring, and *in vitro* onto cells (Ishaug-Riley *et al.*, 1999). To overcome the disadvantages associated with the fiber thickness and random orientation, blends of different biomaterials can be used. Thus, various types of biological materials, synthetic biopolymers and composites have been used to manufacture NFs by electrospinning.

For neural regeneration, the features of the scaffolds are extremely critical in order to mimic the ECM and achieve successful repair. For example, electric fields used during fabrication enhance control of fiber alignment, which is significant for neural regeneration. Electrospun NFs have been shown to enhance and accelerate cellular activities such as proliferation, axon regeneration, growth, and migration (Dong *et al.*, 2006; Liu *et al.*, 2012b) and have been used for neural tissue engineering applications (Wang *et al.*, 2011a; Kijenska *et al.*, 2012). Research to produce nano-structural scaffolds with desired mechanical and surface properties for nerve conduits using electrospinning is continuing.

10.2.2   Self-assembly

Self-assembly (SA) is the reversible and spontaneous formation of a stable organized/ordered structure, from disordered elements through various weak interactions, such as non-covalent bonds and hydrophobic interactions. SA can occur at the molecular level and in the nano- and micro-scale. Numerous types of SA process are carried out in nature. Amino acids, fatty acids, and other molecules come together to form self-assembled nano or larger structures (Toksoz and Guler, 2009).

The SA process has been shown to occur with polypeptide sequences, and di- and tri-block peptide amphiphilic copolymers. Porous NFs of 5–10 nm diameter fabricated by self-assembly from various biodegradable/biocompatible materials have the potential to mimic the three-dimensional (3D) microenvironment of the ECM and promote neural regeneration by enhancing cell attachment, proliferation, differentiation, and migration. Amphiphilic oligopeptides composed of repeating units of hydrophilic and hydrophobic amino acids formed stable β-sheet structures in water and have been used to make scaffolds for neuronal regeneration (Zhang *et al.*, 1995). Zhang *et al.* (1995) reported that with the addition of salts, these oligopeptides self-assembled into stable macroscopic structures of ordered filaments with porous enclosures, and supported cell attachment and synapse formation. Holmes *et al.* (2000) demonstrated the enhancement of neurite outgrowth with self-assembled peptide hydrogels. Various macromolecules such as peptides have the ability to self-assemble and form ordered structures and nanomaterials due to several interactions between the components (Chang *et al.*, 2008). Sur *et al.* (2012) designed a hybrid matrix consisting of collagen type I and a peptide amphiphile (PA). They produced homogenous NFs of 20–30 nm in diameter having the structural properties of the former and epitope-presenting ability of the latter components by self-assembly (Sur *et al.*, 2012). The cellular density was shown to be manipulated by epitope concentration, which could be changed by PA concentration. Furthermore, controllable axon and dendrite growth achieved, as well as enhanced neuronal survival and maturation with epitope concentration adjustment. In another recent study, Angeloni *et al.* (2011) used biodegradable bundles of PA-NFs (Plate XII), which were fabricated by self-assembly and incorporated Sonic hedgehog (SHH) protein molecules, for controlled delivery of the latter, which has an important role in cavernous nerve integrity.

SA is viewed as an approach for changing and controlling the functions of the scaffolds for neural repair (Holmes *et al.*, 2000; Zhang *et al.*, 2005b). However, self-assembled nanomaterials have limited mechanical strength and unfavorable degradation rates. Moreover, fabrication of these structures using this technique for large neural gaps is not very practical (Chang *et al.*, 2008).

10.2.3 Phase separation

Another technique for NF production is phase separation (PS). The rationale for phase separation is the physical incompatibility or immiscibility of the polymer and the solvent. Initially, the polymer is dissolved in a solvent and the resulting solution is kept at the gelation temperature to promote gel formation of the polymer phase. After this, the solvent can be extracted from the gel and a porous nanofibrous skeleton is recovered. There are various types of PS methods for preparing NFs, such as nonsolvent, chemically and thermally induced phase separation (TIPS).

In TIPS, polymer-rich and solvent-rich phases are produced by reducing the temperature of the polymer solution. The solvent-rich phase is removed from the solution by extraction or simple evaporation, leaving behind a solid, skeletal-like polymer matrix with pores. The morphology of the pores depends on various factors like concentration of the polymer solution, solvent properties, PS temperature, and type of solvent removal techniques. TIPS is classified into different methods depending on solvent freezing and phase separation temperatures. If the former is lower than the phase separation temperature, then PS occurs when the solution is cooled to between the lower and upper-critical solution temperatures, that is, in the unstable region of the binary phase diagram. At this temperature, a stable morphological pore network forms in the polymer foam. This is called the liquid–liquid phase separation (LLPS) process. Several synthetic biodegradable polymer scaffolds have been formed using this method in solvents like tetrahydrofuran and dioxane/water (Nam and Park, 1999b; Hua *et al.*, 2002; Pavia *et al.*, 2008).

In the solid–liquid phase separation (SLPS) process, the solvent freezing point is higher than the phase separation temperature of the solution, and the solvent freezes when the temperature is reduced below this point. The polymer comes out of solution with a pore structure corresponding to the crystallization front of the solvent. Similar to LLPS, some of the synthetic biodegradable scaffolds have been produced using this method in solvent mixtures of dioxane and water (Nam and Park, 1999a). Microporous membranes of polypropylene, polyethylene, etc., have been prepared using TIPS and SLPS (Lloyd *et al.*, 1990).

10.3   Nano-structures

Nanomaterials of different morphologies, produced by electrospinning, phase separation, and self-assembly, can enhance axonal outgrowth and facilitate neural regeneration. Most important of these nanomaterials are nanogels (NGs), nanoparticles (NPs) and, in particular, nanotubes (NTs) and nanofibers (NFs), as these can mimic the tubular structures found within cells and tissues (Gilmore *et al.*, 2008).

10.3.1   Nanotubes

Nanotubes are cylindrical structures with diameters of ~1–100 nm (Gallagher *et al.*, 1993; Wang *et al.*, 2004; Jayatissa and Guo, 2009; Tran-Duc *et al.*, 2011). They are capable of simulating various intracellular structures such as microtubules and axons, etc., and can also penetrate cell membranes. They have a very high tensile strength (i.e. high strength to volume ratio) (Yu *et al.*, 2000; Demczyk *et al.*, 2006). With respect to neural regeneration, NGCs with bundles of NTs may provide higher surface area compared to that of conduits alone (Fadel *et al.*, 2008). Moreover, elongation in one-dimensional (1D) can provide better support and orientation to axon regeneration (Huang *et al.*, 2011; Liu *et al.*, 2012a). Carbon nanotubes (CNTs) were tested and recommended for use as NGCs because of their favorable mechanical, elastic, electrical, and rheological properties (Saether *et al.*, 2003; Ashrafi and Hubert, 2006; Schulz *et al.*, 2011). Graphene, the main constituent of

CNTs, has been shown to promote neurite sprouting and outgrowth compared to tissue culture polystyrene substrates (Li *et al.*, 2011). Various functional groups (e.g., carboxylic acid, ethylenediamine, etc.) can be attached to CNTs to modify their chemical and electrical properties to promote controlled neuronal regeneration (Hu *et al.*, 2004). It was reported that human embryonic stem cells differentiated into neural lineages more efficiently on silk-CNT scaffolds with higher expression levels of β-III tubulin and nestin (Chen *et al.*, 2012). CNTs were also shown to promote formation of synaptic contacts and modulation of cell plasticity (Cellot *et al.*, 2011).

10.3.2   Nanofibers

Nanofibers or nanowires are similar to NTs in structure and functionality. They are porous cylinders oriented in 1D with a high interconnectivity and high specific surface area. They can mimic the tubular structure associated with ECM like NTs. Various biomaterials have been used to make NFs using electrospinning (Dong *et al.*, 2006; Meng *et al.*, 2010; Yeganegi *et al.*, 2010). NFs made of polyhydroxyalkanoate (PHA) were observed to promote neural stem cell attachment, synapse formations (synaptogenesis), and CNS regeneration (Xu *et al.*, 2010). Chitosan NF mesh tubes immobilized with electrically charged β-tricalcium phosphate (β-TCP) particles improved nerve regeneration by increasing axon density and area (Wang *et al.*, 2010). Mammadov *et al.* (2012) used a synthetic peptide nanofiber to mimic the activity of laminin and heparan sulfate proteoglycans to enhance axonal growth and induce neuritogenesis, respectively. These self-assembled NF scaffolds were shown to be effective for neurite outgrowth as well as surmounting the inhibitory effect of chondroitin sulfate proteoglycans (Mammadov *et al.*, 2012). Electrospun collagen NFs impregnated with neurotrophin-3 (NT-3) and chondroitinase ABC (ChABC) supported neuronal culture and neurite outgrowth for a longer period than bolus delivery of NT-3. These hold promise for nerve regeneration following spinal cord related injuries, as they provide topographical and biochemical cues and suppress the inhibitory activity during axonal regrowth in the CNS (Liu *et al.*, 2012b).

10.3.3   Nanoparticles

Nanoparticles are tiny particles having at least one dimension under 100 nm. Nanoparticles with lower melting points (Nanda, 2009), higher solar absorption capabilities (Edman Jonsson *et al.*, 2011), and superparamagnetism (Mikhaylova *et al.*, 2004), etc., are typically different from bulk materials. Because of this, NPs have been used in various industrial applications, for example, electronics, biomedical engineering, and optics. NPs have very high surface area-to-volume ratios resulting in high diffusion gradients (Soppimath *et al.*, 2001) and enhanced release of substances such as drugs, proteins, peptides, etc., from NPs to the surrounding environments.

Chitosan-heparin NPs impregnated with nerve growth factors were shown to improve nerve regeneration in mice and to function as a robust NP drug delivery system to the sciatic nerve (Gonçalves *et al.*, 2012). Superparamagnetic NPs functionalized with tropomyosin-receptor-kinases-B (TrkB) receptor antibodies were shown to be endocytosed into signaling endosomes by neurons which in turn promoted neurite outgrowth and activated TrkB-dependent signaling (Steketee *et al.*, 2011). Nano-silver embedded into collagen scaffolds coated with laminin and fibronectin were shown to increase axonal regeneration and a number of the recovered nerves were comparable to an autologous nerve graft (Ding *et al.*, 2011). Nerve regeneration was improved using microstructured polymer filaments in the form of nerve implants containing chitosan/small interfering RNA (siRNA) NPs, which were rapidly internalized by cells (Mittnacht *et al.*, 2010). A nano-structured two-dimensional (2D) substrate comprising gold NPs attached to the surface of a cover glass via an adsorption system has been shown to improve neurite outgrowth of PC12 cells in the presence of electrical stimuli (Park *et al.*, 2008). Such gold NPs can be used as suitable tools for nerve regeneration from neuronal cells.

10.3.4   Nanogels

Nanogels are among the more recent products to be used in nanotechnology. They are nano-scale hydrophilic, cross-linked networks of biocompatible polymers and are mostly used for encapsulating drugs and as efficient drug delivery vehicles (Shidhaye *et al.*, 2008; Peng *et al.*, 2012). Nanogels are prepared using polymerization techniques such as free radical cross-linking (Sanson and Rieger, 2010), free radical precipitation (Blackburn and Lyon, 2008), and nano-emulsion polymerization (Sasaki and Akiyoshi, 2010). Porous scaffolds have been prepared from protein NGs (Zhu *et al.*, 2011) which may be useful for NTE purposes as well. These gels have also been suggested as a promising system for delivering drugs such as oligonucleotides to the brain (Vinogradov *et al.*, 2003).

10.4   Biomaterials for scaffold design

The success of conduit implantation for nerve regeneration depends on engrafting methods and materials as well as cellular activities such as survival, growth, proliferation, and differentiation, and their synergistic use with trophic and growth factors. It is crucial to mimic the 3D architecture of the microenvironment to achieve neural regeneration. Thus, an ideal NGC should possess aspects mimicking the ECM as a critical factor for neuroregeneration. For this purpose, the use of support cells and inducing chemicals on a nano/micro-platform made of appropriate biomaterials and ECM components is essential. NGCs mimicking the ECM can be used to control, direct, and also modify the guidance cues. However, it is a challenging undertaking. As mentioned, biomaterials used for neural regeneration should fulfill requirements such as biodegradability, biocompatibility, porosity, and non-toxicity. Materials with these properties have been tested for neural regeneration (Table 10.1) (Schmidt and Leach, 2003; Rutkowski et al., 2004; Blong et al., 2010; Deumens et al., 2010). Candidates for this role are natural (biologically derived or biological) and synthetic materials. Combinations of these two types of materials have remarkable advantages due to the synergy obtained from the blended composites. Efforts to produce the ideal NGC incorporating ECM using different scaffold materials are likely to provide important benefits.

Table 10.1 Materials produced by electrospinning and most commonly used for neural regeneration scaffolds

Type of materials	Biomaterial name	Nano-product	Reference
Natural	Collagen	NFs	Timnak et al., 2011; Wang et al., 2011b; Liu et al., 2012b
	Chitosan	NF-tubes	Wang et al., 2008
		NTs	Wang et al., 2010
		NPs	Goncalves et al., 2012
	Silk	NFs	Hu et al., 2012
	Fibrin	NGs	Dubey et al., 2001
	Laminin	NFs	Neal et al., 2009
Synthetic	PLLA	NFs	Kakinoki et al., 2011; Prabhakaran et al., 2011; Leach et al., 2011b
	PLGA	NFs	Bini et al., 2004; Panseri et al., 2008; Subramanian et al., 2011
		NFs/Silica NPs	Song et al., 2012
	PCL	NFs	Jin et al., 2011
	PPC	NFs	Wang et al., 2011c
Combined	PCL/Gelatin	NF-Scaffold	Ghasemi-Mobarakeh et al., 2008
	P(LLA-CL)/ Collagen	NF-Scaffold	Kijenska et al., 2012
	SF/ P(LLA-CL)	NFs	Wang et al., 2011a
	PAN-MA/ fibronectin	NFs	Mukhatyar et al., 2011
	SF/PLGA	NFs	Li et al., 2012

10.4.1   Natural materials

Nanofiber scaffolds made from natural (biological or biologically derived) materials represent potential guidance cues to enhance nerve regeneration due to their ability to closely mimic the architecture of endoneurial tubes, which is a critical requirement for peripheral neuroregeneration (Biazar *et al.*, 2010). Natural materials are derived, harvested, or obtained from biological sources and have the advantage of presenting similar features to the host tissues or cells. The ECM is mainly a mixture of proteins and polysaccharides, that is, collagen, proteoglycans, and glycosaminoglycans. Collagen, chitosan, laminin, fibrin, fibronectin, and gelatin have been employed to produce scaffolds by the nano-design techniques for neural tissue engineering applications (Cunha *et al.*, 2011).

Collagen is the main structural component of the ECM, and supports cellular activities and tissue regeneration. It is made of a triple helix and over 20 types of collagen have been identified, with type I being the most common isoform used for biomedical applications. High mechanical strength compared to other biological materials, and good biocompatibility and biodegradability of this protein make it very attractive for NTE applications as well as other tissue engineering therapies. It can also be used with other biomaterials or cross-linked to chemicals to enhance its properties or enable secretion of trophic and growth factors (Liu *et al.*, 2012b). Liu *et al.* (2012b) integrated NT-3 and ChABC onto electrospun collagen NFs. Nearly 50% cross-linking was obtained by microbial transglutaminase and a controlled release of the NT-3 was observed which promoted neurite outgrowth for a longer time than delivery of encapsulated NT-3. Heparin was found to be useful to prevent degradation of ChABC as well as to achieve a controlled NT-3 release. Wang *et al.* (2011b) compared the proliferation of neural progenitor cells on aligned and randomly oriented electrospun NFs made of collagen. The authors found that aligned NFs caused faster expansion and a change in the cell cycle progression (Wang *et al.*, 2011b). Liu *et al.* (2010) produced collagen NFs using a novel photochemical cross-linking method that prevented the denaturation of collagen. Clearly, the use of biofunctional NFs will grow since they supply biochemical cues alongside topographical stimulation and support neural regeneration in the PNS, as well as the CNS. Timnak and co-workers (2011) used collagen and a common type of glycosaminoglycan to fabricate random and aligned electrospun-nanofibrous scaffolds with diameters ranging from 50 to 350 nm that mimicked natural ECM. The biocompatibility of the scaffold was improved by cross-linking with genipin. Fiber diameters were adjusted with viscosity and flow rate of electrospinning solution, and aligned NFs were collected using a rotating collector. The authors stated the positive effect of the scaffold and fiber orientation on cell outgrowth (Timnak *et al.*, 2011). Fast degradation rate and interaction with the integrin ECM receptors are possible disadvantages of collagen and need to be further studied to understand the mechanisms completely before its use *in vivo* (Khaing and Schmidt, 2012).

Chitosan is a linear polysaccharide and has numerous biomedical uses. It has been used for regeneration studies in various forms, such as fibers and sponges (Mi *et al.*, 2001). Wang and co-workers produced electrospun nanofibrous tubes made of chitosan and tested for a sciatic nerve injury in rats. The recovery of sensory function was achieved with enhanced humoral permeation, and cell attachment and migration using the chitosan nano/microfiber mesh tubes (Wang *et al.*, 2008). Wang and co-workers (2010) also tested the ability of electrical charge storage and the effect of β-TCP particles, which were immobilized on chitosan mesh NTs, on a sciatic nerve injury model in rats. Electrophysiological recovery as well as functional recovery of motor and sensory nerves with increased axon density and area with electrical polarization treatment (Wang *et al.*, 2010) was obtained. Gonçalves and co-workers used a novel drug delivery method for a sciatic nerve injury (Goncalves *et al.*, 2012). Chitosan was selected as scaffold material and heparin was bound to growth factors to form nanoparticles. *In vivo* neural regeneration with sensorimotor performance was obtained in mice with the use of these safe nanomedicines (Gonçalves *et al.*, 2012).

Silk is produced naturally by some insects and electrospun silk fibroin NF scaffolds are promising alternative materials for neural regeneration. They are favored due to their significant biocompatibility and mechanical properties (Altman *et al.*, 2003). Hu *et al.* (2012) fabricated electrospun silk fibroin NFs and observed more ordered and aligned SCs with the extended processes with the use of these NFs.

Several natural materials have been used as nano-scaffolds for neural regeneration: fibrinogen, fibrin, gelatin, elastin, laminin, hyaluronic acid, etc. (Khaing and Schmidt, 2012). They are shown to play critical roles in cell adhesion, growth, migration, and differentiation as well as axonal outgrowth and neural regeneration. Biological materials have advantages such as higher biocompatibility, biodegradability, and the ability to mimic the ECM compared to synthetic materials. However, they do pose some significant disadvantages, one being the potential for immunological and/or inflammatory induction (Cunha *et al.*, 2011). Other drawbacks include their weak mechanical properties compared to synthetic materials, and reduced amenability to chemical modification, which is indispensable for neural regeneration applications. Furthermore, variations in sources as well as inherent properties of the sources may produce unexpected responses following implantation. For these reasons, studies on modification techniques and homogeneity are needed to identify suitable biologically derived materials for enhanced neural regeneration.

10.4.2   Synthetic materials

Synthetic materials have been used to mimic the ECM. Their controllable surface and structural properties are critical advantages for NGC fabrication. Biodegradable synthetic materials are favored over non-biodegradable ones since their use eliminates the need for an additional surgery to remove the guidance conduit. Known degradation rates and mechanical properties with reproducibility favor the use of synthetic materials for neural regeneration applications. With their improved features, they are one of the main candidates for NTE conduits and they can be designed to overcome the problems such as immune response and fast degradation. However, lack of mimicking of ECM and possible release of toxic chemicals with the degradation of the NGC material are their main drawbacks. Like biological materials, synthetic materials have been tested as scaffolding material for NGCs. They can be produced in various sizes including nano-scales using the techniques described earlier. These nano-structures made of various biodegradable polymers can be used as NGCs as well as integrated into NGCs. Nano-structured PLA (poly(lactic acid), PGA (poly(glycolic acid)), PLGA (poly(L-lactide-co-glycolide)), PCL (poly(caprolactone)), and PLCL (poly(lactic-co-caprolactone)) are among the most common synthetic materials for NTE applications (Corey *et al.*, 2008; Wang *et al.*, 2009).

PLA is an aliphatic polyester, present in two isoforms, D-(dexorotary) (PDLA) and L-(levorotary) (PLLA) forms, and can also be in a mixed form composed of its two types. The L-form is preferred in biomedical applications since it is present in the body. The crystallinity depends on the form of PLA and is critical for the type of the biomedical use. Leach *et al.* (2011a) fabricated highly aligned electrospun PLLA-NFs and observed that outgrowth of sensory and motor neurons were directed through these nano-scaffolds *in vitro*. PLLA-NFs were modified with oligo-(D-lactic acid) bioactive-peptide conjugates and used for a sciatic nerve injury model in rats (Kakinoki *et al.*, 2011). Kakinoki *et al.* (2011) stated that superior functional reinnervation was obtained with the modified PLLA nanofibrous NGC compared to silicone tube and PLLA conduits. Prabhakaran *et al.* (2011) blended two types of synthetic materials, PLLA and polyaniline, to form NFs with fiber diameter of ~200 nm, and applied an electric field, which resulted in facilitated neurite outgrowth of neural stem cells.

PGA is a similar polymer to PLA, except the repeated unit is glycolic acid instead of lactic acid. It is more hydrophilic and degrades faster than PLA (Table 10.2). To obtain a scaffold material with a desired biodegradability, the copolymer PLGA can be formed by adjusting the ratios of lactic and glycolic acids. Since it is biodegradable and biocompatible, PLGA has been used in many tissue engineering applications. Electrospun PLGA NFs were demonstrated to promote cellular activities and enhance sciatic nerve

Table 10.2 Properties of synthetic biodegradable polymers most commonly used for neural regeneration scaffolds

Polymer	Degradation product	Degradation period (months)	Reference
PLLA	L-lactic acid	6–24	Garlotta, 2001; Zilberman *et al.*, 2005; de Tayrac *et al.*, 2008; Armentano *et al.*, 2010
PGA	Glycolic acid	3–6	Wen and Tresco, 2006; Armentano *et al.*, 2010
PLGA	D,L-lactic and glycolic acids	3–12	Lu *et al.*, 1999, 2000; Armentano *et al.*, 2010;
PCL	Caproic acid	18–36	Nair and Laurencin, 2006; Peña *et al.*, 2006

regeneration in rat models (Bini *et al.*, 2004; Panseri *et al.*, 2008). Lee and co-workers coated PLGA electrospun NFs with polypyrrole to form electrically conductive nano-scaffolds. By applying electrical stimulation, growth and differentiation of rat PC12 cells were shown to be enhanced (Lee *et al.*, 2009). In a recent study, Subramanian *et al.* (2011) manufactured uniaxially aligned electrospun PLGA nanofibrous scaffolds and demonstrated their ability to orient the direction of SCs and this yielded a better proliferation rate than random fibers. Song and co-workers formed dual drug-loaded (PLGA)/mesoporous silica nanoparticles on an electrospun composite mat, with two model drugs (fluorescein (FLU) and rhodamine B (RHB)). The authors stated that most of the FLU was delivered rapidly, whereas RHB release was controlled by the nanoparticles (Song *et al.*, 2012).

PCL is favorable owing to its mechanical properties and slow degradation rate. PCL also has the ability of easy copolymerization with various polymers, such as PGA, PLA, and collagen, which is crucial in forming a desired NGC scaffold. Jin *et al.* (2011) coated electrospun PLCL NFs with multi-walled carbon nanotubes (MWCNTs) and showed that they facilitated neurite outgrowth of rat dorsal root ganglia (DRG) neurons and focal adhesion kinase (FAK) expression of PC-12 cells (Jin *et al.*, 2011). Wang and co-workers used poly(propylene carbonate) (PPC) to fabricate electrospun aligned nanofibrous scaffolds. They showed that the majority of neurite outgrowth and SC migration from DRG extended unidirectionally and parallel to the oriented fibers, while they were randomized on non-aligned fiber films (Wang *et al.*, 2011c).

It is shown that synthetic materials are favored due to their advanced properties for NGC fabrication. Their easy modification and adjustable features for nano-scaffold designs make them one of the most important guidance cues for neural regeneration. However, combined use of synthetic biodegradable materials with the biological ones will overcome their main drawback and the resultant material can closely mimic the ECM.

10.4.3   Combined use of natural and synthetic materials

Combined use of natural and synthetic materials to produce NGCs has been pursued for the purpose of obtaining nano-structures that more closely mimic ECM and enhance nerve regeneration. Overcoming disadvantages of some biomaterials by using other types of biomaterials has resulted in well-developed NGCs. Ghasemi-Mobarakeh and co-workers blended PCL and gelatin, and fabricated nanofibrous scaffolds. The authors reported aligned neurite outgrowth, enhanced neural differentiation, and proliferation with the NFs (Ghasemi-Mobarakeh *et al.*, 2008). Kijenska and colleagues produced electrospun nanofibrous scaffolds with an average diameter of ~250 nm from blends of poly (L-lactic acid)-co-poly(ε-caprolactone) (P(LLA-CL)), collagen I, and collagen III, and investigated the effects of the composite material on cellular activities. Supported cell proliferation was increased by over 20% with the aligned composite scaffold as compared to the polymer alone (Kijenska *et al.*, 2012).

Using electrospinning (Fig. 10.3), Wang and co-workers combined (P(LLA-CL)) and silk fibroin (SF) to fabricate NFs used to treat 10 mm sciatic nerve lesions in rats. Functional recovery and oriented neural regeneration obtained with the composite, SF/P(LLA-CL) was greater than

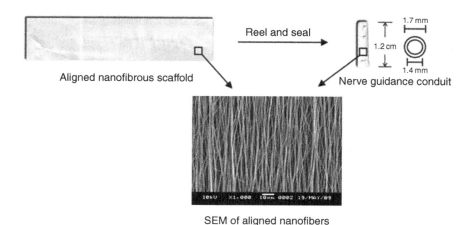

SEM of aligned nanofibers

10.3 The aligned NFs were unreeled from a drum collector onto a stainless steel bar and sealed with nylon monofilament suture stitches. (*Source*: Reprinted from Wang *et al.* (2011a), Copyright 2011, with permission from Elsevier.)

NGCs made of the synthetic polymer itself (Wang *et al.*, 2011a). Fibronectin adsorption on aligned electrospun NFs made of poly-acrylonitrile methyl acrylate (PAN-MA) was shown to be superior compared to smooth films of the same polymer for SC cell migration and neurite outgrowth (Mukhatyar *et al.*, 2011). In another recent study, Li *et al.* (2012) used electrospun NGCs made of PLGA/SF to bridge sciatic nerve gaps in rats and obtained results that were similar to those using autologous nerve grafting.

It is apparent that the use of polymer blends consisting of natural and synthetic materials with nanotechnology provides promising advances for neural regeneration. Researchers are still working on developing the most efficient NGC to repair severe lesions with an enhanced and oriented regeneration. The advantages of composites have prompted researchers to investigate combinations of various biological and synthetic materials for nano-designed conduits.

10.5   Drawbacks of the use of nanomaterials

Nanotechnology is currently one of the most promising and actively pursued areas of research. However, besides the benefits, there may be unexpected drawbacks that will create challenges for scientists in the near future. At present, very little is known about the effects of nanomaterials on biological systems, public health and safety, and the environment.

In neural tissue engineering, not all of the findings on the use of nanomaterials have been positive (Belyanskaya *et al.*, 2009; Dong *et al.*, 2011; Wu *et al.*, 2012). Multi-walled carbon NTs were found to retard the regenerative capacity of axons of DRG neurons without causing the death of cells (Wu *et al.*, 2012). An electrically conducting hydrogel containing single-walled carbon nanotubes (SWCNTs) was shown to reduce SC proliferation in a 2D microenvironment without affecting cell viability. However, a similar hydrogel mimicking a 3D environment seemed to have no significant effect on SC growth (Behan *et al.*, 2011). Effects of SWCNTs were shown to decrease the DNA content of cells derived from avian embryonic spinal cord. Belyanskaya *et al.* (2009) showed that SWCNTs reduced the number of glial cells in both PNS and the CNS, leading to a reduction of total sensory neurons and improved resting membrane potential of cultured DRG neurons as compared to controls.

Magnetic NPs have been used widely in biomedical studies. It was shown that iron oxide nanoparticles decreased the viability of cultured PC12 cells over time (Pisanic II *et al.*, 2007). Moreover, Pisanic II *et al.* (2007) stated that increasing the concentration of NPs caused detachment of the cells and decreased neurite and actin filament counts in mature PC12 cells compared to controls. Nano-alumina, thought to be capable of crossing the blood-brain barrier, has been shown to decrease cell viability and promote loss of membrane integrity of neural stem cells at concentrations >100 µg/ml (Dong *et al.*, 2011).

These drawbacks, along with those that cannot yet be predicted by the investigators, need to be overcome before the use of nano-structures for neural regeneration becomes commonplace in clinical settings. An increase in the number of *in vivo* studies using models which can mimic the human body is necessary before the implantation of NGCs. Promising results obtained up to now are encouraging researchers to investigate the most appropriate and safe biomaterial with the desired size ranges.

10.6   Conclusion and future trends

The cellular and chemical mechanisms essential for functionality of the nervous system are extremely complex and still not entirely understood. This in itself presents formidable challenges to neural tissue engineering efforts. Research to discover and implement neural regeneration strategies has been underway for several decades. Substantial improvements in neuroregeneration with functional recovery have been obtained using NGC implantation. However, technology for NGCs has not advanced sufficiently to allow them to completely replace grafts for their effectiveness, especially in cases of severe injuries and longer gaps. Advances in nanotechnology, along with other NTE strategies, have provided alternatives and implementation of guidance cues to facilitate healing of neural lesions. Several scaffold types have been used to make nano-NGCs from blends of natural or synthetic materials with the desired physical and chemical properties compatible with extracellular matrices. Their use has improved cell survival, adhesion, growth, proliferation, orientation, and differentiation. Composites fabricated using the latest techniques in entubulation and bioengineering have produced encouraging results. Guidance cues are also critical to success, among them the use of specific cell types to maximize reinnervation and promote axonal growth to bridge nerve gaps. Growth factors have been combined with other guidance cues to facilitate nerve regeneration. Recent advances in tissue engineering have enabled the use of cells genetically modified to secrete specific growth factors. Thus, NGCs function not only as guidance cues, but also as platforms for controlled release of therapeutic agents such as trophic factors and cellular constituents such as ECM molecules. Incorporation of these guidance cues may mimic the molecular and cellular components of the microenvironment.

The use of nanotechnology offers many significant and attractive advantages; however, at the molecular level, it poses a number of challenges. Nano-engineered structures can be functionalized for specific use in neural regeneration and other applications. For example, nanoparticles are used in designing devices for the controlled release of medicines, while NFs and

NTs can mimic the tubular structures of the ECM, a desirable feature in designing NGCs for the repair of neural lesions. However, improvements in specificity are still needed to enhance and expand their therapeutic potential for clinical treatments (Silva, 2006).

In spite of the progress made to date, more research on scaffold design and evaluation *in vivo* is still needed. This is a challenging undertaking, especially given the complexities of the nervous system. The ability to duplicate the microenvironment of the ECM with guidance cues must be evaluated in models that mimic the human body. Trials with animals larger than rodents will provide a better understanding of how regenerative therapies will perform in humans. The contributions of nanotechnology should be used for NGCs after overcoming the drawbacks and possible health risks. In spite of the complex challenges, nanotechnology, along with improved biomaterials, drug delivery capability, and tissue engineering will provide alternative therapies with successful clinical applications.

10.7   Acknowledgements

The authors would like to acknowledge US Army Medical Research and Materiel Command under contract W81XWH-11–1–0700 for funding and Rose Perry for the illustration.

10.8   References

Altman, G. H., Diaz, F., Jakuba, C., Calabro, T., Horan, R. L., Chen, J., Lu, H., Richmond, J. and Kaplan, D. L. (2003). Silk-based biomaterials. *Biomaterials*, **24**, 401–16.

Angeloni, N. L., Bond, C. W., Tang, Y., Harrington, D. A., Zhang, S., Stupp, S. I., McKenna, K. E. and Podlasek, C. A. (2011). Regeneration of the cavernous nerve by Sonic hedgehog using aligned peptide amphiphile nanofibers. *Biomaterials*, **32**, 1091–101.

Ashrafi, B. and Hubert, P. (2006). Modeling the elastic properties of carbon nanotube array/polymer composites. *Composites Science and Technology*, **66**, 387–396.

Behan, B. L., Dewitt, D. G., Bogdanowicz, D. R., Koppes, A. N., Bale, S. S. and Thompson, D. M. (2011). Single-walled carbon nanotubes alter Schwann cell behavior differentially within 2D and 3D environments. *Journal of Biomedical Materials Research Part A*, **96A**, 46–57.

Belyanskaya, L., Weigel, S., Hirsch, C., Tobler, U., Krug, H. F. and Wick, P. (2009). Effects of carbon nanotubes on primary neurons and glial cells. *NeuroToxicology*, **30**, 702–711.

Biazar, E., Khorasani, M. T., Montazeri, N., Pourshamsian, K., Daliri, M., Mostafa, R. T., Mahmoud, J. B., Khoshzaban, A., Saeed, H. K., Jafarpour, M. and Roviemiab, Z. (2010). Types of neural guides and using nanotechnology for peripheral nerve reconstruction. *International Journal of Nanomedicine*, **5**, 839–852.

Bini,T. B., Shujun, G.,Ter Chyan,T., Shu,W., Aymeric, L., Lim Ben, H. and Ramakrishna, S. (2004). Electrospun poly(L-lactide-co-glycolide) biodegradable polymer nanofibre tubes for peripheral nerve regeneration. *Nanotechnology*, **15**, 1459.

Blackburn, W. H. and Lyon, L. A. (2008). Size controlled synthesis of monodispersed, core/shell nanogels. *Colloid and Polymer Science*, **286**, 563–9.

Blong, C. C., Jeon, C. J., Yeo, J. Y., Ye, E. A., Oh, J., Callahan, J. M., Law, W. D., Mallapragada, S. K. and Sakaguchi, D. S. (2010). Differentiation and behavior of human neural progenitors on micropatterned substrates and in the developing retina. *Journal of Neuroscience Research*, **88**, 1445–56.

Cellot, G.,Toma, F. M., Kasap Varley, Z., Laishram, J., Villari, A., Quintana, M., cipollone, S., Prato, M. and Ballerini, L. (2011). Carbon nanotube scaffolds tune synaptic strength in cultured neural circuits: novel frontiers in nanomaterial-tissue interactions. *Journal of Neuroscience*, **31**, 12945–53.

Chang, W. C., Kliot, M. and Sretavan, D. W. (2008). Microtechnology and nanotechnology in nerve repair. *Neurological Research*, **30**, 1053–62.

Chen, C.-S., Soni, S., Le, C., Biasca, M., Farr, E., Chen, E.-T. and Chin, W.-C. (2012). Human stem cell neuronal differentiation on silk-carbon nanotube composite. *Nanoscale Research Letters*, **7**, 1–7.

Corey, J. M., Gertz, C. C., Wang, B. S., Birrell, L. K., Johnson, S. L., Martin, D. C. and Feldman, E. L. (2008). The design of electrospun PLLA nanofiber scaffolds compatible with serum-free growth of primary motor and sensory neurons. *Acta Biomaterialia*, **4**, 863–75.

Cunha, C., Panseri, S. and Antonini, S. (2011). Emerging nanotechnology approaches in tissue engineering for peripheral nerve regeneration. *Nanomedicine*, **7**, 50–9.

Demczyk, B. G., Wang, Y. M., Cumings, J., Hetman, M., Han, W., Zettl, A. and Ritchie, R. O. (2006). Direct mechanical measurement of the tensile strength and elastic modulus of multiwalled carbon nanotubes. *Microscopy and Microanalysis*, **12**, 934–5.

Deumens, R., Bozkurt, A., Meek, M. F., Marcus, M. A., Joosten, E. A., Weis, J. and Brook, G. A. (2010). Repairing injured peripheral nerves: Bridging the gap. *Progress in Neurobiology*, **92**, 245–76.

Dillon, G. P., Yu, X. J., Sridharan, A., Ranieri, J. P. and Bellamkonda, R. V. (1998). The influence of physical structure and charge on neurite extension in a 3D hydrogel scaffold. *Journal of Biomaterials Science-Polymer Edition*, **9**, 1049–69.

Ding, T., Lu, W. W., Zheng, Y., Li, Z. Y., Pan, H. B. and Luo, Z. (2011). Rapid repair of rat sciatic nerve injury using a nanosilver-embedded collagen scaffold coated with laminin and fibronectin. *CORD Conference Proceedings*, **6**, 437–47.

Dong, E., Wang, Y., Yang, S.-T., Yuan, Y., Nie, H., Chang, Y., Wang, L., Liu, Y. and Wang, H. (2011). Toxicity of nano gamma alumina to neural stem cells, *Journal of Nanoscience and Nanotechnology*, **11**, 7848–56.

Dong, W., Zhang, T., McDonald, M., Padilla, C., Epstein, J. and Tian, Z. R. (2006). Biocompatible nanofiber scaffolds on metal for controlled release and cell colonization. *Nanomedicine: Nanotechnology, Biology and Medicine*, **2**, 248–52.

Edman Jonsson, G., Fredriksson, H., Sellappan, R. and Chakarov, D. (2011). Nanostructures for Enhanced Light Absorption in Solar Energy Devices. *International Journal of Photoenergy*, Article ID 939807, 11 pages.

Fadel, T. R., Steenblock, E. R., Stern, E., Li, N., Wang, X., Haller, G. L., Pfefferle, L. D. and Fahmy,T. M. (2008). Enhanced Cellular Activation with Single Walled Carbon Nanotube Bundles Presenting Antibody Stimuli. *Nano Letters*, **8**, 2070–6.

Gallagher, M. J., Chen, D., Jacobsen, B. P., Sarid, D., Lamb, L. D., Tinker, F. A., Jiao, J., Huffman, D. R., Seraphin, S. and Zhou, D. (1993). Characterization of carbon nanotubes by scanning probe microscopy. *Surface Science Letters*, **281**, 335–40.

Ghasemi-Mobarakeh, L., Prabhakaran, M. P., Morshed, M., Nasr-Esfahani, M.-H. and Ramakrishna, S. (2008). Electrospun poly(ε-caprolactone)/gelatin nanofibrous scaffolds for nerve tissue engineering. *Biomaterials*, **29**, 4532–9.

Gilmore, J. L., Yi, X., Quan, L. and Kabanov, A. V. (2008). Novel nanomaterials for clinical neuroscience. *Journal of Neuroimmune Pharmacology*, **3**, 83–94.

Gonçalves, N. P., Oliveira, H., Pêgo, A. P. and Saraiva, M. J. (2012). A novel nanoparticle delivery system in vivo targeting of the sciatic nerve: impact on regeneration. *Nanomedicine (London)*, **7**, 1167–80.

Grinyo, J. M., Bestard, O., Torras, J. and Cruzado, J. M. (2010). Optimal immunosuppression to prevent chronic allograft dysfunction. *Kidney International*, **78**, S66-S70.

Gu, X., Ding, F., Yang, Y. and Liu, J. (2011). Construction of tissue engineered nerve grafts and their application in peripheral nerve regeneration. *Progress in Neurobiology*, **93**, 204–30.

Haastert, K., Joswig, H., Jäschke, K.-A., Samii, M. and Grothe, C. (2010). Nerve repair by end-to-side nerve coaptation: histologic and morphometric evaluation of axonal origin in a rat sciatic nerve model. *Neurosurgery*, **66**, 567–76.

Han, Z., Jing, Y., Zhang, S., Liu, Y., Shi, Y. and Wei, L. (2012). The role of immunosuppression of mesenchymal stem cells in tissue repair and tumor growth. *Cell and Bioscience*, **2**, 1–8.

Holmes, T. C., de Lacalle, S., Su, X., Liu, G., Rich, A. and Zhang, S. (2000). Extensive neurite outgrowth and active synapse formation on self-assembling peptide scaffolds. *Proceedings of the National Academy Science of the United States of America*, **97**, 6728–33.

Horner, P. J. and Gage, F. H. (2000). Regenerating the damaged central nervous system. *Nature*, **407**, 963–70.

Hu, A. J., Zuo, B. Q., Zhang, F., Lan, Q. and Zhang, H. X. (2012). Electrospun silk fibroin nanofibers promote Schwann cell adhesion, growth and proliferation. *Neural Regeneration Research*, **7**, 1171–8.

Hu, H., Ni, Y., Montana, V., Haddon, R. C. and Parpura, V. (2004). Chemically Functionalized Carbon Nanotubes as Substrates for Neuronal Growth. *Nano Letters*, **4**, 507–11.

Hua, F. J., Kim, G. E., Lee, J. D., Son, Y. K. and Lee, D. S. (2002). Macroporous poly(L-lactide) scaffold 1. Preparation of a macroporous scaffold by liquid–liquid phase separation of a PLLA–dioxane–water system. *Journal of Biomedical Materials Research*, **63**, 161–7.

Huang, Y.-C., Hsu, S.-H., Kuo, W.-C., Chang-Chien, C.-L., Cheng, H. and Huang, Y.-Y. (2011). Effects of laminin-coated carbon nanotube/chitosan fibers on guided neurite growth. *Journal of Biomedical Materials Research Part A*, **99A**, 86–93.

Ishaug-Riley, S. L., Okun, L. E., Prado, G., Applegate, M. A. and Ratcliffe, A. (1999). Human articular chondrocyte adhesion and proliferation on synthetic biodegradable polymer films. *Biomaterials*, **20**, 2245–56.

Jayatissa, A. H. and Guo, K. (2009). Synthesis of carbon nanotubes at low temperature by filament assisted atmospheric CVD and their field emission characteristics. *Vacuum*, **83**, 853–56.

Jin, G. Z., Kim, M., Shin, U. S. and Kim, H. W. (2011). Neurite outgrowth of dorsal root ganglia neurons is enhanced on aligned nanofibrous biopolymer scaffold with carbon nanotube coating. *Neuroscience Letters*, **501**, 10–14.

Kakinoki, S., Uchida, S., Ehashi, T., Murakami, A. and Yamaoka, T. (2011). Surface modification of poly(L-lactic acid) nanofiber with oligo(D-lactic acid) bioactive-peptide conjugates for peripheral nerve regeneration. *Polymers*, **3**, 820–32.

Karabekmez, F., Duymaz, A. and Moran, S. (2009). Early clinical outcomes with the use of decellularized nerve allograft for repair of sensory defects within the hand. *Hand*, **4**, 245–9.

Khaing, Z. Z. and Schmidt, C. E. (2012). Advances in natural biomaterials for nerve tissue repair. *Neuroscience Letters*, **519**, 103–14.

Kijenska, E., Prabhakaran, M. P., Swieszkowski, W., Kurzydlowski, K. J. and Ramakrishna, S. (2012). Electrospun bio-composite P(LLA-CL)/collagen I/collagen III scaffolds for nerve tissue engineering. *Journal of Biomedical Materials Research Part B: Applied Biomaterials*, **100**, 1093–102.

Langer, R. and Vacanti, J. P. (1993). Tissue engineering. *Science*, **260**, 920–6.

Leach, M. K., Feng, Z.-Q., Gertz, C. C., Tuck, S. J., Regan, T. M., Naim, Y., Vincent, A. M. and Corey, J. M. (2011a). The culture of primary motor and sensory neurons in defined media on electrospun poly-L-lactide nanofiber scaffolds. *Journal of Visualized Experiments*, Feb 15 (48), e2389.

Leach, M. K., Feng, Z. Q., Tuck, S. J. and Corey, J. M. (2011b). Electrospinning fundamentals: optimizing solution and apparatus parameters. *Journal of Visualized Experiments*, Jan 21 (47), e2494.

Lee, J. Y., Bashur, C. A., Goldstein, A. S. and Schmidt, C. E. (2009). Polypyrrole-coated electrospun PLGA nanofibers for neural tissue applications. *Biomaterials*, **30**, 4325–35.

Li, N., Zhang, X., Song, Q., Su, R., Zhang, Q., kong, T., Liu, L., Jin, G., Tang, M. and Cheng, G. (2011). The promotion of neurite sprouting and outgrowth of mouse hippocampal cells in culture by graphene substrates. *Biomaterials*, **32**, 9374–82.

Li, S., Wu, H., Hu, X. D., Tu, C. Q., Pei, F. X., Wang, G. L., Lin, W. and Fan, H. S. (2012). Preparation of electrospun PLGA-silk fibroin nanofibers-based nerve conduits and evaluation in vivo. *Artif Cells Blood Substit Immobil Biotechnol*, **40**, 171–8.

Liu, M., Zhou, G., Song, W., Li, P., Liu, H., Niu, X. and Fan, Y. (2012a). Effect of nano-hydroxyapatite on the axonal guidance growth of rat cortical neurons. *Nanoscale*, **4**, 3201–7.

Liu, T., Teng, W. K., Chan, B. P. and Chew, S. Y. (2010). Photochemical crosslinked electrospun collagen nanofibers: synthesis, characterization and neural stem cell interactions. *J Biomed Mater Res A*, **95**, 276–82.

Liu, T., Xu, J., Chan, B. P. and Chew, S. Y. (2012b). Sustained release of neurotrophin-3 and chondroitinase ABC from electrospun collagen nanofiber scaffold for spinal cord injury repair. *J Biomed Mater Res A*, **100**, 236–42.

Lloyd, D. R., Kinzer, K. E. and Tseng, H. S. (1990). Microporous membrane formation via thermally induced phase separation. I. Solid-liquid phase separation. *Journal of Membrane Science*, **52**, 239–61.

Mammadov, B., Mammadov, R., Guler, M. O. and Tekinay, A. B. (2012). Cooperative effect of heparan sulfate and laminin mimetic peptide nanofibers on the promotion of neurite outgrowth. *Acta Biomaterialia*, **8**, 2077–86.

Marti, M., Sakaguchi, D., Mallapragada, S. (2012). Neural tissue engineering strategies, in *Regenerative Medicine and Cell Therapy, Biomedical and Health Research Series*, Vol. 77, J.F. Stoltz (Ed.), Chap XXII, pp. 275–290, IOS Press BV, Amsterdam.

Meng, Z. X., Zheng, W., Li, L. and Zheng, Y. F. (2010). Fabrication and characterization of three-dimensional nanofiber membrane of PCL–MWCNTs by electrospinning. *Materials Science and Engineering: C*, **30**, 1014–21.

Mi, F. L., Shyu, S. S., Wu, Y. B., Lee, S. T., Shyong, J. Y. and Huang, R. N. (2001). Fabrication and characterization of a sponge-like asymmetric chitosan membrane as a wound dressing. *Biomaterials*, **22**, 165–73.

Mikhaylova, M., Kim, D. K., Bobrysheva, N., Osmolowsky, M., Semenov, V., Tsakalakos, T. and Muhammed, M. (2004). Superparamagnetism of magnetite nanoparticles: dependence on surface modification. *Langmuir*, **20**, 2472–7.

Miller, C., Shanks, H., Witt, A., Rutkowski, G. and Mallapragada, S. (2001). Oriented Schwann cell growth on micropatterned biodegradable polymer substrates. *Biomaterials*, **22**, 1263–9.

Mittnacht, U., Hartmann, H., Hein, S., Oliveira, H., Dong, M., Pêgo, A. P., Kjems, J., Howard, K. A. and Schlosshauer, B. (2010). Chitosan/siRNA nanoparticles biofunctionalize nerve implants and enable neurite outgrowth. *Nano Letters*, **10**, 3933–9.

Mukhatyar, V. J., Salmeron-Sanchez, M., Rudra, S., Mukhopadaya, S., Barker, T. H., Garcia, A. J. and Bellamkonda, R. V. (2011). Role of fibronectin in topographical guidance of neurite extension on electrospun fibers. *Biomaterials*, **32**, 3958–68.

Nam, Y. S. and Park, T. G. (1999a). Biodegradable polymeric microcellular foams by modified thermally induced phase separation method. *Biomaterials*, **20**, 1783–90.

Nam, Y. S. and Park, T. G. (1999b). Porous biodegradable polymeric scaffolds prepared by thermally induced phase separation. *J Biomed Mater Res*, **47**, 8–17.

Nanda, K. 2009. Size-dependent melting of nanoparticles: Hundred years of thermodynamic model. *Pramana*, **72**, 617–28.

Panseri, S., Cunha, C., Lowery, J., del Carro, U., Taraballi, F., Amadio, S., Vescovi, A. and Gelain, F. (2008). Electrospun micro- and nanofiber tubes for functional nervous regeneration in sciatic nerve transections. *BMC Biotechnol*, **8**, 39.

Park, J. S., Park, K., Moon, H. T., Woo, D. G., Yang, H. N. and Park, K.-H. (2008). Electrical Pulsed Stimulation of Surfaces Homogeneously Coated with Gold Nanoparticles to Induce Neurite Outgrowth of PC12 Cells. *Langmuir*, **25**, 451–7.

Pavia, F. C., la Carrubba, V., Piccarolo, S. and Brucato, V. (2008). Polymeric scaffolds prepared via thermally induced phase separation: tuning of structure and morphology. *J Biomed Mater Res A*, **86**, 459–66.

Peng, J., Qi, T., Liao, J., Fan, M., Luo, F., Li, H. and Qian, Z. (2012). Synthesis and characterization of novel dual-responsive nanogels and their application as drug delivery systems. *Nanoscale*, **4**, 2694–704.

Pisanic II, T. R., Blackwell, J. D., Shubayev, V. I., Fiñones, R. R. and Jin, S. (2007). Nanotoxicity of iron oxide nanoparticle internalization in growing neurons. *Biomaterials*, **28**, 2572–81.

Prabhakaran, M. P., Ghasemi-Mobarakeh, L., Jin, G. and Ramakrishna, S. (2011). Electrospun conducting polymer nanofibers and electrical stimulation of nerve stem cells. *J Biosci Bioeng*, **112**, 501–7.

Recknor, J. B., Sakaguchi, D. S. and Mallapragada, S. K. (2006). Directed growth and selective differentiation of neural progenitor cells on micropatterned polymer substrates. *Biomaterials*, **27**, 4098–108.

Rutkowski, G. E., Miller, C. A., Jeftinija, S. and Mallapragada, S. K. (2004). Synergistic effects of micropatterned biodegradable conduits and Schwann cells on sciatic nerve regeneration. *Journal of Neural Engineering*, **1**, 151–7.

Saether, E., Frankland, S. J. V. and Pipes, R. B. (2003). Transverse mechanical properties of single-walled carbon nanotube crystals. Part I: determination of elastic moduli. *Composites Science and Technology*, **63**, 1543–50.

Sanson, N. and Rieger, J. (2010). Synthesis of nanogels/microgels by conventional and controlled radical crosslinking copolymerization. *Polymer Chemistry*, **1**, 965–77.

Sasaki, Y. and Akiyoshi, K. (2010). Nanogel engineering for new nanobiomaterials: from chaperoning engineering to biomedical applications. *The Chemical Record*, **10**, 366–76.

Schmidt, C. E. and Leach, J. B. (2003). Neural tissue engineering: strategies for repair and regeneration. *Annu Rev Biomed Eng*, **5**, 293–347.

Schulz, S. C., Faiella, G., Buschhorn, S. T., Prado, L. A. S. A., Giordano, M., Schulte, K. and Bauhofer, W. (2011). Combined electrical and rheological properties of shear induced multiwall carbon nanotube agglomerates in epoxy suspensions. *European Polymer Journal*, **47**, 2069–77.

Sedaghati, T., Yang, S. Y., Mosahebi, A., Alavijeh, M. S. and Seifalian, A. M. (2011). Nerve regeneration with aid of nanotechnology and cellular engineering. *Biotechnol Appl Biochem*, **58**, 288–300.

Shidhaye, S., Lotlikar, V., Malke, S. and Kadam, V. (2008). Nanogel engineered polymeric micelles for drug delivery. *Current Drug Therapy*, **3**, 209–17.

Siemionow, M. and Brzezicki, G. (2009). Chapter 8 Current techniques and concepts in peripheral nerve repair. *In:* Stefano, G., Pierluigi, T. and Bruno, B. (eds.) *International Review of Neurobiology*. Academic Press.

Silva, G. A. (2006). Neuroscience nanotechnology: progress, opportunities and challenges. *Nat Rev Neurosci*, **7**, 65–74.

Song, B., Wu, C. and Chang, J. (2012). Dual drug release from electrospun poly(lactic-co-glycolic acid)/mesoporous silica nanoparticles composite mats with distinct release profiles. *Acta Biomaterialia*, **8**, 1901–7.

Soppimath, K. S., Aminabhavi, T. M., Kulkarni, A. R. and Rudzinski, W. E. (2001). Biodegradable polymeric nanoparticles as drug delivery devices. *Journal of Controlled Release*, **70**, 1–20.

Steketee, M. B., Moysidis, S. N., Jin, X.-L., Weinstein, J. E., Pita-Thomas, W., Raju, H. B., Iqbal, S. and Goldberg, J. L. (2011). Nanoparticle-mediated signaling endosome localization regulates growth cone motility and neurite growth. *Proceedings of the National Academy of Sciences*, **108**, 19042–7.

Subramanian, A., Krishnan, U. M. and Sethuraman, S. (2011). Fabrication of uniaxially aligned 3D electrospun scaffolds for neural regeneration. *Biomed Mater*, **6**, 025004.

Sur, S., Pashuck, E. T., Guler, M. O., Ito, M., Stupp, S. I. and Launey, T. (2012). A hybrid nanofiber matrix to control the survival and maturation of brain neurons. *Biomaterials*, **33**, 545–55.

Timnak, A., Gharebaghi, F. Y., Shariati, R. P., Bahrami, S. H., Javadian, S., Emami SH, H. and Shokrgozar, M. A. (2011). Fabrication of nano-structured electrospun collagen scaffold intended for nerve tissue engineering. *J Mater Sci Mater Med*, **22**, 1555–67.

Toksoz, S. and Guler, M. O. (2009). Self-assembled peptidic nanostructures. *Nano Today*, **4**, 458–69.

Tran-Duc, T., Thamwattana, N., Cox, B. J. and Hill, J. M. (2011). Encapsulation of a benzene molecule into a carbon nanotube. *Computational Materials Science*, **50**, 2720–6.

Vinogradov, S. V., Batrakova, E. V. and Kabanov, A. V. (2003). Nanogels for Oligonucleotide Delivery to the Brain. *Bioconjugate Chemistry*, **15**, 50–60.

Wang, C. Y., Zhang, K. H., Fan, C. Y., Mo, X. M., Ruan, H. J. and Li, F. F. (2011a). Aligned natural-synthetic polyblend nanofibers for peripheral nerve regeneration. *Acta Biomater*, **7**, 634–43.

Wang, H. B., Mullins, M. E., Cregg, J. M., Hurtado, A., Oudega, M., Trombley, M. T. and Gilbert, R. J. (2009). Creation of highly aligned electrospun poly-L-lactic acid fibers for nerve regeneration applications. *Journal of Neural Engineering*, **6**, 016001.

Wang, W., Itoh, S., Matsuda, A., Ichinose, S., Shinomiya, K., Hata, Y. and Tanaka, J. (2008). Influences of mechanical properties and permeability on chitosan nano/microfiber mesh tubes as a scaffold for nerve regeneration. *Journal of Biomedical Materials Research A*, **84**, 557–66.

Wang, W., Itoh, S., Yamamoto, N., Okawa, A., Nagai, A. and Yamashita, K. (2010). Enhancement of nerve regeneration along a chitosan nanofiber mesh tube on which electrically polarized ß-tricalcium phosphate particles are immobilized. *Acta Biomaterialia*, **6**, 4027–33.

Wang, Y.-H., Chiu, S.-C., Lin, K.-M. and Li, Y.-Y. (2004). Formation of carbon nanotubes from polyvinyl alcohol using arc-discharge method. *Carbon*, **42**, 2535–41.

Wang, Y., Yao, M., Zhou, J., Zheng, W., Zhou, C., Dong, D., Liu, Y., Teng, Z., Jiang, Y., Wei, G. and Cui, X. (2011b). The promotion of neural progenitor cells proliferation by aligned and randomly oriented collagen nanofibers through beta1 integrin/MAPK signaling pathway. *Biomaterials*, **32**, 6737–44.

Wang, Y., Zhao, Z., Zhao, B., Qi, H. X., Peng, J., Zhang, L., Xu, W. J., Hu, P. and Lu, S. B. (2011c). Biocompatibility evaluation of electrospun aligned poly (propylene carbonate) nanofibrous scaffolds with peripheral nerve tissues and cells in vitro. *Chinese Medical Journal (England)*, **124**, 2361–6.

Weinstein, J. and Ralphs, S. C. (2004). External coaptation. *Clinical Techniques in Small Animal Practice*, **19**, 98–104.

Wu, D., Pak, E. S., Wingard, C. J. and Murashov, A. K. (2012). Multi-walled carbon nanotubes inhibit regenerative axon growth of dorsal root ganglia neurons of mice. *Neuroscience Letters*, **507**, 72–7.

Xiong, G., Wang, Y. and Tong, D. (2006). Effects of immunosuppressants on cytokine expressions after repair for nerve injury in a rat model. *Zhongguo Xiu Fu Chong Jian Wai Ke Za Zhi*, **20**, 1163–7.

Xu, X.-Y., Li, X.-T., Peng, S.-W., Xiao, J.-F., Liu, C., Fang, G., Chen, K. C. and Chen, G.-Q. (2010). The behaviour of neural stem cells on polyhydroxyalkanoate nanofiber scaffolds. *Biomaterials*, **31**, 3967–75.

Yeganegi, M., Kandel, R. A. and Santerre, J. P. (2010). Characterization of a biode-gradable electrospun polyurethane nanofiber scaffold: Mechanical properties and cytotoxicity. *Acta Biomaterialia*, **6**, 3847–55.

Yu, M.-F., Lourie, O., Dyer, M. J., Moloni, K., Kelly, T. F. and Ruoff, R. S. (2000). Strength and breaking mechanism of multiwalled carbon nanotubes under ten-sile load. *Science*, **287**, 637–40.

Yucel, D., Kose, G. T. and Hasirci, V. (2010). Tissue engineered, guided nerve tube consisting of aligned neural stem cells and astrocytes. *Biomacromolecules*, **11**, 3584–91.

Zhang, L. J. and Webster, T. J. (2009). Nanotechnology and nanomaterials: Promises for improved tissue regeneration. *Nano Today*, **4**, 66–80.

Zhang, N., Yan, H. and Wen, X. (2005a). Tissue-engineering approaches for axonal guidance. *Brain Reseach Reviews*, **49**, 48–64.

Zhang, S., Holmes, T. C., Dipersio, C. M., Hynes, R. O., Su, X. and Rich, A. (1995). Self-complementary oligopeptide matrices support mammalian cell attach-ment. *Biomaterials*, **16**, 1385–93.

Zhang, Y., Ouyang, H., Lim, C. T., Ramakrishna, S. and Huang, Z. M. (2005b). Electrospinning of gelatin fibers and gelatin/PCL composite fibrous scaffolds. *Journal of Biomedical Materials Research Part B Applied Biomaterials*, **72**, 156–65.

Zheng, L. and Cui, H.-F. (2012). Enhancement of nerve regeneration along a chito-san conduit combined with bone marrow mesenchymal stem cells. *Journal of Materials Science: Materials in Medicine*, **23**(9), 229–302.

Zhu, Q., Yan, M., He, L., Zhu, X., Lu, Y. and Yan, D. (2011). Fabrication of porous scaffolds with protein nanogels. *Science China Chemistry*, **54**, 961–7.

11
Nanomaterials for cartilage tissue engineering

E. J. CHUNG, Northwestern University, USA,
N. SHAH, Parkview Musculoskeletal Institute, USA and
R. N. SHAH, Northwestern University, USA

DOI: 10.1533/9780857097231.2.301

Abstract: This chapter discusses strategies that make use of nanomaterials for cartilage regeneration. In order to understand the necessary requirements for optimal tissue regeneration and design efficacious nanomaterials, cartilage structure and biology are first reviewed and an overview of current clinical strategies for cartilage repair is described. The chapter then discusses three emerging nanomaterials research areas including hydrogels, nanofibers, and nanocomposites. Particular emphasis is given to strategies that aim to regenerate articular cartilage.

Key words: cartilage tissue engineering, hydrogels, nanofibers, nanocomposites, self-assembly.

11.1   Introduction

Cartilage tissues are connective tissues that are present in both vertebrates and invertebrates. Cartilage can function as transitional tissue, as in the case in human bone development, in soft tissue-to-hard tissue junctions (i.e., ligament), or as permanent tissues that provide mechanical rigidity and flexibility, and maintain the shape of a particular structure (i.e., ear, nose, shark skeleton). Cartilage is, however, perhaps most widely known for its role in musculoskeletal movement, providing a low-friction gliding surface to joints. With the aging elderly population and an increase in sports-related injuries and physically active lifestyles, much attention has been drawn to the lack of adequate therapies for proper cartilage regeneration and the need for effective solutions caused by injury or degenerative disease. Damage to cartilage is a source of enormous pain and since cartilage tissues have little intrinsic healing capability, such solutions are essential to restore patient quality of life.

Effective strategies for regenerative medicine include the use of an appropriate cell source, bioactive agents (such as growth factors, drugs, or

301

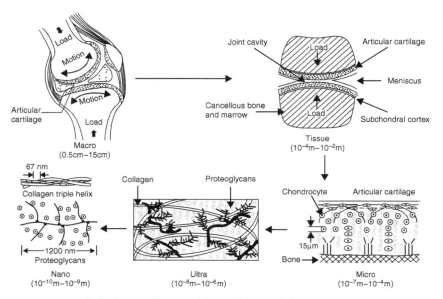

11.1 Articular cartilage resides within the joints and both mechanical and biological properties can be traced back to the nanoscale (Mow *et al.*, 1992).

genes), and/or scaffold technologies. The extracellular matrix (ECM) comprises structural protein and polysaccharide fibers (i.e., collagen, elastin, hyaluronic acid), soluble proteins (i.e., growth factors), and an assortment of glycoproteins and proteoglycans (i.e., fibronectin, laminin, aggrecan), and is considered to be a native scaffold responsible for the maintenance of proper cell function. Therefore, the incorporation of ECM components or ECM-mimetic components into scaffolds for cartilage regeneration provides a familiar environment and regenerative niche that may preserve cell viability, promote cell differentiation and matrix synthesis, and restore damaged cartilage tissues. The use and design of nanomaterials that can emulate the sophisticated architecture and signaling environment of the ECM could potentially achieve enhanced cartilage repair (Fig. 11.1).

In this chapter, we discuss strategies that use nanoscale materials either alone or within composite structures for cartilage tissue engineering applications. Three emerging research areas include the use of hydrogels that have the capability of encapsulating biological agents such as growth factors and cells, nanofibers formed via methods such as self-assembly and electrospinning, and nanocomposites which aim to restore osteochondral defects. In order to understand the necessary requirements for optimal tissue regeneration and design efficacious nanomaterials, cartilage structure and biology are first described and an overview of current clinical strategies for cartilage repair is provided. Particular emphasis is given to strategies that aim to regenerate hyaline cartilage and, specifically, articular cartilage which is found on the surface of long bones of articulating joints that can be damaged due to trauma, disease, or constant mechanical loading over time.

11.2   Cartilage biology and structure

Cartilage is a connective tissue that is predominantly avascular, aneural, and alymphatic and is composed of specialized cells called chondrocytes embedded within a dense ECM. In embryogenesis, cartilage arises from the mesenchyme (Bhosale and Richardson, 2008). At 5 weeks gestation, some mesenchymal chondroprogenitor cells aggregate to form a blastema and differentiate into chondroblasts, which then secrete cartilage matrix. With further development, the extracellular matrix that is produced encases and gradually pushes the cells apart. These mature cells are considered to be chondrocytes. The ratio of the volume of ECM to cells highly favors the matrix with the cell volume averaging only 1–5% of total cartilage volume in human adults (Huber *et al.*, 2000; Poole *et al.*, 2001; Bhosale and Richardson, 2008). Chondrocytes are typically spheroidal in shape, and survive without the need for cell-to-cell contact and in low oxygenated states (Bhosale and Richardson, 2008). In addition, they are non-uniform and vary in number, shape, and metabolic activity in different areas of the cartilage (Huber *et al.*, 2000).

Different types of cartilage are present at various sites throughout the body and can be classified into three types depending on their histological and molecular characterizations: hyaline, elastic, and fibrocartilaginous.

11.2.1   Hyaline (articular) cartilage

Hyaline cartilage is the most abundant form of cartilage and is commonly associated with the skeletal system. It can be found in many locations in the body including the bronchi, larynx, nose, trachea, embryonic skeleton, ribs, and at the ends of long bones. Articular cartilage belongs in this group and is composed primarily of 65–80% by weight fluid containing water with dissolved electrolytes, gases, small proteins, and metabolites; 10–20% collagen with primarily type II collagen; and 10–20% proteoglycans such as aggrecan, decorin, biglycan, and fibromodulin (Buckwalter, 1983; Bhosale and Richardson, 2008). The water content provides nutrition and a medium for lubrication, creating a low-friction gliding surface for joints, as well as load-bearing capability (Bhosale and Richardson, 2008). Collagen fibers are strong in tension and mechanically reinforce the ECM. The orientation of the collagen fibers can be identified by the direction in which the tissue withstands tensile forces and is organized by chondrocytes into a highly ordered and dense structure (Huber *et al.*, 2000). In addition to type II collagen, which makes up 80–95% of the total collagen content, types VI, IX, X, XI, XII, and XIV are also present (Huber *et al.*, 2000; Eyre, 2002; Wachsmuth *et al.*, 2006; Bhosale and Richardson, 2008). These collagen fibers have varying diameters, although type II collagen forms fibrils averaging 20–30 nm in diameter (Montes, 1996). Aggrecans are large, aggregating proteoglycans that constitute 90% of the total cartilage proteoglycan mass. Other proteoglycans found in cartilage include syndecans and glypican, decorin, biglycan, fibromodulin, lumican, epiphycan, and perlecan (Knudson and Knudson, 2001). Aggrecans are elastic, negatively charged macromolecules that are produced by the chondrocytes and are bound to hyaluronan through link proteins. The interaction of water molecules with the high density of negative charges within the aggregating structure provide compressive strength to articular cartilage (Huber *et al.*, 2000). Hyaluronan also binds to the collagen network to provide more stability to the ECM.

Ultra-structure of articular cartilage

Articular cartilage can be divided into four zones depending on the collagen and proteoglycan organization and content: the superficial zone, the transitional zone, the middle (radial) or deep zone, and the calcified cartilage zone (Fig. 11.2) (Umlauf *et al.*, 2010). The superficial zone is the thinnest zone of articular cartilage and composed of flattened ellipsoid chondrocytes that lie parallel to the joint surface. This zone has the highest water content and a thin film of synovial fluid called lubricin covers the cartilage and contributes to the gliding surface (Warman, 2000). Chondrocytes in this zone synthesize a high concentration of collagen and low concentration of proteoglycans. Parallel arrangement of the fibrils is responsible for providing the greatest tensile and shear strength. Disruption of this zone alters the mechanical properties of the articular cartilage and can contribute to the development of degenerative diseases such as osteoarthritis (Bhosale and Richardson, 2008). This layer also acts as a filter for large macromolecules, thereby protecting the cartilage.

The transitional zone is composed of rounded chondrocytes surrounded by abundant ECM. Large diameter collagen fibers are arranged randomly in this zone with a higher proteoglycan content. In the middle or deep zone, groups of chondrocytes are arranged perpendicular to the surface and are spherical in shape. This zone contains the largest diameter collagen fibrils, high concentrations of proteoglycan, and the lowest water and cellular content (Bhosale and Richardson, 2008).

An irregular line called the tidemark separates the deep zone from the calcified zone. The calcified zone is characterized by a small number

(a)   (b)

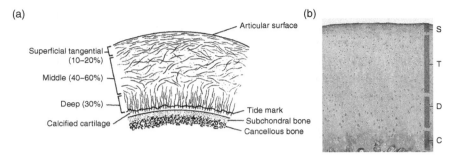

11.2 The four zones of articular cartilage. (a) Cross-sectional depiction (Mow *et al.*, 1992) and (b) azan staining of articular cartilage identifying the four zones (Slomianka, 2009).

of rounded, hypertrophic chondrocytes with very little metabolic activity located in uncalcified lacunae. Chondrocytes in this zone are unique in that they synthesize type X collagen, which contributes to providing structural integrity and shock absorbance (Umlauf *et al.*, 2010). This zone is also marked by the absence of proteoglycans and the presence of large collagen fibers that are arranged perpendicular to the articular surface and anchored in a calcified matrix. The calcified zone contacts the underlying subchondral cortical bone, known as the articular end plate. The organization of the subchondral bone is characteristic for any joint and contains blood vessels. This specific organization of articular cartilage in which the cartilage interfaces with the ends of long bones results from complex developmental processes called endochondral ossification (DeLise *et al.*, 2000). In addition to chondrogenesis derived from early mesenchyme as described earlier, mineralization of the matrix is followed by the invasion of bone cells.

Ultra-structure of extracellular matrix

In addition to zonal organization, the matrix surrounding chondrocytes of articular cartilage is also organized into three different zones by structural differences and a distribution of proteoglycans, link proteins, and hyaluronic acid: pericellular, territorial, and inter-territorial (Poole *et al.*, 1982; Buckwalter, 1983; Poole, 1997). The pericellular matrix surrounds the chondrocytes and is approximately 2 mm wide. This zone is rich in proteoglycans such as aggrecan and decorin, type VI collagen, and hyaluronic acid (Poole *et al.*, 1982; Keene *et al.*, 1988). The territorial matrix surrounds the pericellular region and is present throughout the four zones of cartilage. It can surround individual chondrocytes, a cluster of chondrocytes, or columns of chondrocytes. The collagen fibrils in this region are large, organized in a criss-cross manner, and form a fibrillar basket around the cells, protecting them from mechanical impacts (Bhosale and Richardson, 2008). Proteoglycans in this region are rich in chondroitin sulfate (Meachim and Stockwell, 1979). Most of the matrix volume is composed of the inter-territorial matrix which is made up of collagen fibrils with the largest diameter and aggregates of proteoglycan molecules rich in keratin sulfate (Meachim and Stockwell, 1979).

11.2.2   Elastic cartilage

Elastic cartilage or yellow cartilage is located in the ear and the larynx (epiglottis). Elastic cartilage is histologically similar to hyaline cartilage but contains an abundant amount of yellow elastic fibers. Elastic fibers are derived from elastin proteins, ranging from 10 to 12 nm in diameter, and provide elasticity to the tissue (Montes, 1996). In addition, elastic cartilage differs from hyaline cartilage in the spatial arrangement of cells in which chondrocytes are found lying between the elastic fibers of the network (Nabzdyk *et al.*, 2009). Furthermore, elastic cartilage is more compact than hyaline cartilage and provides flexibility as well as maintains the overall shape of tissues.

11.2.3   Fibrocartilage

Fibrocartilage, which is composed of type I collagen, is found in the intervertebral discs, menisci, tendons, ligaments, pubic symphysis, and in healing articular cartilage. Unlike hyaline and elastic cartilage, fibrocartilage is known to be a transitional tissue that lacks a perichondrium (Benjamin and Evans, 1990). Fibroblasts are dispersed throughout the bundles of large collagen fibers and chondrocytes with little surrounding ECM are sparsely scattered. Therefore, the structural and functional properties of fibrocartilage are intermediate between those of dense fibrous connective tissue and hyaline cartilage. In the case of ligaments and tendons, fibrocartilage functions to directly merge hyaline cartilage and connective tissue. Fibrocartilage is also found in scar tissue that is formed upon hyaline cartilage damage and is widely known as an inadequate replacement for the smooth, glassy, articular cartilage that covers the surface of the knee joint (Minas and Nehrer, 1997).

11.2.4   Mechanical properties of cartilage tissues

The mechanical properties of hyaline, elastic, and fibrocartilage differ. Since fibrocartilage contains structural cartilaginous components in addition to dense fibrous components, their mechanical properties are superior to both hyaline and elastic cartilage but inferior to tendon (Benjamin and Evans,

1990). The tensile strength of fibrocartilage is approximately 10 MPa which is less than that of tendon (55 MPa) but greater than hyaline cartilage (4 MPa), while maximum elongation of fibrocartilage is approximately 13% which is less than that of hyaline cartilage (about 25%), but greater than that of tendon (9–5%) (Yamada, 1976). Under compression, the strength of fibrocartilage is similar to that of hyaline cartilage (10–59 MPa) (Hodge *et al.*, 1986; Kerin *et al.*, 1998; Huber *et al.*, 2000), although the elastic modulus is approximately half of articular cartilage (410 KPa vs. 700 KPa) (Proctor *et al.*, 1989).

Ultimately, the ultra-structure of the ECM in cartilage tissues allows for the viscoelastic properties of the material. For example, in articular cartilage, the ability to undergo reversible deformation and act as a weight-bearing tissue depends on the specific arrangement of macromolecules in the ECM (Kempson *et al.*, 1973; Jeffery *et al.*, 1991). In many ways, the ECM in cartilage can be considered to be a natural hydrogel that is composed primarily of proteoglycans and water, embedded with chondrocytes, and reinforced with collagen fibers (Maroudas, 1976; Huber *et al.*, 2000). The volume and concentration of the water is determined by the interaction of water with the aggregating proteoglycans (Linn and Sokoloff, 1965; Lai *et al.*, 1981). Proteoglycans provide negatively charged sites derived mainly from chondroitin sulfate, and bind firmly to collagen fibrils. Cations dissolved in tissue fluid balance the negatively charged proteoglycans, contributing to the overall mechanical properties of cartilage (Yoshikawa *et al.*, 1997).

The mechanical loading of articular cartilage also directly translates into the deformation of the matrix and can cause a variety of downstream effects, including increased proteoglycans in the pericellular matrix, changes in ion concentrations and pH, and deformation of chondrocytes (Gray *et al.*, 1988). The deformation of chondrocytes depends on the viscoelastic properties of the surrounding pericellular matrix (Guilak *et al.*, 1999). Mechanical and physicochemical cues determine the response of the cells, and can cause changes in the flow of tissue fluid and the charge density around chondrocytes. The consequential electrostatic repulsive forces and osmotic pressure differences determine the hydration state of the tissue and contribute to the overall mechanical response to cartilage deformation (Mow *et al.*, 1999).

11.3 Clinical approaches in the treatment of cartilage defects

Clinical strategies to repair injuries to articular cartilage tissues range from bone marrow stimulation to the use of biomimetic biomaterials.

11.3.1   Articular cartilage

Injury to articular cartilage leads to pain, joint dysfunction, and degenerative joint disease. The treatment of articular cartilage injuries has evolved over time to multiple surgical strategies. The ultimate goal of articular cartilage repair is to recreate the three-dimensional nanostructure of the ECM with viable chondrocytes and to preserve joint integrity.

Injury to articular cartilage can be categorized into two groups: partial thickness or full thickness cartilage defects (Ahmed and Hincke, 2010). Partial thickness defects are restricted within the cartilage region only and do not have access to bone marrow-derived stem cells, therefore limiting the healing capacity which can result in an increase of lesion size over time (Buckwalter, 1983). On the other hand, full thickness defects span the entire cartilage region and may include injury to the subchondral bone. These types of defects may partially heal spontaneously if bleeding occurs through the subchondral bone, exposing the site to stem cells from the bone marrow. The resulting repair tissue, however, is made up of scar tissue that lacks the biomechanical, biochemical, and architectural characteristics of native articular cartilage (Temenoff and Mikos, 2000; McCormick *et al.*, 2008).

Despite the limitations of self-regeneration or repair of articular cartilage, several current clinical techniques are available in an attempt to improve function and relieve symptoms of joint dysfunction, pain, and swelling. Current clinical treatments include chondral debridement, marrow stimulation (microfracture, subchondral drilling, and abrasion arthroplasty), screw or pin fixation, osteochondral autografting or allografting procedures, and cell-based therapies using cultured autologous chondrocytes (Alford and Cole, 2005; Brittberg, 2010).

Reparative or bone marrow stimulation techniques such as microfracture are simple arthroscopic, one-stage surgical techniques that aim to initiate bleeding from the subchondral bone by perforating the subchondral plate, allowing the migration of bone marrow stromal cells to the lesion site. These pluripotent stem cells differentiate into fibrochondrocytes and lead to the formation of reparative tissue. This fibrocartilage (as described in Section 11.2.3) does not resemble the surrounding hyaline cartilage, has less type II collagen, and is mostly composed of type I collagen (Nehrer and Minas, 2000; Steinwachs *et al.*, 2008).

Restoration techniques of articular cartilage include osteochondral autograft/allograft plug transplantation, autologous chondrocyte implantation, and more recently, allograft juvenile minced cartilage implantation. Variations of these techniques involve the use of scaffolds and tissue engineering. Osteochondral grafting utilizes either autograft or fresh-frozen allograft bone dowel(s) composed of bone and cartilage that are impacted into a defect. These techniques are often utilized for defects of larger sizes as well as for defects that have injury to both the articular cartilage and sub-chondral bone. While clinical studies have shown improvement in function and decreased pain, issues still exist with donor site morbidity, cost, graft mismatch, graft orientation, and the ability of the cartilage on the end of the plug to integrate with the normal surrounding articular cartilage (Recht *et al.*, 2003; Evans *et al.*, 2004; Kessler *et al.*, 2008).

Cell-based therapies are either 1- or 2-stage surgical procedures that use harvested chondroctyes from the patient or minced juvenile chondrocytes. Autologous chondrocyte implantation is a 2-stage procedure where chon-drocytes are harvested at the index procedure and then grown or expanded in monolayer. During the second procedure, the cells are injected into the defective site underneath a flap partially sutured to the surrounding native cartilage and made up of either a type I/III collagen patch or the patient's own harvested periosteum. The flap is subsequently sealed with fibrin glue and sutures. This procedure, which is primarily used in larger defect sizes, has yielded acceptable clinical results with regard to improved function and decreased pain. This procedure, however, has had limited success in restoring native articular cartilage tissue both in its structure and biome-chanical properties. Furthermore, there are concerns regarding the ability of the implanted cells to incorporate and produce extracellular matrix in the correct cell-to-matrix ratio and in the appropriate collagen fibril align-ment to restore the native ultrastructure and function of articular cartilage. Additionally, the surgical procedure has had documented complications of periosteal hypertrophy and delamination, inadequate cell supply, donor site morbidity, and arthrofibrosis (Kessler *et al.*, 2008).

Recently, a single-stage technique has been adopted clinically. This technique employs the use of clinically available fibrin glue to secure tissue-engineered cartilage. In this procedure, aseptically minced cartilage obtained from a juvenile allograft donor joint is implanted while mixed with fibrin glue, which serves as a scaffold carrier (McCormick *et al.*, 2008; Ahmed and Hincke, 2010).

Emerging technologies utilize nanotechnology advances as surgical adjuncts to accelerate healing, decrease surgical co-morbidity, and enhance and more closely resemble the native articular cartilage ECM nanostruc-ture. Both cell-based and non-cell-based strategies have been employed with biomimetic scaffolds. By utilizing bio-mimicking scaffolds, the goal is to enhance chondrocyte proliferation and viability within a matrix that closely resembles the three-dimensional (3D) architecture of articular car-tilage. Multiple scaffolding materials including, but not exclusive to, natural polymers, carbohydrates (hyaluronan, agarose, alginate, chitosan), proteins (collagen, gelatin, fibrin), and synthetic polymers (polylactic acid, polygly-colic acid, polyethylene oxide) are currently being tested (Temenoff *et al.*, 2002; Kessler and Grande, 2008; Kessler *et al.*, 2008; Ahmed and Hincke,

2010). Additionally, growth factors that have been either co-implanted with the scaffolds or concentrated from the surrounding normal tissue have been shown to enhance chondrocyte proliferation and viability.

11.3.2   Other cartilage tissues

Due to birth defects, damage, disease, and for cosmetic reasons, cartilage tissues that are commonly repaired are located in the auricle and the nose. Birth defects such as microtia can result in an underdeveloped pinna (the external ear) (Nagata, 1993). Aurical reconstruction uses grafts derived from synthetic polymers (i.e., polyethylene, Medpor) or that are sculpted from the patient's own rib cartilage (costal cartilage) (Romo et al., 2006). Furthermore, a customized ear prosthesis made out of silicone can be implanted into the skull to resemble a normal ear (dos Santos et al., 2010; Walton and Beahm, 2010). Surgical reconstruction of the nose also requires the use of cartilage grafts that are typically harvested from the septum of the patient's nose or the rib (Kridel et al., 2009). The use of porous and high density polyethylene grafts (Medpor) is also common (Kridel et al., 2009).

11.4   Nanomaterials: strategies for cartilage regeneration

Cartilage regeneration requires the restoration of both mechanical and biological properties, and an appropriate scaffold is an essential component to such restoration. It is logical that scaffolds for cartilage regeneration mimic the ECM as much as possible since native ECM itself is a 3D scaffold that simultaneously provides physical support as well as regulates a variety of cellular functions. The functionality of the ECM can be traced back to its nanocomponents, and the use and design of nanomaterials is desirable to achieve proper cartilage regeneration.

The innate advantages of nanoscaled features on cellular behavior and mechanical properties have been known since the 1960s (Rosenberg, 1963). A variety of studies have shown nanometer-sized roughness and surface topography allow for better cell attachment and ECM secretion (Laurencin et al., 1999; Webster et al., 1999, 2001; Fan et al., 2002). Furthermore, nanoscaled surface architecture can affect the adsorption of proteins, ultimately affecting cell behavior (Webster et al., 2000; Wei and Ma, 2004). In terms of mechanical properties, it is widely accepted that the incorporation of nanometer-sized components over micrometer-sized components can provide additional structural support and yields higher stiffness and strength to the resulting composite (Radford, 1971; Bartczak et al., 1999; Teo et al., 1999; Lau et al., 2006; Fu et al., 2008). This is especially important when considering the full repair of articular cartilage: while the ECM in cartilage is referred to as a natural hydrogel that is composed primarily of glycosaminogly-cans (GAGs) and collagen fibers (Maroudas, 1976; Huber *et al.*, 2000), the underlying subchondral bone is often considered a natural nanocomposite in which an elastic collagen matrix is reinforced with hydroxyapatite nano-crystals (Murugan and Ramakrishna, 2005).

Biomaterial approaches to restoring cartilage structure and function that take advantage of the biomimetic properties of nanoscaled features include hydrogels, nanofiber systems, and nanocomposites.

11.4.1   Hydrogels

Hydrogels mimic the natural aqueous and nanofibrous environment of car-tilage ECM. Hydrogels are particularly useful because of their ability to deliver and encapsulate chondrocytes in a 3D environment that maintain their rounded phenotype. Furthermore, hydrogels support the transfer of nutrients and waste, and are injectable scaffolds that can complement exist-ing and emerging, minimally invasive surgical procedures to fill defects of varying size and shape.

Hydrogels are composed of a network of natural and/or synthetic poly-mers or proteins that are cross-linked physically or chemically. Hydrogels that are physically cross-linked consist of molecular entanglements and/or secondary forces such as ionic or hydrogen bonding or hydrophobic interac-tions. Chemically cross-linked hydrogels are covalently bonded. Parameters including molecular weight, polymer concentration, cross-linking method, cross-linking density, and network morphology dictate the physical and chemical properties of the hydrogel such as swelling ratio, degradation rate, mechanical properties, and biocompatibility (Chung and Burdick, 2008). Since the ECM of cartilage is considered to be an example of a natural hydrogel, we provide examples of both general hydrogels as well as those with nanodesign.

Synthetic (general) hydrogels

Hydrogels that are composed of synthetic polymers include poly(vinyl alco-hol) (PVA) and poly(ethylene glycol) (PEG) hydrogels. Due to the harsh manufacturing techniques of traditional PVA hydrogel, PVA hydrogels can-not encapsulate cells until made porous after preparation (Bichara *et al.*, 2011; Spiller *et al.*, 2011). Nonetheless, PVA hydrogels have been shown to support chondrogenesis and can be modified with methacrylate (Ifkovits and Burdick, 2007), chondroitin sulfate (Crispim *et al.*, 2012), and argin-ine-gycine-aspartic acid (RGD) domains (Bryant *et al.*, 2008) to enhance mechanical or biological properties. PEG hydrogels are relatively inert and are the most prevalent class of synthetic hydrogels for cartilage tissue engineering. PEG hydrogels can be modified to be photo-crosslinkable (Elisseeff *et al.*, 1999) or thermoreversible (Liang *et al.*, 2011), and can support cartilage tissue formation via chondrocytes (Elisseeff *et al.*, 1999), mesenchymal stem cells (MSCs) (Elisseeff *et al.*, 1999), and embryonic stem cells (ESCs) (Hwang *et al.*, 2006). The material properties of PEG hydrogels can easily be altered by incorporating other components including lactic acid (Bryant *et al.*, 2004), caprolactone (Park *et al.*, 2007), trimethylene carbonate (Zara *et al.*, 2011), fumarate (Temenoff *et al.*, 2002), and methacrylate (Elisseeff *et al.*, 1999). PEG hydrogels can also be functionalized with biomimetic and natural components such as peptides (i.e., RGD, collagen-mimetic) (Lee *et al.*, 2006; Liu *et al.*, 2010), agarose (Roberts *et al.*, 2011), chitosan (Bhattarai *et al.*, 2010), collagen (Roberts *et al.*, 2011), chondroitin sulfate (Villanueva *et al.*, 2010), and hyaluronic acid (Jha *et al.*, 2009) for enhanced bioactivity and biocompatibility. The potential for PEG hydrogels to induce specific zonal architectures of articular cartilage have been recently reported by Nguyen *et al.* (2011) and was achieved by varying the mechanical properties and concentrations of chondroitin sulfate, matrix metalloproteinase-sensitive peptide, and hyaluronic acid within the hydrogel.

Natural (general) hydrogels

Hydrogels that are composed of natural materials have been fabricated using polysaccharides (i.e., chitosan, agarose, alginate, and hyaluronic acid) and proteins (i.e., silk, collagen, and fibrin). Although some batch-to-batch variability exists from such natural sources, natural materials are commonly used for cartilage tissue engineering due to their abundance and intrinsic pro-chondrogenic biochemical properties. Furthermore, most natural polymers can be degraded by endogenous enzymes. Chitosan-based hydrogels are used for cartilage tissue engineering, because of the structural similarity to cartilage GAGs, and are prepared by inducing ionic or chemical cross-linking of partially N-deacetlyated chitin derived from the exoskeleton of arthropods (Berger *et al.*, 2004). The crystallinity and the degree of acetylation of chitosan can be used to control the material properties of the resulting hydrogel (Ganji *et al.*, 2007). Furthermore, like synthetic hydrogels, chitosan can be further modified to increase bioactivity (Park *et al.*, 2008; Tan *et al.*, 2009; Hu *et al.*, 2011).

Agarose and alginate, both polysaccharide-based and derived from marine algae, were two of the first materials used as hydrogels for cartilage tissue engineering due to the ease of gelation and cell encapsulation (Spiller *et al.*, 2011). Alginate is generally cross-linked with divalent ions while agarose spontaneously gels due to hydrogen bonding (Awad *et al.*, 2004; Spiller *et al.*, 2011). Agarose alone can effectively achieve chondrogenesis, resulting in the highest GAG to DNA ratio when compared to type I collagen, alginate, fibrin, and polyglycolic acid scaffolds (Mouw *et al.*, 2005).

Hyaluronic acid-based hydrogels are one of the most extensively studied natural materials for cartilage tissue engineering. Hyaluronic acid is a linear, nanofibrous polysaccharide found natively in articular cartilage and synovial fluid, and is involved in the regulation of wound healing (Xie *et al.*, 2011), cell migration (Lei *et al.*, 2011), ECM deposition and organization (Fraser *et al.*, 1997), and cell differentiation (Jha *et al.*, 2011). These hydrogels can also be cross-linked and further functionalized into formats that are both photopolymerizable (Smeds and Grinstaff, 2001; Nettles *et al.*, 2004), hydrolytically degradable (Sahoo *et al.*, 2008), and proteinase-sensitive (Kim *et al.*, 2010). Hyaluronic acid hydrogels support chondrogenic differentiation of ESCs and MSCs (Chung and Burdick, 2009; Toh *et al.*, 2010), and have been found to promote cartilage tissue formation by chondrocytes and MSCs better than fibrin and PEG hydrogels, emphasizing the importance of the native biochemical and architectural niche (Park *et al.*, 2005; Chung and Burdick, 2009).

Silk-based hydrogels have more recently been developed and are derived from insect origins or by recombinant methods. Although traditional methods of forming silk hydrogels consisted of high temperature, low pH, and long gelation periods that are not suitable for incorporation of live cells, Chao and colleagues developed a novel method utilizing sonication that allows for chondrocyte encapsulation (Wang *et al.*, 2008; Chao *et al.*, 2010). Such hydrogels have been shown to support cell proliferation and yield cartilaginous constructs that successfully produced type I and II collagen and GAGs (Chao *et al.*, 2010), confirming their potential for cartilage applications (Plate XIII, see colour section between pages 264 and 265).

Hydrogels with nanodesign

Natural hydrogels based on proteins, such as collagen and fibrin, are also common for cartilage regeneration. Collagen hydrogels provide intrinsic cell-binding motifs and nanoarchitecture, although type II collagen hydrogels have been shown to promote more efficient chondrogenesis of encapsulated MSCs compared to type I collagen gels (Bosnakovski *et al.*, 2006; Lu *et al.*, 2010). Fibrin is prepared from fibrinogen and is another commonly used natural and fibrous protein that has pro-chondrogenic properties. Fibrin hydrogels exhibit cell adhesive properties and have been shown to be more favorable in the production of cartilage tissue over alginate hydrogels (Ho *et al.*, 2010a). Moreover, incorporation and release of growth factors such as transforming growth factor beta (TGFβ) is a common strategy to enhance the chondrogenic potential of fibrin-based hydrogels (Bosnakovski *et al.*, 2006; Park *et al.*, 2011).

Self-assembly is a phenomenon inspired by nature and involves the spontaneous association of disordered components into well-organized structures (Hartgerink *et al.*, 2001a, 2002; Hwang *et al.*, 2002; Klok *et al.*, 2002; Silva *et al.*, 2004; Stupp, 2005; Sargeant *et al.*, 2008). Self-assembly implies that molecules are designed to spontaneously organize into supramolecular structures held together through noncovalent interactions such as electrostatic or ionic interactions, hydrogen bonding, hydrophobic interactions, and van der Waals interactions. Large collections of these bonds, though relatively weak compared to covalent bonds, can result in very stable structures. Self-assembly can be particularly useful in cartilage regeneration because starting liquid materials containing dissolved molecules can be injected arthroscopically and form a solid matrix *in situ* that can mechanically and biologically support tissue formation.

Peptide-based, self-assembling materials are most commonly explored for cartilage regeneration. Ionic self-complementary peptides incorporate alternating positive and negative amino acid repeats within the peptide sequence to form stable fibers that gel through β-sheet formation in aqueous solution. These gels have been shown to allow for chondrocyte encapsulation and maintenance, as well as chondrogenic differentiation upon incorporating bone marrow stromal cells into TGFβ-1-containing hydrogels (Kisiday *et al.*, 2002; Kopesky *et al.*, 2010). Fluorenylmethyloxycarbonyl-protected di- and tri-peptides can also self-assemble to form hydrogels for cartilage applications. *In vitro* studies indicate that these hydrogels can support chondrocyte survival and proliferation (Jayawarna *et al.*, 2007).

Hydrogels for cartilage applications can also be formed through the use of β-hairpin peptides. These peptides are composed of alternating hydrophilic and hydrophobic residues flanking an intermittent tetrapetide and dissolve in aqueous solutions in random coil conformations. Under specific stimuli, molecules can be triggered to fold into a β-hairpin conformation that undergoes rapid self-assembly into a β-sheet-rich, highly cross-linked hydrogel (Schneider *et al.*, 2002; Pochan *et al.*, 2003; Ozbas *et al.*, 2004, 2007, Haines *et al.*, 2005; Haines-Butterick *et al.*, 2007). These hydrogels supported cell growth and can be further tailored to have specific and controlled biodegradation rates by incorporation of sequences cleaveable by matrix metalloproteinase-13 (Giano *et al.*, 2011).

Peptide amphiphiles (PAs) are self-assembling molecules that also use hydrophobic and hydrophilic elements to drive self-assembly. PAs self-assemble into nanofibers that are <10 nm in diameter and can form hydrogels (Hartgerink *et al.*, 2001b). PA gels containing a TGFβ-1 binding sequence can efficiently release TGFβ-1 and differentiate human MSCs *in vitro* (Shah *et al.*, 2010) (Plate XIV). Its use as a biomaterial adjunct to current microfracture procedures has been investigated in a full thickness articular cartilage rabbit model. Shah *et al.* (2010) found that treating the defects with the TGFβ-1-binding PA, with or without the addition of exogenous TGFβ-1 growth factor, led to a significant enhancement in cartilage regeneration compared to defects treated with TGFβ-1 growth factor alone or with a non-growth factor-binding PA mixed with TGFβ-1. These results indicated that this growth factor-binding PA system was effective at binding TGFβ-1 present in the joint space, and did not require the addition of exogenous growth factor to see an enhanced regenerative effect.

11.4.2   Nanofiber systems using top-down approaches

The use of nanofiber systems is appealing because their dimensions are structurally similar to native ECM, providing essential cues for cell adhesion, proliferation, differentiation, and migration among many other essential cellular functions. Conventional processing techniques have difficulty producing fibers smaller than 10 μm in diameter, which are several orders of magnitude larger than the fiber components of the ECM (10–500 nm). Other techniques (besides self-assembly) that have proven successful in simulating biomimetic nanofibers for cartilage regeneration include electrospinning and phase separation.

Electrospinning as a technique to produce nanofibrous scaffolds

Electrospinning consists of an electric field between a target and a positively charged capillary filled with a polymer solution. When the electrostatic charge becomes larger than the surface tension of the polymer solution at the capillary tip, a polymer jet is created. The polymer jet travels from the charged capillary to the grounded collector and allows for the production of continuous nanofibers, which are collected to create unique structures with specific composition and mechanical properties (Fig. 11.3).

Nanofibrous scaffolds for cartilage regeneration have been electrospun from natural materials such as chitosan (Subramanian *et al.*, 2005), type II collagen (Shields *et al.*, 2004), and hyaluronic acid (Lee *et al.*, 2009), as well as typical biodegradable, synthetic polymers used in tissue engineering such as poly(lactic-co-glycolic acid) (PLGA) (Shin *et al.*, 2006), poly(lactic acid) (PLA) (Chen and Su, 2011), and poly(caprolactone) (PCL) (Alves da Silva *et al.*, 2010; Ho *et al.*, 2010b). Nanofibers derived by electrospinning result in larger diameter fibers than fibers formed via self-assembly and range from 50 to 700 nm (Nisbet *et al.*, 2009). Electrospun materials are advantageous for tissue engineering applications not only because of their nanoarchitecture, but because the resulting, highly porous scaffolds allow for nutrient and waste exchange as well as cell ingrowth and tissue integration *in vivo*. Furthermore, electrospun scaffolds have been shown to support

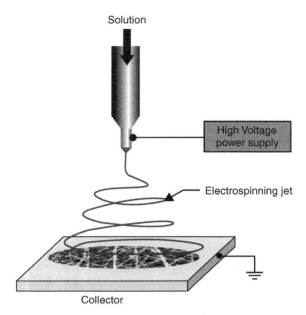

Solution

High Voltage
power supply

Electrospinning jet

Collector

11.3 A typical electrospinning set-up (Teo and Ramakrishna, 2006).

chondrogenesis using chondrocytes or MSCs, and TGFβ can be incorporated to enhance their chondrogenic potential (Zhang *et al.*, 2005).

Electrospun PCL nanofibers have been extensively studied for cartilage regeneration. Li and colleagues have fabricated nanofibrous scaffolds consisting of 700 nm diameter PCL fibers (Li *et al.*, 2005). The structural integrity was maintained over a 21-day culture period in which MSCs differentiated into chondrocytes with TGFβ loading. Electrospun type II collagen scaffolds have also been shown to provide a suitable environment for chondrocyte growth and infiltration *in vitro* (Shields *et al.*, 2004). In addition to providing native cues favorable for cartilage regeneration, electrospun type II collagen scaffolds can be further cross-linked for enhanced mechanical properties.

Other methods of producing nanofibrous scaffolds

Although phase separation has been used for several years as a technique to create porous polymer membranes, Ma and Zhang recently pioneered the use of this technique as a means of producing nanofibrous, 3D scaffolds with porosities as high as 98.5% (Ma and Zhang, 1999). Phase separation is a thermodynamic separation of a polymer solution into a polymer-rich component and a polymer-poor/solvent-rich component. Scaffolds can be created from a variety of biodegradable polymers, and are composed of fibers with diameters that range from 50 to 500 nm (Krishnan *et al.*, 2004; Smith and Ma, 2004). One of the main benefits of phase separation is the relative ease with which nanofibers can be made using this technique without the use of specialized equipment. Although this is a relatively novel technique for fabricating nanofibers for cartilage regeneration, Hu and colleagues have shown that PLLA scaffolds fabricated by this method support chondrogenesis with TGFβ-1 incorporation (Hu *et al.*, 2009a). An increase in type II collagen and aggrecan expression and GAG production was found after culturing MSCs on PLLA nanofibrous scaffolds for 3 weeks.

Carbon nanotubes (CNTs) are another promising biomaterial for cartilage regeneration. CNTs are essentially rolled-up graphene sheets with capped ends. CNTs can be single-walled carbon nanotubes, which are made up of a single cylindrical graphite sheet, or multi-walled carbon nanotubes, which are made up of concentric cylinders of graphite. CNTs have 0.4–2 nm diameters (Smart *et al.*, 2006) and possess mechanical strength similar to collagen fibers (Ribeiro *et al.*, 2012). CNTs have been combined with polymers such as PLLA (Hu *et al.*, 2009b), polycarbonate urethane (PCU), polyimides (Ribeiro *et al.*, 2012), and natural proteins such as type I collagen (Lu *et al.*, 2008), as well as combined with self-assembly and electrospinning techniques for cartilage applications (Chen *et al.*, 2010).

11.4.3   Nanocomposites in osteochondral repair

The structure of native articular cartilage gradually transitions from cartilaginous tissue into bone tissue as described under Section 11.2.1 in 'Ultra-structure of articular cartilage'. Therefore, biomaterials for osteochondral defects are being developed in which both cartilage and subchondral bone repair are targeted. The underlying subchondral bone, or trabecular bone, is a highly porous material (50–90%) (Temenoff *et al.*, 2000; Sikavitsas *et al.*, 2001) with compression moduli ranging from 1 to 1524 MPa (Carter and Hayes, 1977; Goulet *et al.*, 1994; Morgan *et al.*, 2003). The mechanical properties of bone can be traced back to its hierarchical structure and organization provided through the inorganic and organic phases at the nanoscale (Fig. 11.4). The inorganic phase constitutes 60–70% of bone by weight and contains spindle-shaped hydroxyapatite (HAp, $Ca_{10}(PO_4)_6(OH)_2$) crystals with an average length of 50 nm, width of 25 nm, and thickness of 2–5 nm. HAp is secreted by osteoblasts, provides stiffness to bone, and is embedded in the structural framework of the organic phase composed primarily of collagen (Meunier and Boivin, 1997). Therefore, bone is a natural nanocomposite in which a reinforcement phase of HAp nanocrystals is embedded in an elastic collagen matrix, acting as the 'glue' that holds the two phases together (Murugan and Ramakrishna, 2005). A composite, by definition, is

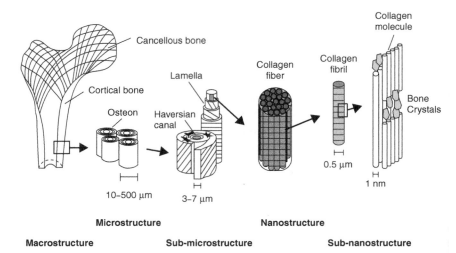

11.4 Hierarchical structural organization of bone (Rho *et al.*, 1998).

a material that consists of two chemically identifiable phases: a particulate, filler phase (i.e., HAp) and a binding matrix phase (i.e., collagen) (Supova, 2009).

Scaffolds that aim to repair both cartilage and subchondral bone aim to mimic the nanostructure of both tissues. The majority of the scaffolds incorporate HAp nanoparticles, emulating the particulate phase of native bone. The incorporation of HAp is advantageous because of its osteoconductive properties, enhancing the bone-bonding capability of the material. In addition to HAp, other nanosized reinforcements composed of cellulose, clay, and titania are also considered.

Nanohydroxyapatite (HAp)-reinforced biomaterials

Synthetic polymers reinforced with nanohydroxyapatite for osteochondral repair include PVA (Pan *et al.*, 2008; Pan and Xiong, 2009; Yusong, 2010; Qu *et al.*, 2011), PLGA (Xue *et al.*, 2010), poly(diol-co-citrate) (POC) (Chung *et al.*, 2011), and PCL (Raghunath *et al.*, 2009). As discussed under Section 11.4.1 in 'Synthetic (general) hydrogels', PVA hydrogels mimic the natural aqueous and nanofibrous environment of cartilage ECM and have been shown to support chondrogenesis. By reinforcing PVA gels with HAp, the resulting nanocomposites have the potential to mimic both the cartilage and bone tissues necessary for osteochondral repair. Pan and colleagues incorporated up to 9% nanoHAp particles and showed that the mechanical properties can be tailored for cartilage or bone by altering the HAp content (Pan *et al.*, 2008; Pan and Xiong, 2009). Furthermore, Xue *et al.* (2010) showed similar

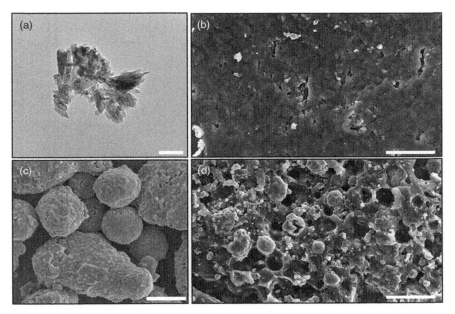

11.5 Electron microscopy images of individual HAp particles and POC-HAp microcomposites and nanocomposites. (a) TEM image of HAp nanocrystals (scale bar (SB): 100 nm) and SEM images of (b) POC-HAp nanocomposites, (c) HAp micropartices, and (d) POC-HAp microcomposites (SB: 20 μm) (Chung *et al.*, 2011).

results upon reinforcing PLGA with 11% by weight HAp, and implanted MSC-loaded porous PLGA scaffolds into articular cartilage defects in a rat model. Osteochondral defects developed mature hyaline tissues, abundant in GAGs and type II collagen with little fibrous tissue after 12 weeks. Chung and colleagues reinforced POC with HAp and compared the efficiency of osteo-chondral repair of nanocomposites over microcomposites in a rabbit model (Chung *et al.*, 2011) (Fig. 11.5). By incorporating 65% by weight HAp, the min-eral content in bone was achieved (Chung *et al.*, 2011). Although both micro-composites and nanocomposites supported articular cartilage development, nanocomposites were mechanically superior and achieved better host tissue integration at the bone-implant interface after 6 weeks post-implantation.

As with synthetic polymers, HAp can reinforce natural materials such as chitosan (Peter *et al.*, 2010; Shi *et al.*, 2012) and collagen (Kon *et al.*, 2009, 2010a, 2010b, 2011) for osteochondral repair. Shi *et al.* (2012) fabricated porous nanocomposite scaffolds based on N-carboxyethylchitosan and HAp via vacuum freeze-drying. Scaffolds were >85% porous with pore sizes ranging from 200 to 500 μm. Chondrocytes were isolated from New Zealand white rabbit trachea and after 8 weeks of culture on scaffolds, neonatal

Cartilage region
Collagen I (100%)

Transition region
Hydroxyapatite (40%)
Collagen I (60%)

Bone region
Hydroxyapatite (70%)
Collagen I (30%)

11.6 Multi-layered collagen-HAp nanocomposite with varying HAp content (Kon *et al.*, 2009).

cartilage-like tissue was formed that exhibited type II collagen staining and supported chondrocyte viability.

While nanocomposites described thus far have been used in bulk to fill in and repair osteochondral defects, Kon and colleagues have investigated the use of a multi-layered collagen-HAp nanocomposite that accounts for the varying HAp concentrations which exist in the ultra-structure of articular cartilage (Kon *et al.*, 2009, 2010a, 2010b, 2011). The collagen-HAp scaffold consists of a cartilaginous layer which is composed of type I collagen and has a smooth surface to account for the tribological functions of the articular cartilage. The intermediate layer accounts for the tidemark and consists of a combination of type I collagen (60%) and HAp (40%). The lower layer is primarily mineralized with a 30:70 blend of collagen and HAp. Although each layer is synthesized separately, the final construct is physically combined via freeze-drying (Fig. 11.6). Clinical trials including 13 patients and 15 defects showed 13 out of 15 defects had completely integrated with the surrounding tissue and filled the osteochondral defect (Kon *et al.*, 2010b). Furthermore, cartilage repair was significantly improved upon using the International Cartilage Repair Society visual scoring system.

Other types of nanoparticle reinforcement

Although reinforcement with HAp is prevalent, other types of nanoparticle reinforcement include cellulose (Karaaslan *et al.*, 2011), clay (Zhang *et al.*, 2010), and titania (Kay *et al.*, 2002). Karaaslan *et al.* (2011) fabricated poly(2-hydroxyethylmthacrylate) (PHEMA) hydrogels reinforced with cellulose nanowhiskers, or long pointed rods in the shape of a whisker derived from aspen wood, for cartilage applications (Fig. 11.7). These hydrogels have mechanical properties that are favorable for load-bearing tissues and show potential for applications in cartilage regeneration. Another class of novel hydrogels is formed by *in situ* free radical polymerization with N-vinylpyrrolidone (NVP) and clay particles (Zhang *et al.*, 2010). Zhang

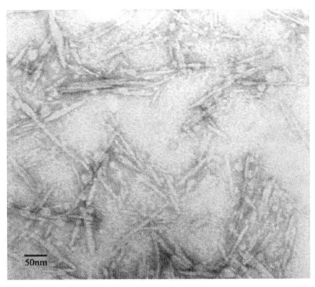

11.7 Transmission electron microscopy image of cellulose whiskers isolated from aspen wood pulp used to reinforce hydrogels for cartilage regeneration (Karaaslan *et al.*, 2011).

and colleagues (2010) showed how clay particles are arranged in an orderly fashion within hydrogels that support chondrocyte attachment. Lastly, PLGA reinforced with 30% by weight 32 nm titania nanoparticles has also shown evidence of chondrocyte biocompatibility and its potential for cartilage applications (Kay *et al.*, 2002).

11.5 Conclusion

With the rise in cartilage degeneration due to disease and injury, research into the development of nanomaterials for cartilage regeneration is growing. Designing biomaterials to mimic the structure of native cartilage at the nanoscale is advantageous because it may allow for restoration of normal tissue function at the mechanical, structural, and cellular level. Ultimately, such therapies have the potential to enhance quality of life, restoring active lifestyles that are characteristic of many patients. Although many advances have been made in scaffold fabrication technology in which the ECM-mimetic biomaterials fill in articular cartilage defects in bulk, recent trends in scaffold fabrication aim to restore the multi-layered organization that resides within cartilage. Furthermore, as medicine turns to non-invasive measures, a rise in techniques such as self-assembly that can complement minimally invasive surgical procedures is likely.

11.6   References

Ahmed, T. A. and Hincke, M. T. (2010). Strategies for articular cartilage lesion repair and functional restoration. *Tissue Engineering Part B Review*, **16**, 305–329. 10.1089/ten.TEB.2009.0590

Alford, J. W. and Cole, B. J. (2005). Cartilage restoration, Part 1. *The American Journal of Sports Medicine*, **33**, 295–306. 10.1177/0363546504273510

Alves da Silva, M. L., Martins, A., Costa-Pinto, A. R., Costa, P., Faria, S., Gomes, M., Reis, R. L. and Neves, N. M. (2010). Cartilage tissue engineering using electrospun PCL nanofiber meshes and MSCs. *Biomacromolecules*, **11**, 3228–3236. 10.1021/bm100476r

Awad, H. A., Wickham, M. Q., Leddy, H. A., Gimble, J. M. and Guilak, F. (2004). Chondrogenic differentiation of adipose-derived adult stem cells in agarose, alginate, and gelatin scaffolds. *Biomaterials*, **25**, 3211–3222. 10.1016/j.biomaterials.2003.10.045

Bartczak, Z., Argon, A. S., Cohen, R. E. and Weinberg, M. (1999). Toughness mechanism in semi-crystalline polymer blends: II. High-density polyethylene toughened with calcium carbonate filler particles. *Polymer*, **40**, 2347–2365.

Benjamin, M. and Evans, E. J. (1990). Fibrocartilage. *Journal of Anatomy*, **171**, 1–15.

Berger, J., Reist, M., Mayer, J. M., Felt, O. and Gurny, R. (2004). Structure and interactions in chitosan hydrogels formed by complexation or aggregation for biomedical applications. *European Journal of Pharmaceutics and Biopharmaceutics*, **57**, 35–52. 10.1016/s0939–6411(03)00160–7

Bhattarai, N., Gunn, J. and Zhang, M. (2010). Chitosan-based hydrogels for controlled, localized drug delivery. *Advanced Drug Delivery Reviews*, **62**, 83–99. 10.1016/j.addr.2009.07.019

Bhosale, A. M. and Richardson, J. B. (2008). Articular cartilage: structure, injuries and review of management. *British Medical Bulletin*, **87**, 77–95. 10.1093/bmb/ldn025

Bichara, D. A., Zhao, X., Bodugoz-Senturk, H., Ballyns, F. P., Oral, E., Randolph, M. A., Bonassar, L. J., Gill, T. J. and Muratoglu, O. K. (2011). Porous poly(vinyl alcohol)-hydrogel matrix-engineered biosynthetic cartilage. *Tissue Engineering Part A*, **17**, 301–309. 10.1089/ten.tea.2010.0322

Bosnakovski, D., Mizuno, M., Kim, G., Takagi, S., Okumura, M. and Fujinaga, T. (2006). Chondrogenic differentiation of bovine bone marrow mesenchymal stem cells (MSCs) in different hydrogels: Influence of collagen type II extracellular matrix on MSC chondrogenesis. *Biotechnology and Bioengineering*, **93**, 1152–1163. 10.1002/bit.20828

Brittberg, M. (2010). Cell carriers as the next generation of cell therapy for cartilage repair. *The American Journal of Sports Medicine*, **38**, 1259–1271. 10.1177/0363546509346395

Bryant, S. J., Bender, R. J., Durand, K. L. and Anseth, K. S. (2004). Encapsulating chondrocytes in degrading PEG hydrogels with high modulus: engineering gel structural changes to facilitate cartilaginous tissue production. *Biotechnology and Bioengineering*, **86**, 747–755. 10.1002/bit.20160

Bryant, S. J., Nicodemus, G. D. and Villanueva, I. (2008). Designing 3D photopolymer hydrogels to regulate biomechanical cues and tissue growth for cartilage tissue engineering. *Pharmaceutical Research*, **25**, 2379–2386. 10.1007/s11095–008–9619-y

Buckwalter, J. A. (1983). Articular cartilage. *Instructional Course Lectures*, **32**, 349–370.

Carter, D. and Hayes, W. 1977. The compressive behavior of bone as a two-phase porous structure. *Journal of Bone and Joint Surgery America*, **59**, 954–962.

Chao, P.-H. G., Yodmuang, S., Wang, X., Sun, L., Kaplan, D. L. and Vunjak-Novakovic, G. (2010). Silk hydrogel for cartilage tissue engineering. *Journal of Biomedical Materials Research Part B: Applied Biomaterials*, **95B**, 84–90. 10.1002/jbm.b.31686

Chen, J.-P. and Su, C.-H. (2011). Surface modification of electrospun PLLA nanofibers by plasma treatment and cationized gelatin immobilization for cartilage tissue engineering. *Acta Biomaterialia*, **7**, 234–243. 10.1016/j.actbio.2010.08.015

Chen, Y., Bilgen, B., Pareta, R. A., Myles, A. J., Fenniri, H., Ciombor, D. M., Aaron, R. K. and Webster, T. J. (2010). Self-assembled rosette nanotube/hydrogel composites for cartilage tissue engineering. *Tissue Engineering. Part C, Methods*, **16**, 1233–1243.

Chung, C. and Burdick, J. A. (2008). Engineering cartilage tissue. *Advanced Drug Delivery Reviews*, **60**, 243–262. 10.1016/j.addr.2007.08.027

Chung, C. and Burdick, J. A. (2009). Influence of three-dimensional hyaluronic acid microenvironments on mesenchymal stem cell chondrogenesis. *Tissue Engineering Part A*, **15**, 243–254. 10.1089/ten.tea.2008.0067

Chung, E. J., Qiu, H., Kodali, P., Yang, S., Sprague, S. M., Hwong, J., Koh, J. and Ameer, G. A. (2011). Early tissue response to citric acid-based micro- and nanocomposites. *Journal of Biomedical Materials Research Part A*, **96**, 29–37. 10.1002/jbm.a.32953

Crispim, E. G., Piai, J. F., Fajardo, A. R., Ramos, E. R. F., Nakamura, T. U., Nakamura, C. V., Rubira, A. F. and Muniz, E. C. (2012). Hydrogels based on chemically modified poly(vinyl alcohol) (PVA-GMA) and PVA-GMA/chondroitin sulfate: preparation and characterization. *Express Polymer Letters*, **6**, 383–395. 10.3144/expresspolymlett.2012.41

Delise, A. M., Fischer, L. and Tuan, R. S. (2000). Cellular interactions and signaling in cartilage development. *Osteoarthritis and Cartilage*, **8**, 309–334. 10.1053/joca.1999.0306

Dos Santos, D. M., Goiato, M. C., Pesqueira, A. A., Bannwart, L. C., Rezende, M. C. R. A., Magro-Filho, O. and Moreno, A. (2010). Prosthesis auricular with osseointegrated implants and quality of life. *Journal of Craniofacial Surgery*, **21**, 94–96. 10.1097/SCS.0b013e3181c4651a.

Elisseeff, J., Anseth, K., Sims, D., McIntosh, W., Randolph, M., Yaremchuk, M. and Langer, R. (1999). Transdermal photopolymerization of poly(ethylene oxide)-based injectable hydrogels for tissue-engineered cartilage. *Plastic and Reconstructive Surgery*, **104**, 1014–1022. 10.1097/00006534–199909040–00017

Evans, P. J., Miniaci, A. and Hurtig, M. B. (2004). Manual punch versus power harvesting of osteochondral grafts. *Arthroscopy: The Journal of Arthroscopic and Related Surgery*, **20**, 306–310. 10.1016/j.arthro.2004.01.012

Eyre, D. (2002). Collagen of articular cartilage. *Arthritis Research*, **4**, 30–35.

Fan, Y. W., Cui, F. Z., Hou, S. P., Xu, Q. Y., Chen, L. N. and Lee, I. S. (2002). Culture of neural cells on silicon wafers with nano-scale surface topograph. *Journal of Neuroscience Methods*, **120**, 17–23. 10.1016/s0165–0270(02)00181–4

Fraser, J. R. E., Laurent, T. C. and Laurent, U. B. G. (1997). Hyaluronan: its nature, distribution, functions and turnover. *Journal of Internal Medicine*, **242**, 27–33. 10.1046/j.1365–2796.1997.00170.x

Fu, S.-Y., Feng, X.-Q., Lauke, B. and Mai, Y.-W. (2008). Effects of particle size, particle/matrix interface adhesion and particle loading on mechanical properties of particulate-polymer composites. *Composites Part B: Engineering*, **39**, 933–961.

Ganji, F., Abdekhodaie, M. J. and Ramazani, A. (2007). Gelation time and degradation rate of chitosan-based injectable hydrogel. *Journal of Sol-Gel Science and Technology*, **42**, 47–53. 10.1007/s10971–006–9007–1

Giano, M. C., Pochan, D. J. and Schneider, J. P. (2011). Controlled biodegradation of self-assembling β-hairpin peptide hydrogels by proteolysis with matrix metalloproteinase-13. *Biomaterials*, **32**, 6471–6477. 10.1016/j.biomaterials.2011.05.052

Goulet, R. W., Goldstein, S. A., Ciarelli, M. J., Kuhn, J. L., Brown, M. B. and Feldkamp, L. A. (1994). The relationship between the structural and orthogonal compressive properties of trabecular bone. *Journal of Biomechanics*, **27**, 375–377, 379–389.

Gray, M. L., Pizzanelli, A. M., Grodzinsky, A. J. and Lee, R. C. (1988). Mechanical and physicochemical determinants of the chondrocyte biosynthetic response. *Journal of Orthopaedic Research*, **6**, 777–792. 10.1002/jor.1100060602

Guilak, F., Jones, W. R., Ting-Beall, H. P. and Lee, G. M. (1999). The deformation behavior and mechanical properties of chondrocytes in articular cartilage. *Osteoarthritis and Cartilage*, **7**, 59–70. 10.1053/joca.1998.0162

Haines-Butterick, L., Rajagopal, K., Branco, M., Salick, D., Rughani, R., Pilarz, M., Lamm, M. S., Pochan, D. J. and Schneider, J. P. (2007). Controlling hydrogelation kinetics by peptide design for three-dimensional encapsulation and injectable delivery of cells. *Proceedings of the National Academy of Sciences of the United States America*, **104**, 7791–7796.

Haines, L. A., Rajagopal, K., Ozbas, B., Salick, D. A., Pochan, D. J. and Schneider, J. P. (2005). Light-activated hydrogel formation via the triggered folding and self-assembly of a designed peptide. *Journal of the American Chemical Society*, **127**, 17025–17029.

Hartgerink, J. D., Beniash, E. and Stupp, S. I. (2001a). Self-assembly and mineralization of peptide-amphiphile nanofibers. *Science*, **294**, 1684–1688.

Hartgerink, J. D., Beniash, E. and Stupp, S. I. (2001b). Self-assembly and mineralization of peptide-amphiphile nanofibers. *Science*, **294**, 1684–1688. 10.1126/science.1063187

Hartgerink, J. D., Beniash, E. and Stupp, S. I. (2002). Peptide-amphiphile nanofibers: a versatile scaffold for the preparation of self-assembling materials. *Proceedings of the National Academy of Sciences of the United States of America*, **99**, 5133–5138.

Ho, S. T. B., Cool, S. M., Hui, J. H. and Hutmacher, D. W. (2010a). The influence of fibrin based hydrogels on the chondrogenic differentiation of human bone marrow stromal cells. *Biomaterials*, **31**, 38–47. 10.1016/j.biomaterials.2009.09.021

Ho, S. T. B., Ekaputra, A. K., Hui, J. H. and Hutmacher, D. W. (2010b). An electrospun polycaprolactone–collagen membrane for the resurfacing of cartilage defects. *Polymer International*, **59**, 808–817. 10.1002/pi.2792

Hodge, W. A., Fijan, R. S., Carlson, K. L., Burgess, R. G., Harris, W. H. and Mann, R. W. (1986). Contact pressures in the human hip joint measured in vivo. *Proceedings of the National Academy of Sciences of the United States of America*, **83**, 2879–2883.

Hu, J., Feng, K., Liu, X. and Ma, P. X. (2009a). Chondrogenic and osteogenic differentiations of human bone marrow-derived mesenchymal stem cells on a nanofibrous scaffold with designed pore network. *Biomaterials*, **30**, 5061–5067. 10.1016/j.biomaterials.2009.06.013

Hu, J., Feng, K., Liu, X. H. and Ma, P. X. (2009b). Chondrogenic and osteogenic differentiations of human bone marrow-derived mesenchymal stem cells on a nanofibrous scaffold with designed pore network. *Biomaterials*, **30**, 5061–5067. 10.1016/j.biomaterials.2009.06.013

Hu, X. H., Li, D. and Gao, C. Y. (2011). Chemically cross-linked chitosan hydrogel loaded with gelatin for chondrocyte encapsulation. *Biotechnology Journal*, **6**, 1388–1396. 10.1002/biot.201100017

Huber, M., Trattnig, S. and Lintner, F. (2000). Anatomy, biochemistry, and physiology of articular cartilage. *Investigative Radiology*, **35**, 573–580.

Hwang, J. J., Iyer, S. N., Li, L. S., Claussen, R., Harrington, D. A. and Stupp, S. I. (2002). Self-assembling biomaterials: liquid crystal phases of cholesteryl oligo(L-lactic acid) and their interactions with cells. *Proceedings of the National Academy of Sciences of the United States of America*, **99**, 9662–9667. 10.1073/pnas.152667399

Hwang, N. S., Varghese, S., Theprungsirikul, P., Canver, A. and Elisseeff, J. (2006). Enhanced chondrogenic differentiation of murine embryonic stem cells in hydrogels with glucosamine. *Biomaterials*, **27**, 6015–6023. 10.1016/j. biomaterials.2006.06.033

Ifkovits, J. L. and Burdick, J. A. (2007). Review: photopolymerizable and degradable biomaterials for tissue engineering applications. *Tissue Engineering*, **13**, 2369–2385. 10.1089/ten.2007.0093

Jayawarna, V., Smith, A., Gough, J. E. and Ulijn, R. V. (2007). Three-dimensional cell culture of chondrocytes on modified di-phenylalanine scaffolds. *Biochemical Society Transactions*, **35**, 535–537.

Jeffery, A., Blunn, G., Archer, C. and Bentley, G. (1991). Three-dimensional collagen architecture in bovine articular cartilage. *Journal of Bone and Joint Surgery, British Volume*, **73B**, 795–801.

Jha, A. K., Xu, X., Duncan, R. L. and Jia, X. (2011). Controlling the adhesion and differentiation of mesenchymal stem cells using hyaluronic acid-based, doubly crosslinked networks. *Biomaterials*, **32**, 2466–2478. 10.1016/j. biomaterials.2010.12.024

Jha, A. K., Yang, W. D., Kirn-Safran, C. B., Farach-Carson, M. C. and Jia, X. Q. (2009). Perlecan domain I-conjugated, hyaluronic acid-based hydrogel particles for enhanced chondrogenic differentiation via BMP-2 release. *Biomaterials*, **30**, 6964–6975. 10.1016/j.biomaterials.2009.09.009

Karaaslan, M. A., Tshabalala, M. A., Yelle, D. J. and Buschle-Diller, G. (2011). Nanoreinforced biocompatible hydrogels from wood hemicelluloses and cellulose whiskers. *Carbohydrate Polymers*, **86**, 192–201. 10.1016/j. carbpol.2011.04.030

Kay, S., Thapa, A., Haberstroh, K. M. and Webster, T. J. (2002). Nanostructured polymer/nanophase ceramic composites enhance osteoblast and chondrocyte adhesion. *Tissue Engineering*, **8**, 753–761.

Keene, D. R., Engvall, E. and Glanville, R. W. (1988). Ultrastructure of type VI collagen in human skin and cartilage suggests an anchoring function for this filamentous network. *The Journal of Cell Biology*, **107**, 1995–2006. 10.1083/jcb.107.5.1995

Kempson, G. E., Muir, H., Pollard, C. and Tuke, M. (1973). The tensile properties of the cartilage of human femoral condyles related to the content of collagen and glycosaminoglycans. *Biochimica et Biophysica Acta (BBA) – General Subjects*, **297**, 456–472. 10.1016/0304–4165(73)90093–7

Kerin, A. J., Wisnom, M. R. and Adams, M. A. (1998). The compressive strength of articular cartilage. *Proceedings of the Institution of Mechanical Engineers, Part H: Journal of Engineering in Medicine*, **212**, 273–280. 10.1243/0954411981534051

Kessler, M. W., Ackerman, G., Dines, J. S. and Grande, D. (2008). Emerging technologies and fourth generation issues in cartilage repair. *Sports Medicine and Arthroscopy Review*, **16**, 246–254 10.1097/JSA.0b013e31818d56b3

Kessler, M. W. and Grande, D. A. (2008). Tissue engineering and cartilage. *Organogenesis*, **4**, 28–32.

Kim, J., Kim, I. S., Cho, T. H., Kim, H. C., Yoon, S. J., Choi, J., Park, Y., Sun, K. and Hwang, S. J. (2010). In vivo evaluation of MMP sensitive high-molecular weight HA-based hydrogels for bone tissue engineering. *Journal of Biomedical Materials Research Part A*, **95A**, 673–681. 10.1002/jbm.a.32884

Kisiday, J., Jin, M., Kurz, B., Hung, H., Semino, C., Zhang, S. and Grodzinsky, A. J. (2002). Self-assembling peptide hydrogel fosters chondrocyte extracellular matrix production and cell division: implications for cartilage tissue repair. *Proceedings of the National Academy of Sciences*, **99**, 9996–10001. 10.1073/pnas.142309999

Klok, H. A., Hwang, J. J., Iyer, S. N. and Stupp, S. I. (2002). Cholesteryl-(L-Lactic Acid) n⁻ building blocks for self-assembling biomaterials. *Macromolecules*, **35**, 746–759.

Knudson, C. B. and Knudson, W. (2001). Cartilage proteoglycans. *Seminars in Cell and Developmental Biology*, **12**, 69–78. 10.1006/scdb.2000.0243

Kon, E., Delcogliano, M., Filardo, G., Altadonna, G. and Marcacci, M. (2009). Novel nano-composite multi-layered biomaterial for the treatment of multifocal degenerative cartilage lesions. *Knee Surgery Sports Traumatology Arthroscopy*, **17**, 1312–1315. 10.1007/s00167–009–0819–8

Kon, E., Delcogliano, M., Filardo, G., Busacca, M., Di Martino, A. and Marcacci, M. (2011). Novel nano-composite multilayered biomaterial for osteochondral regeneration. *The American Journal of Sports Medicine*, **39**, 1180–1190. 10.1177/0363546510392711

Kon, E., Delcogliano, M., Filardo, G., Fini, M., Giavaresi, G., Francioli, S., Martin, I., Pressato, D., Arcangeli, E., Quarto, R., Sandri, M. and Marcacci, M. (2010a). Orderly osteochondral regeneration in a sheep model using a novel nano-composite multilayered biomaterial. *Journal of Orthopaedic Research*, **28**, 116–124. 10.1002/jor.20958

Kon, E., Delcogliano, M., Filardo, G., Pressato, D., Busacca, M., Grigolo, B., Desando, G. and Marcacci, M. (2010b). A novel nano-composite multi-layered biomaterial for treatment of osteochondral lesions: Technique note and an early stability pilot clinical trial. *Injury*, **41**, 693–701. 10.1016/j.injury.2009.11.014

Kopesky, P. W., Vanderploeg, E. J., Sandy, J. S., Kurz, B. and Grodzinsky, A. J. (2010). Self-assembling peptide hydrogels modulate in vitro chondrogenesis of bovine bone marrow stromal cells. *Tissue Engineering Part A*, **16**, 465–477. 10.1089/ten.TEA.2009.0158

Kridel, R. W. H., Ashoori, F., Liu, E. S. and Hart, C. G. (2009). Long-term use and follow-up of irradiated homologous costal cartilage grafts in the nose. *Archives of Facial Plastic Surgery*, **11**, 378–394. 10.1001/archfacial.2009.91

Krishnan, J., Kotaki, M., Yanzhong, Z., Xiumei, M. and Ramakrishna, S. (2004). Recent advances in polymer nanofibers. *Journal of Nanoscience and Nanotechnology*, **4**, 52–65. 10.1166/jnn.2004.078

Lai, W. M., Mow, V. C. and Roth, V. (1981). Effects of nonlinear strain-dependent permeability and rate of compression on the stress behavior of articular cartilage. *Journal of Biomechanical Engineering*, **103**, 61–66.

Lau, K.-T., Gu, C. and Hui, D. (2006). A critical review on nanotube and nanotube/nanoclay related polymer composite materials. *Composites Part B: Engineering*, **37**, 425–436.

Laurencin, C. T., Ambrosio, A. M., Borden, M. D. and Cooper, J. A. (1999). Tissue engineering: orthopedic applications. *Annual Review of Biomedical Engineering*, **1**, 19–46.

Lee, H. J., Lee, J. S., Chansakul, T., Yu, C., Elisseeff, J. H. and Yu, S. M. (2006). Collagen mimetic peptide-conjugated photopolymerizable PEG hydrogel. *Biomaterials*, **27**, 5268–5276. 10.1016/j.biomaterials.2006.06.001

Lee, K. Y., Jeong, L., Kang, Y. O., Lee, S. J. and Park, W. H. (2009). Electrospinning of polysaccharides for regenerative medicine. *Advanced Drug Delivery Reviews*, **61**, 1020–1032. 10.1016/j.addr.2009.07.006

Lei, Y., Gojgini, S., Lam, J. and Segura, T. (2011). The spreading, migration and proliferation of mouse mesenchymal stem cells cultured inside hyaluronic acid hydrogels. *Biomaterials*, **32**, 39–47. 10.1016/j.biomaterials.2010.08.103

Li, W. J., Tuli, R., Okafor, C., Derfoul, A., Danielson, K. G., Hall, D. J. and Tuan, R. S. (2005). A three-dimensional nanofibrous scaffold for cartilage tissue engineering using human mesenchymal stem cells. *Biomaterials*, **26**, 599–609. 10.1016/j.biomaterials.2004.03.005

Liang, Y., Deng, L., Chen, C., Zhang, J., Zhou, R., Li, X., Hu, R. and Dong, A. (2011). Preparation and properties of thermoreversible hydrogels based on methoxy poly(ethylene glycol)-grafted chitosan nanoparticles for drug delivery systems. *Carbohydrate Polymers*, **83**, 1828–1833. 10.1016/j.carbpol.2010.10.048

Linn, F. C. and Sokoloff, L. (1965). Movement and composition of interstitial fluid of cartilage. *Arthritis and Rheumatism*, **8**, 481–494. 10.1002/art.1780080402

Liu, S. Q., Tian, Q. A., Wang, L., Hedrick, J. L., Hui, J. H. P., Yang, Y. Y. and Ee, P. L. R. (2010). Injectable biodegradable poly(ethylene glycol)/RGD PEPTIDE hybrid hydrogels for in vitro chondrogenesis of human mesenchymal stem cells. *Macromolecular Rapid Communications*, **31**, 1148–1154. 10.1002/marc.200900818

Lu, G. Y., Sheng, B. Y., Wei, Y. J., Wang, G., Zhang, L. H., Ao, Q., Gong, Y. D. and Zhang, X. F. (2008). Collagen nanofiber-covered porous biodegradable carboxymethyl chitosan microcarriers for tissue engineering cartilage. *European Polymer Journal*, **44**, 2820–2829. 10.1016/j.eurpolymj.2008.06.021

Lu, Z., Doulabi, B. Z., Huang, C., Bank, R. A. and Helder, M. N. (2010). Collagen type II enhances chondrogenesis in adipose tissue-derived stem cells by affecting cell shape. *Tissue Engineering Part A*, **16**, 81–90. 10.1089/ten.tea.2009.0222

Ma, P. X. and Zhang, R. (1999). Synthetic nano-scale fibrous extracellular matrix. *Journal of Biomedical Materials Research*, **46**, 60–72.

Maroudas, A. (1976). Balance between swelling pressure and collagen tension in normal and degenerate cartilage. *Nature*, **260**, 808–809.

McCormick, F., Yanke, A., Provencher, M. T. and Cole, B. J. (2008). Minced articular cartilage – basic science, surgical technique, and clinical application. *Sports Medicine and Arthroscopy Review*, **16**, 217–220. 10.1097/JSA.0b013e31818e0e4a

Meachim, G. and Stockwell R.A. (1979). The matrix. In: Freeman MAR, ed. *Adult Articular Cartilage*. Second edition. Pitman Medical.

Meunier, P. J. and Boivin, G. (1997). Bone mineral density reflects bone mass but also the degree of mineralization of bone: therapeutic implications. *Bone*, **21**, 373–377.

Minas, T. and Nehrer, S. (1997). Current concepts in the treatment of articular cartilage defects. *Orthopedics*, **20**, 525–538.

Montes, G. S. (1996). Structural biology of the fibres of the collagenous and elastic systems. *Cell Biology International*, **20**, 15–27. 10.1006/cbir.1996.0004

Morgan, E. F., Bayraktar, H. H. and Keaveny, T. M. (2003). Trabecular bone modulus-density relationships depend on anatomic site. *Journal of Biomechanics*, **36**, 897–904. S002192900300071X [pii]

Mouw, J. K., Case, N. D., Guldberg, R. E., Plaas, A. H. K. and Levenston, M. E. (2005). Variations in matrix composition and GAG fine structure among scaffolds for cartilage tissue engineering. *Osteoarthritis and Cartilage*, **13**, 828–836. 10.1016/j.joca.2005.04.020

Mow, V. C., Ratcliffe, A. and Robin Poole, A. (1992). Cartilage and diarthrodial joints as paradigms for hierarchical materials and structures. *Biomaterials*, **13**, 67–97.

Mow, V. C., Wang, C. C. and Hung, C. T. (1999). The extracellular matrix, interstitial fluid and ions as a mechanical signal transducer in articular cartilage. *Osteoarthritis and Cartilage*, **7**, 41–58. 10.1053/joca.1998.0161

Murugan, R. and Ramakrishna, S. (2005). Development of nanocomposites for bone grafting. *Composites Science and Technology*, **65**, 2385–2406.

Nabzdyk, C., Pradhan, L., Molina, J., Emerson, P., Paniagua, D. and Rosenstrauch, D. (2009). Auricular chondrocytes – from benchwork to clinical applications. *In Vivo*, **23**, 369–380.

Nagata, S. (1993). A new method of total reconstruction of the auricle for microtia. *Plastic and Reconstructive Surgery*, **92**, 187–201.

Nehrer, S. and Minas, T. (2000). Treatment of articular cartilage defects. *Investigative Radiology*, **35**, 639–646.

Nettles, D. L., Vail, T. P., Morgan, M. T., Grinstaff, M. W. and Setton, L. A. (2004). Photocrosslinkable hyaluronan as a scaffold for articular cartilage repair. *Annals of Biomedical Engineering*, **32**, 391–397. 10.1023/b:abme.0000017552.65260.94

Nguyen, L. H., Kudva, A. K., Guckert, N. L., Linse, K. D. and Roy, K. (2011). Unique biomaterial compositions direct bone marrow stem cells into specific chondrocytic phenotypes corresponding to the various zones of articular cartilage. *Biomaterials*, **32**, 1327–1338. 10.1016/j.biomaterials.2010.10.009

Nisbet, D. R., Forsythe, J. S., Shen, W., Finkelstein, D. I. and Horne, M. K. (2009). Review paper: a review of the cellular response on electrospun nanofibers for tissue engineering. *Journal of Biomaterials Applications*, **24**, 7–29. 10.1177/0885328208099086

Ozbas, B., Rajagopal, K., Haines-Butterick, L., Schneider, J. P. and Pochan, D. J. (2007). Reversible stiffening transition in beta-hairpin hydrogels induced by ion complexation. *Journal of Physical Chemistry B*, **111(50)**, 13901–13908.

Ozbas, B., Rajagopal, K., Schneider, J. P. and Pochan, D. J. (2004). Semiflexible chain networks formed via self-assembly of beta-hairpin molecules. *Physics Review Letters*, **93**, 268106.

Pan, Y. and Xiong, D. (2009). Study on compressive mechanical properties of nano-hydroxyapatite reinforced poly(vinyl alcohol) gel composites as biomaterial. *Journal of Materials Science: Materials in Medicine*, **20**, 1291–1297. 10.1007/s10856-008-3679-8

Pan, Y., Xiong, D. and Gao, F. (2008). Viscoelastic behavior of nano-hydroxyapatite reinforced poly(vinyl alcohol) gel biocomposites as an articular cartilage. *Journal of Materials Science: Materials in Medicine*, **19**, 1963–1969. 10.1007/s10856-007-3280-6

Park, J. S., Shim, M. S., Shim, S. H., Yang, H. N., Jeon, S. Y., Woo, D. G., Lee, D. R., Yoon, T. K. and Park, K. H. (2011). Chondrogenic potential of stem cells derived from amniotic fluid, adipose tissue, or bone marrow encapsulated in fibrin gels containing TGF-beta 3. *Biomaterials*, **32**, 8139–8149. 10.1016/j.biomaterials.2011.07.043

Park, J. S., Woo, D. G., Sun, B. K., Chung, H. M., Im, S. J., Choi, Y. M., Park, K., Huh, K. M. and Park, K. H. (2007). In vitro and in vivo test of PEG/PCL-based hydrogel scaffold for cell delivery application. *Journal of Controlled Release*, **124**, 51–59. 10.1016/j.jconrel.2007.08.030

Park, K. M., Joung, Y. K., Park, K. D., Lee, S. Y. and Lee, M. C. (2008). RGD-conjugated chitosan-Pluronic hydrogels as a cell supported scaffold for articular cartilage regeneration. *Macromolecular Research*, **16**, 517–523.

Park, S. H., Park, S. R., Chung, S. I., Pai, K. S. and Min, B. H. (2005). Tissue-engineered cartilage using fibrin/hyaluronan composite gel and its in vivo implantation. *Artificial Organs*, **29**, 838–845. 10.1111/j.1525-1594.2005.00137.x

Peter, M., Ganesh, N., Selvamurugan, N., Nair, S. V., Furuike, T., Tamura, H. and Jayakumar, R. (2010). Preparation and characterization of chitosan-gelatin/nanohydroxyapatite composite scaffolds for tissue engineering applications. *Carbohydrate Polymers*, **80**, 687–694. 10.1016/j.carbpol.2009.11.050

Pochan, D. J., Schneider, J. P., Kretsinger, J., Ozbas, B., Rajagopal, K. and Haines, L. (2003). Thermally reversible hydrogels via intramolecular folding and consequent self-assembly of a de novo designed peptide. *Journal of the American Chemical Society*, **125**, 11802–11803.

Poole, A. R., Kojima, T., Yasuda, T., Mwale, F., Kobayashi, M. and Laverty, S. (2001). Composition and structure of articular cartilage: a template for tissue repair. *Clinical Orthopaedics and Related Research*, **391**, S26–S33.

Poole, A. R., Pidoux, I., Reiner, A. and Rosenberg, L. (1982). An immunoelectron microscope study of the organization of proteoglycan monomer, link protein, and collagen in the matrix of articular cartilage. *The Journal of Cell Biology*, **93**, 921–937. 10.1083/jcb.93.3.921

Poole, C. A. (1997). Review. Articular cartilage chondrons: form, function and failure. *Journal of Anatomy*, **191**(Pt 1), 1–3.

Proctor, C. S., Schmidt, M. B., Whipple, R. R., Kelly, M. A. and Mow, V. C. (1989). Material properties of the normal medial bovine meniscus. *Journal of Orthopaedic Research*, **7**, 771–782. 10.1002/jor.1100070602

Qu, D., Li, J., Li, Y., Khadka, A., Zuo, Y., Wang, H., Liu, Y. and Cheng, L. (2011). Ectopic osteochondral formation of biomimetic porous PVA-n-HA/PA6 bilayered scaffold and BMSCs construct in rabbit. *Journal of Biomedical Materials Research Part B: Applied Biomaterials*, **96B**, 9–15. 10.1002/jbm.b.31697

Radford, K. C. (1971). The mechanical properties of an epoxy resin with a second phase dispersion. *Journal of Materials Science*, **6**, 1286–1291. 10.1007/bf00552042

Raghunath, J., Georgiou, G., Armitage, D., Nazhat, S. N., Sales, K. M., Butler, P. E. and Seifalian, A. M. (2009). Degradation studies on biodegradable nanocomposite based on polycaprolactone/polycarbonate (80:20%) polyhedral oligomeric silsesquioxane. *Journal of Biomedical Materials Research Part A*, **91A**, 834–844. 10.1002/jbm.a.32335

Recht, M., White, L. M., Winalski, C. S., Miniaci, A., Minas, T. and Parker, R. D. (2003). MR imaging of cartilage repair procedures. *Skeletal Radiology*, **32**, 185–200. 10.1007/s00256–003–0631–3

Rho, J.-Y., Kuhn-Spearing, L. and Zioupos, P. (1998). Mechanical properties and the hierarchical structure of bone. *Medical Engineering and Physics*, **20**, 92–102.

Ribeiro, R., Banda, S., Ounaies, Z., Ucisik, H., Usta, M. and Liang, H. (2012). A tribological and biomimetic study of PI-CNT composites for cartilage replacement. *Journal of Materials Science*, **47**, 649–658. 10.1007/s10853–011–5835–7

Roberts, J. J., Earnshaw, A., Ferguson, V. L. and Bryant, S. J. (2011). Comparative study of the viscoelastic mechanical behavior of agarose and poly(ethylene glycol) hydrogels. *Journal of Biomedical Materials Research Part B-Applied Biomaterials*, **99B**, 158–169. 10.1002/jbm.b.31883

Romo, T., Presti, P. M. and Yalamanchili, H. R. (2006). Medpor alternative for microtia repair. *Facial Plastic Surgery Clinics of North America*, **14**, 129–36, vi.

Rosenberg, M. D. (1963). Cell guidance by alterations in monomolecular films. *Science*, **139**, 411. 10.1126/science.139.3553.411

Sahoo, S., Chung, C., Khetan, S. and Burdick, J. A. (2008). Hydrolytically degradable hyaluronic acid hydrogels with controlled temporal structures. *Biomacromolecules*, **9**, 1088–1092. 10.1021/bm800051m

Sargeant, T. D., Guler, M. O., Oppenheimer, S. M., Mata, A., Satcher, R. L., Dunand, D. C. and Stupp, S. I. (2008). Hybrid bone implants: self-assembly of peptide amphiphile nanofibers within porous titanium. *Biomaterials*, **29**, 161–171.

Schneider, J. P., Pochan, D. J., Ozbas, B., Rajagopal, K., Pakstis, L. and Kretsinger, J. (2002). Responsive hydrogels from the intramolecular folding and self-assembly of a designed peptide. *Journal of the American Chemical Society*, **124**, 15030–15037.

Shah, R. N., Shah, N. A., del Rosario Lim, M. M., Hsieh, C., Nuber, G. and Stupp, S. I. (2010). Supramolecular design of self-assembling nanofibers for cartilage regeneration. *Proceedings of the National Academy of Sciences*, **107**, 3293–3298. 10.1073/pnas.0906501107

Shi, H. C., Lu, D., Wang, W. P., Lu, Y. and Zeng, Y. J. (2012). N-carboxyethylchitosan/nanohydroxyapatite composites scaffold for tracheal cartilage tissue-engineering applications. *Micro and Nano Letters*, **7**, 76–79. 10.1049/mnl.2011.0628

Shields, K. J., Beckman, M. J., Bowlin, G. L. and Wayne, J. S. (2004). Mechanical properties and cellular proliferation of electrospun collagen type II. *Tissue Engineering*, **10**, 1510–1517. 10.1089/ten.2004.10.1510

Shin, H. J., Lee, C. H., Cho, I. H., Kim, Y.-J., Lee, Y.-J., Kim, I. A., Park, K.-D., Yui, N. and Shin, J.-W. (2006). Electrospun PLGA nanofiber scaffolds for articular cartilage reconstruction: mechanical stability, degradation and cellular responses under mechanical stimulation in vitro. *Journal of Biomaterials Science, Polymer Edition*, **17**, 103–119. 10.1163/156856206774879126

Sikavitsas, V. I., Temenoff, J. S. and Mikos, A. G. (2001). Biomaterials and bone mechanotransduction. *Biomaterials*, **22**, 2581–2593.

Silva, G. A., Czeisler, C., Niece, K. L., Beniash, E., Harrington, D. A., Kessler, J. A. and Stupp, S. I. (2004). Selective differentiation of neural progenitor cells by high-epitope density nanofibers. *Science*, **303**, 1352–1355.

Slomianka, L. (2009). *Blue Histology – Skeletal Tissues – Cartilage* [Online]. Available: http://www.lab.anhb.uwa.edu.au/mb140/CorePages/Cartilage/Cartil. htm [Accessed 05/03/2012 2012].

Smart, S. K., Cassady, A. I., Lu, G. Q. and Martin, D. J. (2006). The biocompatibility of carbon nanotubes. *Carbon*, **44**, 1034–1047. 10.1016/j.carbon.2005.10.011

Smeds, K. A. and Grinstaff, M. W. (2001). Photocrosslinkable polysaccharides for in situ hydrogel formation. *Journal of Biomedical Materials Research*, **54**, 115–121. 10.1002/1097–4636(200101)54:1<115::aid-jbm14>3.0.co;2-q

Smith, L. A. and Ma, P. X. (2004). Nano-fibrous scaffolds for tissue engineering. *Colloids and Surfaces B: Biointerfaces*, **39**, 125–131. 10.1016/j. colsurfb.2003.12.004

Spiller, K. L., Holloway, J. L., Gribb, M. E. and Lowman, A. M. (2011). Design of semi-degradable hydrogels based on poly(vinyl alcohol) and poly(lactic-co-glycolic acid) for cartilage tissue engineering. *Journal of Tissue Engineering and Regenerative Medicine*, **5**, 636–647. 10.1002/term.356

Steinwachs, M. R., Guggi, T. and Kreuz, P. C. (2008). Marrow stimulation techniques. *Injury*, **39**, 26–31. 10.1016/j.injury.2008.01.042

Stupp, S. I. (2005. Biomaterials for regenerative medicine. *MRS Bulletin*, **30**, 546.

Subramanian, A., Vu, D., Larsen, G. F. and Lin, H.-Y. (2005). Preparation and evaluation of the electrospun chitosan/PEO fibers for potential applications in cartilage tissue engineering. *Journal of Biomaterials Science, Polymer Edition*, **16**, 861–873. 10.1163/1568562054255682

Supova, M. (2009). Problem of hydroxyapatite dispersion in polymer matrices: a review. *Journal of Materials Science: Materials in Medicine*, **20**, 1201–1213. 10.1007/s10856–009–3696–2

Tan, H. P., Chu, C. R., Payne, K. A. and Marra, K. G. (2009). Injectable in situ forming biodegradable chitosan-hyaluronic acid based hydrogels for cartilage tissue engineering. *Biomaterials*, **30**, 2499–2506. 10.1016/j.biomaterials.2008.12.080

Temenoff, J. S., Athanasiou, K. A., Lebaron, R. G. and Mikos, A. G. (2002). Effect of poly(ethylene glycol) molecular weight on tensile and swelling properties of oligo(poly(ethylene glycol) fumarate) hydrogels for cartilage tissue engineering. *Journal of Biomedical Materials Research*, **59**, 429–437. 10.1002/ jbm.1259

Temenoff, J. S., Lu, L., Mikos, A. G. (2000). Bone tissue engineering using synthetic biodegradable polymer scaffolds. In Davies, J. E., ed., *Bone Engineering*, Em Squared, Toronto.

Temenoff, J. S. and Mikos, A. G. (2000). Review: tissue engineering for regeneration of articular cartilage. *Biomaterials*, **21**, 431–440. 10.1016/ s0142–9612(99)00213–6

Teo, W. E., Liao, S., Chan, C. and Ramakrishna, S. (2011). Fabrication and characterization of hierarchically organized nanoparticle-reinforced nanofibrous composite scaffolds. *Acta Biomaterialia*, **7**, 193–202. 10.1016/j.actbio.2010.07.041

Teo, W. E. and Ramakrishna, S. (2006). A review on electrospinning design and nano-fibre assemblies. *Nanotechnology*, IOP Publishing Ltd, **17**, R89.

Toh, W. S., Lee, E. H., Guo, X. M., Chan, J. K. Y., Yeow, C. H., Choo, A. B. and Cao, T. (2010). Cartilage repair using hyaluronan hydrogel-encapsulated human embryonic stem cell-derived chondrogenic cells. *Biomaterials*, **31**, 6968–6980. 10.1016/j.biomaterials.2010.05.064

Umlauf, D., Frank, S., Pap, T. and Bertrand, J. (2010). Cartilage biology, pathology, and repair. *Cellular and Molecular Life Sciences*, **67**, 4197–4211. 10.1007/s00018–010–0498–0

Villanueva, I., Gladem, S. K., Kessler, J. and Bryant, S. J. (2010). Dynamic loading stimulates chondrocyte biosynthesis when encapsulated in charged hydrogels prepared from poly(ethylene glycol) and chondroitin sulfate. *Matrix Biology*, **29**, 51–62. 10.1016/j.matbio.2009.08.004

Wachsmuth, L., Soder, S., Fan, Z., Finger, F. and Aigner, T. (2006). Immunolocalization of matrix proteins in different human cartilage subtypes. *Histology and Histopathology*, **21**, 477–485.

Walton, R. L. and Beahm, E. K. (2010). Auricular reconstruction for microtia plastic and reconstructive surgery. In: Siemionow, M. Z. and Eisenmann-Klein, M. (eds.), *Plastic and Reconstructive Surgery*. Springer, London.

Wang, X., Kluge, J. A., Leisk, G. G. and Kaplan, D. L. (2008). Sonication-induced gelation of silk fibroin for cell encapsulation. *Biomaterials*, **29**, 1054–1064. 10.1016/j.biomaterials.2007.11.003

Warman, M. L. (2000). Human genetic insights into skeletal development, growth, and homeostasis. *Clinical Orthopaedics and Related Research*, **379**, S40–S54.

Webster, T. J., Ergun, C., Doremus, R. H., Siegel, R. W. and Bizios, R. (2000). Specific proteins mediate enhanced osteoblast adhesion on nanophase ceramics. *Journal of Biomedical Materials Research*, **51**, 475–483. 10.1002/1097–4636(20000905)51:3<475::aid-jbm23>3.0.co;2–9

Webster, T. J., Schadler, L. S., Siegel, R. W. and Bizios, R. (2001). Mechanisms of enhanced osteoblast adhesion on nanophase alumina involve vitronectin. *Tissue engineering*, **7**, 291–301.

Webster, T. J., Siegel, R. W. and Bizios, R. (1999). Osteoblast adhesion on nanophase ceramics. *Biomaterials*, **20**, 1221–1227. 10.1016/s0142–9612(99)00020–4

Wei, G. and Ma, P. X. (2004). Structure and properties of nano-hydroxyapatite/polymer composite scaffolds for bone tissue engineering. *Biomaterials*, **25**, 4749–4757.

Xie, Y., Upton, Z., Richards, S., Rizzi, S. C. and Leavesley, D. I. (2011). Hyaluronic acid: evaluation as a potential delivery vehicle for vitronectin:growth factor complexes in wound healing applications. *Journal of Controlled Release*, **153**, 225–232. 10.1016/j.jconrel.2011.03.021

Xue, D., Zheng, Q., Zong, C., Li, Q., Li, H., Qian, S., Zhang, B., Yu, L. and Pan, Z. (2010). Osteochondral repair using porous poly(lactide-co-glycolide)/nano-hydroxyapatite hybrid scaffolds with undifferentiated mesenchymal stem cells in a rat model. *Journal of Biomedical Materials Research Part A*, **94A**, 259–270. 10.1002/jbm.a.32691

Yamada, M. (1976). Ultrastructural and cytochemical studies on the calcification of the tendon-bone joint. *Archives of Histology Japan*, **39**, 347–378.

Yoshikawa, T., Nishida, K., Doi, T., Inoue, H., Ohtsuka, A., Taguchi, T. and Murakami, T. (1997). Negative charges bound to collagen fibrils in the rabbit articular cartilage: a light and electron microscopic study using cationic colloidal iron. *Archives of Histology Japan*, **60**, 435–443.

Yusong, P. (2010). Swelling properties of nano-hydroxyapatite reinforced poly(vinyl alcohol) gel biocomposites. *Micro and Nano Letters, IET*, **5**, 237–240. 10.1049/mnl.2010.0061

Zara, J. N., Siu, R. K., Zhang, X., Shen, J., Ngo, R., Lee, M., Li, W., Chiang, M., Chung, J., Kwak, J., Wu, B. M., Ting, K. and Soo, C. (2011). High doses of bone morphogenetic protein 2 induce structurally abnormal bone and inflammation in vivo. *Tissue Engineering Part A*, **17**, 1389–1399. 10.1089/ten.TEA.2010.0555

Zhang, J. J., Li, W. D. and Rong, J. H. (2010). Preparation and characterization of PVP/clay nanocomposite hydrogels. *Chemical Journal of Chinese Universities-Chinese*, **31**, 2081–2087.

Zhang, Y., Lim, C., Ramakrishna, S. and Huang, Z.-M. (2005). Recent development of polymer nanofibers for biomedical and biotechnological applications. *Journal of Materials Science: Materials in Medicine*, **16**, 933–946. 10.1007/s10856–005–4428-x

12

Biomaterials and nano-scale features for ligament regeneration

A. F. CIPRIANO and H. LIU, University of California, USA

DOI: 10.1533/9780857097231.2.334

Abstract: This chapter focuses on the use of biomaterials, nano-scale features, and scaffold enhancements for the construction of successful biological scaffolds for ligament regeneration. Ligament composition, structure, physiology, function, and properties are first introduced in order to identify key design parameters for scaffold design. The second part of the chapter describes injury, healing, and treatment of the anterior cruciate ligament focusing on current techniques utilized clinically for reconstruction. Subsequent content focuses on the materials, techniques, and features being used to develop scaffolds for ligament regeneration. Finally, the approaches (e.g., addition of growth factors and platelet rich plasma, and induced mechanical strain) which have been used to enhance engineered scaffolds for ligament regeneration are discussed.

Key words: anterior cruciate ligament healing, ligament primary regeneration, engineered scaffolds, nanomaterials, carbon nanomaterials, nanocomposites, biodegradable, growth factors.

12.1 Introduction

Ligaments are bone–bone connective tissues whose function is primarily associated with the transfer of loads within the musculoskeletal system. Fully developed ligaments experience some of the highest induced mechanical forces in the body despite requiring only small amounts of nutrients and having relatively low oxygen concentration and low cell density (Woo et al., 2006). The anterior cruciate ligament (ACL) demonstrates poor healing potential and limited vascularization compared to other ligaments in the human body (Lu et al., 2005). Therefore, ligament regeneration studies have mainly focused on the regeneration of the ACL and lessons learned from these studies have been extrapolated to develop methods to assist regeneration of other types of ligaments in the body.

The ACL plays an important role in restraining tibial translational and rotational motion. It is also the most commonly injured ligament of the

334

knee with more than 175 000 affected patients every year (Myer *et al.*, 2004). Ligament injury occurs when the induced mechanical forces exceed a critical threshold generating permanent tissue damage that results in impaired joint function and mobility. A poor healing mechanism associated with the ACL usually necessitates use of reconstructive surgery using various types of graft materials to treat injuries (Fu *et al.*, 2000; Lu *et al.*, 2005). Graft reconstruction methods remain as the gold standard, but recent long-term studies showed that patients who sustain an ACL tear have a 78% risk of developing premature osteoarthritis within only 14 years following injury (von Porat *et al.*, 2004). Additional long-term studies showed that, although grafts provide initial functional stability, normal knee kinematics are not restored (Murray *et al.*, 2010).

As an alternative to surgical reconstruction, recent biomechanical studies showed that conserving injured ACL remnants contributes to joint stability for up to several months after injury (Nakamae *et al.*, 2010), thus sparking an interest once again in the long abandoned ligament regeneration approach. Additionally, Murray *et al.* (2006) recently demonstrated that the premature loss of the provisional scaffold in the wound site after ACL rupture is a key factor that limits ligament regeneration in animal models. Other results reported by Murray *et al.* (2010) showed that placement of a substitute, engineered, degradable scaffold in animal ACL injury sites initiated and helped promote healing of the ruptured tissue after initial suturing of the torn ligament. Nanomaterials are of special interest in tissue engineering because results have demonstrated cell preference towards materials with nano-scaled features that mimic physiological surfaces and structures. Therefore, advancements in the synthesis of biological nanomaterials and results from recent studies have led researchers to pursue ligament regeneration through the development of novel provisional scaffolds to address the anatomic and biomechanical requirements for ligament repair.

12.2 Anterior cruciate ligament (ACL) composition, structure and properties

In order to determine the materials, features and/or regeneration-enhancing components that are needed for scaffold design, it is important to fully understand the nature of the condition being addressed. For ligaments, it is important to understand composition, structure, biomechanics, injury mechanism, and healing mechanism. This information will point to key design parameters that will serve as guidelines for the development of new methods for ligament regeneration.

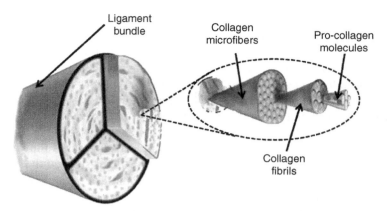

12.1 Hierarchical structure of ligaments. Ligaments in their simplest form are synthesized as pro-collagen molecules, then nano-scale fibrils, micro-scale fibers, and finally into macro-scale bundles.

12.2.1   ACL composition, structure and physiology

The composition of ligaments is approximately two-thirds water and one-third solid components (Frank, 2004). The solid fraction (70–80% by mass) is mainly composed of closely packed triple-helical type-I collagen fiber bundles oriented axially in a parallel fashion. Located circumferentially around these collagen fibers are interspersed ligament fibroblasts. Ligament fibroblasts are the major cell type found in ligaments. The primary function of ligament fibroblasts is to maintain ligament matrix density and composition (Woo *et al.*, 2006). The ligament matrix is also composed of elastin (5 wt.%), collagen (approximately 20 wt.%), proteoglycans, and glycoproteins (Frank, 2004; Woo *et al.*, 2006).

Ligaments are organized structures composed of small building blocks that assemble into larger units (Fig. 12.1). The smallest building blocks found in ligament tissue are pro-collagen molecules (approximately one nanometer in diameter) (Kuo *et al.*, 2010). Pro-collagen molecules are subsequently organized into parallel collagen nano-scale fibrils (approximately 50–280 nm diameter) (McBride *et al.*, 1988; Silver *et al.*, 2003) and finally into micro-scale collagen fibers. Both nano-scale fibrils and microfibers are highly aligned axially along the direction of the applied load. Furthermore, some ligaments have a zonal architecture which transitions near the area of insertion to the bone. These zones include ligament tissue, fibrocartilage, mineralized fibrocartilage, and bone (Woo *et al.*, 1987). Such transitional composition allows for a gradient in force distribution and applied differential tension between zones (Iwahashi *et al.*, 2010).

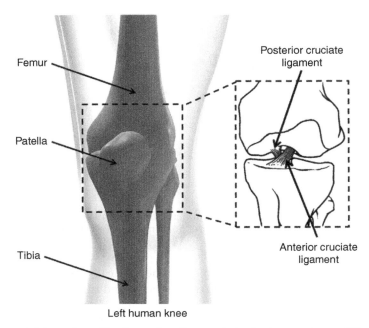

12.2 Location of both cruciate ligaments in the human knee. ACL and PCL excerpt shown resides behind the patella.

Physiologically there are two cruciate ligaments in the human knee, named anterior and posterior according to their site of attachment to the tibia. Both cruciate ligaments are located intra-articularly within the knee capsule and cross each other at an oblique angle (Fig. 12.2). The ACL arises from the anterior part of the intercondylar eminentia of the tibia and extends to the posterolateral (PL) aspect of the intercondylar fossa of the femur where it is surrounded by a synovial layer (Woo *et al.*, 2006). Furthermore, the ACL is composed of two separate bundles, the anteromedial (AM) and PL, and studies have shown that each bundle functions individually even under minimal loads (Sakane *et al.*, 1997).

12.2.2 ACL function, motion mechanics and mechanical properties

The main function of ligaments is to transfer uniaxial forces between musculoskeletal tissues (bone-to-bone) to achieve joint movement and stability while retaining responsiveness to many local and systemic factors (Kuo *et al.*, 2010). Under low loading conditions ligaments are subject to deformations that help dissipate induced stresses and achieve joint homeostasis.

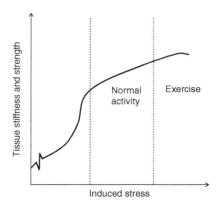

12.3 Diagram illustrating ligament stiffness and strength response to different levels of stress corresponding to physical activities.

As load is increased, an improvement in mechanical properties is observed until a point is reached where the tissue yields and ruptures. A load versus elongation graph can be plotted as a non-linear, concave-upward curve. As shown in Fig. 12.3, the normal range of physiological activities is represented by the relatively constant middle region of the curve, whereas long term exercise results in a slight increase in mechanical properties (Woo *et al.*, 2006). In the knee, the AM and PL bundles of the ACL prevent hyperextension by providing anterior–posterior translation restraint and tibial rotational stability, respectively (Butler *et al.*, 1980; Kanamori *et al.*, 2002; Yagi *et al.*, 2002). During flexion at the knee joint the AM is taut while the PL is loose, and during extension the roles of the AM and PL are reversed (Woo *et al.*, 2006).

Ligament biomechanics are limited by the structural and mechanical properties of the tissue and the maximum stresses they can support. Ligament structural properties of interest include stiffness, ultimate load, ultimate elongation, and absorbed energy at failure. Mechanical properties of interest include tangent modulus, ultimate tensile strength, ultimate strain, and strain energy density (Woo *et al.*, 2006). Reported values for non-injured ACL mechanical and structural properties include: tangent modulus of 283 ± 114 MPa and 154 ± 120 MPa (for AM and PL bundles, respectively) (Butler *et al.*, 1992), maximum tensile load of 2160 ± 157 N, and linear stiffness of 242 ± 28 N/mm (Woo *et al.*, 1991).

12.3   Injury, healing and treatment of the ACL

Understanding the healing response mechanism of ligaments to injury is critical in developing regeneration strategies capable of emulating the natural healing response. The following section discusses the mechanisms that lead to ligament rupture, subsequent body response to such disruptions, and current techniques used to fix ligament ruptures. These concepts will help the reader identify the desired physiological responses to injury which will serve as guidelines for biomaterial-assisted ligament regeneration techniques.

12.3.1   ACL injury

ACL injuries can result in either partial or complete rupture. The main cause of ruptures can be attributed to large strains induced by traumatic forces applied to the knee in a twisting moment by means of direct or indirect forces. Non-contact, sudden deceleration with a change of direction or single limb landing that results in a rotation-induced collapse of the knee account for nearly half of ACL rupture cases (Boden *et al.*, 2000). Increased loads on the ACL result in constantly increasing stiffness and absorbed energy and culminate in tensile failure (Frank, 2004). Cyclical loads induce ligament 'creep', or deformation by means of permanent elongation, where excessive creep can result in joint laxity and ultimately predispose the joint to further injury. Figure 12.4 shows magnetic resonance imaging (MRI) scans of torn ligaments where the severed ends are outlined and the gap of the torn ligament can be observed.

A secondary cause of ACL rupture can be attributed to microscopic cellular changes and damage, such as altered organization of the extracellular matrix (ECM), that induce a higher propensity to injury. In such cases, ligament tissue with microstructural irregularities can rupture under even moderate levels of strain (Provenzano *et al.*, 2002). Microscopic cellular alterations can be correlated with conditions that lead to incorrectly formed ligaments, narrow intercondylar notches, and knee joint deformities (Bradbury *et al.*, 1998).

12.3.2   Extra-articular and intra-articular ligament healing mechanisms

The self-initiating healing process in extra-articular ligaments can be roughly divided into three phases (Frank *et al.*, 1983; Weiss *et al.*, 1991): the inflammatory, reparative, and remodeling phases. Histology results of the different phases during the healing of extra-articular ligaments are illustrated in Fig. 12.5a–12.5f. First, the inflammatory phase is marked by the retraction of disrupted ligament ends and hematoma formation which is initiated after injury and lasts for a few weeks. Second, the reparative phase, which has a duration of about six weeks, is characterized by fibroblast

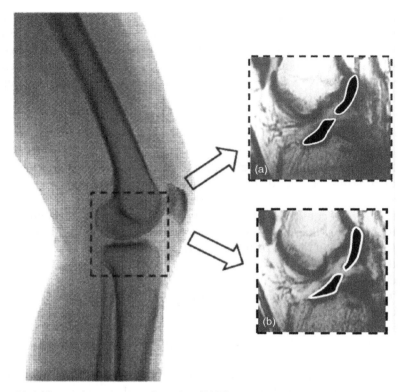

12.4 Magnetic resonance imaging (MRI) scans demonstrating discontinuities in the ACL. Highlighted areas in (a) and (b) represent complete ruptures of the ACL. (*Source*: Adapted from Barry *et al.*, 1996. Reproduced with permission from the *Journal of Skeletal Radiology*.)

migration, proliferation, cell contraction, and by the synthesis of an organized glycoprotein-collagen matrix. Lastly, the remodeling phase is marked by alignment and further organization of collagen fibers and matrix maturation which can be years in duration (Frank, 2004).

Intra-articular ligaments such as the ACL do not heal on their own due to differences in the ligament healing cascade of events. In the intra-articular environment, formation of a stabilized fibrin clot is not observed (Harrold, 1961; Andersen and Gormsen, 1970). Figure 12.6 illustrates the injury response of the extra-articular medial collateral ligament (MCL) and intra-articular ACL, where the main difference resides in the ability of the MCL to form a stable fibrin clot. Figure 12.6a shows a schematic diagram comparing the healing step of both MCL and ACL. Figure 12.6b and 12.6c show micrographs of 7 day-old slit wounds in MCL and ACL, respectively, in a canine model.

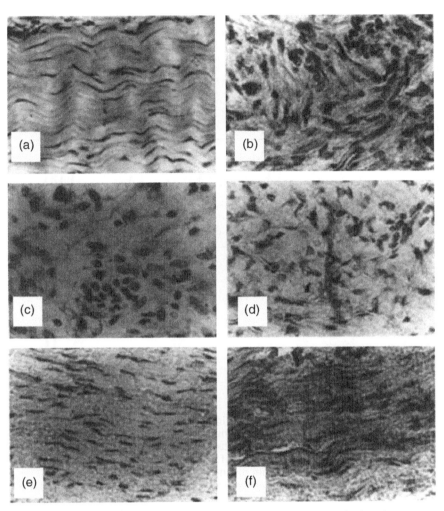

12.5 Histology results from extra-articular ligaments during the healing process. (a) normal ligament; (b) neoligament filled by vascular inflammatory tissue (after 10 days); (c) and (d) decrease in fibroblast numbers and size, and some evidence of longitudinal alignment of nuclei (after 3 and 4 weeks, respectively); (e) continued remodeling with improving realignment and further decrease in cell numbers (after 14 weeks); (f) cells remained larger and more numerous than those seen in normal tissue (after 40 weeks). (*Source*: Reproduced with permission from *The American Journal of Sports Medicine*.)

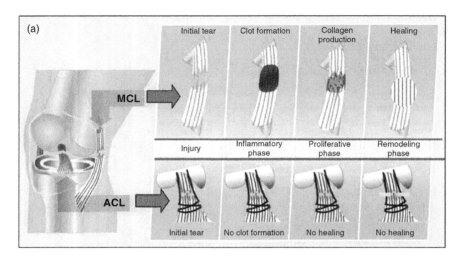

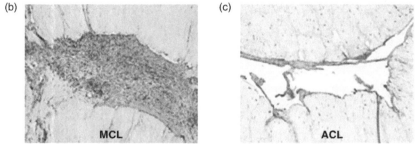

12.6 Healing in extra-articular ligaments compared with healing in intra-articular ligaments. (a) Schematic diagram comparing failure of ACL healing with the autocatalytic healing of the medial collateral ligament (MCL). Micrographs of 7 day-old slit wounds in canine (b) MCL and (c) ACL. Main distinction resides in the ability of the extra-articular MCL to generate a provisional scaffolding material; the ACL wound, however, remains unfilled. (*Source*: Reproduced with permission from the *Journal of Clinical Sports Medicine*.)

The micrographs shown in Fig. 12.6 indicate the lack of a stabilized fibrin clot in the ACL compared with the fibrin clot formed in the extra-articular MCL. The lack of formation of a stable fibrin-platelet clot at the ligament tear site can be attributed to blood dissipation into the synovial fluid caused by disruption to the synovial sheath of the ACL following injury. The lack of clot formation can also be linked to the presence of urokinase plasminogen activator (UPA) in the synovial fluid after injury. UPA promotes plasmin circulation that leads to fibrin consumption and inhibits formation of a functional clot (Rosc *et al.*, 2002). Due to the lack of a blood clot, the supply of inflammatory mediators that secrete cytokines and growth factors is diverted towards the synovial fluid and thus fails to assist in healing of the injured ligament (Haus *et al.*, 2012). It is therefore hypothesized that the lack of functional healing inherent to the ACL may be attributed to the premature loss of the fibrin-platelet provisional clot (Murray *et al.*, 2006).

12.3.3   Torn ACL treatment: reconstruction through graft insertion

Over the past several decades, the gold standard method of ACL reconstruction has relied on the use of biologic tissue grafts (85–95% success rate) (West and Harner, 2005) mainly due to their positive integration into the joint (Fu *et al.*, 1999). Reconstruction methods can be classified into three main categories depending on the source material: autograft, where graft material is harvested from the injured patient; allograft, where graft material is obtained from a donor; and finally, synthetic grafts, where graft materials are biocompatible materials rather than biological tissue. Graft reconstruction procedures involve removal of the torn ACL remnants followed by surgical insertion of the preferred ligament graft as determined by the operating surgeon. Rehabilitation from surgical ligament reconstruction procedures is incremental and results in restored mobility. Post-surgical rehabilitation is usually completed about six months following the surgical intervention (Shelbourne and Nitz, 1990).

The most common type of biological tissue graft used in ligament reconstruction is the autograft. Traditionally, autografts are harvested from the patient's patellar-bone tendon but can also be obtained from the patient's hamstring tendon. Studies have shown the latter to be associated with faster recovery and less anterior knee pain (van Eck *et al.*, 2010). Benefits of using autografts include low risk of adverse inflammatory reactions and virtually no risk of disease transmission (Fu *et al.*, 1999). On the other hand, allografts have been harvested from tissue donors at sites such as the patellar tendon, the Achilles tendon, and the fascia lata (Nikolaou *et al.*, 1986). Allografts are advantageous as they avoid donor site morbidity, decrease surgical time, and reduce post-operative pain (Fu *et al.*, 1999). Lastly, synthetic grafts have been fabricated from various polymers and FDA-approved products including Proplast (Vitek Inc, Houston, TX), Leeds-Keio (Neoligaments Ltd, UK), Gore-Tex (W. L. Gore & Associates, Flagstaff, AZ), Stryker (Stryker, Kalamazoo, MI), and prosthesis (Guidoin *et al.*, 2000). Synthetic grafts were first introduced in the 1970s and sought to perform as stand-alone ligament replacements. Although synthetic materials provide adequate mechanical strength, these benefits are often offset by major complications such as foreign-body inflammation and synovitis (Guidoin *et al.*, 2000).

A major shortcoming associated with graft reconstruction methods is the relation to premature osteoarthritis (von Porat *et al.*, 2004). Overcoming premature osteoarthritis has been the focus of current studies and has shifted graft reconstruction research toward the restoration of normal knee kinematics by means of techniques that reinstate native ligament anatomy (van Eck *et al.*, 2010). Ongoing research in ligament reconstruction via graft insertion also seeks to address post-operative shortcomings such as anterior knee pain, decreased range of motion, and mechanically inferior tissue (Petrigliano *et al.*, 2006).

12.4 Engineered scaffold materials for ligament regeneration

Recent studies demonstrated that conservation of the injured ACL remnants improved knee biomechanics, accelerated revascularization, and enhanced joint proprioceptive functioning (Adachi *et al.*, 2002; Nakamae *et al.*, 2010). Interest in conserving the injured ACL remnants, as opposed to ACL reconstruction methods, has fueled research studies that utilize advancements in biomaterial synthesis and characterization in order to develop engineered tissue scaffolds capable of regenerating the torn ACL remnants. Further motivation for research studies that preserve ACL remnants are based on the findings reported by Murray *et al.* (2006) which indicated that the lack of functional healing inherent to the ACL may be attributed to the premature loss of the fibrin-platelet provisional scaffold. Termed 'primary regeneration', these repair techniques of the torn ligament remnants seek to restore native function with the aid of a temporary, engineered scaffold.

In order to effectively assist in repairing and restoring functionality to the torn ACL, these engineered scaffolds should have the following characteristics: biodegradable and biocompatible; porous, to facilitate cell infiltration and proliferation; exhibit sufficient mechanical strength for immediate load bearing post-implantation; and promote the formation of ligament tissue by providing an environment that facilitates nutrient transport and appropriate regulatory stimuli. An effectively engineered scaffold will result in collagen synthesis in an organized manner similar to the native ligament structure to achieve mechanical and biochemical properties required for load bearing (Petrigliano *et al.*, 2006). Additionally, as the desired physiological response is obtained, the engineered scaffold should degrade at a carefully examined rate to avoid accumulation of by-products within the body. Figure 12.7 illustrates a schematic diagram of the main requirements and the relationship between the required characteristics leading to the successful development of a regenerated functional ligament. This section will consider materials that have been used in ligament regeneration studies, and the benefits and limitations of each material.

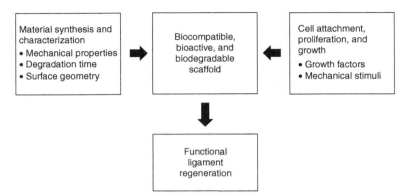

12.7 Schematic diagram indicating the main requirements and relation between the required characteristics leading to development of successful engineered scaffolds for ligament regeneration.

12.4.1   Natural biopolymers

Considerable interest in the application of natural, protein-based fibers as scaffold materials for ligament regeneration has been motivated by the fact that these materials often provide functional chemical groups for cellular binding and generate non-toxic by-products upon degradation (Kuo *et al.*, 2010). Of these, collagen has been of particular interest since ligaments are mainly composed of this protein. Results from previous studies have shown fibroblast proliferation, both *in vitro* and *in vivo*, in collagen matrices (Dunn *et al.*, 1995; Bellincampi *et al.*, 1998). Further motivation to utilize collagen as a biomaterial for engineered scaffolds has been sparked by recent animal studies on ACL regeneration which showed ligament fibroblast proliferation in collagen matrices when combined with platelet rich plasma (PRP) (Murray *et al.*, 2007a). While collagen scaffolds have been shown to support cell attachment, proliferation, and ECM coverage, the limitations associated with mechanical properties such as poor tensile strength have offset the benefits leading to a consideration of alternate materials.

Fibrin has also been considered as a scaffolding material for ligament regeneration due to its natural occurrence and its biodegradable and biocompatible properties. Fibrin participates in natural wound healing by entrapping activated platelets in the wound site (Vavken *et al.*, 2011a). The ability to entrap activated platelets allows fibrin-based scaffolds to reap the benefits of growth factors that are expressed during the inflammatory healing phase. Results from fibrin-based scaffolds showed enhanced cellular migration and growth factor delivery capabilities (Wong *et al.*, 2003). Other studies showed that positive fibroblast response and collagen synthesis were highly dependent on fibrin concentration and the best results were obtained with low fibrin concentrations (Vavken *et al.*, 2011a). Fibrin concentration dependency was also observed in a study involving human mesenchymal stem cells (hMSCs) where Ho *et al.* (2006) concluded that both cell proliferation and migration are equally important indicators of healing. Positive cell response and collagen production make fibrin seem like an adequate scaffold material for ACL regeneration (Vavken *et al.*, 2011a). However, fibrin has poor mechanical properties, similar to collagen (Perka *et al.*, 2000; Ho *et al.*, 2006).

In response to the need for a biopolymer with enhanced mechanical properties, silk has been considered as a scaffold material for ligament regeneration. The remarkable tensile strength and toughness of silk, which is unmatched by natural proteins, has made it a popular biopolymer in tissue engineering. Further characteristics that make silk an adequate material for ligament scaffold include: the capability of exposing surface amino acids for cell adhesion; the ability to conserve structural stability in solution while providing a slow degradation rate; and the ability to be fabricated into gels, films, braided fibers, or nano-fibers (Altman *et al.*, 2002a). These beneficial characteristics allowed for the development of a braided silk scaffold seeded with adult human progenitor bone marrow stromal cells (BMSCs). These scaffolds provided the required mechanical strength and enabled *in vitro* cell differentiation toward ligament regeneration, as shown in the scanning electron microscope (SEM) images in Fig. 12.8 (Altman *et al.*, 2002a).

Other natural biopolymers that have been considered for ligament scaffolds include chitosan, alginate, and hyaluronic acid. Both chitosan and alginate are inherently biocompatible materials that have been shown to promote *in vitro* fibroblast growth and type-I collagen production (Majima *et al.*, 2005). Results by Majima and co-workers also demonstrated that a coating of arginine-glycine-aspartic acid peptide sequence on chitosan and alginate improved *in vitro* fibroblast proliferation (Chen *et al.*, 2003; Majima *et al.*, 2005). Separate experimental results showed that modified hyaluronic acid-based scaffolds seeded with BMSCs also produced type-I collagen and other fibroblast markers (Cristino *et al.*, 2005). Although studies using natural biopolymers for constructing three-dimensional (3D) matrices have shown promising results in terms of promoting cell proliferation, the main limitation still resides in the inability to modify the mechanical and degradation properties of these materials. The ability to control mechanical and degradation properties of a temporary scaffold is important for a successful scaffold design for ligament regeneration.

12.4.2   Synthetic degradable polymers

The ease with which mechanical and chemical properties can be reproduced, ease of fabrication into different shapes and sizes, the ability to degrade by

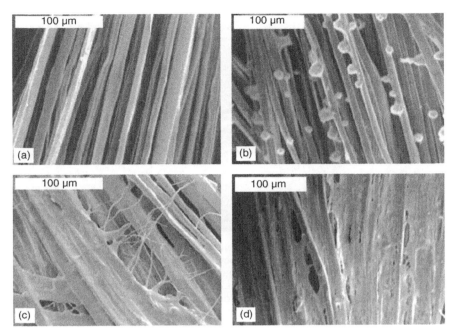

12.8 Scanning electron microscope (SEM) images showing adherence, proliferation, and cell sheet formation by human BMSCs on a silk cord matrix: (a) prior to seeding; (b) time zero of cell culture; (c) 1 day after cell culture; and (d) 14 days after cell culture. Scale bars = 100 μm. (*Source*: Reproduced with permission from the *Journal of Biomaterials*.)

hydrolysis, and the ability to provide vascularization have led researchers to explore synthetic biodegradable polyesters as alternative scaffold materials for ligament regeneration (Perka *et al.*, 2000). An additional advantage of using synthetic biomaterials resides in the controlled biodegradable property and the ability to incorporate controlled drug/growth factor delivery to further promote cell proliferation and functionality. Of these synthetic polymers, polyhydroxyesters encompass a family of materials that have been used extensively as ligament scaffolds. Advantages include the ability to degrade by hydrolysis, an ability to provide fibroblast adhesion and proliferation at rates comparable to tissue-culture plastic, and FDA approval (Kuo *et al.*, 2010). Examples of polyhydroxyesters include poly(ε-caprolactone) (PCL), poly(D, L-lactic acid) (PLA), poly(D, L-lactic-co-glycolic acid) (PLGA), and blends of different percentages between these polymers. An *in vitro* comparative study between fabricated 3D braided scaffolds with human recombinant fibronectin (Fn)-coated polyglycolic acid (PGA-Fn), poly-L-lactic acid (PLLA-Fn), and polylactic-co-glycolic acid 82:18 (PLAGA-Fn) concluded that PLLA-Fn scaffolds were the most appropriate choice for ligament

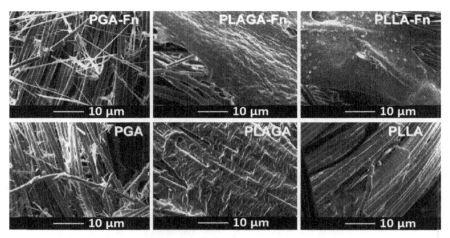

12.9 SEM images of ACL fibroblast growth after 14 days of culture on braided polymer scaffolds covered with fibronectin (PGA-Fn, PLAGA-Fn, and PLLA-Fn) compared with non-covered braided polymer scaffolds (PGA, PLAGA, and PLLA). Scale bars = 10 µm. (*Source:* Reproduced with permission from the *Journal of Biomaterials.*)

regeneration due to higher levels of matrix production, slower degradation rates, and improved cell adhesion (Lu *et al.*, 2005). The SEM images in Fig. 12.9 show ACL fibroblast growth on Fn-coated and non-Fn-coated polymer braided scaffolds (PGA, PLAGA, and PLLA) after 14 days of culture; the SEM image corresponding to the PLLA-Fn scaffold showed the most cell growth. Results from this study also showed that mechanical properties comparable to those of non-injured ligaments could be obtained by biodegradable polyesters.

In a similar *in vitro* comparative study, a comparison between PCL and PLLA pointed towards PCL as a better choice over PLLA for ligament regeneration due to the slower degradation rates and higher ductility of PCL (Vieira *et al.*, 2011). Taken together, these studies showed a promising future for functionally enhanced synthetic scaffolds (e.g., Fn-coated polymers) due to the versatile mechanical and degradation properties and the improved positive cell response.

12.4.3   Carbon nanomaterials

Advantageous properties of carbon fibers include high strength-to-weight ratio, flexibility, biological inertness and compatibility, and long *in vivo* stability (Saito *et al.*, 2011). Additionally, electrical conductive properties inherent to carbon nanomaterials have been shown to direct cell growth within the scaffolds when applied current is induced through aligned carbon nano-tubes (Harrison and Atala, 2007). Previous studies have shown adequate carbon fiber biocompatibility and suitability for ligament regeneration in sheep models (Neugebauer and Claes, 1983) and canine models (Aragona *et al.*, 1983) along with adequate biocompatibility with various cell types, especially osteoblasts (Zanello *et al.*, 2006). Although not much work has been done at the interface of carbon nanomaterials and ligament regeneration, the properties mentioned combined with advancement in material synthesis indicate that carbon nanomaterials are a promising candidate for ligament scaffolds.

12.4.4 Nano-fibers and porosity in engineered ligament scaffolds

The ability to process synthetic biopolymers into nano-fibers that closely resemble native collagen fibrils is an additional benefit of using synthetic materials for ligament scaffold construction. Nano-fiber scaffolds offer a significant increase in the surface-area-to-volume ratio, leading to improved cellular interaction with individual fibrils. Nano-fiber and nano-porous scaffolds are also known to improve cell attachment leading to proliferation, migration, and differentiation. Studies have evaluated the ideal pore size to favor cell response and it was shown that pores of about 150 μm promote cell growth in soft tissue (Lu *et al.*, 2005). Other studies explored *in vitro* cell response on different polyhydroxyester nano-fiber scaffolds and obtained robust fibroblast growth on PCL (Reed *et al.*, 2009), PLA, and PGA substrates (Lin *et al.*, 1999). Furthermore, cell growth comparisons on PCL nano-fiber constructs between 3D PCL scaffolds and two-dimensional (2D) surfaces demonstrated improved growth and increased longevity of fibroblasts on 3D scaffolds.

Synthetic polymers can be fabricated into nano-fibers by electrospinning, self-assembly, or phase separation. Of these, electrospinning is the most studied because the parameters such as fiber diameter, composition, and porosity can be controlled precisely. Electrospinning allows for nano-fiber fabrication from degradable polymers (both natural and synthetic) such as collagen, gelatin, hyaluronic acid, silk fibroin, carbon, PLA, polyurethane, PCL, PLGA, and PLLA with fiber diameters ranging from about three nanometers to over one micrometer (Vasita and Katti, 2006). Additionally, it has been reported that aligned nano-fibers significantly increased collagen production and provided cell morphologies that had better resemblance to native tissue, in contrast to randomly oriented fibers (Lee *et al.*, 2005).

An additional aspect that has been shown to favor cell proliferation and differentiation is the increased homogeneous dispersion of nano-features (Liu and Webster, 2007). The experimental evidence of enhanced cell response associated with incorporation of nano-sized features into scaffolds can be rationalized and explained by the fact that natural tissues and related extracellular matrices have native hierarchical nanostructures (Liu and Webster, 2007). The ability of these nano-scale surface features to promote desired cell responses, independent of the biomaterial choice, certainly justifies incorporation of these features into ligament scaffold designs to further enhance the systemic healing response.

12.4.5   Composites for engineered ligament scaffolds

Both natural polymers and degradable synthetic polymers are often combined as composites to fabricate scaffolds that are able to enhance the benefits of each individual component. In one study, a PLGA-collagen composite was synthesized into rolled microsponges to combine the mechanical strength of PLGA with the cellular binding affinity of collagen (Chen *et al.*, 2004). The results from this study revealed a large population of ACL cells, newly deposited collagen, and scaffold integration after 8 weeks of subcutaneous implantation into the dorsum of a mouse. In addition to microsponges, nanocomposites have been synthesized into well-ordered three-dimensional porous structures for tissue engineering (Liu and Webster, 2011). The results showed improved cell infiltration into these 3D porous structures with nano-sized surface features. Composite nano-fibers have also been synthesized to induce porous structures for tissue engineering scaffolds. For example, deposition of composite nano-fiber-based scaffold composed of electrospun PLGA (65:35) nano-fibers (300–900 nm diameter) into a PLGA (10:90) microfiber mesh (25 µm diameter) resulted in a scaffold with 2–50 µm pores. This composite scaffold improved mechanical strength and degradation resistance of the micro-fibers and increased the surface area for cell attachment. Results for this *in vitro* study showed a high level of matrix production relative to other scaffold 3D structures (Sahoo *et al.*, 2006). Although current research trends have focused on exploring synthetic polymers for composite scaffold fabrication, considerations have also been given to natural biopolymer composites. For example, a silk-collagen (type-I) composite scaffold showed rapid tissue ingrowth and favorable genetic expression results (Chen *et al.*, 2008). The extensive availability of degradable polymers, synthesis methods, fabrication methods for porous 3D scaffolds, and surface feature geometry allows for a wide range of combinations of these elements to develop biocompatible and bioactive scaffolds for ligament regeneration.

Due to the ease of fabrication, carbon nano-fibers and nanotubes have also been combined with polyhydroxyesters to produce composites with enhanced mechanical properties (Chrissafis *et al.*, 2010; Mackle *et al.*, 2011) and improved degradation rates (Chrissafis *et al.*, 2010). *In vitro* cell response studies to aligned carbon nanotube composites showed fibroblast morphology resembling that of native tissues (Yuen *et al.*, 2008). Although not much work has been done specifically for ACL regeneration, adequate biocompatibility has been shown *in vitro* with periodontal ligament cells (Mei *et al.*, 2007) and hMSCs (Mackle *et al.*, 2011) seeded on carbon nano-composites. These results indicated that carbon nanocomposites should be further researched for ligament scaffold construction.

12.5   Methods for enhancing engineered scaffolds for ligament regeneration

In addition to the material selection, an array of enhancement methods that promote ligament regeneration has been developed and should be considered in the design of ligament scaffolds. Growth factors, platelet rich plasma (PRP), and induced mechanical stimuli have all been shown to induce positive results for ligament regeneration and will be discussed in the following sections.

12.5.1   Growth factors for engineered ligament scaffolds

Five growth factors have shown promise for ligament regeneration due to their notable upregulation during natural ligament healing (Molloy *et al.*, 2003): insulin-like growth factor-I (IGF-I), transforming growth factor-β (TGF-β), vascular endothelial growth factor (VEGF), platelet-derived growth factor (PDGF), and basic fibroblastic growth factor (bFGF) (Kuo *et al.*, 2010). Studies focusing on the effects of growth factors suggest that growth factor receptors might be a bottleneck in the healing response (Vavken *et al.*, 2011b). Additional growth factor studies suggest that the lack of a functional clot in the intra-articular ligament is associated with the decrease in concentration of growth factors PDGF, TGF-β, and bFGF that results from the diverted translation of these growth factors into the synovial fluid (Murray *et al.*, 2007b). Together, these results favor the incorporation of growth factors and receptor enhancers in scaffold synthesis in order to improve the ligament healing response.

Both bFGF (Murakami *et al.*, 1999) and VEGF (Everts *et al.*, 2006) have been of interest for tissue engineering applications due to their key role in the natural response to wound healing (e.g., influence on revascularization) and to their association in strengthening of healing ligaments (e.g., enhanced cell proliferative response) (Vavken *et al.*, 2011b). Studies have shown that VEGF receptors 1 and 2 are expressed by ligament fibroblasts at high levels during the inflammatory response phase. This increased VEGF receptor response has been further correlated with a higher degree of tissue repair (Vavken *et al.*, 2011b). In terms of functional results obtained from VEGF infused scaffolds, studies have shown increased anterior–posterior laxity and reduced linear stiffness when VEGF is included in the scaffold (Yoshikawa *et al.*, 2006).

12.5.2   Platelet rich plasma (PRP)

It has been hypothesized that healing in the ACL may be limited by the environment in which the cells reside rather than an inherent inability for ACL regeneration (Petrigliano *et al.*, 2006). Due to the limited intrinsic healing properties of intra-articular ligaments and the lack of formation of a functional clot (refer to Fig. 12.6), growth factors secreted during the inflammatory phase of the healing process are not utilized effectively for ligament healing. Studies have shown that timing of clot formation is critical during the healing process (Marx, 2001; Robert, 2004) mainly because platelets secrete 95% of their pre-synthesized growth factors between 10 and 60 min following activation (Illingworth *et al.*, 2010). In a key study, Fallouh *et al.* (2010) were able to synthesize a collagen-PRP scaffold to function as an exogenous clot to torn ACL remnants. In this study platelets were activated with type-I collagen in order to form the fibrin clot and secrete growth factors while simultaneously preventing plasmin degradation (Murray *et al.*, 2007b). Results showed an improved response in production of ECM and increased collagen proliferation (Fallouh *et al.*, 2010) during the remodeling phase, leading to restored ligament strength and functionality (Murray *et al.*, 2007a).

The key benefits of PRP resulted from the high concentration of platelets available for natural fibrin-platelet clot formation (phase 1 of ligament healing), which in turn led to collagen production and subsequent healing phases. Platelets play a key role in this process because they secrete a number of biologically active molecules and growth factors, such as PDGF, TGF-β, and VEGF, all of which facilitate blood coagulation. Normal human physiological platelet levels are approximately 3×10^5 platelets/μL-plasma while 'rich' levels are approximately 1×10^6 platelets/μL-plasma (Foster *et al.*, 2009). Naturally, with an increased concentration of platelets, there is an expected increase in concentration of the fundamental protein growth factors (Vavken *et al.*, 2011b). Some studies showed as much as thirteen-fold increase in growth factors in the PRP condition (Fallouh *et al.*, 2010). It is important to mention that although there is an increase in platelet concentration, the ratios of the growth factors induced by PRP are consistent with normal biological ratios (Foster *et al.*, 2009).

Results from animal studies using a PRP-collagen scaffold on torn ACL remnants showed that concentration levels of fibronectin, fibrinogen, pro-collagen, and a number of growth factors were similar to those observed in extra-articular ligament wounds at 3 and 6 weeks post-injury (Murray *et al.*, 2007b). Other *in vitro* studies on PRP reported increased expression of type-III collagen, which has been identified as playing an important role in the early stages of ligament healing (Fallouh *et al.*, 2010). It has also been suggested that a collagen-PRP hydrogel is capable of stimulating fibroblast ingrowth and endothelial cell invasion, both of which are important functions of the fibrin clot (Murray *et al.*, 2007b). PRP has also been used to induce tissue regeneration in various clinical applications such as wound healing, bone regeneration, maxillofacial surgery, cardiovascular surgery, and orthopedic surgery, among others (Fallouh *et al.*, 2010). It is important to keep in mind that although increased erythrocyte proliferation is desired for successful development of a clot, studies have found that high erythrocyte concentrations can inhibit fibroblast formation (Harrison *et al.*, 2011), leading to an induced constraint on the healing cascades. Therefore, in order to obtain an improved scaffold design, erythrocyte concentration response must be kept in mind and held under a determined threshold in order to avoid the possibility of canceling the positive effects induced by PRP.

12.5.3 Effects of induced *in vitro* mechanical strain on scaffold functionality

In recent years, the use of *in vitro* dynamic experiments has become widely accepted in the tissue engineering community (Garvin *et al.*, 2003). External mechanical forces can positively affect cell proliferation, density, and differentiation within engineered scaffolds (Toyoda *et al.*, 1998). In an effort to replicate native knee kinematics, studies focused on the response of seeded ligament scaffolds to applied external tensile and torsional cyclical strains to determine collagen production and messenger ribonucleic acid (mRNA) expression levels (Altman *et al.*, 2002b). In a similar study, cellular alignment in a fibroblast-seeded PLGA scaffold was shown to increase after application of cyclic strain over a period of 2 weeks (Kyaw *et al.*, 2005). A comparative study between induced mechanical strains and immobilized injured ligaments resulted in a greater percentage of organized collagen fibrils and increased mechanical properties for the mechanically strained samples, suggesting the importance of induced strains for proper ligament healing (Woo *et al.*, 1987).

Further fibroblast alignment on scaffolds can be achieved by utilizing patterned scaffold structures, a concept termed 'contact guidance'. Cellular alignment and proliferation depends on the size and type of the channeling structure/pattern (Kuo *et al.*, 2010). In a study investigating the combined effect of cyclical strain with contact guidance on rat MCL fibroblasts, the results showed increased alignment in response to 3.5% cyclic strain at 1 Hz for 2 h when compared with cultures on smooth and static polydimethylsiloxane (PDMS) (Jones *et al.*, 2005). Figure 12.10 shows fluorescence images from this study where cells cultured on smooth PDMS (Fig. 12.10a and 12.10c) showed randomly aligned cells, whereas the cultures on the patterned substrates induced with cyclical strain (Fig. 12.10b and 12.10d) showed cells which were highly aligned along the direction of the patterns. These results show the potential benefits of utilizing bioreactors where external mechanical forces can be induced on the seeded scaffolds in order to obtain enhanced cellular alignment and potential to mimic the cellular alignment of native tissue.

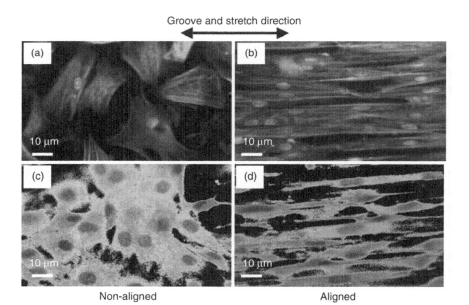

12.10 Fluorescence images from rat MCL fibroblasts cultured on patterned substrates (contact guidance) and induced with 3.5% cyclic strain at 1 Hz for 2 h. (a) and (c): Cells cultured on smooth PDMS substrates showed randomly aligned cytoskeletal and nuclear morphology; (b) and (d): cells induced with contact guidance and cyclic strain were substantially more aligned, based on cytoskeletal and nuclear morphology. Scale bars = 10 μm. (*Source*: Reproduced with permission from the *Journal of Biomechanics*.)

12.6   Conclusion and future trends

Despite the long history of ACL healing via graft reconstruction, research in nano-scaled biomaterials and understanding of the healing mechanism has led to a shift in focus towards ligament regeneration. The main advantage of primary regeneration over current treatment methods resides in an enhanced biomechanical response due to the preservation of native anatomical features and structures. Although the exact reasons pertaining to the limited healing mechanism of the ACL remains elusive, results such as those obtained by Murray *et al.* (2006, 2007, 2010) have revealed paths that enhance the healing capabilities of intra-articular ligaments such as the ACL. Furthermore, the advancement of synthesis and characterization of biopolymers at the nano-scale (nano-fibers, nano-pores, etc.) has opened new avenues through which complete and timely ACL regeneration may be achieved without long-term negative effects. Although most of the topics covered in this chapter introduce individual solutions to ligament regeneration, it is likely that the ultimate solution leading to total regeneration will be one which fully utilizes the benefits of composites, nano-features, and growth factors.

12.7   References

Adachi, N., Ochi, M., Uchio, Y., Iwasa, J., Ryoke, K. and Kuriwaka, M. (2002). Mechanoreceptors in the anterior cruciate ligament contribute to the joint position sense. *Acta Orthop Scand*, **73**, 330–4.

Altman, G. H., Horan, R. L., Lu, H. H., Moreau, J., Martin, I., Richmond, J. C. and Kaplan, D. L. (2002a). Silk matrix for tissue engineered anterior cruciate ligaments. *Biomaterials*, **23**, 4131–41.

Altman, G. H., Horan, R. L., Martin, I., Farhadi, J., Stark, P. R., Volloch, V., Richmond, J. C., Vunjak-Novakovic, G. and Kaplan, D. L. (2002b). Cell differentiation by mechanical stress. *FASEB J*, **16**, 270–2.

Andersen, R. B. and Gormsen, J. (1970). Fibrin dissolution in synovial fluid. *Acta Rheumatol Scand*, **16**, 319–33.

Aragona, J., Parsons, J. R., Alexander, H. and Weiss, A. B. (1983). Medial collateral ligament replacement with a partially absorbable tissue scaffold. *Am J Sports Med*, **11**, 228–33.

Bellincampi, L. D., Closkey, R. F., Prasad, R., Zawadsky, J. P. and Dunn, M. G. (1998). Viability of fibroblast-seeded ligament analogs after autogenous implantation. *J Orthop Res*, **16**, 414–20.

Boden, B. P., Dean, G. S., Feagin, J. A., Jr. and Garrett, W. E., Jr. (2000). Mechanisms of anterior cruciate ligament injury. *Orthopedics*, **23**, 573–8.

Bradbury, N., Borton, D., Spoo, G. and Cross, M. J. (1998). Participation in sports after total knee replacement. *Am J Sports Med*, **26**, 530–5.

Butler, D. L., Guan, Y., Kay, M. D., Cummings, J. F., Feder, S. M. and Levy, M. S. (1992). Location-dependent variations in the material properties of the anterior cruciate ligament. *J Biomech*, **25**, 511–8.

Butler, D. L., Noyes, F. R. and Grood, E. S. (1980). Ligamentous restraints to anterior-posterior drawer in the human knee. A biomechanical study. *J Bone Joint Surg Am*, **62**, 259–70.

Chen, G., Sato, T., Sakane, M., Ohgushi, H., Ushida, T., Tanaka, J. and Tateishi, T. (2004). Application of PLGA-collagen hybrid mesh for three-dimensional culture of canine anterior cruciate ligament cells. *Mater Sci Eng C*, **24**, 861–6.

Chen, J., Altman, G. H., Karageorgiou, V., Horan, R., Collette, A., Volloch, V., Colabro, T. and Kaplan, D. L. (2003). Human bone marrow stromal cell and ligament fibroblast responses on RGD-modified silk fibers. *J Biomed Mater Res A*, **67**, 559–70.

Chen, X., Qi, Y. Y., Wang, L. L., Yin, Z., Yin, G. L., Zou, X. H. and Ouyang, H. W. (2008). Ligament regeneration using a knitted silk scaffold combined with collagen matrix. *Biomaterials*, **29**, 3683–92.

Chrissafis, K., Paraskevopoulos, K. M., Jannakoudakis, A., Beslikas, T. and Bikiaris, D. (2010). Oxidized multiwalled carbon nanotubes as effective reinforcement and thermal stability agents of poly(lactic acid) ligaments. *J Appl Polym Sci*, **118**, 2712–21.

Cristino, S., Grassi, F., Toneguzzi, S., Piacentini, A., Grigolo, B., Santi, S., Riccio, M., Tognana, E., Facchini, A. and Lisignoli, G. (2005). Analysis of mesenchymal stem cells grown on a three-dimensional HYAFF 11-based prototype ligament scaffold. *J Biomed Mater Res A*, **73**, 275–83.

Dunn, M. G., Liesch, J. B., Tiku, M. L. and Zawadsky, J. P. (1995). Development of fibroblast-seeded ligament analogs for ACL reconstruction. *J Biomed Mater Res*, **29**, 1363–71.

Everts, P. A., Knape, J. T., Weibrich, G., Schonberger, J. P., Hoffmann, J., Overdevest, E. P., Box, H. A. and van Zundert, A. (2006). Platelet-rich plasma and platelet gel: a review. *J Extra Corpor Technol*, **38**, 174–87.

Fallouh, L., Nakagawa, K., Sasho, T., Arai, M., Kitahara, S., Wada, Y., Moriya, H. and Takahashi, K. (2010). Effects of autologous platelet-rich plasma on cell viability and collagen synthesis in injured human anterior cruciate ligament. *J Bone Joint Surg Am*, **92**, 2909–16.

Foster, T. E., Puskas, B. L., Mandelbaum, B. R., Gerhardt, M. B. and Rodeo, S. A. (2009). Platelet-rich plasma: from basic science to clinical applications. *Am J Sports Med*, **37**, 2259–72.

Frank, C., Woo, S. L., Amiel, D., Harwood, F., Gomez, M. and Akeson, W. (1983). Medial collateral ligament healing. A multidisciplinary assessment in rabbits. *Am J Sports Med*, **11**, 379–89.

Frank, C. B. (2004). Ligament structure, physiology and function. *J Musculoskelet Neuronal Interact*, **4**, 199–201.

Fu, F. H., Bennett, C. H., Lattermann, C. and Ma, C. B. (1999). Current trends in anterior cruciate ligament reconstruction. Part 1: Biology and biomechanics of reconstruction. *Am J Sports Med*, **27**, 821–30.

Fu, F. H., Bennett, C. H., Ma, C. B., Menetrey, J. and Lattermann, C. (2000). Current trends in anterior cruciate ligament reconstruction. Part II. Operative procedures and clinical correlations. *Am J Sports Med*, **28**, 124–30.

Garvin, J., Qi, J., Maloney, M. and Banes, A. J. (2003). Novel system for engineering bioartificial tendons and application of mechanical load. *Tissue Eng*, **9**, 967–79.

Guidoin, M. F., Marois, Y., Bejui, J., Poddevin, N., King, M. W. and Guidoin, R. (2000). Analysis of retrieved polymer fiber based replacements for the ACL. *Biomaterials*, **21**, 2461–74.

Harrison, B. S. and Atala, A. (2007). Carbon nanotube applications for tissue engineering. *Biomaterials*, **28**, 344–53.

Harrison, S. L., Vavken, P. and Murray, M. M. (2011). Erythrocytes inhibit ligament fibroblast proliferation in a collagen scaffold. *J Orthop Res*, **29**, 1361–6.

Harrold, A. J. (1961). The defect of blood coagulation in joints. *J Clin Pathol*, **14**, 305–8.

Haus, B. M., Mastrangelo, A. N. and Murray, M. M. (2012). Effect of anterior cruciate healing on the uninjured ligament insertion site. *J Orthop Res*, **30**, 86–94.

Ho, W., Tawil, B., Dunn, J. C. and Wu, B. M. (2006). The behavior of human mesenchymal stem cells in 3D fibrin clots: dependence on fibrinogen concentration and clot structure. *Tissue Eng*, **12**, 1587–95.

Illingworth, K. D., Musahl, V., Lorenz, S. G. F. and Fu, F. H. (2010). Use of fibrin clot in the knee. *Oper Tech Orthop*, **20**, 90–7.

Iwahashi, T., Shino, K., Nakata, K., Otsubo, H., Suzuki, T., Amano, H. and Nakamura, N. (2010). Direct anterior cruciate ligament insertion to the femur assessed by histology and 3-dimensional volume-rendered computed tomography. *Arthroscopy*, **26**, S13–20.

Jones, B. F., Wall, M. E., Carroll, R. L., Washburn, S. and Banes, A. J. (2005). Ligament cells stretch-adapted on a microgrooved substrate increase intercellular communication in response to a mechanical stimulus. *J Biomech*, **38**, 1653–64.

Kanamori, A., Zeminski, J., Rudy, T. W., Li, G., Fu, F. H. and Woo, S. L. (2002). The effect of axial tibial torque on the function of the anterior cruciate ligament: a biomechanical study of a simulated pivot shift test. *Arthroscopy*, **18**, 394–8.

Kuo, C. K., Marturano, J. E. and Tuan, R. S. (2010). Novel strategies in tendon and ligament tissue engineering: advanced biomaterials and regeneration motifs. *Sports Med Arthrosc Rehabil Ther Technol*, **2**, 20.

Kyaw, M., Tay, T. E., Goh, J. H. C., Ouyang, H. W. and Toh, S. L. (2005). *Cyclic uniaxial strains on Fibroblasts-seeded PLGA scaffolds for tissue engineering of Ligaments*, Bellingham, WA, International Society of Photo-Optical Instrumentation Engineers.

Lee, C. H., Shin, H. J., Cho, I. H., Kang, Y. M., Kim, I. A., Park, K. D. and Shin, J. W. (2005). Nanofiber alignment and direction of mechanical strain affect the ECM production of human ACL fibroblast. *Biomaterials*, **26**, 1261–70.

Lin, V. S., Lee, M. C., O'Neal, S., McKean, J. and Sung, K. L. (1999). Ligament tissue engineering using synthetic biodegradable fiber scaffolds. *Tissue Eng*, **5**, 443–52.

Liu, H. and Webster, T. J. (2007). Nanomedicine for implants: a review of studies and necessary experimental tools. *Biomaterials*, **28**, 354–69.

Liu, H. and Webster, T. J. (2011). Enhanced biological and mechanical properties of well-dispersed nanophase ceramics in polymer composites: From 2D to 3D printed structures. *Mater Sci Eng C*, **31**, 77–89.

Lu, H. H., Cooper, J. A., Manuel, S., Freeman, J. W., Attawia, M. A., Ko, F. K. and Laurencin, C. T. (2005). Anterior cruciate ligament regeneration using braided biodegradable scaffolds: in vitro optimization studies. *Biomaterials*, **26**, 4805–16.

Mackle, J. N., Blond, D. J., Mooney, E., McDonnell, C., Blau, W. J., Shaw, G., Barry, F. P., Murphy, J. M. and Barron, V. (2011). In vitro characterization of an electroactive carbon-nanotube-based nanofiber scaffold for tissue engineering. *Macromol Biosci*, **11**, 1272–82.

Majima, T., Funakosi, T., Iwasaki, N., Yamane, S. T., Harada, K., Nonaka, S., Minami, A. and Nishimura, S. (2005). Alginate and chitosan polyion complex hybrid fibers for scaffolds in ligament and tendon tissue engineering. *J Orthop Sci*, **10**, 302–7.

Marx, R. E. (2001). Platelet-rich plasma (PRP): what is PRP and what is not PRP? *Implant Dent*, **10**, 225–8.

Marx, R. E. (2004). Platelet-rich plasma: evidence to support its use. *J Oral Maxil Surg*, **62**, 489–96.

McBride Jr, D. J., Trelstad, R. L. and Silver, F. H. (1988). Structural and mechanical assessment of developing chick tendon. *Int J Biol Macromol*, **10**, 194–200.

Mei, F., Zhong, J., Yang, X., Ouyang, X., Zhang, S., Hu, X., Ma, Q., Lu, J., Ryu, S. and Deng, X. (2007). Improved biological characteristics of poly(L-lactic acid) electrospun membrane by incorporation of multiwalled carbon nanotubes/hydroxyapatite nanoparticles. *Biomacromolecules*, **8**, 3729–35.

Molloy, T., Wang, Y. and Murrell, G. (2003). The roles of growth factors in tendon and ligament healing. *Sports Med*, **33**, 381–94.

Murakami, S., Takayama, S., Ikezawa, K., Shimabukuro, Y., Kitamura, M., Nozaki, T., Terashima, A., Asano, T. and Okada, H. (1999). Regeneration of periodontal tissues by basic fibroblast growth factor. *J Periodontal Res*, **34**, 425–30.

Murray, M. M., Magarian, E., Zurakowski, D. and Fleming, B. C. (2010). Bone-to-bone fixation enhances functional healing of the porcine anterior cruciate ligament using a collagen-platelet composite. *Arthroscopy*, **26**, S49–57.

Murray, M. M., Spindler, K. P., Abreu, E., Muller, J. A., Nedder, A., Kelly, M., Frino, J., Zurakowski, D., Valenza, M., Snyder, B. D. and Connolly, S. A. (2007a). Collagen-platelet rich plasma hydrogel enhances primary repair of the porcine anterior cruciate ligament. *J Orthop Res*, **25**, 81–91.

Murray, M. M., Spindler, K. P., Ballard, P., Welch, T. P., Zurakowski, D. and Nanney, L. B. (2007b). Enhanced histologic repair in a central wound in the anterior cruciate ligament with a collagen-platelet-rich plasma scaffold. *J Orthop Res*, **25**, 1007–17.

Murray, M. M., Spindler, K. P., Devin, C., Snyder, B. S., Muller, J., Takahashi, M., Ballard, P., Nanney, L. B. and Zurakowski, D. (2006). Use of a collagen-platelet rich plasma scaffold to stimulate healing of a central defect in the canine ACL. *J Orthop Res*, **24**, 820–30.

Myer, G. D., Ford, K. R. and Hewettt, T. E. (2004). Rationale and clinical techniques for anterior cruciate ligament injury prevention among female athletes. *J Athl Training*, **39**, 352–64.

Nakamae, A., Ochi, M., Deie, M., Adachi, N., Kanaya, A., Nishimori, M. and Nakasa, T. (2010). Biomechanical function of anterior cruciate ligament remnants: how long do they contribute to knee stability after injury in patients with complete tears? *Arthroscopy*, **26**, 1577–85.

Neugebauer, R. and Claes, L. (1983). The biological reaction of the tissues to carbon fibre ligament prosthesis in sheep-knees. *Aktuelle Probl Chir Orthop*, **26**, 96–100.

Nikolaou, P. K., Seaber, A. V., Glisson, R. R., Ribbeck, B. M. and Bassett, F. H., 3RD (1986). Anterior cruciate ligament allograft transplantation. Long-term function, histology, revascularization, and operative technique. *Am J Sports Med*, **14**, 348–60.

Perka, C., Schultz, O., Spitzer, R. S., Lindenhayn, K., Burmester, G. R. and Sittinger, M. (2000). Segmental bone repair by tissue-engineered periosteal cell transplants with bioresorbable fleece and fibrin scaffolds in rabbits. *Biomaterials*, **21**, 1145–53.

Petrigliano, F. A., McAllister, D. R. and Wu, B. M. (2006). Tissue engineering for anterior cruciate ligament reconstruction: a review of current strategies. *Arthroscopy*, **22**, 441–51.

Provenzano, P. P., Heisey, D., Hayashi, K., Lakes, R. and Vanderby, R., Jr. (2002). Subfailure damage in ligament: a structural and cellular evaluation. *J Appl Physiol*, **92**, 362–71.

Reed, C. R., Han, L., Andrady, A., Caballero, M., Jack, M. C., Collins, J. B., Saba, S. C., Loboa, E. G., Cairns, B. A. and van Aalst, J. A. (2009). Composite tissue engineering on polycaprolactone nanofiber scaffolds. *Ann Plast Surg*, **62**, 505–12.

Rosc, D., Powierza, W., Zastawna, E., Drewniak, W., Michalski, A. and Kotschy, M. (2002). Post-traumatic plasminogenesis in intraarticular exudate in the knee joint. *Med Sci Monit*, **8**, CR371–8.

Sahoo, S., Ouyang, H., Goh, J. C., Tay, T. E. and Toh, S. L. (2006). Characterization of a novel polymeric scaffold for potential application in tendon/ligament tissue engineering. *Tissue Eng*, **12**, 91–9.

Saito, N., Aoki, K., Usui, Y., Shimizu, M., Hara, K., Narita, N., Ogihara, N., Nakamura, K., Ishigaki, N., Kato, H., Haniu, H., Taruta, S., Kim, Y. A. and Endo, M. (2011). Application of carbon fibers to biomaterials: a new era of nano-level control of carbon fibers after 30-years of development. *Chem Soc Rev*, **40**, 3824–34.

Sakane, M., Fox, R. J., Woo, S. L., Livesay, G. A., Li, G. and Fu, F. H. (1997). In situ forces in the anterior cruciate ligament and its bundles in response to anterior tibial loads. *J Orthop Res*, **15**, 285–93.

Shelbourne, K. D. and Nitz, P. (1990). Accelerated rehabilitation after anterior cruciate ligament reconstruction. *Am J Sports Med*, **18**, 292–9.

Silver, F. H., Freeman, J. W. and Seehra, G. P. (2003). Collagen self-assembly and the development of tendon mechanical properties. *J Biomech*, **36**, 1529–53.

Toyoda, T., Matsumoto, H., Fujikawa, K., Saito, S. and Inoue, K. (1998). Tensile load and the metabolism of anterior cruciate ligament cells. *Clin Orthop Relat Res*, 247–55.

van Eck, C. F., Schreiber, V. M., Mejia, H. A., Samuelsson, K., van Dijk, C. N., Karlsson, J. and Fu, F. H. (2010). 'Anatomic' anterior cruciate ligament reconstruction: A systematic review of surgical techniques and reporting of surgical data. *Arthroscopy*, **26**, S2–12.

Vasita, R. and Katti, D. S. (2006). Nanofibers and their applications in tissue engineering. *Int J Nanomedicine*, **1**, 15–30.

Vavken, P., Joshi, S. M. and Murray, M. M. (2011a). Fibrin concentration affects ACL fibroblast proliferation and collagen synthesis. *Knee*, **18**, 42–6.

Vavken, P., Saad, F., Fleming, B. and Murray, M. (2011b). VEGF receptor mRNA expression by ACL fibroblasts is associated with functional healing of the ACL. *Knee Surg Sport Traumat Arthrosc*, **19**, 1675–82.

Vieira, A. C., Vieira, J. C., Ferra, J. M., Magalhaes, F. D., Guedes, R. M. and Marques, A. T. (2011). Mechanical study of PLA-PCL fibers during in vitro degradation. *J Mech Behav Biomed Mater*, **4**, 451–60.

von Porat, A., Roos, E. M. and Roos, H. (2004). High prevalence of osteoarthritis 14 years after an anterior cruciate ligament tear in male soccer players: a study of radiographic and patient relevant outcomes. *Ann Rheum Dis*, **63**, 269–73.

Weiss, J. A., Woo, S. L., Ohland, K. J., Horibe, S. and Newton, P. O. (1991). Evaluation of a new injury model to study medial collateral ligament healing: primary repair versus nonoperative treatment. *J Orthop Res*, **9**, 516–28.

West, R. V. and Harner, C. D. (2005). Graft selection in anterior cruciate ligament reconstruction. *J Am Acad Orthop Surg*, **13**, 197–207.

Wong, C., Inman, E., Spaethe, R. and Helgerson, S. (2003). Fibrin-based biomaterials to deliver human growth factors. *Thromb Haemost*, **89**, 573–82.

Woo, S. L.-Y., Hollis, J. M., Adams, D. J., Lyon, R. M. and Takai, S. (1991). Tensile properties of the human femur-anterior cruciate ligament-tibia complex. *Am J Sport Med*, **19**, 217–25.

Woo, S. L., Gomez, M. A., Sites, T. J., Newton, P. O., Orlando, C. A. and Akeson, W. H. (1987). The biomechanical and morphological changes in the medial collateral ligament of the rabbit after immobilization and remobilization. *J Bone Joint Surg Am*, **69**, 1200–11.

Woo, S. L. Y., Abramowitch, S. D., Kilger, R. and Liang, R. (2006). Biomechanics of knee ligaments: injury, healing, and repair. *J Biomech*, **39**, 1–20.

Yagi, M., Wong, E. K., Kanamori, A., Debski, R. E., Fu, F. H. and Woo, S. L. (2002). Biomechanical analysis of an anatomic anterior cruciate ligament reconstruction. *Am J Sports Med*, **30**, 660–6.

Yoshikawa, T., Tohyama, H., Katsura, T., Kondo, E., Kotani, Y., Matsumoto, H., Toyama, Y. and Yasuda, K. (2006). Effects of local administration of vascular endothelial growth factor on mechanical characteristics of the semitendinosus tendon graft after anterior cruciate ligament reconstruction in sheep. *Am J Sports Med*, **34**, 1918–25.

Yuen, F. L., Zak, G., Waldman, S. D. and Docoslis, A. (2008). Morphology of fibroblasts grown on substrates formed by dielectrophoretically aligned carbon nanotubes. *Cytotechnology*, **56**, 9–17.

Zanello, L. P., Zhao, B., Hu, H. and Haddon, R. C. (2006). Bone cell proliferation on carbon nanotubes. *Nano Lett*, **6**, 562–7.

Part III
Application of nanomaterials in hard tissue engineering

13

Nanomaterials for hard–soft tissue interfaces

E. C. BECK and M. S. DETAMORE,
University of Kansas, USA

DOI: 10.1533/9780857097231.3.363

Abstract: The field of interfacial tissue engineering is striving to restore the structural and functional characteristics of hard–soft tissue interfaces, which include osteochondral and bone–tendon/ligament interfaces. This chapter first discusses the structural and functional characteristics of these interfaces and then describes the two main types of nanomaterials emerging in interfacial tissue engineering strategies: nanoparticles and nanofibers. Additionally, the chapter discusses approaches used to employ these nanomaterials in interfacial constructs.

Key words: tissue engineering, interface, gradient, nanoparticle, nanofiber.

13.1 Introduction

Engineering of interfacial tissues is emerging as one of the grandest challenges of tissue engineering. This emergence is in part because of the complexity of the interface itself, but also because interfacial tissues are in high clinical demand. Hard–soft tissue interfaces (e.g., bone–cartilage, bone–tendon, bone–ligament), which are the focus of this chapter, are especially prone to injury. Considering the bone–tendon/ligament interface alone, repair methods using grafts are not necessarily focused on regenerating the tissue interface, which can lead to compromised graft function and poor long-term clinical outcome (Lu and Jiang, 2006; Mikos *et al.*, 2006). Therefore, the field of interfacial tissue engineering is striving to restore the structural and functional characteristics of the interface itself to form a seamless transition from hard to soft tissues.

Bone, cartilage, and ligament tissues are discussed in other chapters of this book. Therefore, for further information on these individual tissues, the reader is directed to chapters 11, 12, and 14. However, the structural and functional characteristics of the interfaces of these tissues will be described here. These characteristics are highly complex because the interface is responsible for the transfer of mechanical loads between two dissimilar

363

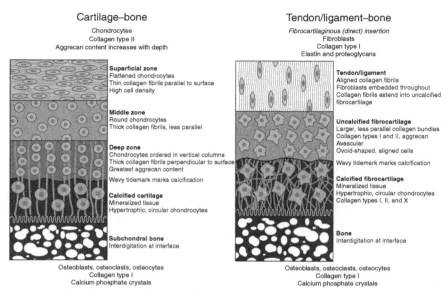

13.1 Depiction of the cartilage–bone and tendon/ligament–bone interfaces and their compositions. (*Source*: Used with permission from Yang and Temenoff (2009).)

and heterogeneous tissues (Woo *et al.*, 1983; Mikos *et al.*, 2006). Therefore, the physiology and function of each interface must be understood prior to designing interfacial constructs. The bone–cartilage interface, also known as the osteochondral interface, unites hyaline cartilage with subchondral bone and is the most extensively studied interface in tissue engineering. Cartilage can be further divided into three zones (Fig. 13.1). The superficial zone at the articular surface is characterized by flattened chondrocytes and thin collagen II fibrils that run parallel to the joint surface (Ulrich-Vinther *et al.*, 2003). The middle zone contains more rounded chondrocytes and slightly larger and less parallel collagen II fibers (Poole *et al.*, 2001). The deep zone contains chondrocytes and collagen II fibrils that run perpendicular to the articular surface (Ulrich-Vinther *et al.*, 2003). Underlying the three cartilage zones is a wavy tidemark that marks the beginning of the calcified cartilage zone, which extends the collagen fibrils from the deep zone to the subchondral bone. Finally, the subchondral bone is primarily composed of collagen I and hydroxyapatite (HAp) crystals (Yang and Temenoff, 2009). The subchondral bone is secured to the calcified cartilage by interdigitation at the interface of these two zones. In addition to providing a secure attachment, this interdigitation reduces stress concentrations at the bone–cartilage interface to allow for efficient transfer of mechanical loading (Lu and Jiang, 2006).

Although osteochondral constructs are more widely explored in interfacial tissue engineering, the bone–tendon/ligament interfaces are gaining attention as well. Tendons transfer loads from muscle to bone, while ligaments link bone to bone and help ensure joint stability (Martin *et al.*, 1998). The insertions of tendons and ligaments to bones can vary drastically, but they can generally be classified as direct or indirect (Martin *et al.*, 1998). Direct insertions are characterized by four transition zones (Fig. 13.1). The first zone is fibrous connective tissue that contains fibroblasts with aligned collagen fibrils. These collagen fibrils extend to the next zone, the uncalcified fibrocartilage zone, which contains larger collagen fibrils that are less parallel than the previous region. A wavy tidemark separates the uncalcified fibrocartilage zone from the calcified cartilage zone. Finally, interdigitation secures the calcified cartilage zone to the fourth zone, which is the underlying bone (Benjamin and Ralphs, 1998). Indirect insertions do not contain the fibrocartilage zones, but instead the tendon is directly secured to the bone through Sharpey's fibers, which are collagen fibers that extend directly into the underlying bone (Martin *et al.*, 1998).

Primarily, tissue engineering strategies to regenerate osteochondral and bone–tendon/ligament interfacial tissues are stratified in nature. These stratified designs are considered 'graded' designs, as they combine two or more distinct layers that aim to hone in on the characteristics of the various interfacial zones. More recently, continuously graded approaches have been considered, where instead of containing discrete layers, the constructs provide a more gradual tissue interface transition, similar to the native interface itself (Dormer *et al.*, 2010a). These interfacial tissue engineering approaches and strategies have been thoroughly reviewed in the literature (Lu and Jiang, 2006; Mikos *et al.*, 2006; Grayson *et al.*, 2008; Lim and Temenoff, 2009; Lu and Spalazzi, 2009; Moffat *et al.*, 2009; Yang and Temenoff, 2009; Dormer *et al.*, 2010a; Lu *et al.*, 2010; Deng *et al.*, 2011; Panseri *et al.*, 2012; Sahoo *et al.*, 2011; Seidi *et al.*, 2011; Smith *et al.*, 2012; Seidi and Ramalingam, 2012; Zhang *et al.*, 2012b), however the quest for the best approach continues.

An emerging trend in interfacial tissue engineering is the incorporation of nanomaterials in the interfacial design. Prior to 2009, few publications existed that incorporated nanomaterials for interfacial constructs. However, the number of publications in this area continues to increase steadily each year, with just over five publications in 2009 alone, doubling to more than ten publications in 2011 alone. Additionally, the number of publications in this area so far for 2012 is on a similar pace to that of 2011, emphasizing that nanomaterials are rapidly gaining attention and generating interest in the field of interfacial tissue engineering. Nanomaterials are typically classified as materials that are 1–100 nm in size, although many groups consider materials smaller than 1 μm to be nanoscale. Nanomaterials can take many forms, including nanoparticles, nanofibers, nanocrystals, nanorods, nanotubes, etc. In a relatively short period, nanomaterials have exhibited extraordinary potential to advance medical therapies. In the field of tissue engineering, nanomaterials have become of growing interest recently because the extracellular matrices of tissues are composed of hierarchically nanostructured materials that can regulate cellular functions such as differentiation, morphogenesis, adhesion, proliferation, and migration (Tsang *et al.*, 2010). Thus, nanomaterials have the potential to advance current strategies for interfacial tissue engineering.

The following sections review the two nanomaterial components currently being used in hard–soft tissue engineering: nanoparticles and nanofibers. In addition, a section is included that discusses strategies incorporating nanomaterials in hard–soft tissue interfaces. Although this text is organized by nanomaterial component, because of overlap in nanomaterial use in some strategies, the tables are arranged by interface type. Table 13.1 provides the nanomaterials employed in osteochondral strategies and Table 13.2 provides the nanomaterials used in bone–tendon/ligament strategies.

13.2   Nanoparticles

Recent interfacial tissue engineering constructs have incorporated nanoparticles of various types of materials. Because all hard–soft tissues include bone, it is not surprising that the majority of nanoparticles used in interfacial tissue engineering constructs are ceramic. Not only are ceramics part of the native extracellular matrix (ECM) of bone, they can also increase the mechanical strength, biocompatibility, and osteoconductivity of tissue engineered scaffolds (Cool *et al.*, 2007; Dorozhkin, 2010). In the following sections, we will discuss interfacial constructs that incorporate ceramic nanoparticles of hydroxyapatite (HAp), the most commonly incorporated ceramic nanoparticle, beta-tricalcium phosphate (β-TCP), and other mineral-based materials. Although most of the nanoparticle materials in the following sections are mineral-based, newly emerging nanoparticle materials, including magnetic and fluorescent nanoparticles, will also be discussed in the following sections.

13.2.1   Hydroxyapatite (HAp) nanoparticles

It is well known that the nanostructure of bone, including HAp, plays an important role in the overall mechanical properties of bone (Rho *et al.*, 1998). HAp nanocrystals constitute approximately 70% of native bone matrix (Kaplan *et al.*, 1994). These HAp nanocrystals are roughly 50 nm long, 25 nm wide, and 2–5 nm thick (Robinson, 1952; Johansen and Parks,

Table 13.1 Applications of nanomaterials in osteochondral interfaces

Reference(s)	Nanomaterial	Additional material(s)	Biological model	Scaffold formulation	Highlighted finding
Bian et al. (2012)	β-TCP nanoparticle	Collagen I acrylamide	In vitro: scaffold cultured with rabbit BMSCs for 7 days	Triphasic	Gel casting avoided delamination between phases
Cheng et al. (2011)	Collagen I nanofibers	NA	In vitro: scaffold cultured with rabbit BMSCs for 5 weeks	Triphasic	Calcified cartilage layer formed in triphasic constructs as compared to biphasic controls
Chung et al. (2011a, 2011b)	HAp nanoparticles	POC	In vivo: New Zealand White rabbits osteochondral defect for 26 weeks	Nanocomposite	Nanocomposites induced more trabecular bone formation and had superior mechanical properties compared to microcomposites
Erisken et al. (2008, 2010)	PCL nanofibers, HAp nanoparticles	NA	In vitro: cultured with MC3T3-E1 cells for 4 weeks	Continuous gradient	Tissue constructs formed a gradient in ECM deposition and construct mechanical properties approached native tissue as culture period increased
Erisken et al. (2011)	PCL nanofibers, β-GP nanoparticles	Insulin	In vitro: cultured with human adipose-derived stromal cells for 8 weeks	Opposing continuous gradients	Differentiation of seeded cells towards chondrogenic or osteogenic lineage depended upon position in the scaffold
Kon et al. (2009, 2010b, 2011); Filardo et al. (2010)	Collagen I nanofibers, HAp nanoparticles	NA	In vivo: Human osteochondral defect	Triphasic	The triphasic constructs demonstrate safety and potential in human clinical trials

(Continued)

Table 13.1 Continued

Reference(s)	Nanomaterial	Additional material(s)	Biological model	Scaffold formulation	Highlighted finding
Kon et al. (2010a, 2010c)	Collagen I nanofibers, HAp nanoparticles	NA	In vivo: Sheep and horse osteochondral defect for 6 months	Triphasic	Comparable extent of regeneration observed with or without autologous chondrocytes suggesting the constructs recruit native cells
Grogan et al. (2012)	Iron oxide nanoparticles, maghemite nanoparticles, magnetite nanoparticles	Magnetite, alginate, agarose	In vivo: New Zealand White rabbits osteochondral defect for 4 weeks	Biphasic	Manipulation of the magnetic field in vitro and in situ results in multi-directional cellular arrangements
Harley et al. (2010a)	CaP nanoparticles	Collagen I, collagen II, CS	Not directly evaluated	Biphasic with continuous interface	Liquid phase co-synthesis successfully resulted in a continuous interface between phases
Hu et al. (2009)	PLLA nanofibers	NA	In vitro: cultured with human BMSCs for 6 weeks	Nanocomposite	PLLA scaffold supports both chondrogenic and osteogenic differentiation when exposed to the respective growth signals
Khanarian et al. (2012)	HAp nanoparticles	Agarose, micron-sized HAp	In vitro: cultured with bovine chondrocytes for 2 weeks	Nanocomposite	Superior ECM deposition with gels composed of micron-sized HAp in comparison to nanoHAp
Lee et al. (2012)	Fluorescent silica nanoparticles	Heparin-conjugated fibrin	In vivo: male nude rat osteochondral defect for 21 days	NA	IOMTS successfully monitored MSC migration and can be used the effect of growth factors on cartilage repair

Reference	Material	Matrix	In vivo evaluation	Type	Results
Liu et al. (2009)	HAp nanoparticles	Collagen I	Not directly evaluated	Continuous gradient	Opposing gradients of Ca and P successfully formed with HAp nucleation in interior of scaffold
Liu et al. (2011)	Star-shaped PLLA nanoparticles	NA	In vivo: New Zealand White rabbit osteochondral defect for 8 weeks	Nanocomposite	Microspheres assembled from PLLA nanoparticles resulted in enhanced cartilage regeneration compared to control microspheres
Matsusaki et al. (2008)	HAp nanoparticles	Rabbit synovial MSCs	In vivo: Japanese White rabbit osteochondral defect for 6 weeks	Nanocomposite	Scaffolds containing HAp exhibited accelerated osteoinduction as compared to scaffolds without HAp
Mohan et al. (2011)	HAp nanoparticles	PLGA microspheres containing chondrogenic or osteogenic factors	In vivo: New Zealand White rabbit osteochondral defect for 12 weeks	Continuous gradient	Greatest extent of regeneration achieved when both material and signal gradients were present
Nirmala et al. (2010)	PCL nanofibers, HAp and Ag nanoparticles	NA	Not directly evaluated	Nanocomposite	Incorporating Ag resulted in superior mechanical properties and mineral deposition
Qu et al. (2011)	HAp nanoparticles	PVA, PA6	In vivo: Implanted intramuscularly in rabbit for 12 weeks	Biphasic	Osteogenesis and chondrogenesis observed in respective layers after 12 week implantation
Rodrigues et al. (2012)	Carbon nanotubes or nanofibers	PVA	In vivo: Wistar rat osteochondral defect for 12 weeks	Nanocomposite	Significantly more calcium and phosphorus deposition occurred in constructs with carbon nanofibers compared to constructs with carbon nanotubes and gels of PVA only

(Continued)

Table 13.1 Continued

Reference(s)	Nanomaterial	Additional material(s)	Biological model	Scaffold formulation	Highlighted finding
Tampieri et al. (2008)	Collagen I nanofibers, HAp nanoparticles	HA	In vivo: ID mice for 8 weeks subcutaneously	Triphasic	Scaffold layers selectively support cell differentiation towards osteogenic or chondrogenic lineages
Tampieri et al. (2011)	Collagen I nanofibers, HAp nanoparticles, magnetite nanoparticles	NA	Not directly evaluated	Triphasic	Scaffolds containing magnetite nanoparticles had superior mechanical properties compared to scaffolds without magnetite
Xue et al. (2010)	HAp nanoparticles	PLGA	In vivo: SD rats osteochondral defect for 12 weeks	Nanocomposite	Greatest extent of regeneration observed in PLGA scaffolds containing HAp compared to PLGA control
Yunos et al. (2011)	PDLLA nanofibers, HAp nanoparticles	Bioglass	In vitro: cultured with ATDC5 chondrocyte cells line for 10 days	Biphasic	Chondrocyte cells were able to attach, proliferate, and migrate into the scaffolds

β-GP: β-glycerophosphate; β-TCP: β-tricalcium phosphate; Ag: silver; BMSCs: bone marrow stem cells; Ca: calcium; CaP: calcium phosphate; CS: chondroitin sulfate; ECM: extracellular matrix; HA: hyaluronic acid; HAp: hydroxyapatite; ID: immunodeficient; IOMTS: in vivo osteochondral mesenchymal stem cells tracking system; MSCs: mesenchymal stem cells; NA: not applicable; P: phosphorous; PA6: polyamide-6; PCL: polycaprolactone; PDLLA: poly(D,L-lactide); PLGA: poly(lactic-co-glycolic acid); PLLA: poly(L-lactic acid); POC: poly(1,8-octanediol-co-citrate); PVA: polyvinyl alcohol; SD: Sprague-Dawley.

Table 13.2 Applications of nanomaterials in bone–tendon and bone–ligament interfaces

Reference(s)	Nanomaterial	Additional material(s)	Biological model	Scaffold formulation	Highlighted finding
Li *et al.* (2009)	PLGA nanofibers, CaP nanoparticles	NA	*In vitro*: cultured with MC3T3 cells for 3 days	Continuous gradient	Mineral gradient resulted in a gradient of stiffness and cell density
Ramalingam *et al.* (2012)	PCL nanofibers, amorphous CaP nanoparticles	NA	*In vitro*: cultured with MC3T3-E1 cells for 7 days	Continuous gradient	Adhesion and proliferation of seeded cells were enhanced as CaP concentration increased along length of scaffold
Samavedi *et al.* (2011)	PCL and PEU nanofibers, HAp nanoparticles	NA	*In vitro*: cultured with MC3T3-E1 cells for 7 days	Continuous gradient	Mechanical gradient with differences in tensile moduli observed along length of scaffold
Spalazzi *et al.* (2008)	PLGA nanofibers	PLGA/bioactive glass microspheres	*In vitro*: cultured with bovine tendons	Biphasic	Scaffold induced compression of tendons led to increased ECM deposition compared to tendon control
Xie *et al.* (2010)	PLGA nanofibers	NA	*In vitro*: cultured with rat tendon fibroblasts for 7 days	Biphasic	Tendon fibroblasts and deposited collagen I were aligned in aligned portion of scaffold and were randomly distributed in randomly oriented part of scaffold

CaP: calcium phosphate; HAp: hydroxyapatite; NA: not applicable; PCL: polycaprolactone; PEU: poly(ester urethane); PLGA: poly(lactic-co-glycolic acid).

1960). One method used in interfacial tissue engineering is to directly incorporate synthetic HAp in the constructs (Xue *et al.*, 2010; Chung *et al.*, 2011a, 2011b; Mohan *et al.*, 2011; Qu *et al.*, 2011; Samavedi *et al.*, 2011; Khanarian *et al.*, 2012). Samavedi *et al.* (2011) fabricated a continuously graded mesh by co-electrospinning nano-HAp/polycaprolactone (PCL) with poly(ester urethane). This approach for the bone–ligament/tendon interface resulted in a continuous mineral gradient as well as a gradient in tensile moduli throughout the scaffold. Mohan *et al.* (2011) also constructed a continuous mineral gradient for the osteochondral interface by incorporating nano-HAp into poly(D,L-lactic-co-glycolic acid) (PLGA) microspheres. The gradient was formed by filling a mold with opposing types of microspheres, PLGA with or without nano-HAp, and microspheres loaded with transforming growth factor-β1 for chondrogenesis. Scaffolds containing both mineral and signal gradients were found to promote the greatest extent of regeneration. Another group used a bi-layered approach, composed of a poly vinyl alcohol (PVA)/gelatin/nano-HAp layer and a polyamide-6 layer (Qu *et al.*, 2011). Osteogenically induced bone marrow stem cells (BMSCs) were seeded onto the HAp layer and chondrogenically induced BMSCs were seeded onto the polyamide-6 layer, and the constructs were implanted intramuscularly. After 12 weeks, the layers retained the corresponding osteogenic and chondrogenic gene expression (Qu *et al.*, 2011). Other groups have used non-stratified nanocomposites that have shown promise for osteochondral tissue engineering. These nanocomposites incorporated HAp and synthetic polymers, including PLGA (Xue *et al.*, 2010) and poly(1,8-octanediol-co-citrate) (POC) (Chung *et al.*, 2011a, 2011b). Xue *et al.* (2010) implanted PLGA constructs containing HAp into rat osteochondral defects and observed superior osteochondral regeneration in the scaffolds incorporating HAp compared to PLGA control scaffolds.

In addition to solely incorporating nano-HAp, some groups compared both HAp microparticle and nanoparticle incorporation (Chung *et al.*, 2011b; Khanarian *et al.*, 2012). For bone tissue engineering, a debate currently exists on the use of HAp microparticles versus nanoparticles (Wang *et al.*, 2011). In comparison to microparticles, nanoparticles are hypothesized to better mimic the native bone environment and enhance osteogenic differentiation (Ngiam *et al.*, 2011). However, drawbacks of nanoparticles include reduced scaffold porosity and particle aggregation (Wang *et al.*, 2011). In interfacial tissue engineering, this microscale versus nanoscale debate exists as well. Chung *et al.* (2011b) directly compared HAp microcomposites versus nanocomposites and found that nanocomposites displayed the highest strength and stiffness and allowed for more trabecular bone formation at the implant–tissue interface. However, Khanarian *et al.* (2012) compared HAp microparticle and nanoparticle incorporation in agarose gels and found superior mechanical properties and matrix production with HAp microparticles relative to HAp nanoparticles. These studies suggest the need for further research and characterization of microscale and nanoscale interfacial constructs.

Not all studies that incorporate HAp used synthetic HAp. A newer alternative technique to incorporate HAp is by inducing the nucleation and growth of HAp crystals (Zhang and Ma, 1999). This approach mimics native HAp deposition by the use of simulated body fluid (SBF), which is a solution that contains the approximate concentrations of ions found *in vivo*. Scaffolds can be soaked in this solution and HAp crystals are directly nucleated on the surface, where the size of HAp crystals can be regulated by the amount of soak time (Smith and Ma, 2011). In native bone, HAp is deposited on collagen I fibers, and the size and orientation of the deposited HAp crystals are regulated by the collagen fibers (Rho *et al.*, 1998). Although this nucleation technique may be used for bone tissue engineering (Zhang *et al.*, 2012a), several interfacial tissue engineering studies also tried to directly mimic this native mineralization by allowing collagen I nanofibers to soak in SBF (Tampieri *et al.*, 2008, 2011; Kon *et al.*, 2009, 2010a, 2010b, 2010c, 2011; Liu *et al.*, 2009; Filardo *et al.*, 2010). Tampieri *et al.* (2008) developed a tri-layered osteochondral scaffold, which consisted of a biomineralized collagen layer to mimic native subchondral bone, an intermediate layer composed of collagen and less mineral, and an upper layer composed of hyaluronic acid and collagen to mimic the articular cartilage. These constructs were implanted subcutaneously in mice for 8 weeks and the layers of the construct were found to selectively support osteogenesis or chondrogenesis. From the same group, Kon *et al.* (2010a) studied a similar tri-layered scaffold in sheep, except the cartilaginous layer lacked hyaluronic acid. Blank scaffolds and scaffolds seeded with autologous chondrocytes were implanted in sheep osteochondral defects. The extent of regeneration in the seeded constructs was similar to that of the blank scaffolds, suggesting the scaffold induced local cellular infiltration *in vivo* (Kon *et al.*, 2010a). This scaffold has even been further tested in horses and in human clinical trials (Kon *et al.*, 2010b, 2010c, 2011). In another study, Liu *et al.* (2009) created a continuous gradient of HAp on a collagen construct by a controlled diffusion-precipitate method that allowed calcium and phosphate solutions to diffuse into the construct via opposing gradients. This resulted in a calcium rich and a calcium depleted side, where nano-HAp precipitation was the most prominent in the interior of the scaffold.

HAp nucleation has also been performed on PCL (Li *et al.*, 2009; Nirmala *et al.*, 2010; Samavedi *et al.*, 2011) and poly(D,L-lactide) (PDLLA) (Yunos *et al.*, 2011; Zou *et al.*, 2012). Li *et al.* (2009) created a continuously graded HAp scaffold by locally varying the immersion time of PCL nanofibers in SBF. The PCL scaffold was placed in a tilted position in a glass beaker and a mineral solution was fed into the beaker at a constant rate. Thus, because the scaffold region at the bottom of the beaker was exposed to SBF for a longer period, there was more HAp deposition at the bottom than in the scaffold region at the top of the beaker. After seeding with mouse preosteoblasts for 3 days *in vitro*, the cell density within the scaffold was found to increase with increasing mineral content along the mineral gradient (Li *et al.*, 2009). HAp has also been nucleated onto scaffoldless osteochondral constructs, which were composed of ECM directly secreted by mesenchymal stem cells (MSCs) (Matsusaki *et al.*, 2008). HAp was deposited onto the constructs with an alternate soaking process with calcium and phosphate solutions. Scaffolds containing HAp exhibited accelerated osteoinduction *in vivo* as compared to scaffolds without HAp (Matsusaki *et al.*, 2008).

13.2.2   β-tricalcium phosphate (TCP) nanoparticles

In addition to being osteoconductive and biocompatible, β-TCP has received great attention from the interfacial and bone tissue engineering communities due to its tunable bioresorption rates (Porter *et al.*, 2009). Erisken *et al.* (2008) recently created a functionally graded β-TCP scaffold via a hybrid twin-screw extrusion/electrospinning (TSEE) process for osteochondral interfaces (Erisken *et al.*, 2008). This process allowed the time-dependent feeding of β-TCP nanoparticles and PCL solution resulting in a continuous gradation of β-TCP in an electrospun nanofibrous mesh. The same group further tested the scaffold for its linear viscoelastic and compressive properties by testing scaffolds seeded with mouse preosteoblasts and unseeded scaffolds (Erisken *et al.*, 2010). They found that the viscoelastic and biomechanical properties increased (i.e., became closer to the native osteochondral interface) by increasing the culture period. Another group fabricated a multilayer scaffold that employed β-TCP, in which the scaffold contained a bone phase, a cartilage phase, and a transition phase in between. The bone phase was created by ceramic stereolithography and the cartilage phase, consisting of collagen I, was added to the bone phase by gel casting. The gel casting method was also proposed to aid in the reduction of delamination between bone and cartilage layers (Bian *et al.*, 2012).

13.2.3   Other mineral nanoparticles

Although the primary mineral components used in tissue engineering are HAp and β-TCP, other mineral nanomaterial components such as amorphous calcium phosphate and β-glycerophosphate (β-GP) have also been incorporated into interfacial constructs (Harley *et al.*, 2010a, 2010b; Erisken *et al.*, 2011; Ramalingam *et al.*, 2012). Harley *et al.* (2010a) developed a layered scaffold with a continuous interface via a liquid-phase co-synthesis technique. The bone phase was composed of a mineralized type I collagen/chondroitin sulfate and the cartilage phase was composed of unmineralized collagen II/chondroitin sulfate. The layers were then allowed to diffuse, creating a graded interface. Erisken *et al.* (2011) used the previously described TSEE process to create scaffolds with opposing gradients of insulin and β-GP. The scaffolds were seeded with human adipose-derived stromal cells and it was found that the cells differentiated selectively towards chondrogenic or osteogenic lineages depending upon the location in the scaffold (Erisken *et al.*, 2011). As an alternative electrospinning technique, Ramalingam *et al.* (2012) used a two-spinneret approach to create a scaffold of continuously graded amorphous calcium phosphate nanoparticles. The spinnerets were set 2.5 cm apart oriented vertically above a spinning mandrel, where each spinneret was connected to a syringe loaded with either PCL or PCL with calcium phosphate nanoparticles. Preosteoblast cells were seeded onto the constructs for 7 days and it was observed that cell adhesion and proliferation increased as the calcium phosphate concentration increased along the length of scaffold (Ramalingam *et al.*, 2012).

13.2.4   Emerging nanoparticle materials

Although most nanoparticle materials used in interfacial tissue engineering are mineral-based, it is important to consider the potential benefits of other nanoparticle materials as well. Some emerging nanoparticle materials in interfacial tissue engineering are magnetic (Tampieri *et al.*, 2011; Grogan *et al.*, 2012), fluorescent (Lee *et al.*, 2012), or are composed of silver (Nirmala *et al.*, 2010). Grogan *et al.* (2012) fabricated a graded scaffold by mixing alginate hydrogels with cells labeled with magnetic iron-oxide nanoparticles and exposing the cells to varying external magnetic fields. The resulting scaffold produced a scaffold with varying cellular orientations that closely mimicked that of the osteochondral interface. Lee *et al.* (2012) developed another interesting use for fluorescent nanoparticles in interfacial tissue engineering. They created an *in vivo* cell tracking system to monitor the migration of mesenchymal stem cells in osteochondral defects. Nirmala *et al.* (2010) introduced the use of silver nanoparticles for osteochondral constructs. They incorporated hydroxyapatite and silver nanoparticles in PCL electrospun nanofibers and then submerged the constructs in SBF to allow for further mineralization. Incorporating the silver nanoparticles resulted in superior mechanical properties and apatite deposition. Additionally, silver nanoparticles are known for their antimicrobial properties, providing a further advantage of their incorporation (Sondi and Salopek-Sondi, 2004).

13.2.5   Summary of nanoparticles

Overall, nanoparticle use in interfacial tissue engineering is primarily focused on the incorporation of natural materials like hydroxyapatite and β-TCP. The most common methods utilize the nucleation of hydroxyapatite onto collagen scaffolds, which closely mimics the natural mineral deposition *in vivo*. In general, scaffolds incorporating nano-HAp and other mineral gradients appear to result in superior interface mechanical properties as well as enhance the selective cellular response within the constructs. Although there is some debate on whether nanoscale or microscale minerals are superior for interfacial constructs, nanomaterials certainly show promise for the field. Additionally, further developments in the newly emerging nanoparticle materials also have the potential to advance the interfacial tissue engineering field. These emerging nanomaterials provide new methods to produce gradient scaffolds (e.g., using magnetic fields) and fluorescent nanoparticles also provide a new method to evaluate the constructs *in vivo*.

13.3   Nanofibers

Nanofibers are another promising nanomaterial used in interfacial tissue engineering due to their high surface-to-volume ratio and their ability to mimic the fibrous organization of the ECM (Seidi *et al.*, 2011). Nanofibers also possess unique mechanical properties whereby the tensile modulus, shear modulus, and tensile strength have shown the capacity to increase as the diameter of the nanofiber decreases (Chew *et al.*, 2006; Yang *et al.*, 2008). These properties are especially useful to the bone–tendon/ligament interface, where the native mechanical loading transitions from primarily compressive in bone to tensile loading in the tendon/ligament. Three common methods to produce nanofibers are self-assembly, electrospinning, and phase separation. Additionally, the materials that constitute these nanofibers in interfacial tissue engineering are primarily collagen and synthetic polymers. Thus, the following sections will discuss the collagen and synthetic nanofibers for interfacial tissue applications and each section will also discuss how the materials are incorporated.

13.3.1   Collagen nanofibers

As the most common bodily protein, accounting for 30% of all proteins in the body, and giving rise to the structural support and tensile strength of the native ECM (Kew *et al.*, 2011), collagen is also the most commonly employed nanofibrous material in interfacial tissue engineering. Collagen I is the most prevalent type of collagen in bone, while collagen II is more prevalent in hyaline cartilage. However, collagen I is primarily the only type of collagen currently being incorporated into interfacial constructs. Additionally, all publications currently employing collagen I nanofibers into their interfacial constructs use collagen self-assembly to create the nanofibers (Tampieri *et al.*, 2008, 2011; Kon *et al.*, 2009, 2010a, 2010b, 2010c, 2011; Filardo *et al.*, 2010; Cheng *et al.*, 2011). Cheng *et al.* (2011) encapsulated rabbit MSCs into collagen microspheres that were fabricated out of self-assembled collagen I nanofibers. The encapsulated microspheres were then either chondrogenically or osteogenically induced and the microspheres were brought together to form three layers, a chondrogenic and osteogenic layer, and an intermediate layer in between containing undifferentiated MSCs in collagen microspheres. Compared to biphasic control scaffolds that did not contain an intermediate layer, the triphasic scaffolds successfully formed a calcified cartilage layer (Cheng *et al.*, 2011). As previously mentioned in the HAp nanoparticle section, the other group incorporating collagen nanofibers via self-assembly developed a tri-layered osteochondral scaffold, which consisted of a biomineralized collagen layer to mimic native subchondral bone, an intermediate layer composed of collagen and less mineral, and an upper layer composed of hyaluronic acid and collagen to mimic the articular cartilage (Tampieri *et al.*, 2008). This group went on to study a similar tri-layered scaffold in sheep, horses, and even in human clinical trials (Kon *et al.*, 2010a, 2010b, 2010c, 2011).

13.3.2   Synthetic nanofibers

Although incorporation of natural material more closely mimics the native ECM of tissues, synthetic nanofibers have also received a lot of attention in interfacial tissue engineering due to the ability to control degradation and tune mechanical properties (Vavken *et al.*, 2009). Primarily synthetic nanofibers are made via electrospinning, however one interfacial tissue engineering strategy used phase separation to make nanofibers (Hu *et al.*, 2009). Through this phase separation technique, Hu *et al.* (2009) created nanofibrous scaffolds out of poly(L-lactic acid) (PLLA). The PLLA scaffolds supported the differentiation of human bone marrow stem cells along osteogenic or chondrogenic lineages when exposed to the respective growth signals.

As previously described, several studies have incorporated synthetic nanofibers with mineral-based nanoparticles, where the fibers were electrospun PCL, PLGA, or PDLLA (Erisken *et al.*, 2008, 2010, 2011; Li *et al.*, 2009; Nirmala *et al.*, 2010; Samavedi *et al.*, 2011; Yunos *et al.*, 2011; Ramalingam *et al.*, 2012). Other electrospun interfacial constructs incorporate only the synthetic nanofiber itself (Spalazzi *et al.*, 2008; Xie *et al.*, 2010; Liu *et al.*,

2011). Xie *et al.* (2010) fabricated a unique 'aligned-to-random' oriented nanofibrous scaffold by specifically designing a collector with varying electric fields. These scaffolds were created to mimic the change in collagen fiber orientation in the bone–tendon insertion site. Tendon fibroblasts cultured on these scaffolds were observed to be aligned in the aligned fiber portion and were oriented more randomly in the random fiber portion. Although collagen II was not observed on any portion of the scaffolds after 7 days of culture, collagen I deposition was oriented in the direction of the nanofibers in the aligned portion of the scaffold and was randomly distributed in the random portion (Xie *et al.*, 2010). Aside from electrospinning and phase separation, Liu *et al.* (2011) fabricated nanofibrous hollow microspheres via self-assembled star-shaped PLLA. These microspheres were designed to be injectable for cartilage and osteochondral defects, mimic the nanoscale topography of native ECM, and they were also designed to be highly porous. In comparison to solid microspheres composed of the same material, the nanofibrous microspheres supported a significantly larger amount of cartilage regeneration in an ectopic model (Liu *et al.*, 2011).

Alternative synthetic materials that have emerged are carbon nanofibers and nanorods (Rodrigues *et al.*, 2012). Rodrigues *et al.* (2012) tested both pure polyvinyl alcohol (PVA) hydrogels and carbon-reinforced hydrogels in osteochondral defects. The reinforced gels contained either carbon nanofibers or carbon nanorods. These carbon nanomaterials are desirable because they are lightweight and possess high stiffness and axial strength (Treacy *et al.*, 1996). After 12 weeks of implantation, the carbon nanofiber-reinforced PVA accumulated the most calcium and phosphorus suggesting carbon nanofibers may have potential for treating osteochondral defects (Rodrigues *et al.*, 2012).

13.3.3   Summary of nanofibers

Although self-assembled collagen fibers are the most commonly employed nanofiber for interfacial tissue engineering applications, other promising synthetic nanofiber materials are being explored, including carbon nanofibers. The unique mechanical properties as well as the ability to mimic the native ECM of interfacial tissues make these materials highly desirable for interfacial constructs. Specifically, electrospinning has been successfully used to spatially control the alignment of nanofibers, which resulted in a gradation in cellular alignment and ECM secretion. Overall, it appears nanofibers are used primarily in conjunction with nanoparticles and are used as a tool to create gradients in mineral content. Thus, nanofiber use may create gradients in mechanical properties and cellular responses that mimic native tissue interfaces.

13.4 Strategies incorporating nanomaterials in hard–soft tissue interfaces

Though it is certainly important to consider types of nanomaterials to be used in interfacial tissue engineering, it is also essential to consider the overall interfacial strategy used that employs these materials. As stated in the introduction, interfacial tissue engineering strategies as a whole are primarily stratified in nature. This trend certainly remains evident (Tables 13.1 and 13.2) in strategies that employ nanomaterials, where the majority of the strategies are stratified (i.e., biphasic or triphasic) or homogeneous (i.e., nanocomposite). The stratified approaches more closely mimic the native tissue components and/or structure, however, they may risk delamination between phases. The use of homogeneous constructs reduces the risk of delamination (Harley et al., 2010a; Xue et al., 2010), although homogeneous approaches lack to regenerate the structural/functional anisotropy that exists in tissue interfaces. Consequently, continuous interfaces have received recent attention in the literature as they provide a more seamless transition between tissues (Dormer et al., 2010a). Continuous interfaces that employ nanomaterials are certainly a minority (Tables 13.1 and 13.2) as compared to the stratified and homogeneous approaches, although continuous approaches are more common in bone–tendon/ligament interfaces. Interestingly, the main continuous approaches involve electrospinning techniques. However, Detamore and Berkland et al. have created continuous interfaces by loading a mold with opposing gradients of osteogenic and chondrogenic microspheres (Dormer et al., 2010b, 2011, 2012; Singh et al., 2010; Mohan et al., 2011). Additionally, strategies that incorporate a continuous or stratified design primarily create a gradient of ceramic components. However, other gradient approaches exist, including gradients in fiber (Xie et al., 2010) and cellular (Grogan et al., 2012) alignment.

13.5 Conclusion and future trends

As the field of interfacial tissue engineering progresses, the underlying goal remains, which is to regenerate a structural and functional interface that transitions between two dissimilar tissues. This chapter has provided an overview of the structural and functional characteristics of the most widely explored tissue interfaces, specifically the osteochondral and bone–tendon/ligament interfaces. However, tissue engineering studies of other tissue interfaces (e.g., muscle–tendon interface) are beginning to emerge as well (Ladd et al., 2011).

Highlighted in this chapter is the burgeoning integration of nanomaterial use in interfacial tissue engineering strategies. The nanomaterials currently being used in interfacial applications are nanoparticles and nanofibers.

Interestingly, the most common nanomaterials used in interfacial strategies are mineral-based nanoparticles and collagen nanofibers, which are components found in the native interfaces. This use of raw materials further emphasizes the potential of using raw materials in tissue engineering strategies (Detamore and Renth, 2012). However, nanomaterial use in general provides many benefits to interfacial strategies, including superior mechanical properties of the interface, and providing a more biomimetic environment.

In addition to the type of nanomaterial employed, the overall strategy to regenerate the interface is crucial as well. As previously mentioned, interfacial strategies that employ nanomaterials are primarily stratified or homogeneous, although continuous gradient approaches are gaining attention in the field. It is also important to note that, although these strategies are defined as osteochondral or bone–tendon/ligament strategies, approaches from either interface can be tailored for the design of a desired interface. Because interface structure and function are both crucial entities for successful regeneration of interfaces, future successful interfacial designs are likely to incorporate a strategy that mimics the gradient of natural interface as well as incorporate the nanoscale components of the interface. Thus, continuous gradients that incorporate gradients of minerals, collagen, and/or other nanoscale elements may become the next generation of interfacial designs.

13.6   Acknowledgments

The authors would like to acknowledge the National Science Foundation Graduate Research Fellowship for the support of Emily Beck.

13.7   References

Benjamin, M. and Ralphs, J. (1998). Fibrocartilage in tendons and ligaments – an adaptation to compressive load. *Journal of Anatomy*, **193**, 481–494.

Bian, W., Li, D., Lian, Q., Li, X., Zhang, W., Wang, K. and Jin, Z. (2012). Fabrication of a bio-inspired beta-Tricalcium phosphate/collagen scaffold based on ceramic stereolithography and gel casting for osteochondral tissue engineering. *Rapid Prototyping Journal*, **18**, 68–80.

Cheng, H., Luk, K. D. K., Cheung, K. and Chan, B. P. (2011). In vitro generation of an osteochondral interface from mesenchymal stem cell–collagen microspheres. *Biomaterials*, **32**, 1526–1535.

Chew, S. Y., Hufnagel, T. C., Lim, C. T. and Leong, K. W. (2006). Mechanical properties of single electrospun drug-encapsulated nanofibres. *Nanotechnology*, **17**, 3880.

Chung, E. J., Kodali, P., Laskin, W., Koh, J. L. and Ameer, G. A. (2011a). Long-term in vivo response to citric acid-based nanocomposites for orthopaedic tissue engineering. *Journal of Materials Science: Materials in Medicine*, **32**, 2131–2138.

Chung, E. J., Qiu, H., Kodali, P., Yang, S., Sprague, S. M., Hwong, J., Koh, J. and Ameer, G. A. (2011b). Early tissue response to citric acid–based micro and nanocomposites. *Journal of Biomedical Materials Research Part A*, **96**, 29–37.

Cool, S., Kenny, B., Wu, A., Nurcombe, V., Trau, M., Cassady, A. and Grøndahl, L. (2007). Poly (3-hydroxybutyrate-co-3-hydroxyvalerate) composite biomaterials for bone tissue regeneration: In vitro performance assessed by osteoblast proliferation, osteoclast adhesion and resorption, and macrophage proinflammatory response. *Journal of Biomedical Materials Research Part A*, **82**, 599–610.

Deng, M., James, R., Laurencin, C. and Kumbar, S. (2011). Nanostructured polymeric scaffolds for orthopaedic regenerative engineering. *NanoBioscience, IEEE Transactions on*, **11**, 3–14.

Detamore, M. S. and Renth, A. N. (2012). Leveraging 'raw materials' as building blocks and bioactive signals in regenerative medicine. *Tissue Engineering Part B: Reviews*, **18**, 341–362.

Dormer, N. H., Berkland, C. J. and Detamore, M. S. (2010a). Emerging techniques in stratified designs and continuous gradients for tissue engineering of interfaces. *Annals of Biomedical Engineering*, **38**, 2121–2141.

Dormer, N. H., Busaidy, K., Berkland, C. J. and Detamore, M. S. (2011). Osteochondral interface regeneration of rabbit mandibular condyle with bioactive signal gradients. *Journal of Oral and Maxillofacial Surgery*, **69**, e50–e57.

Dormer, N. H., Singh, M., Wang, L., Berkland, C. J. and Detamore, M. S. (2010b). Osteochondral interface tissue engineering using macroscopic gradients of bioactive signals. *Annals of Biomedical Engineering*, **38**, 2167–2182.

Dormer, N. H., Singh, M., Zhao, L., Mohan, N., Berkland, C. J. and Detamore, M. S. (2012). Osteochondral interface regeneration of the rabbit knee with macroscopic gradients of bioactive signals. *Journal of Biomedical Materials Research Part A*, **100A**, 162–170.

Dorozhkin, S. V. (2010). Bioceramics of calcium orthophosphates. *Biomaterials*, **31**, 1465–1485.

Erisken, C., Kalyon, D. M. and Wang, H. (2008). Functionally graded electrospun polycaprolactone and β-tricalcium phosphate nanocomposites for tissue engineering applications. *Biomaterials*, **29**, 4065–4073.

Erisken, C., Kalyon, D. M. and Wang, H. (2010). Viscoelastic and biomechanical properties of osteochondral tissue constructs generated from graded polycaprolactone and beta-tricalcium phosphate composites. *Journal of Biomechanical Engineering*, **132**, 091013.

Erisken, C., Kalyon, D. M., Wang, H., Ornek-Ballanco, C. and Xu, J. (2011). Osteochondral tissue formation through adipose-derived stromal cell differentiation on biomimetic polycaprolactone nanofibrous scaffolds with graded insulin and beta-glycerophosphate concentrations. *Tissue Engineering: Part A*, **17**, 1239–1252.

Filardo, G., Kon, E., Delcogliano, M., Giordano, G., Bonanzinga, T., Marcacci, M. and Zaffagnini, S. (2010). Novel nano-composite multilayered biomaterial for the treatment of patellofemoral cartilage lesions. *Patellofemoral Pain, Instabilty, and Arthritis: Clinical Presentation, Imaging, and Treatment*, **12**, 255.

Grayson, W. L., Chao, P. H. G., Marolt, D., Kaplan, D. L. and Vunjak-Novakovic, G. (2008). Engineering custom-designed osteochondral tissue grafts. *Trends in Biotechnology*, **26**, 181–189.

Grogan, S. P., Pauli, C., Chen, P., Du, J., Chung, C. B., Kong, S. D., Colwell Jr, C. W., Lotz, M., Jin, S. and D'lima, D. (2012). In situ tissue engineering using magnetically guided 3D cell patterning. *Tissue Engineering Part C*, **18**, 1–11.

Harley, B. A., Lynn, A. K., Wissner-Gross, Z., Bonfield, W., Yannas, I. V. and Gibson, L. J. (2010a). Design of a multiphase osteochondral scaffold III: Fabrication of layered scaffolds with continuous interfaces. *Journal of Biomedical Materials Research Part A*, **92**, 1078–1093.

Harley, B. A., Lynn, A. K., Wissner-Gross, Z., Bonfield, W., Yannas, I. V. and Gibson, L. J. (2010b). Design of a multiphase osteochondral scaffold. II. Fabrication of a mineralized collagen–glycosaminoglycan scaffold. *Journal of Biomedical Materials Research Part A*, **92**, 1066–1077.

Hu, J., Feng, K., Liu, X. and Ma, P. X. (2009). Chondrogenic and osteogenic differentiations of human bone marrow-derived mesenchymal stem cells on a nanofibrous scaffold with designed pore network. *Biomaterials*, **30**, 5061–5067.

Johansen, E. and Parks, H. F. (1960). Electron microscopic observations on the three-dimensional morphology of apatite crystallites of human dentine and bone. *The Journal of Biophysical and Biochemical Cytology*, **7**, 743–746.

Kaplan, F., Hayes, W., Keaveny, T., Boskey, A., Einhorn, T. and Iannotti, J. 1994. Form and function of bone. In Simon SP, ed. *Orthopedic Basic Science*. Columbus, OH: American Academy of Orthopedic Surgeons, 127–185.

Kew, S., Gwynne, J., Enea, D., Abu-Rub, M., Pandit, A., Zeugolis, D., Brooks, R., Rushton, N., Best, S. and Cameron, R. (2011). Regeneration and repair of tendon and ligament tissue using collagen fibre biomaterials. *Acta Biomaterialia*, **7**, 3237–3247.

Khanarian, N. T., Haney, N. M., Burga, R. A. and Lu, H. H. (2012). A functional agarose-hydroxyapatite scaffold for osteochondral interface regeneration. *Biomaterials*, **33**, 5247–5258.

Kon, E., Delcogliano, M., Filardo, G., Altadonna, G. and Marcacci, M. (2009). Novel nano-composite multi-layered biomaterial for the treatment of multifocal degenerative cartilage lesions. *Knee Surgery, Sports Traumatology, Arthroscopy*, **17**, 1312–1315.

Kon, E., Delcogliano, M., Filardo, G., Busacca, M., Di Martino, A. and Marcacci, M. (2011). Novel nano-composite multilayered biomaterial for osteochondral regeneration. *The American Journal of Sports Medicine*, **39**, 1180–1190.

Kon, E., Delcogliano, M., Filardo, G., Fini, M., Giavaresi, G., Francioli, S., Martin, I., Pressato, D., Arcangeli, E. and Quarto, R. 2010a. Orderly osteochondral regeneration in a sheep model using a novel nano-composite multilayered biomaterial. *Journal of Orthopaedic Research*, **28**, 116–124.

Kon, E., Delcogliano, M., Filardo, G., Pressato, D., Busacca, M., Grigolo, B., Desando, G. and Marcacci, M. (2010b). A novel nano-composite multi-layered biomaterial for treatment of osteochondral lesions: Technique note and an early stability pilot clinical trial. *Injury*, **41**, 693–701.

Kon, E., Mutini, A., Arcangeli, E., Delcogliano, M., Filardo, G., Nicoli Aldini, N., Pressato, D., Quarto, R., Zaffagnini, S. and Marcacci, M. (2010c). Novel nano-structured scaffold for osteochondral regeneration: Pilot study in horses. *Journal of Tissue Engineering and Regenerative Medicine*, **4**, 300–308.

Ladd, M. R., Lee, S. J., Stitzel, J. D., Atala, A. and Yoo, J. J. (2011). Co-electrospun dual scaffolding system with potential for muscle–tendon junction tissue engineering. *Biomaterials*, **32**, 1549–1559.

Lee, J. M., Kim, B. S., Lee, H. and Im, G. I. (2012). In vivo tracking of mesenchymal stem cells using fluorescent nanoparticles in an osteochondral repair model. *Molecular Therapy*, **20**, 1434–1442.

Li, X., Xie, J., Lipner, J., Yuan, X., Thomopoulos, S. and Xia, Y. (2009). Nanofiber scaffolds with gradations in mineral content for mimicking the tendon-to-bone insertion site. *Nano Letters*, **9**, 2763–2768.

Lim, J. and Temenoff, J. (2009). Tendon and ligament tissue engineering: Restoring tendon/ligament and its interfaces. *Fundamentals of Tissue Engineering and Regenerative Medicine*, **20**, 255.

Liu, C., Han, Z. and Czernuszka, J. (2009). Gradient collagen/nanohydroxyapatite composite scaffold: Development and characterization. *Acta Biomaterialia*, **5**, 661–669.

Liu, X., Jin, X. and Ma, P. X. (2011). Nanofibrous hollow microspheres self-assembled from star-shaped polymers as injectable cell carriers for knee repair. *Nature Materials*, **10**, 398–406.

Lu, H. and Jiang, J. (2006). Interface tissue engineering and the formulation of multiple-tissue systems. *Advances in Biochemical Engineering/Biotechnology*, **102**, 91–111.

Lu, H. H. and Spalazzi, J. P. (2009). Biomimetic stratified scaffold design for ligament-to-bone interface tissue engineering. *Combinatorial Chemistry 38; High Throughput Screening*, **12**, 589–597.

Lu, H. H., Subramony, S. D., Boushell, M. K. and Zhang, X. (2010). Tissue engineering strategies for the regeneration of orthopedic interfaces. *Annals of Biomedical Engineering*, **38**, 2142–2154.

Martin, R. B., Burr, D. B. and Sharkey, N. A. (1998). *Skeletal Tissue Mechanics*, Springer Verlag.

Matsusaki, M., Kadowaki, K., Tateishi, K., Higuchi, C., Ando, W., Hart, D. A., Tanaka, Y., Take, Y., Akashi, M. and Yoshikawa, H. (2008). Scaffold-free tissue-engineered construct–hydroxyapatite composites generated by an alternate soaking process: Potential for repair of bone defects. *Tissue Engineering Part A*, **15**, 55–63.

Mikos, A. G., Herring, S. W., Ochareon, P., Elisseeff, J., Lu, H. H., Kandel, R., Schoen, F. J., Toner, M., Mooney, D. and Atala, A. (2006). Engineering complex tissues. *Tissue Engineering*, **12**, 3307–3339.

Moffat, K. L., Wang, I., Rodeo, S. A. and Lu, H. H. (2009). Orthopedic interface tissue engineering for the biological fixation of soft tissue grafts. *Clinics in Sports Medicine*, **28**, 157–176.

Mohan, N., Dormer, N. H., Caldwell, K. L., Key, V. H., Berkland, C. J. and Detamore, M. S. (2011). Continuous gradients of material composition and growth factors for effective regeneration of the osteochondral interface. *Tissue Engineering Part A*, **17**, 2845–2855.

Ngiam, M., Nguyen, L., Liao, S., Chan, C. and Ramakrishna, S. (2011). Biomimetic nanostructured materials: Potential regulators for osteogenesis. *Annals of the Academy of Medicine, Singapore*, **40**, 213–220.

Nirmala, R., Nam, K. T., Park, D. K., Woo-Il, B., Navamathavan, R. and Kim, H. Y. (2010). Structural, thermal, mechanical and bioactivity evaluation of silver-loaded bovine bone hydroxyapatite grafted poly (ε-caprolactone) nanofibers via electrospinning. *Surface and Coatings Technology*, **205**, 174–181.

Panseri, S., Russo, A., Cunha, C., Bondi, A., Di Martino, A., Patella, S. and Kon, E. (2012). Osteochondral tissue engineering approaches for articular cartilage and subchondral bone regeneration. *Knee Surgery, Sports Traumatology, Arthroscopy*, **20**, 1182–1191.

Poole, A. R., Kojima, T., Yasuda, T., Mwale, F., Kobayashi, M. and Laverty, S. (2001). Composition and structure of articular cartilage: A template for tissue repair. *Clinical Orthopaedics and Related Research*, **391**, S26.

Porter, J. R., Ruckh, T. T. and Popat, K. C. (2009). Bone tissue engineering: A review in bone biomimetics and drug delivery strategies. *Biotechnology Progress*, **25**, 1539–1560.

Qu, D., Li, J., Li, Y., Khadka, A., Zuo, Y., Wang, H., Liu, Y. and Cheng, L. (2011). Ectopic osteochondral formation of biomimetic porous PVA-n-HA/PA6 bilayered scaffold and BMSCs construct in rabbit. *Journal of Biomedical Materials Research Part B: Applied Biomaterials*, **96**, 9–15.

Ramalingam, M., Young, M. F., Thomas, V., Sun, L., Chow, L. C., Tison, C. K., Chatterjee, K., Miles, W. C. and Simon Jr, C. G. (2012). Nanofiber scaffold gradients for interfacial tissue engineering. *Journal of Biomaterials Applications*, **27**, 695–705.

Rho, J. Y., Kuhn-Spearing, L. and Zioupos, P. (1998). Mechanical properties and the hierarchical structure of bone. *Medical Engineering and Physics*, **20**, 92–102.

Robinson, R. A. (1952). An electron-microscopic study of the crystalline inorganic component of bone and its relationship to the organic matrix. *The Journal of Bone and Joint Surgery (American)*, **34**, 389–476.

Rodrigues, A. A., Batista, N. A., Bavaresco, V. P., Baranauskas, V., Ceragioli, H. J., Peterlevitz, A. C., Mariolani, J. R. L., Santana, M. H. A. and Belangero, W. D. (2012). In vivo evaluation of hydrogels of polyvinyl alcohol with and without carbon nanoparticles for osteochondral repair. *Carbon*, **50**, 2091–2099.

Sahoo, S., Teh, T. K. H., He, P., Toh, S. L. and Goh, J. C. H. (2011). Interface Tissue Engineering: Next Phase in Musculoskeletal Tissue Repair. *Annals of the Academy of Medicine, Singapore*, **40**, 245.

Samavedi, S., Olsen Horton, C., Guelcher, S. A., Goldstein, A. S. and Whittington, A. R. (2011). Fabrication of a model continuously graded co-electrospun mesh for regeneration of the ligament-bone interface. *Acta Biomaterialia*, **7**, 4131–4138.

Seidi, A. and Ramalingam, M. (2012). Impact of gradient biomaterials on interface tissue engineering. *Journal of Biomaterials and Tissue Engineering*, **2**, 89–99.

Seidi, A., Ramalingam, M., Elloumi-Hannachi, I., Ostrovidov, S. and Khademhosseini, A. (2011). Gradient biomaterials for soft-to-hard interface tissue engineering. *Acta Biomaterialia*, **7**, 1441–1451.

Singh, M., Dormer, N., Salash, J. R., Christian, J. M., Moore, D. S., Berkland, C. and Detamore, M. S. (2010). Three-dimensional macroscopic scaffolds with a gradient in stiffness for functional regeneration of interfacial tissues. *Journal of Biomedical Materials Research Part A*, **94**, 870–876.

Smith, I. O. and Ma, P. X. (2011). Biomimetic scaffolds in tissue engineering. In: Pallua, N. and Suscheck, C. V. (eds), *Tissue Engineering*, Springer, Berlin Heidelberg. 31–39.

Smith, L., Xia, Y., Galatz, L. M., Genin, G. M. and Thomopoulos, S. (2012). Tissue engineering strategies for the tendon/ligament-to-bone insertion. *Connective Tissue Research*, **53**, 95–105.

Sondi, I. and Salopek-Sondi, B. (2004). Silver nanoparticles as antimicrobial agent: A case study on *E. coli* as a model for gram-negative bacteria. *Journal of Colloid and Interface Science*, **275**, 177–182.

Spalazzi, J. P., Vyner, M. C., Jacobs, M. T., Moffat, K. L. and Lu, H. H. (2008). Mechanoactive scaffold induces tendon remodeling and expression of fibrocartilage markers. *Clinical Orthopaedics and Related Research*, **466**, 1938–1948.

Tampieri, A., Landi, E., Valentini, F., Sandri, M., D'alessandro, T., Dediu, V. and Marcacci, M. (2011). A conceptually new type of bio-hybrid scaffold for bone regeneration. *Nanotechnology*, **22**, 015104.

Tampieri, A., Sandri, M., Landi, E., Pressato, D., Francioli, S., Quarto, R. and Martin, I. (2008). Design of graded biomimetic osteochondral composite scaffolds. *Biomaterials*, **29**, 3539–3546.

Treacy, M., Ebbesen, T. and Gibson, J. (1996). Exceptionally high Young's modulus observed for individual carbon nanotubes. *Nature*, **381**, 678–680.

Tsang, K. Y., Cheung, M. C. H., Chan, D. and Cheah, K. S. E. (2010). The developmental roles of the extracellular matrix: Beyond structure to regulation. *Cell and Tissue Research*, **339**, 93–110.

Ulrich-Vinther, M., Maloney, M. D., Schwarz, E. M., Rosier, R. and O'keefe, R. J. (2003). Articular cartilage biology. *Journal of the American Academy of Orthopaedic Surgeons*, **11**, 421–430.

Vavken, P., Meyer, U., Meyer, T., Handschel, J. and Wiesman, H. (2009). Tissue engineering of ligaments and tendons. *Fundamentals of Tissue Engineering and Regenerative Medicine*, **17**, 317–327.

Wang, H., Leeuwenburgh, S. C. G., Li, Y. and Jansen, J. A. (2011). The use of micro-and nanospheres as functional components for bone tissue regeneration. *Tissue Engineering Part B: Reviews*, **18**, 24–39.

Woo, S., Gomez, M., Seguchi, Y., Endo, C. and Akeson, W. (1983). Measurement of mechanical properties of ligament substance from a bone-ligament-bone preparation. *Journal of Orthopaedic Research*, **1**, 22–29.

Xie, J., Li, X., Lipner, J., Manning, C. N., Schwartz, A. G., Thomopoulos, S. and Xia, Y. 2010. 'Aligned-to-random' nanofiber scaffolds for mimicking the structure of the tendon-to-bone insertion site. *Nanoscale*, **2**, 923–926.

Xue, D., Zheng, Q., Zong, C., Li, Q., Li, H., Qian, S., Zhang, B., Yu, L. and Pan, Z. (2010). Osteochondral repair using porous poly (lactide-co-glycolide)/nano-hydroxyapatite hybrid scaffolds with undifferentiated mesenchymal stem cells in a rat model. *Journal of Biomedical Materials Research Part A*, **94**, 259–270.

Yang, L., Fitié, C. F. C., Van Der Werf, K. O., Bennink, M. L., Dijkstra, P. J. and Feijen, J. (2008). Mechanical properties of single electrospun collagen type I fibers. *Biomaterials*, **29**, 955–962.

Yang, P. J. and Temenoff, J. S. (2009). Engineering orthopedic tissue interfaces. *Tissue Engineering Part B: Reviews*, **15**, 127–141.

Yunos, D., Ahmad, Z., Salih, V. and Boccaccini, A. (2011). Stratified scaffolds for osteochondral tissue engineering applications: Electrospun PDLLA nanofibre coated Bioglass®-derived foams. *Journal of Biomaterials Applications*, **27**, 537–551.

Zhang, Q., Mochalin, V. N., Neitzel, I., Hazeli, K., Niu, J., Kontsos, A., Zhou, J. G., Lelkes, P. I. and Gogotsi, Y. (2012a). Mechanical properties and biomineralization of multifunctional nanodiamond-PLLA composites for bone tissue engineering. *Biomaterials*, **33**, 5067–5075.

Zhang, R. and Ma, P. X. (1999). Porous poly (L-lactic acid)/apatite composites created by biomimetic process. *Journal of Biomedical Materials Research*, **45**, 285–293.

Zhang, X., Bogdanowicz, D., Erisken, C., Lee, N. M. and Lu, H. H. (2012b). Biomimetic scaffold design for functional and integrative tendon repair. *Journal of Shoulder and Elbow Surgery*, **21**, 266–277.

Zou, B., Liu, Y., Luo, X., Chen, F., Guo, X. and Li, X. (2012). Electrospun fibrous scaffolds with continuous gradations in mineral contents and biological cues for manipulating cellular behaviors. *Acta Biomaterialia*, **8**, 1576–1585.

14
Mineralization of nanomaterials for bone tissue engineering

B. MARELLI, C. E. GHEZZI and
S. N. NAZHAT, McGill University, Canada

DOI: 10.1533/9780857097231.3.387

Abstract: As a major component of the body, type I collagen has long been considered as a biomaterial with an enormous potential. In bone, type I collagen constitutes 95 weight % (wt%) of the total organic matter, which in combination with carbonated hydroxyapatite constitutes a highly organised natural nanobiocomposite. Therefore, the mineralization of collagen based biomaterials constitutes an interesting challenge for bone tissue engineering. This chapter reviews some of the recent key trends in acellular based approaches to the mineralization of nanostructured matrices, *in vitro*.

Key words: collagen, plastic compression, mineralization, bioactive glass, silk fibroin.

14.1 Bone: a nanobiocomposite material

Mineralized skeletal tissues contribute to maintaining the body shape, and are involved in motion and in the control of several physiological activities. Bone in particular is a vascularized composite tissue that is composed of a mineral phase, an organic phase and water (Glimcher, 2006). The bulk composition of these three components depends on the bone type and age (Lees, 1987). As an average, wet bone can be considered to be composed of 65 weight% (wt%) mineral (carbonated hydroxyapatite, CHA), 23 wt% organic (95 wt% of which is type I collagen) and 12 wt% water (Currey, 2002). In the dry state, the composition of bone ranges between 60 and 85 wt% inorganic and 15 and 40 wt% organic.

The understanding of the collagen framework and its role in the nucleation and organization of CHA would prove beneficial in the design of bone tissue engineered scaffolds as well as for the prevention of pathological processes (i.e., ectopic calcification).

The organic phase of bone includes water, collagens (mainly type I), non-collagenous proteins (NCPs) and glycosaminoglycans (GAGs) (Weiner *et al.*, 1999). NCPs make up approximately 9 wt% of the total protein content

387

in bone (Boskey, 1989). The principal functions of NCPs relate to mineralization, that is, crystal nucleation, orientation and growth as well as cell signalling and ion homeostasis (Boskey, 1989; Feng *et al.*, 2006). GAGs, on the other hand are highly anionic hetero-polysaccharide chains that exist alone or as components of proteoglycans (PGs). In bone they represent less than 1 wt% of the organic phase and may have an active role in its formation (Vejlens, 1971; Mania *et al.*, 2009).

Bone is a dynamic environment, where cells are constantly responding to external stimuli (Owan *et al.*, 1997). Three unique cell types can be identified in bone, which are responsible for its homeostasis: osteoblasts, osteocytes and osteoclasts. Osteoblasts are derived from bone-lining cells, and are responsible for its formation. These cells initially produce an osteoid in the form of seams of some micrometres thickness, made by a gel of fairly randomly oriented collagen fibrils and proteoglycans. The collagenous matrix is then aligned and minerals are later formed (Currey, 2002; Vuong and Hellmich, 2011) (Fig. 14.1).

Osteocytes are derived from osteoblasts: they are imprisoned in the mineralized matrix and have much less bone deposition capability than active osteoblasts. These are mechanical deformation sensing cells within bone, and

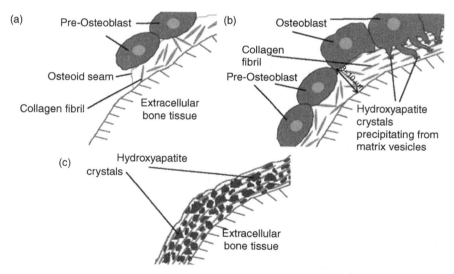

14.1 Working mode of osteoblast. (a) Pre-osteoblastic cells deposit in the extracellular space and produce an osteoid seam reinforced by randomly oriented fibrils. (b) Primary mineralization: osteoblasts order the collagen fibrils through cell-mediated stretch of the matrix and mediate hydroxyapatite formation. (c) Secondary mineralization: crystals grow without the control of local cells. (*Source*: Adapted from Vuong and Hellmich (2011). Copyright © 2011 Elsevier Ltd.)

connect to other osteocytes in a network via cell processes running through channels (*canaliculi*) of about 0.03–0.2 μm in diameter (Guyton and Hall, 2006). Osteoclasts are macrophage lineage-derived, giant multi-nucleated cells, which resorb bone and are required for bone remodelling (Guyton and Hall, 2006).

14.1.1   Collagen

A collagen molecule is generally defined as a protein that contains at least one triple helical domain, which is able to form supramolecular aggregates, and is deposited within the extracellular matrix (ECM) (Kadler *et al.*, 1996; Gelse *et al.*, 2003). Collagens are trimeric molecules composed of three α-chains, which contain the sequence repeat (G-X-Y)n, G being glycine, X frequently proline and Y hydroxyproline (Ricard-Blum *et al.*, 2005). The three α-chains stabilize themselves in a triple helix quaternary structure, which is the characteristic feature of the collagen superfamily (Wess, 2008). The side chains of each X and Y residues are exposed at the surface of the triple helix, giving the collagen molecule a significant capacity for lateral interactions with other molecules of the ECM and resulting in the formation of various supramolecular assemblies (Gelse *et al.*, 2003).

To date, more than 29 types of collagen have been identified. Of these, collagen types I, II, III, V and XI are the only collagens that can be organized in highly ordered supramolecular aggregates (i.e., fibrils) (Kadler *et al.*, 1996). Collagen fibrils have a distinctive suprastructure (with a periodicity of 67 nm, *D*-period) given by the quarter-staggered arrangement of individual collagen monomers in fibril-arrays (around 100 nm in diameter) (Gelse *et al.*, 2003). Within the fibrillar collagen sub-family, type I collagen is considered as the 'construction unit' of the body, forming approximately 95% of the organic mass of bone and is the major component in tendons, ligaments, skin, cornea and vessels, with the exception of cartilage and brain (Kadler, 1995). The type I collagen triple helix is a heterotrimer of two identical α1(I)-chains and one α2(I)-chain (Li, 2000). In tendon and ligament, type I collagen provides tensile strength, while in bone and tooth its mineralization combines strength with toughness and stiffness (Vanderby and Provenzano, 2003). In addition, type I collagen is involved in a plethora of additional biological functions, the accomplishment of which requires the maintenance of the natural triple helical structure of the protein (Prockop and Kivirikko, 1995; Gelse *et al.*, 2003; Ricard-Blum *et al.*, 2005).

14.1.2   Carbonated hydroxyapatite (CHA)

The mammalian mineral phase is a particular type of apatite, namely dahlite (Lowenstam and Weiner, 1989; Landis, 1999). The term apatite indicates a mineral structure with the chemical formula $A_4B_6(MO_4)_6X_2$, where, in vertebrates, both A and B are calcium, MO_4 is a phosphate group and X is a hydroxide ion (Brown and Chow, 1976). Hydroxyapatite $(Ca_{10}(PO_4)_6(OH)_2)$ is the mineral component of bones and teeth. In biological hydroxyapatite, the substitution of lattice ions takes place through surface ionic exchanges with those present in the body fluids (Nancollas and Tomson, 1976). In general, biological apatite contains substantial carbonate substitutions, OH^- deficiencies and other imperfections in the crystal lattice (i.e., presence of magnesium and fluoride ions), making it a low crystalline CHA (Boskey, 2007). These imperfections occur because the body utilizes bone as a reservoir to maintain the homeostasis of several ions (Guyton and Hall, 2006). Substitutions of carbonate for hydroxide are identified as A-type CHA, while carbonate substitutions for phosphate are named as B-type CHA (Barralet *et al.*, 2002). These two substitutions can be distinguished, for example, by Fourier transform infrared (FTIR) spectroscopy (Koutsopoulos, 2002). CHA crystals of bone are very small platelets with irregular and imperfect shapes, which are 30–50 nm long, 20–25 nm wide and 1.5–4 nm thick (Fernandez-Morán and Engström, 1957; Wachtel and Weiner, 1994; Ziv and Weiner, 1994; Hassenkam *et al.*, 2004) and are embedded within the collagen fibrils.

14.2   Collagen as a biomaterial

As a major component of the body, type I collagen has long been considered as a biomaterial with an enormous potential (Ramshaw, 1996; Ramshaw *et al.*, 2009; Mandal *et al.*, 2010). Various forms of collagen are extensively used as medical devices (Meena *et al.*, 1999; John *et al.*, 2001; Lee *et al.*, 2001) such as in bioprosthetic heart valves, arteriovenous shunts, suture threads, haemostatic membranes, wound dressings, tissue augmentation, ophthalmic barriers and drug delivery (Table 14.1) (Ramshaw, 1996; Ramshaw *et al.*, 2009).

In addition, the uses of collagen in numerous tissue engineering applications are currently under investigation (Parenteau-Bareil *et al.*, 2010) as it offers the following advantages (Lee *et al.*, 2001):

* it is abundant and can be easily and cheaply isolated with a high grade of purity (medical grade >99% purity);
* it can be processed and reconstituted in various forms useful in surgery;
* it can be hybridized with other materials to tailor their properties;
* it can be processed into tissue-equivalent constructs; and
* its products have been approved as a safe material for clinical use by the major health agencies (United States Food and Drug Administration, (FDA) and European Medicines Agency, (EMA)).

Table 14.1 Commercially available collagen-based matrices

Product	Company	Source	Application
Alloderm®	LifeCell	Human skin	Dermis repair
Apligraf®	Organogenesis	Human neonatal male foreskin	Dermis repair
Allopatch®	Musculoskeletal Transplant Foundation	Human skin	Dermis repair
Axis™ Dermis	Mentor	Human dermis	Dermis repair and female pelvic floor repair
Bard® Dermal Allograft	Bard	Cadaveric human dermis	Dermis repair and female pelvic floor repair
CuffPatch™	Arthrotek	Porcine small intestinal submucosa	Tendon repair
DurADAPT™	Pegasus Biologics	Horse pericardium	Dura mater repair
Dura-Guard®	Synovis Surgical	Bovine pericardium	Dura mater repair
Durasis®	Cook Biotech	Porcine small intestinal submucosa	Dura mater repair
Durepair®	TEI Biosciences	Foetal bovine skin	Dura mater repair
FasLata®	Bard	Cadaveric fascia lata	Soft tissue repair
Graft Jacket®	Wright Medical Technology, Inc.	Human skin	Soft tissue repair
Integra®	Integra Life Science	Bovine tendon	Dermis repair
IntegraOS™	Integra Life Science	Bovine tendon	Bone repair
Oasis®	Healthpoint	Porcine small intestinal submucosa	Wound repair
OrthADAPT™	Pegasus Biologics	Horse pericardium	Soft tissue repair
Pelvicol®	Bard	Porcine Dermis	Pelvic repair
Peri-Guard®	Synovis Surgical	Bovine pericardium	Cardiac, thoracic and soft tissue repair
Permacol™	Tissue Science Laboratories	Porcine skin	Hernia and abdominal wall repair
PriMatrix™	TEI Biosciences	Foetal bovine skin	Dermis repair
Restore®	DePuy Orthopaedics, Inc	Porcine small intestinal submucosa	Soft tissue repair
Stratasis®	Cook Biotech	Porcine small intestinal submucosa	Urethra repair
SurgiMend™	TEI Biosciences	Foetal bovine skin	Soft tissue repair

(*Continued*)

Table 14.1 Continued

Product	Company	Source	Application
SurgiSIS®	Cook Biotech	Porcine small intestinal submucosa	Urethra repair
Suspend™	Mentor	Human fascia lata	Urethra repair
Synergraft®	CryoLife, Inc	porcine heart valve	heart valve
TissueMend®	TEI Biosciences	Foetal bovine skin	Soft tissue repair
Vascu-Guard®	Synovis Surgical	Bovine pericardium	Vascular patch
Veritas®	Synovis Surgical	Bovine pericardium	Soft tissue repair (abdominal wall)
Xelma™	Mölnlycke	ECM proteins	Venous leg ulcers treatment
Xenform™	TEI Biosciences	Foetal bovine skin	Soft tissue repair (urethra)
Zimmer Collagen Patch®	Tissue Science Laboratories	Porcine dermis	Soft tissue repair

Source: Adapted from Badylak *et al.* (2009).

14.2.1 Collagen origin and variability

Type I collagen is generally extracted from collagen-rich tissues in living animals, particularly mammals (Ramshaw *et al.*, 2009). Common sources are human skin, rat-tail tendons, bovine dermis and *ligamentum nuchae*, as well as swine skin and intestine due to the ease of the purification process required to isolate type I collagen from other ECM components (Gilbert *et al.*, 2006; Mandal *et al.*, 2010). The source of collagen is fundamental since collagen properties differ from one animal to another and are influenced by the processing steps that the protein undergoes during extraction (Lin and Liu, 2006). This makes it difficult to compare the results obtained from one source to another, especially in the case of cellular responses to the material. Since human collagen is available at high cost and at a low extent, Fibrogen® Inc has recently started to produce and distribute recombinant human collagen (Yang *et al.*, 2004; Liu *et al.*, 2008). This collagen is potentially less immunogenic than animal sources and, more importantly, its composition is reproducible (Yang *et al.*, 2004; Badylak and Gilbert, 2008). However, this is produced by yeast engineered with human DNA sequences, which lack the fundamental extracellular post-translational modifications that impart the unique properties of the protein. For non-medical applications, cheaper sources of collagen originate from marine life forms such as sponges, fish and jellyfish (Meena *et al.*, 1999). In addition, the fear of transmissible spongiform encephalopathy derived from mammal sources has led to new efforts to obtain collagen derived from marine species as an alternative and safer source of collagen for biomedical applications (Song *et al.*, 2006).

14.2.2 Types of collagen-based biomaterials

Two main approaches are used to produce collagen-based biomaterials: decellularization of ECM and reconstitution of solubilized collagen (Ramshaw, 1996; Lee *et al.*, 2001; Parenteau-Bareil *et al.*, 2010). In the first methodology, tissues generally undergo decellularization and purification to reduce immunogenic responses (Gilbert *et al.*, 2006) and result in an insoluble, intact collagen matrix. In addition, a subsequent chemical fixation may be used to strengthen and increase the durability of the collagenous matrix (Johnson *et al.*, 1999). Physical, chemical or enzymatic techniques are currently used to produce implantable acellular collagen matrices (Gilbert *et al.*, 2006). Physical methods involve snap freezing (Roberts *et al.*, 1991), high pressure (Lin *et al.*, 2004) and mechanical agitation (Dahl *et al.*, 2003) in order to induce cell membrane lysis and tissue removal. Chemical treatments include the use of acid or alkaline solutions (De Filippo *et al.*, 2002; Falke *et al.*, 2003), chelating agents (e.g., ethylenediamine tetra-acetic acid) (Vyavahare *et al.*, 1997; Goissis *et al.*, 2000), ionic or non-ionic (Dahl *et al.*, 2003; Ketchedjian *et al.*, 2005) and zwitterionic detergents (Dahl *et al.*, 2003) to remove the cellular content of ECM.

Enzymatically, ECM can be purified by trypsin (Gamba *et al.*, 2002), exo- and endo-nuclease treatments (Woods and Gratzer, 2005), which specifically cleave proteins and nucleases to remove DNA and RNA. However, while protocols based on a combination of physical, chemical and enzymatic treatments appear to remove most of the cellular material responsible for immunogenic responses, resulting in the production of biologic scaffold materials that are safe for implantation (Gilbert *et al.*, 2006), some concerns are still present in terms of fully removing cellular debris from a tissue. Although the number of failed grafts due to immunogenic rejection is limited, it does not exclude the necessity of immunosuppressive drugs (Allman *et al.*, 2002; Badylak, 2004). In addition, purification processes alter the ECM composition, structure, mechanical properties and cell responses, with significant implications for subsequent *in vitro* and *in vivo* performance (Gilbert *et al.*, 2006). In addition, the properties of decellularized matrices are ultimately dictated by the animal and tissue source; they are not bioengineered. Nevertheless, it is important to underline that in a sector where technology transfer is generally very difficult, a number of naturally occurring ECM-derived matrices and related decellularization protocols have received regulatory approval from the FDA and EMA to be used clinically, including human dermis (Alloderm®, LifeCell, Corp.), porcine intestine (SurgiSIS®, Cook Biotech, Inc.; Restore®, DePuy Orthopaedics, Inc.), porcine urinary bladder (ACell, Inc.) and porcine heart valves (Synergraft®, CryoLife, Inc.).

Collagen-based biomaterials may also be obtained through extraction, purification and solubilization of collagen from tissues (Rajan *et al.*, 2007).

Modern extraction methods are based on three basic principles of solubilization: acid solutions (Steven and Jackson, 1967; Rajan *et al.*, 2007), neutral salt solutions (Gross *et al.*, 1955) and proteolytic solutions (Drake *et al.*, 1966; Miller and Kent Rhodes, 1982). Proteolytic extraction, however, alters the collagen molecular structure by cleaving the terminal telopeptide regions and results in a proportional decrease in tropocollagen self-assembled fibrils (Rubin *et al.*, 1963). To avoid this effect, endogenous proteases can be inhibited during acid solubilization (Miller, 1972). Nonetheless, the acid extraction using a slight pepsin solubilization, is the most effective technique in terms of yield, albeit some telopeptides are cleaved or partially denatured (Miller and Kent Rhodes, 1982). Various protocols have been reported to solubilize collagen and several sources are available in the market (Chandrakasan *et al.*, 1976).

Tropocollagen solutions are used to produce reconstituted collagen by changing the pH, temperature and ionic concentration to physiological levels. Tropocollagen polymerization may also be achieved between 15°C and 40°C and at pH values between 5 and 9.2 (Harris and Reiber, 2007; Li *et al.*, 2009; Achilli *et al.*, 2010). The nanostructured type I collagen gels are reconstituted *in vitro* through the neutralization of dilute acidic solutions of collagen. Adjustment to physiological pH, temperature and ionic concentrations induces fibril formation to create porous, native ECM-like gels. The nano-fibrils, ranging from 30 to 100 nm in diameter consist of staggered collagen molecules assembled end-to-end and laterally. In addition, tropocollagen polymerization in physiological conditions results in a protein concentration of <0.85 wt%, which implies the formation of highly hydrated collagen (HHC) hydrogels with low collagen fibrillar density (CFD) and poor mechanical properties (strength of 4–6 kPa and elastic modulus of 20–30 kPa) (Fig. 14.2a) (Roeder *et al.*, 2002; Boccafoschi *et al.*, 2007).

Several types of collagen constructs are manufactured by post-processing HHC gels to make reconstituted collagen usable as a biomaterial. Of these, cross-linking of the collagen fibrils is the most common route to increasing the mechanical properties, achieving an elastic modulus of up to 5 MPa and the stability of reconstituted collagen constructs (Duan and Sheardown, 2006; Avery and Bailey, 2008). Three types of cross-linking methods have been developed to strengthen collagen constructs: physical, chemical and enzymatic (Wahl and Czernuszka, 2006; Avery and Bailey, 2008). Collagen films or sheets can also be produced by the formation of thin HHC gels, which are then cross-linked and dehydrated in a desiccator (Lu *et al.*, 2007). These have been used as drug delivery systems in the treatment of tissue infection, such as cornea or in liver cancer, as a corneal shield or in the engineering of skin and cornea (Lee *et al.*, 2001; Ruszczak and Friess, 2003).

Collagen sponges are the most successful regenerated-collagen based biomaterials and they can be manufactured in different ways, such as

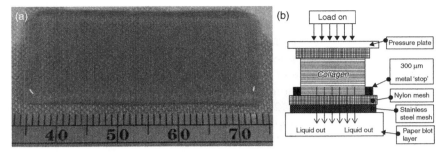

14.2 Plastic compression technique. (a) As made highly hydrated collagen (HHC) gel. (b) Schematic of plastic compression technique applied to HHC gel to obtained denser constructs with higher stability and enhanced mechanical properties. Cast HHC gels are placed on a porous support consisting of (bottom to top) absorbent paper blot layers, a stainless steel mesh and a polymer mesh and a compressive stress is applied for 5 min to expel the excess water from the collagen gel. (*Source*: Reproduced from Brown *et al.* (2005). Copyright © 2005 WILEY-VCH.)

freeze-drying of cross-linked HHC gels containing porogenic material (Chvapil, 1977). These are commercially available for different applications (e.g., BICOL® Collagen Sponge, DePuy Orthopaedics, Inc.). Collagen sponges have been very useful in the treatment of severe burns and as wound dressings, such as pressure sores, donor sites, and leg and decubitus ulcers, as well as for *in vitro* test systems (Ramshaw, 1996; Meena *et al.*, 1999; Lee *et al.*, 2001). They are also used in drug delivery, for example as *in situ* short-term delivery (3–7 days) of antibiotics (Friess, 1998). In addition, collagen sponges have been proposed as scaffolds for tissue engineering applications (e.g., small intestine tissue engineering) (Nakase *et al.*, 2006).

Despite a plethora of techniques to increase collagen stability, none can guarantee a homogenous increase in the collagen construct stability and mechanical properties without the side effects of cytotoxicity and prolonged time lag in preparation. Recently, Brown *et al.* (2005) developed a processing technique based on plastic compression (PC), to rapidly fabricate dense collagen (DC) gels, which mimic the ECM fibrillar density, microstructure and biological properties (Fig. 14.2b).

Through PC, DC is achieved within a 5 min window by simply applying a constant stress (approximately 1 kPa) on top of the HHC gel in order to expel the excess casting fluid. The mechanical properties of the matrices are significantly increased achieving a tensile strength value of 0.6 MPa and an elastic modulus of 1.5 MPa via the controlled increase in nano-fibrillar density to greater than 10 wt% (Abou Neel *et al.*, 2006). Ghezzi *et al.* (2011) have shown that PC does not perturb seeded cells within DC scaffolds and DC supports an increase in cell proliferation and distribution over time relative to HHC gels. Bitar *et al.* (2007, 2008) assessed the suitability of DC scaffolds for bone tissue engineering purposes. Osteoid-like DC scaffolds seeded with human osteosarcoma cells (ATCC HOS TE85) were capable of maintaining cell viability and function for up to 10 days, with preservation of the scaffold initial mechanical properties. Furthermore, Buxton *et al.* (2008) cultivated primary pre-osteoblasts in the presence of osteogenic supplements for 14 days, resulting in their mineralization. More recently, Pedraza *et al.* (2010) assessed the *in vitro* cellular biomineralization of PC collagen scaffolds with MC3T3-E1 murine calvarial osteoblasts cultivated up to 7 weeks (Fig. 14.3a). Furthermore, DC gels have been implemented as an *in vitro* model, in order to contribute to the understanding of collagen mineralization. In fact, the organization of collagen in mineralized tissues provides a hierarchically structured framework in which nanoscopic CHA crystals are nucleated, embedded and organized within the collagen nano-fibrils structure (Rho *et al.*, 1998). By applying PC, Marelli *et al.* (2011) produced biomimetic nano-fibrillar collagen gels with tailored properties (Plate XV, see colour section between pages 264 and 265), resulting in a microenvironment more favourable to crystal formation, due to the greater compacted nano-fibrillar structures and, consequently, altered gel electrostatic properties (Marelli *et al.*, 2011b).

14.2.3 Cell-mediated remodelling and biodegradability of collagen

The relationship between cells and the scaffold where they are situated is bidirectional: the mechanical, chemical and biological properties of the construct influence cell gene expression and metabolism, while cells reorganize their extracellular space (Fig. 14.3a) (Ma *et al.*, 2008; Geckil *et al.*, 2010; Pedraza *et al.*, 2010). Collagenous scaffold environments provide seeded cells with biochemical and mechanical cues to direct their function. Seeded cells use proteolytic enzymes (e.g., metalloproteases) to remodel the collagenous matrix (Birkedal-Hansen, 1995). These enzymes facilitate cells to regulate their relationship with the environment by directly cleaving structural macromolecules of the ECM and consequently determining a mechanical breakdown (Birkedal-Hansen, 1995). At the same time, cells produce new collagen, which substitute the matrix (Birkedal-Hansen, 1995), resulting in a distortion of the original scaffold geometry (Fig. 14.3b) (Mio *et al.*, 1998; Serpooshan *et al.*, 2010). *In vivo*, the bidirectional remodelling of the ECM during morphogenetic events is required for structure development and function of the forming tissue, but represents a big drawback in the use of collagen as biomaterials, since the structural properties of the original construct may be compromised before the formation of the

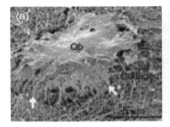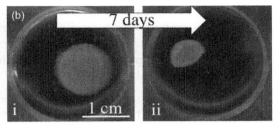

14.3 Cell-induced remodelling of collagen gels. (a) Electron micrograph of a single cell embedded within the collagen matrix. White arrows indicate osteoblast (Ob) cell processes. (*Source*: Reproduced from Pedraza *et al.* (2010). Copyright © 2010 Mary Ann Liebert, Inc.) (b) Effects of NIH/3T3 fibroblast cell-induced remodelling on collagen gel geometry: (i) As made gel, and (ii) at day 7 in culture.

new tissue or the completion of the healing processes (Ramshaw, 1996). This phenomenon is particularly evident in the use of collagen gel and is decelerated by the use of a dense gel or by the stabilization of the matrix through cross-linking (Zhu *et al.*, 2001; Brinkman *et al.*, 2003; Serpooshan *et al.*, 2010; Ghezzi *et al.*, 2011). At the same time, collagen biodegradability is an attractive aspect for collagen-based biomaterials and tissue engineering applications, as it constitutes a fundamental step toward the formation of a new tissue (Parenteau-Bareil *et al.*, 2010).

In vivo, collagen is digested by proteolytic enzymes such as collagenases (Holton *et al.*, 1983; Mott and Werb, 2004). Collagen biodegradability may also be reproduced *in vitro* with *ad hoc* protocols, in order to be studied and predicted (Holton *et al.*, 1983). In addition, the degradation product of type I collagen has also been shown to induce a chemotactic attraction of human fibroblasts (Postlethwaite *et al.*, 1978).

14.3 Approaches to the mineralization of collagenous constructs

Several attempts have been made in order to achieve the *in vitro* acellular mineralization of collagenous constructs for potential bone tissue engineering purposes. Herein, a summary of the most recent attempts is given. The rationale of all of these strategies has been to obviate the poor mineralization properties of collagenous structures. Generally, collagen does not mineralize *in vivo* (Habibovic *et al.*, 2010) and the hybridization of collagenous constructs with supraphysiological doses of recombinant proteins (e.g., NCPs) has been associated with high costs (Guyer *et al.*, 2007) and potential health risks (Shields *et al.*, 2006). Attention has therefore been focused on either the phosphorylation of collagen (e.g., with sodium trimetaphosphate)

(Liu *et al.*, 2011a), incorporating bioactive ceramics in the collagen matrix such as hydroxyapatite (HA) and bioactive glass particles (Marelli *et al.*, 2010, 2011a) in order to develop *ad hoc* mineralizing solutions, or to incorporate functional groups in the collagen macromolecule, in order to increase its potential as a crystal nucleator.

14.3.1   Incorporation of hydroxyapatite

The most immediate approach to achieve collagen mineralization is the incorporation of HA crystals within the collagenous framework. Although it is possible to achieve a fast and extensive mineralization, the structure of the resulting collagen-HA hybrid is not biomimetic and the presence of slowly degrading nanoparticles may be toxic to cells and the host. The following methods have been proposed.

Microsphere of collagen-hydroxyapatite

Yin Hsu *et al.* (1999) have developed microspheres composed of particulate HA dispersed in a fibrous collagen matrix (type I, extracted from rat-tail tendon) and tested their biological behaviour with rat derived osteoblasts. The microspheres were formed by dispersing HA powder in a collagen solution, which was exposed to an olive oil-water emulsion to form micro-beads. Due to the density of the gel, the beads were retained at the interface between oil and water, forming spherical droplets with a diameter ranging from 75 to 300 μm. Osteoblastic phenotype was evaluated by measuring alkaline phosphatase (ALP) activity. An increase in ALP activity with incubation time (up to day 14) was registered. Further analyses studied the microstructure of the spheres and the osteoblastic response to this material (Wu *et al.*, 2004). The collagen nanofibres were 30–90 nm in diameter and the HA particles had an average diameter of 5 μm. Moreover, confocal microscopy images coupled with three different stains showed that the osteoblast cells grew confluent on the microspheres and displayed prominent focal adhesion via stress fibres. DNA staining demonstrated the presence of cellular mitosis of the osteoblasts grown on microspheres. However, the mechanical properties of the microspheres were not investigated. Thus, it is only possible to hypothesize their potential use can be as carriers for osteoblasts, but not as bone-filling material.

Thermally triggered liposome

A cutting edge attempt to mineralize collagen matrix was proposed by Pederson *et al.* (2003) using thermally triggered liposome. Calcium- and phosphate-loaded liposomes were prepared by using the interdigitation-fusion method (Ahl *et al.*, 1994). Liposome lipid chains were engineered to become softer and permeable to calcium and phosphate species at 35–37°C. Liposomes were added to the solution used to polymerize collagen (type I, lyophilized and then solved in acetic acid) at pH 7.4. As an *in vitro* test, the liposome/collagen mixture was placed in an incubator at 35°C to induce collagen self-assembly and then at 37°C to induce its mineralization. The scanning microcalorimetry performed on the liposome/collagen mixture revealed that the liposomal mineralization in the presence of collagen was an endothermic reaction, probably indicating the formation of amorphous calcium phosphate (ACP) as initial reaction product. However, there was no study on cellular interactions with the liposomal mineralized collagen scaffold.

Critical point drying and solid free form (SFF)

Solid free form (SFF) is a scaffolding technique used to control construct properties such as internal microstructure, porosity, mechanical properties and biodegradation. Wahl and co-workers used a multi-stage technique for SFF (Wahl *et al.*, 2007). At first, biocompatible moulds were generated with different external shapes and internal microstructures using SFF fabrication method. Secondly, different type I collagen (from Achilles tendon) and HA particle suspensions were poured in the moulds, frozen at −30 or −80°C, dehydrated in ethanol and dried with critical point drying (CPD) using liquid CO_2 in order to create dry porous scaffolds. Then, a dehydrothermal treatment (DHT, dehydration of the constructs for three days at 120°C under vacuum) was applied to cross-link the collagen fibres. The morphological characterization (by SEM) of these scaffolds (30 wt% collagen–70 wt% HA) indicated a 300 µm pore size structure, which was considered as sufficient for the osteoblastic infiltration. The SFF fabrication also permitted the formation of microchannels ($\oslash$ = 135 µm) within the scaffolds. Although CPD did not alter the collagen triple helix, DHT treatment heavily degraded the protein, showing the major limitation of this technique.

Electrospinning of collagen-HA solution

Electrospinning (ES) is a well-known technique used to obtain nano-sized fibres (Wahl *et al.*, 2007). ES forces a polymeric solution through a needle by a syringe-pump via the application of a high voltage electric potential (range of kV). The consequent acceleration causes the evaporation of the solvent and the formation of nanoscopic fibres, which are collected onto a grounded collector-plate. Venugopal *et al.* (2008) have used this technique to form electrospun collagen/nano-HA matrices. Although the use of ES with type I collagen appears pointless since the collagen matrix is already composed of fibres with an average diameter of 100 nm, the proliferation of human foetal osteoblast was demonstrated for 10 days. However, this time period was too short to investigate the possible cytotoxic effect of the HA nanoparticles. Furthermore, no study was conducted on the denaturation effect of ES on collagen protein (a phenomenon often observed in the literature) (Zeugolis *et al.*, 2008).

14.3.2   Incorporation of bioactive glass

Bioactive glasses (BGs) are considered an important category of materials used in bone regeneration attributable to their osteoinductive and osteoconductive properties (Jones *et al.*, 2007). The incorporation of BG in collagenous matrices has therefore been proposed as a route to produce scaffolds for bone tissue engineering applications. In 1995, an isolated attempt was made to implant a hydrogel composed of collagen and 45S5 Bioglass® particles (Pohunkovà and Adam, 1995). Marelli *et al.* (2010) hybridized dense nano-fibrillar collagen scaffolds and 45S5 Bioglass® to produce scaffolds for bone regeneration therapy. The plastic compression technique was used to rapidly produce the constructs, while the presence of 45S5 Bioglass® within the dense collagenous 'osteoid-like' environment resulted in an accelerated three-dimensional mineralization of the matrix when exposed to an *in vitro* physiological environment. Hybrid scaffolds biomineralized as early as day 3 in simulated body fluid (SBF) (Fig. 14.4, Plate XVI).

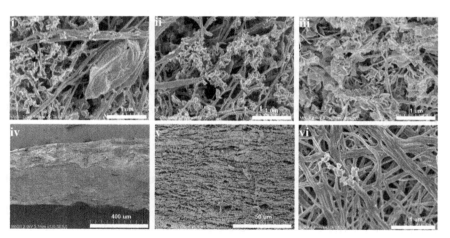

14.4 SEM micrographs of DC–μ BG 40–60 hybrid scaffolds and DC. Images (i), (ii), (iii) refer to days 3, 7 and 14 in SBF, respectively (iv,v) SEM micrographs of the cross section of DC-μ BG 40–60 at day in SBF, confirming the presence of pores (dimensional = 5–10 μm). (vi) SEM micrograph of the DC after 14 days in SBF. (*Source*: Copyright © 2010 American Chemical Society (ACS).)

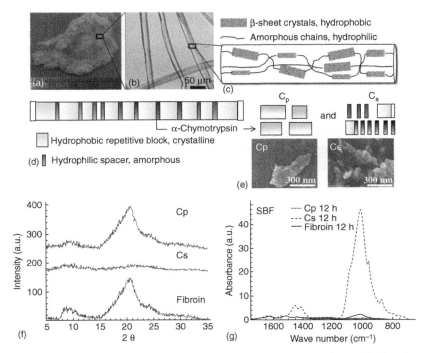

14.5 Multi-scale structure of SF and preparation of Cp and Cs fractions.
(a) SF extracted from *Bombyx Mori* cocoons; (b) optical microscopy
image of SF fibres; (c) each fibre is composed of repeated hydrophobic
crystalline (β-sheets) polypeptides linked together by hydrophilic
amorphous amino-acidic sequences; (d) α-chymotrypsin was used to
selectively separate the crystalline (Cp) and the amorphous (Cs) regions
of the protein; (e) SEM micrographs of Cp and Cs fragments; (f) X-ray
diffraction (XRD) diffractograms of SF, Cs and Cp fragments. SF and Cp
fragments revealed a crystalline structure, while Cs fragments were
amorphous in nature; (g) Attenuated total reflectance (ATR)-FTIR spectra
of SF, Cp and Cs at 12 h in SBF. Cs fragments templated the formation
of carbonated apatite as their spectra were characterized by strong
carbonate (1440 and 872 cm⁻¹) and phosphate (1078, 1035 and 957 cm⁻¹)
absorptions. SF and Cp fragments exhibited only minor or no indications
of apatite formation, respectively. (*Source*: Copyright © 2012 Elsevier Ltd.)

Through tensile mechanical analysis, it was possible to relate the progression of mineralization to changes in scaffold mechanical properties,
which resulted in a ductile-to-brittle transition behaviour, through an
increase in the elastic modulus and a corresponding decrease in the strain
at Uniaxial Tensile Strength (UTS). In addition, the hybridization of DC
gels with nanosized BGs resulted in an accelerated osteoblastic differentiation of pre-osteoblast homogenously seeded at the point of collagen fibrillogenesis in the hybrid gels (Marelli *et al.*, 2011a). With a similar approach,

Xu and co-workers incorporated 20% by weight of phosphatidylserine in freeze-dried collagen disks to enucleate calcium phosphate while in the presence of 45S5 bioactive glass (Xu *et al.*, 2011). They showed that the hybrid scaffold was bioactive and supported bone regeneration in a rat model.

14.3.3   Incorporation of fibroin derived polypeptides

Silk fibroin (SF) is an intriguing biomaterial with outstanding mechanical properties and good biocompatibility (Vepari and Kaplan, 2007). It has been shown that SF can direct the formation of CHA nanocrystals, but only when taken outside of physiological boundaries. In this scenario, it has been reported that the chymotryptic digestion of SF generates soluble polypeptides (Cs) with characteristics similar to the NCPs found in the mineralizing osteoid milieu (Marelli *et al.*, 2012). When immersed in SBF, Cs fragments demonstrated the formation of carbonated apatite, suggesting their potential in bone regenerative medicine (Fig. 14.5).

The potential of Cs to conceptually mimic the role of anionic non-collagenous proteins in biomineralization processes was further investigated via their incorporation (up to 10% by weight) in bulk osteoid-like dense collagen (DC) gels. Within 6 h in SBF, apatite was formed in DC–Cs hybrid gels, and by day 7, CHA crystals were extensively formed. This accelerated three-dimensional (3D) mineralization resulted in a nine-fold increase in the compressive modulus of the hydrogel.

14.3.4   Supersaturated calcium phosphate solutions

An alternative strategy to induce apatite formation within the collagen matrix is the soaking of the scaffold in supersaturated calcium phosphate solutions. Different techniques have been used in this strategy.

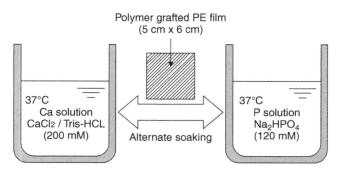

14.6 Schematic of the alternate soaking process of a collagen matrix in calcium and phosphate solutions. (*Source*: Adapted from Gòes *et al.* (2007). Copyright © 2007 Elsevier Ltd.)

Alternate soaking process

Gòes *et al.* (2007) evaluated the formation of apatite on hydrolyzed and natural type I collagen matrix alternately soaked (100 times) in $CaCl_2$ and K_2HPO_4 solutions at 25°C for 120 s (Fig. 14.6) (Taguchi *et al.*, 2000). An extensive mineral phase (60 wt%) was rapidly formed, but only on the surface of the matrix. In addition, FTIR revealed the altered collagen triple helical structure due to the sudden and repeated pH changes during the alternate soaking.

Simultaneous titration

Kikuchi and co-workers tried to reconstruct a bone-like nanostructure using a simultaneous titration system. An aqueous suspension of $Ca(OH)_2$ and an aqueous solution of H_3PO_4 with the addition of collagen (from porcine dermis) were stored in two different reservoirs (Fig. 14.7) (Kikuchi *et al.*, 2004). By using two tube pumps, the solutions were mixed in a bath at 37°C, maintaining the pH at 9.0. The process led to the formation of a collagen–HA matrix, where the *c*-axes of the HA nanocrystals were aligned along the collagen fibres. However, the *in vivo* testing of the constructs was unsuccessful; after 2 weeks, the debris of the composites was encapsulated by fibrotic tissue and macrophages were identified in the implant *situ*.

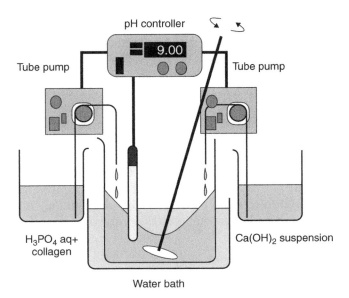

14.7 Schematic of the apparatus used for simultaneous titration. (*Source*: Reproduced from Kikuchi *et al.* (2004). Copyright © 2004 Elsevier Ltd.)

Dual membrane diffusion method

In a recent study, Ehrlich *et al.* (2009) used a dual diffusion membrane system to orient crystal growth of octacalcium phosphate (OCP)/HA on a collagen matrix. The method was based on the system where a cation-selective membrane and an anion-selective dialysis membrane were used to control ionic diffusion into a small reaction volume. Crystal growth experiments were carried out at 37°C for 7 days using $CaCl_2$ and NaH_2PO_4 solutions. Mineralization was extensive but superficial.

Simulated body fluid

In 1991, Kokubo proposed a method to reproduce the *in vivo* apatite formation *in vitro* by using a simulated body fluid (SBF) with ionic concentrations nearly equal to those of human blood plasma (Kokubo, 1991). During the last two decades, SBF has not only been used to study the bioactivity of materials, but also to mineralize scaffolds for bone tissue engineering purposes (Kokubo and Takadama, 2006). In the literature there is a lively interest in the use of SBF to mineralize different scaffolds (natural and artificial polymers) and different recipes have also been proposed (Kokubo and Takadama, 2006; Muller and Muller, 2006). However, the Kokubo recipe is still considered valid and the proposed modifications seem only to tailor the solution to better interact with the studied substrate. More recently, while Bohner and Lemaitre (2009) have questioned the use of SBF to investigate the bioactivity of materials, no alternative solutions have been proposed and SBF may still be regarded as a standard method to study acellular mineralization processes.

The soaking of a collagenous construct in SBF can be considered as a bivalent technique, which investigates the bioactivity and mineralizes the scaffolds. However, only a few studies have been conducted to investigate collagen mineralization in SBF. Zhang and co-workers used 1.5 × concentrated SBF to mineralize slices of commercial haemostatic collagen sponges for 7 days (Zhang *et al.*, 2004) and qualitatively showed the mineralization of the sponges with HA formation. Honda *et al.* (2007) conditioned collagen/OCP hybrid in SBF for 10 days, and showed a higher extent of mineralization when compared to collagen alone, probably through the OCP-apatite conversion. However, no HA formation was observed.

14.3.5   Collagen functionalization

Chelant agent

Rhee and Tanaka (1999) have proposed to use citric acid to increase the nucleation of HA within a collagen membrane where SBF was used as the mineralizing environment. The rationale was to bind citric acid to collagen and to use its negative charge to chelate calcium ions. Then, a critical size cluster can be formed by adsorbing calcium and phosphate ions, and/or another citric acid molecules, which acts as a nucleus for HA crystals. Although FTIR and the X-ray diffraction analysis suggested the formation of HA on the collagen membrane, no biological study was conducted.

Acidified collagen

It has been shown that the introduction of negatively charged groups on collagen molecules provides nucleation sites for HA, due to its close similarity with the *in vivo* role of anionic NCPs. Several groups have proposed acidifying collagen molecules (Liu *et al.*, 2011b). For example, collagen phosphorylation has been shown to be an attractive option (Li and Chang, 2008). It is usually achieved with the same method used for the phosphorylation of food (usually soy-derived) proteins. Briefly, collagen is dispersed in a Na_2HPO_4 solution at pH 9.0. Then, sodium trimetaphosphate is added and the solution is kept for 5 h at room temperature. The mixture is then dialyzed in distilled water to remove excess ions such as Na^+, $P_2O_7^{4-}$ and $P_3O_{10}^{5-}$. Although the efficacy of the method has been largely demonstrated, phosphorylation requires the processing of collagen outside of physiological conditions for long time periods, with consequent alteration of its structure.

The enrichment of collagen with poly-l-aspartic acid (polyAsp) is another possibility (Deshpande and Beniash, 2008). Bradt *et al.* (1999) carried out mineralization of collagen fibrils in the presence of polyAsp involving simultaneous assembly of collagen fibrils and calcium phosphate precipitation. The resulting attachment of needle-shaped crystals to collagen fibrils, did not show any specific orientation with respect to collagen fibres. By using a similar methodology, but with a different buffer solution, Zhang and co-workers were able to form HA crystals on the collagen fibrils, oriented along the fibril axis (Zhang *et al.*, 2003). In addition, Olszta *et al.* (2007) reported the mineralization of collagen fibrils via homogeneous precipitation of amorphous calcium phosphate from a highly supersaturated solution followed by polyAsp induced crystallization. Collagen fibrils were mineralized with the crystalline *c*-axes oriented along the axis of the fibrils. Polyvinylphosphonic acid (PVPA) (Tay and Pashley, 2008; Kim *et al.*, 2010; Gu *et al.*, 2011) and polyacrylic acid (PAA) (Tay and Pashley, 2008) have also been investigated as anionic matrix proteins analogues, with results similar to those obtained with polyAsp. Although promising, all of these results have been achieved in two-dimensional (2D) models and their homogeneous reproduction in 3D represents the next step of the research.

Enzymatic approach

In vivo, collagen mineralization is enzymatically triggered and catalysed. Therefore, the same enzymes could be used *in vitro* to enhance and influence HA formation within the collagen fibres. Tomomatsu and co-workers prepared a film of collagen/calcium phosphate by treating the collagen with ALP in an aqueous phase. $CaCl_2$ and Na_2HPO_4 solutions were used to induce the formation of HA crystals. MC3T3 pre-osteoblast cell line differentiation was promoted in osteogenic medium (Tomomatsu *et al.*, 2008). Moreover, the group recently extended their work by developing a collagen/CaP multilayered sheet (Yamauchi *et al.*, 2004).

Baht *et al.* (2008) have recently investigated the effect of anchoring bone sialoprotein (BSP) to a collagen matrix. The results were very encouraging since BSP revealed a high affinity to the triple helixes of fibrillar collagen and acted as a nucleator of HA, once perfused at 37°C for 5 days with buffers containing $Ca(NO_3)_2$ or Na_2HPO_4.

14.4   Conclusion

The *in vitro* replica of a homogenously mineralized three-dimensional nano-fibrillar collagenous matrix is actively being sought in order to engineer a bone-like tissue to be used in bone regenerative therapy. Although numerous strategies have been developed to pursue this aim, only recently has the *in vitro* design of a collagenous tissue-like construct ready to be implanted been achieved. The maintenance of the nanofibrillar, triple-helical structure of native collagen with a collagen fibrillar density within the physiological values has been proven effective in comparison to freeze-dried scaffolds. In addition, the hybridization of dense collagenous constructs with micron- and nano-sized particles that inoculate calcium phosphates, and anionic fibroin derived polypeptides, which mimic the role of NCPs in native osteoid, both resulted in a rapid and extensive mineralization of the dense collagenous constructs upon conditioning in simulated body fluids.

14.5   References

UniProt Database [Online]. Available: http://www.uniprot.org/ [Accessed April 15th (2011)].

Abou Neel, E. A., Cheema, U., Knowles, J. C., Brown, R. A. and Nazhat, S. N. (2006). Use of multiple unconfined compression for control of collagen gel scaffold density and mechanical properties. *Soft Matter*, **2**, 986–992.

Achilli, M., Lagueux, J. and Mantovani, D. (2010). On the effects of UV-C and pH on the mechanical behavior, molecular conformation and cell viability of collagen-based scaffold for vascular tissue engineering. *Macromolecular Bioscience*, **10**, 307–316.

Ahl, P. L., Chen, L., Perkins, W. R., Minchey, S. R., Boni, L. T., Taraschi, T. F. and Janoff, A. S. (1994). Interdigitation-fusion: A new method for producing lipid vesicles of high internal volume. *Biochimica et Biophysica Acta (BBA) – Biomembranes*, **1195**, 237–244.

Allman, A. J., Mcpherson, T. B., Merrill, L. C., Badylak, S. F. and Metzger, D. W. (2002). The Th2-restricted immune response to xenogeneic small intestinal submucosa does not influence systemic protective immunity to viral and bacterial pathogens. *Tissue Engineering*, **8**, 53–62.

Avery, N. C. and Bailey, A. J. (2008). Restraining cross-links responsible for the mechanical properties of collagen fibers: Natural and artificial. In: FRATZL, P. (ed.) *Collagen: Structure and Mechanics*. New York, NY: Springer LCC.

Badylak, S. F. (2004). Xenogeneic extracellular matrix as a scaffold for tissue reconstruction. *Transplant Immunology*, **12**, 367–377.

Badylak, S. F., Freytes, D. O. and Gilbert, T. W. (2009). Extracellular matrix as a biological scaffold material: Structure and function. *Acta Biomaterialia*, **5**, 1–13.

Badylak, S. F. and Gilbert, T. W. (2008). Immune response to biologic scaffold materials. *Seminars in Immunology*, **20**, 109–116.

Baht, G. S., Hunter, G. K. and Goldberg, H. A. (2008). Bone sialoprotein-collagen interaction promotes hydroxyapatite nucleation. *Matrix Biology*, **27**, 600–608.

Barralet, J. E., Aldred, S., Wright, A. J. and Coombes, A. G. A. (2002). In vitro behavior of albumin-loaded carbonate hydroxyapatite gel. *Journal of Biomedical Materials Research*, **60**, 360–367.

Birkedal-Hansen, H. (1995). Proteolytic remodeling of extracellular matrix. *Current Opinion in Cell Biology*, **7**, 728–735.

Bitar, M., Brown, R. A., Salih, V., Kidane, A. G., Knowles, J. C. and Nazhat, S. N. (2008). Effect of cell density on osteoblastic differentiation and matrix degradation of biomimetic dense collagen scaffolds. *Biomacromolecules*, **9**, 129–135.

Bitar, M., Salih, V., Brown, R. A. and Nazhat, S. N. (2007). Effect of multiple unconfined compression on cellular dense collagen scaffolds for bone tissue engineering. *Journal of Materials Science: Materials in Medicine*, **18**, 237–244.

Boccafoschi, F., Rajan, N., Habermehl, J. and Mantovani, D. (2007). Preparation and characterization of a scaffold for vascular tissue engineering by direct-assembling of collagen and cells in a cylindrical geometry. *Macromolecular Bioscience*, **7**, 719–726.

Bohner, M. and Lemaitre, J. (2009). Can bioactivity be tested in vitro with SBF solution? *Biomaterials*, **30**, 2175–2179.

Boskey, A. L. (1989). Noncollagenous matrix proteins and their role in mineralization. *Bone and Mineral*, **6**, 111–123.

Boskey, A. L. (2007). Mineralization of bones and teeth. *Elements*, **3**, 385–391.

Bradt, J.-H., Mertig, M., Teresiak, A. and Pompe, W. (1999). Biomimetic mineralization of collagen by combined fibril assembly and calcium phosphate formation. *Chemistry of Materials*, **11**, 2694–2701.

Brinkman, W.T., Nagapudi, K., Thomas, B. S. and Chaikof, E. L. (2003). Photo-cross-linking of type I collagen gels in the presence of smooth muscle cells: Mechanical properties, cell viability, and function. *Biomacromolecules*, **4**, 890–895.

Brown, R. A., Wiseman, M., Chuo, C. B., Cheema, U. and Nazhat, S. N. (2005). Ultrarapid engineering of biomimetic materials and tissues: Fabrication of nano- and microstructures by plastic compression. *Advanced Functional Material*, **15**, 1762–1770.

Brown, W. E. and Chow, L. C. (1976). Chemical properties of bone-mineral. *Annual Review of Materials Science*, **6**, 213–236.

Buxton, P. G., Bitar, M., Gellynck, K., Parkar, M., Brown, R. A., Young, A. M., Knowles, J. C. and Nazhat, S. N. (2008). Dense collagen matrix accelerates osteogenic differentiation and rescues the apoptotic response to MMP inhibition. *Bone*, **43**, 377–385.

Chandrakasan, G., Torchia, D. A. and Piez, K. A. (1976). Preparation of intact monomeric collagen from rat tail tendon and skin and the structure of the nonhelical ends in solution. *Journal of Biological Chemistry*, **251**, 6062–6067.

Chvapil, M. (1977). Collagen sponge: Theory and practice of medical applications. *Journal of Biomedical Materials Research*, **11**, 721–741.

Currey, J. D. (2002). The structure of bone tissue. In J. D. Currey (ed.) *Bones: Structure and Mechanics*. Princeton: Princeton University Press.

Dahl, S. L. M., Koh, J., Prabhakar, V. and Niklason, L. E. (2003). Decellularized native and engineered arterial scaffolds for transplantation. *Cell Transplantation*, **12**, 659–666.

De Filippo, R. E., Yoo, J. J. and Atala, A. (2002). Urethral replacement using cell seeded tubularized collagen matrices. *The Journal of Urology*, **168**, 1789–1793.

Deshpande, A. S. and Beniash, E. (2008). Bio-inspired synthesis of mineralized collagen fibrils. *Crystal Growth Design*, **8**, 3084–3090.

Drake, M. P., Davison, P. F., Bump, S. and Schmitt, F. O. (1966). Action of proteolytic enzymes on tropocollagen and insoluble collagen. *Biochemistry*, **5**, 301–312.

Duan, X. and Sheardown, H. (2006). Dendrimer crosslinked collagen as a corneal tissue engineering scaffold: Mechanical properties and corneal epithelial cell interactions. *Biomaterials*, **27**, 4608–4617.

Ehrlich, H., Hanke, T., Born, R., Fischer, C., Frolov, A., Langrock, T., Hoffmann, R., Schwarzenbolz, U., Henle, T., Simon, P., Geiger, D., Bazhenov, V. V. and Worch, H. (2009). Mineralization of biomimetically carboxymethylated collagen fibrils in a model dual membrane diffusion system. *Journal of Membrane Science*, **326**, 254–259.

Falke, G., Yoo, J. J., Kwon, T. G., Moreland, R. and Atala, A. (2003). Formation of corporal tissue architecture in vivo using human cavernosal muscle and endothelial cells seeded on collagen matrices. *Tissue Engineering*, **9**, 871–879.

Feng, J. Q., Ward, L. M., Liu, S., Lu, Y., Xie, Y., Yuan, B., Yu, X., Rauch, F., Davis, S. I., Zhang, S., Rios, H., Drezner, M. K., Quarles, L. D., Bonewald, L. F. and White, K. E. (2006). Loss of DMP1 causes rickets and osteomalacia and identifies a role for osteocytes in mineral metabolism. *Nature Genetics*, **38**, 1310–1315.

Fernandez-Morán, H. and Engström, A. (1957). Electron microscopy and X-ray diffraction of bone. *Biochimica et Biophysica Acta*, **23**, 260–264.

Friess, W. (1998). Collagen – biomaterial for drug delivery. *European Journal of Pharmaceutics and Biopharmaceutics*, **45**, 113–136.

Gamba, P. G., Conconi, M. T., Lo Piccolo, R., Zara, G., Spinazzi, R. and Parnigotto, P. P. (2002). Experimental abdominal wall defect repaired with acellular matrix. *Pediatric Surgery International*, **18**, 327–331.

Geckil, H., Xu, F., Zhang, X. H., Moon, S. and Demirci, U. (2010). Engineering hydrogels as extracellular matrix mimics. *Nanomedicine*, **5**, 469–484.

Gelse, K., Pöschl, E. and Aigner, T. (2003). Collagens – Structure, function, and biosynthesis. *Advanced Drug Delivery Reviews*, **55**, 1531–1546.

Ghezzi, C. E., Muja, N., Marelli, B. and Nazhat, S. N. (2011). Real time responses of fibroblasts to plastically compressed fibrillar collagen hydrogels. *Biomaterials*, **32**, 4761–4772.

Gilbert, T. W., Sellaro, T. L. and Badylak, S. F. (2006). Decellularization of tissues and organs. *Biomaterials*, **27**, 3675–3683.

Glimcher, M. J. (2006). Bone: Nature of the calcium phosphate crystals and cellular, structural, and physical chemical mechanisms in their formation. *Reviews in Mineralogy and Geochemistry*, **64**, 223–282.

Gòes J. C., Figueirò S. D., Oliveira A. M., Macedo A. A. M., Silva C. C., Ricardo N. M. P. S. and Sombraa A. S. B. (2007). Apatite coating on anionic and native collagen films by an alternate soaking process. *Acta Biomaterialia*, **3**, 773–778.

Goissis, G., Suzigan, S., Parreira, D. R., Maniglia, J. V., Braile, D. M. and Raymundo, S. (2000). Preparation and characterization of collagen-elastin matrices from blood vessels intended as small diameter vascular grafts. *Artificial Organs*, **24**, 217–223.

Gross, J., Highberger, J. H. and Schmitt, F. O. (1955). Extraction of collagen from connective tissue by neutral salt solutions. *Proceedings of the National Academy of Sciences*, **41**, 1–7.

Gu, L. S., Kim, Y. K., Liu, Y., Takahashi, K., Arun, S., Wimmer, C. E., Osorio, R., Ling, J. Q., Looney, S. W., Pashley, D. H. and Tay, F. R. (2011). Immobilization of a phosphonated analog of matrix phosphoproteins within cross-linked collagen as a templating mechanism for biomimetic mineralization. *Acta Biomaterialia*, **7**, 268–277.

Guyer, R. D., Tromanhauser, S. G. and Regan, J. J. (2007). An economic model of one-level lumbar arthroplasty versus fusion. *The Spine Journal: Official Journal of the North American Spine Society*, **7**, 558–562.

Guyton, A. C. and Hall, J. E. (2006). Parathyroid hormone, calcitonin, calcium and phosphate metabolism, vitamin D, bone, and teeth. In: INC., E. (ed.) *TextBook of Medical Physiology*. 11th ed. Philadelphia: W.B. Saunders.

Habibovic, P., Bassett, D. C., Doillon, C. J., Gerard, C., McKee, M. D. and Barralet, J. E. (2010). Collagen biomineralization in vivo by sustained release of inorganic phosphate ions. *Advanced Materials*, **22**, 1858–1862.

Harris, J. R. and Reiber, A. (2007). Influence of saline and pH on collagen type I fibrillogenesis in vitro: Fibril polymorphism and colloidal gold labelling. *Micron*, **38**, 513–521.

Hassenkam, T., Fantner, G. E., Cutroni, J. A., Weaver, J. C., Morse, D. E. and Hansma, P. K. (2004). High-resolution AFM imaging of intact and fractured trabecular bone. *Bone*, **35**, 4–10.

Holton, J. S., Litt, M. H., Kohn, R. R. and Hamlin, C. R. (1983). Human collagen digestion: Nature of the digestion kinetics as a function of age and structure. *Experimental Gerontology*, **18**, 477–483.

Honda, Y., Kamakura, S., Sasaki, K. and Suzuki, O. (2007). Formation of bone-like apatite enhanced by hydrolysis of octacalcium phosphate crystals deposited in collagen matrix. *Journal of Biomedical Materials Research Part B: Applied Biomaterials*, **80B**, 281–289.

John, A. M. R., Paul, R. V. and Jerome, A. W. (2001). Applications of collagen in medical devices. *Biomedical Engineering Applications, Basis and Communications*, **13**, 14–26.

Johnson, K. A., Rogers, G. J., Roe, S. C., Howlett, C. R., Clayton, M. K., Milthorpe, B. K. and Schindhelm, K. (1999). Nitrous acid pretreatment of tendon xeno-grafts cross-linked with glutaraldehyde and sterilized with gamma irradiation. *Biomaterials*, **20**, 1003–1015.

Jones, J. R., Gentleman, E. and Polak, J. (2007). Bioactive glass scaffolds for bone regeneration. *Elements*, **3**, 393–399.

Kadler, K. (1995). *Extracellular Matrix 1: Fibril-forming Collagens*. London: Academic Press.

Kadler, K. E., Holmes, D. F., Trotter, J. A. and Chapman, J. A. (1996). Collagen fibril formation. *Biochemical Journal*, **316**, 1–11.

Ketchedjian, A., Jones, A. L., Krueger, P., Robinson, E., Crouch, K., Wolfinbarger, L., Jr. and Hopkins, R. (2005). Recellularization of decellularized allograft scaffolds in ovine great vessel reconstructions. *Annals of Thoracic Surgery*, **79**, 888–896.

Kikuchi, M., Ikoma, T., Itoh, S., Matsumoto, H. N., Koyama, Y., Takakuda, K., Shinomiya, K. and Tanaka, J. (2004). Biomimetic synthesis of bone-like nano-composites using the self-organization mechanism of hydroxyapatite and collagen. *Composites Science and Technology*, **64**, 819–825.

Kim, Y. K., Gu, L. S., Bryan, T. E., Kim, J. R., Chen, L. A., Liu, Y., Yoon, J. C., Breschi, L., Pashley, D. H. and Tay, F. R. (2010). Mineralisation of reconstituted collagen using polyvinylphosphonic acid/polyacrylic acid templating matrix protein analogues in the presence of calcium, phosphate and hydroxyl ions. *Biomaterials*, **31**, 6618–6627.

Kokubo, T. (1991). Bioactive glass ceramics: Properties and applications. *Biomaterials*, **12**, 155–163.

Kokubo, T. and Takadama, H. (2006). How useful is SBF in predicting in vivo bone bioactivity? *Biomaterials*, **27**, 2907–2915.

Koutsopoulos, S. (2002). Synthesis and characterization of hydroxyapatite crystals: A review study on the analytical methods. *Journal of Biomedical Materials Research*, **62**, 600–612.

Landis, W. J. (1999). The nature of calcium phosphate in mineralizing vertebrate tissues. *Phosphorus, Sulfur, and Silicon and the Related Elements*, **144**, 185–188.

Lee, C. H., Singla, A. and Lee, Y. (2001). Biomedical applications of collagen. *International Journal of Pharmaceutics*, **221**, 1–22.

Lees, S. (1987). Considerations regarding the structure of the mammalian mineralized osteoid from viewpoint of the generalized packing model. *Connective Tissue Research*, **16**, 281–303.

Li, S.-T. (2000). Biologic biomaterials: Tissue-derived biomaterials. In: Bronzino, J. D. (ed.), *The Biomedical Engineering Handbook*, 2nd ed. CRC Press LLC.

Li, X. and Chang, J. (2008). Preparation of bone-like apatite–collagen nanocomposites by a biomimetic process with phosphorylated collagen. *Journal of Biomedical Materials Research Part A*, **85A**, 293–300.

Li, Y., Asadi, A., Monroe, M. R. and Douglas, E. P. (2009). pH effects on collagen fibrillogenesis in vitro: Electrostatic interactions and phosphate binding. *Materials Science and Engineering: C*, **29**, 1643–1649.

Lin, P., Chan, W. C. W., Badylak, S. F. and Bhatia, S. N. (2004). Assessing porcine liver-derived biomatrix for hepatic tissue engineering. *Tissue Engineering*, **10**, 1046–1053.

Lin, Y. K. and Liu, D. C. (2006). Comparison of physical-chemical properties of type I collagen from different species. *Food Chemistry*, **99**, 244–251.

Liu, W., Merrett, K., Griffith, M., Fagerholm, P., Dravida, S., Heyne, B., Scaiano, J. C., Watsky, M. A., Shinozaki, N., Lagali, N., Munger, R. and Li, F. (2008). Recombinant human collagen for tissue engineered corneal substitutes. *Biomaterials*, **29**, 1147–1158.

Liu, Y., Kim, Y.-K., Dai, L., Li, N., Khan, S. O., Pashley, D. H. and Tay, F. R. (2011a). Hierarchical and non-hierarchical mineralisation of collagen. *Biomaterials*, **32**, 1291–1300.

Liu, Y., Li, N., Qi, Y.-P., Dai, L., Bryan, T. E., Mao, J., Pashley, D. H. and Tay, F. R. (2011b). Intrafibrillar collagen mineralization produced by biomimetic hierarchical nanoapatite assembly. *Advanced Materials*, **23**, 975–980.

Lowenstam, H. A. and Weiner, S. (1989). *On Biomineralization.* Oxford: Oxford University Press.

Lu, J. T., Lee, C. J., Bent, S. F., Fishman, H. A. and Sabelman, E. E. (2007). Thin collagen film scaffolds for retinal epithelial cell culture. *Biomaterials*, **28**, 1486–1494.

Ma, W., Tavakoli, T., Derby, E., Serebryakova, Y., Rao, M. and Mattson, M. (2008). Cell-extracellular matrix interactions regulate neural differentiation of human embryonic stem cells. *BMC Developmental Biology*, **8**, 90.

Mandal, A., Panigrahi, S. and Zhang, W. (2010). Collagen as biomaterial for medical application – drug delivery and scaffolds for tissue regeneration: A review. *Biological Engineering, ASABE*, **2**, 63–88.

Mania, V. M., Kallivokas, A. G., Malavaki, C., Asimakopoulou, A. P., Kanakis, J., Theocharis, A. D., Klironomos, G., Gatzounis, G., Mouzaki, A., Panagiotopoulos, E. and Karamanos, N. K. (2009). A comparative biochemical analysis of glycosaminoglycans and proteoglycans in human orthotopic and heterotopic bone. *IUBMB Life*, **61**, 447–452.

Marelli, B., Ghezzi, C., Mohn, D., Stark, W., Barralet, J., Boccaccini, A. and Nazhat, S. (2011a). Accelerated mineralization of dense collagen-nano bioactive glass hybrid gels increases scaffold stiffness and regulates osteoblastic function. *Biomaterials*, **32**, 8915–8926.

Marelli, B., Ghezzi, C. E., Alessandrino, A., Barralet, J. E., Freddi, G. and Nazhat, S. N. (2012). Silk fibroin derived polypeptide-induced biomineralization of collagen. *Biomaterials*, **33**, 102–108.

Marelli, B., Ghezzi, C. E., Barralet, J. E., Boccaccini, A. R. and Nazhat, S. N. (2010). Three-dimensional mineralization of dense nanofibrillar collagen-bioglass hybrid scaffolds. *Biomacromolecules*, **11**, 1470–1479.

Marelli, B., Ghezzi, C. E., Barralet, J. E. and Nazhat, S. N. (2011b). Collagen gel fibrillar density dictates the extent of mineralization in vitro. *Soft Matter*, **7**, 9898–9907.

Meena, C., Mengi, S. and Deshpande, S. (1999). Biomedical and industrial applications of collagen. *Journal of Chemical Sciences*, **111**, 319–329.

Miller, E. J. (1972). Structural studies on cartilage collagen employing limited cleavage and solubilization with pepsin. *Biochemistry*, **11**, 4903–4909.

Miller, E. J. and Kent Rhodes, R. (1982). Preparation and characterization of the different types of collagen. In: Leon W. Cunningham, D. W. F. (ed.), *Methods in Enzymology.* Academic Press.

Mio, T., Liu, X.-D., Adachi, Y., Striz, I., Sk√∂ld, C. M., Romberger, D. J., Spurzem, J. R., Illig, M. G., Ertl, R. and Rennard, S. I. (1998). Human bronchial epithelial cells modulate collagen gel contraction by fibroblasts. *American Journal of Physiology – Lung Cellular and Molecular Physiology*, **274**, L119-L126.

Mott, J. D. and Werb, Z. (2004). Regulation of matrix biology by matrix metallopro-teinases. *Current Opinion in Cell Biology*, **16**, 558–564.

Muller, L. and Muller, F. A. (2006). Preparation of SBF with different content and its influence on the composition of biomimetic apatites. *Acta Biomaterialia*, **2**, 181–189.

Nakase, Y., Hagiwara, A., Nakamura, T., Kin, S., Nakashima, S., Yoshikawa, T., Fukuda, K.-I., Kuriu, Y., Miyagawa, K., Sakakura, C., Otsuji, E., Shimizu, Y., Ikada, Y. and Yamagishi, H. (2006). Tissue engineering of small intestinal tis-sue using collagen sponge scaffolds seeded with smooth muscle cells. *Tissue Engineering*, **12**, 403–412.

Nancollas, G. H. and Tomson, M. B. (1976). The precipitation of biological minerals. *Faraday Discussions of the Chemical Society*, **61**, 175–183.

Olszta, M. J., Cheng, X., Jee, S. S., Kumar, R., Kim, Y.-Y., Kaufman, M. J., Douglas, E. P. and Gower, L. B. (2007). Bone structure and formation: A new perspective. *Materials Science and Engineering: R: Reports*, **58**, 77–116.

Owan, I., Burr, D. B., Turner, C. H., Qiu, J., Tu, Y., Onyia, J. E. Duncan, R. L. (1997). Mechanotransduction in bone: Osteoblasts are more responsive to fluid forces than mechanical strain. *American Journal of Physiology – Cell Physiology*, **273**, C810–C815.

Parenteau-Bareil, R. M., Gauvin, R. and Berthod, F. S. (2010). Collagen-based bio-materials for tissue engineering applications. *Materials*, **3**, 1863–1887.

Pederson, A. W., Ruberti, J. W. and Messersmith, P. B. (2003). Thermal assembly of a biomimetic mineral/collagen composite. *Biomaterials*, **24**, 4881–4890.

Pedraza, C. E., Marelli, B., Chicatun, F., McKee, M. D. and Nazhat, S. N. (2010). An in vitro assessment of a cell-containing collagenous extracellular matrix-like scaf-fold for bone tissue engineering. *Tissue Engineering Part A*, **16**, 781–793.

Pohunková, H. and Adam, M. (1995). Reactivity and the fate of some composite bio-implants based on collagen in connective tissue. *Biomaterials*, **16**, 67–71.

Postlethwaite, A. E., Seyer, J. M. and Kang, A. H. (1978). Chemotactic attraction of human fibroblasts to type I, II, and III collagens and collagen-derived pep-tides. *Proceedings of the National Academy of Sciences of the United States of America*, **75**, 871–875.

Prockop, D. J. and Kivirikko, K. I. (1995). Collagens: Molecular biology, diseases, and potentials for therapy. *Annual Review of Biochemistry*, **64**, 403–434.

Rajan, N., Habermehl, J., Cote, M. -F., Doillon, C. J. and Mantovani, D. (2007). Preparation of ready-to-use, storable and reconstituted type I collagen from rat tail tendon for tissue engineering applications. *Nature Protocols*, **1**.

Ramshaw, J., Peng, Y., Glattauer, V. and Werkmeister, J. (2009). Collagens as bioma-terials. *Journal of Materials Science: Materials in Medicine*, **20**, 3–8.

Ramshaw, J. A. M. (1996). Collagen-based biomaterials. *Biotechnology and genetic engineering reviews*, **13**, 370–382.

Rhee, S.-H. and Tanaka, J. (1999). Effect of citric acid on the nucleation of hydroxy-apatite in a simulated body fluid. *Biomaterials*, **20**, 2155–2160.

Rho, J.-Y., Kuhn-Spearing, L. and Zioupos, P. (1998). Mechanical properties and the hierarchical structure of bone. *Medical Engineering and Physics*, **20**, 92–102.

Ricard-Blum, S., Ruggiero, F. and van der Rest, M. (2005). The collagen superfam-ily. In: Brinckmann, J., Notbohm, H. and Müller, P. K. (eds.) *Collagen*. Springer Berlin / Heidelberg.

Roberts, T. S., Drez, D., McCarthy, W. and Paine, R. (1991). Anterior cruciate ligament reconstruction using freeze-dried, ethylene oxide-sterilized, bone-patellar
tendon-bone allografts. *The American Journal of Sports Medicine*, **19**, 35–41.

Roeder, B. A., Kokini, K., Sturgis, J. E., Robinson, J. P. and Voytik-Harbin, S. L. (2002).
Tensile mechanical properties of three-dimensional type I collagen extracellular matrices with varied microstructure. *Journal of Biomechanical Engineering*,
124, 214–222.

Rubin, A. L., Pfahl, D., Speakman, P. T., Davidson, P. F. and Schmitt, F. O. (1963).
Tropocollagen: Significance of protease-induced alterations. *Science*, **139**,
37–39.

Ruszczak, Z. and Friess, W. (2003). Collagen as a carrier for on-site delivery of antibacterial drugs. *Advanced Drug Delivery Reviews*, **55**, 1679–1698.

Serpooshan, V., Julien, M., Nguyen, O., Wang, H., Li, A., Muja, N., Henderson, J. E.
and Nazhat, S. N. (2010). Reduced hydraulic permeability of three-dimensional
collagen scaffolds attenuates gel contraction and promotes the growth and differentiation of mesenchymal stem cells. *Acta Biomaterialia*, **6**, 3978–3987.

Shields, L. B. E., Raque, G. H., Glassman, S. D., Campbell, M., Vitaz, T., Harpring,
J. and Shields, C. B. (2006). Adverse effects associated with high-dose recombinant human bone morphogenetic protein-2 use in anterior cervical spine
fusion. *Spine*, **31**, 542–547. 10.1097/01.brs.0000201424.27509.72.

Song, E., Yeon Kim, S., Chun, T., Byun, H.-J. and Lee, Y. M. (2006). Collagen scaffolds derived from a marine source and their biocompatibility. *Biomaterials*, **27**,
2951–2961.

Steven, F. S. and Jackson, D. S. (1967). Purification and amino acid composition of
monomeric and polymeric collagens. *Biochemical Journal*, **104**, 534.

Taguchi, T., Muraoka, Y., Matsuyama, H., Kishida, A. and Akashi, M. (2000). Apatite
coating on hydrophilic polymer-grafted poly(ethylene) films using an alternate
soaking process. *Biomaterials*, **22**, 53–58.

Tay, F. R. and Pashley, D. H. (2008). Guided tissue remineralisation of partially
demineralised human dentine. *Biomaterials*, **29**, 1127–1137.

Tomomatsu, O., Tachibana, A., Yamauchi, K. and Tanabe, T. (2008). A film of colla-
gen/calcium phosphate composite prepared by enzymatic mineralization in an
aqueous phase. *Journal of the Ceramic Society of Japan*, **116**, 10–13.

Vanderby, R. and Provenzano, P. P. (2003). Collagen in connective tissue: From tendon to bone. *Journal of Biomechanics*, **36**, 1523–1527.

Vejlens, L. (1971). Glycosaminoglycans of human bone tissue. *Calcified Tissue
International*, **7**, 175–190.

Venugopal, J., Low, S., A., Sampath, T. and Ramakrishna, S. (2008). Mineralization
of osteoblasts with electrospun collagen/hydroxyapatite nanofibers. *Journal of
Materials Science: Materials in Medicine*, **19**, 2039–2046.

Vepari, C. and Kaplan, D. L. (2007). Silk as a biomaterial. *Progress in Polymer
Science*, **32**, 991–1007.

Vuong, J. and Hellmich, C. (2011). Bone fibrillogenesis and mineralization:
Quantitative analysis and implications for tissue elasticity. *Journal of Theoretical
Biology*, **287**, 115–130.

Vyavahare, N., Hirsch, D., Lerner, E., Baskin, J. Z., Schoen, F. J., Bianco, R., Kruth, H. S.,
Zand, R. and Levy, R. J. (1997). Prevention of bioprosthetic heart valve calcification by ethanol preincubation: Efficacy and mechanisms. *Circulation*, **95**, 479–488.

Wachtel, E. and Weiner, S. (1994). Small-angle x-ray scattering study of dispersed crystals from bone and tendon. *Journal of Bone and Mineral Research*, **9**, 1651–1655.

Wahl, D., Sachlos, E., Liu, C. and Czernuszka, J. (2007). Controlling the processing of collagen-hydroxyapatite scaffolds for bone tissue engineering. *Journal of Materials Science: Materials in Medicine*, **18**, 201–209.

Wahl, D. A. and Czernuszka, J. T. (2006). Collagen-hydroxyapatite composites for hard tissue repair. *European Cells and Materials*, **11**, 43–56.

Weiner, S., Traub, W. and Wagner, H. D. (1999). Lamellar bone: Structure-function relations. *Journal of Structural Biology*, **126**, 241–255.

Wess, T. J. (2008). Collagen fibrillar structure and hierarchies. In: Fratzl, P. (ed.), *Collagen: Structures and Mechanics*. New York, NY: Springer LCC.

Woods, T. and Gratzer, P. F. (2005). Effectiveness of three extraction techniques in the development of a decellularized bone-anterior cruciate ligament-bone graft. *Biomaterials*, **26**, 7339–7349.

Wu, T. J., Huang, H. H., Lan, C. W., Lin, C. H., Hsu, F. Y. and Wang, Y. J. (2004). Studies on the microspheres comprised of reconstituted collagen and hydroxyapatite. *Biomaterials*, **25**, 651–658.

Xu, C., Su, P., Chen, X., Meng, Y., Yu, W., Xiang, A. P. and Wang, Y. (2011). Biocompatibility and osteogenesis of biomimetic bioglass-collagen-phospha tidylserine composite scaffolds for bone tissue engineering. *Biomaterials*, **32**, 1051–1058.

Yamauchi, K., Goda, T., Takeuchi, N., Einaga, H. and Tanabe, T. (2004). Preparation of collagen/calcium phosphate multilayer sheet using enzymatic mineraliza- tion. *Biomaterials*, **25**, 5481–5489.

Yang, C., Hillas, P. J., B√°ez, J. A., Nokelainen, M., Balan, J., Tang, J., Spiro, R. and Polarek, J. W. (2004). The application of recombinant human collagen in tissue engineering. *BioDrugs*, **18**, 103–119.

Yin Hsu, F., Chueh, S. C. and Wang, Y. J. (1999). Microspheres of hydroxyapatite/ reconstituted collagen as supports for osteoblast cell growth. *Biomaterials*, **20**, 1931–1936.

Zeugolis, D. I., Khew, S. T., Yew, E. S. Y., Ekaputra, A. K., Tong, Y. W., Yung, L.-Y. L., Hutmacher, D. W., Sheppard, C. and Raghunath, M. (2008). Electro-spinning of pure collagen nano-fibres – Just an expensive way to make gelatin? *Biomaterials*, **29**, 2293–2305.

Zhang, L.-J., Feng, X.-S., Liu, H.-G., Qian, D.-J., Zhang, L., Yu, X.-L. and Cui, F.-Z. (2004). Hydroxyapatite/collagen composite materials formation in simulated body fluid environment. *Materials Letters*, **58**, 719–722.

Zhang, W., Liao, S. S. and Cui, F. Z. (2003). Hierarchical self-assembly of nano-fibrils in mineralized collagen. *Chemistry of Materials*, **15**, 3221–3226.

Zhu, Y. K., Umino, T., Liu, X. D., Wang, H. J., Romberger, D. J., Spurzem, J. R. and Rennard, S. I. (2001). Contraction of fibroblast-containing collagen gels: Initial collagen concentration regulates the degree of contraction and cell survival. *In Vitro Cellular and Developmental Biology-Animal*, **37**, 10–16.

Ziv, V. and Weiner, S. (1994). Bone crystal sizes: A comparison of transmission electron microscopic and x-ray diffraction line width broadening techniques. *Connective Tissue Research*, **30**, 165–175.

15

Nanomaterials for dental and craniofacial tissue engineering

S. H. ZAKY, S. YOSHIZAWA and C. SFEIR,
University of Pittsburgh School of Dental Medicine, USA

DOI: 10.1533/9780857097231.3.415

Abstract: This chapter highlights the auspicious prospects that nanotechnology has for engineering craniofacial and dental tissues. We discuss current and promising strategies applying nano-alteration to scaffolding materials to biomimetically regenerate craniofacial bone, tooth tissues as well as the compelling potential for whole tooth regeneration. The chapter also considers nanotechnology applications to enhance tissues vascularization and approaches to control spatiotemporal release of biological factors by integrated gene and drug nano-delivery systems. We finally include the potential of micro/nano-arrays libraries with high-throughput characterization to funnel nanotechnology strategies and efforts into the optimum biomimetic approach for tissues regeneration.

Key words: craniofacial dental tissue engineering, nanotechnology tooth regeneration, nanotechnology scaffolds engineering, nano gene drug delivery, nano vascularisation.

15.1 Introduction

The last two decades have brought significant advancement in the field of tissue regeneration. This progress, fueled largely by the expanding understanding of cellular, molecular and developmental biology, is evidenced by a vast increase in tissue engineering (TE) innovations. Somewhere along this steeply ascending learning curve, a whole new world of *nanos* has unfolded, with regenerative strategies increasingly built around the actual molecular paradigms orchestrating cellular behavior, function and fate. Although first predicted in the 1960s,[1] it is only now that nanotechnology is being effectively applied to tissue regeneration, most visibly through the deliberate fine tuning of molecular particles in the cell microenvironment. Cells, the regeneration key players, can in turn be 'tuned' to recognize and feel their way toward attachment, migration, proliferation, differentiation and the secretion of factors which dictate tissue specificity.

415

Nanotechnology is of particular relevance in strategizing the regeneration of the craniofacial complex. With highly specialized tissues residing within a small physical space, craniofacial interventions mandate minimal invasiveness for the preservation of aesthetics. Application of nanotechnology in this region must also effectively support mechanical strength of the occlusal load-bearing structures, the need for high vascularization and mobilization against microbial assault in the oral cavity.[2]

Although craniofacial TE approaches (e.g., mainly employing a substrate or scaffold, with cell and growth factor delivery)[3] have already advanced tremendously, these advancements are often confounded by clinical translation hurdles.[4–6] Particularly troublesome aspects can be sufficient cell harvests, immunogenicity in allogeneic settings and challenges of *ex vivo* manipulation,[6] and for tissue-specific scaffolds, achieving control of the resorption rate, desirable mechanical properties, growth factor availability, steady mass transfer into the core and hierarchical vascular growth are all ongoing challenges that can be confronted with nanotechnological techniques.[2,7]

In the following section, the prospects for nanotechnology in craniofacial regeneration will be considered. Within such a frame, we will consider current and potential strategies for scaffolding materials, nano-alteration for rebuilding specific craniofacial structures, and approaches to enhance vascularization and control of spatiotemporal release of biological factors crucial for regeneration.

15.2   Nanotechnology for engineered substrates

As the study of tissue regeneration is still in its relative infancy, myriad scaffolding materials are currently being tested for their efficacy in tissue regeneration. Innovation and creativity have quickly driven into the background the notion of scaffolds solely as physical support to cells. One pivotal requirement of a scaffold in supporting cellular function is to allow necessary signals from the extracellular environment to affect cellular physiology. Nanoarchitecturally, extracellular matrix (ECM) comprises interwoven, hierarchically organized fibers (10 to several 100s of nanometers) involved in the storage, activation and release of a wide range of factors that provide cell support, and direct cell behavior through cell–cell and cell–soluble factor interactions.[8,9] These interactions form a conduit for transmission of key physical (topographical and biochemical) information, and via matrix interaction, the cell can recognize, feel or 'Braille read' its bed through an environment-sensing cytoskeleton and associated integrin receptors.

Indeed, biomimetic approaches increasingly leverage nanotechnology for tailoring biomaterial architecture and properties, thereby emulating the complexity and functionality of ECM (Fig. 15.1). These biomimetic approaches ultimately alter gene expression, thus guiding cell and

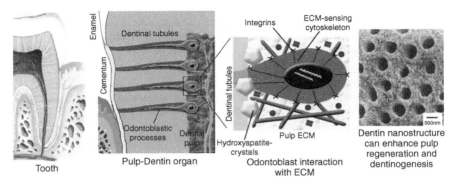

Tooth — **Pulp-Dentin organ** — **Odontoblast interaction with ECM** — **Dentin nanostructure can enhance pulp regeneration and dentinogenesis**

Enamel · Dentinal tubules · Cementum · Dentinal tubules · Odontoblastic processes · Dental pulp · Hydroxyapatite-crystals · Integrins · ECM-sensing cytoskeleton · Pulp ECM · 500nm

15.1 Illustration of a tooth: at the organ level (macro-level), the odontoblast cell position in the pulp-dentin (micro-level), followed by an enlarging illustration for the odontoblast interaction with its pulpal (non-mineralized) and its dentinal (extracellularly mineralized) matrices (nano-level).

tissue-specific spatiotemporal behavior through processes such as adhesion, contraction, migration, proliferation, differentiation, self-renewal and apoptosis.[10] Prevalent views support that, nanoscopically, the surface topography of ECM has a more profound effect on single cell adhesion, morphogenesis, migration, alignment and differentiation than do pore size and microscopic geometry.[11–14] Compared to conventional micron-structured materials, nanotopographic materials boast higher surface areas and concomitant increases in surface energy and wettability, factors which affects anisotropic stresses and modulations at the cell-surface/protein interface.[15–17]

If dentin is examined macroscopically, it is a calcified tissue forming the stress-bearing body of the tooth and odontoblastic processes. The properties and mechanics of the microenvironment around the odontoblast cell are, however, likely to be very different from the macroscopic properties of the tissue it weaves. An odontoblast, in contact with the dental pulp soft matrix and the dentin hard surface, provides an example of a hierarchically structured microenvironment where mechanically and biologically different matrices impact cellular behavior.

Recent advances in nanotechnology have also enabled the design and fabrication of nano-scale ECM-analog materials.[18,19] To maintain tissue-specific architecture and function, the fidelity of ECMs of various tissues in the craniofacial complex must be maintained, specifically with respect to the composition and spatial organization of collagens, elastins, proteoglycans and adhesion molecules (Table 15.1, assembled from Ten Cate's Oral Histology, 7th edition).[20] These differences are of crucial consideration when designing tissue-specific scaffolds.

Table 15.1 Composition and organization of ECM in various craniofacial structures

	Nanostructure (ECM) (for cell sensing)	Microstructure (at best duplicated in a TE scaffold)	Gross macrostructure
Bone	Protein-based soft hydrogel template of collagen (coll.) type I, laminin, fibronectin and vitronectin, structural proteins bone sialoprotein (BSP), osteocalcin (OC), osteopontin (OP), osteonectin (ON), proteoglycans, growth factors, serum proteins and water (33%) Hard inorganic components of nanocrystalline hydroxyapatite (HA) (20–80 nm long and 2–5 nm thick) (67%)	Lamellae, osteons, and Haversian systems	Trabecular cancellous
Enamel	Carbon apatite crystals 60–70 nm wide, 25–30 nm thick (96%) Ameloblastin protein (rods sheath) (4%)	Rods, interred enamel and rod sheath	Ceramic layer covering the coronal tooth structure
Dentin	Coll. types I, III and V, dentin phosphoprotein (DPP), dentin sialoprotein (DSP), dentin glycoprotein (DGP), dentin matrix protein 1 (DMP-1), ON, OC, BSP, OP, matrix phosphoglycoproteins, proteoglycans and serum proteins. Coll. I acts as a scaffold for the crystal structures. Coll. assembled in form of fibrils are 50–200 nm in diameter Uniform small plates of bone like HA minerals (70%)	Dentinal tubules	Resilient core of tooth structure
Cementum	Coll. I mainly and coll. III, XII; traces of V, VI and XIV. ALP, BSP, DMP-1, DSP, fibronectin, OC, ON, proteoglycans, tenascin, cementum attachment protein. Coll. is assembled in form of bundles or sheets (50%) HA in the form of uniform small plates (45–50%)	Mineralized cellular connective tissue	Layer covering root dentin

Table 15.1 Continued

	Nanostructure (ECM) (for cell sensing)	Microstructure (at best duplicated in a TE scaffold)	Gross macrostructure
PDL	Coll. types I, III and XII Dermatan sulfate	Bundles of collage fibers and Highly cellular and vascularized tissue	Soft tissue between cementum and alveolar bone
Dental pulp	Coll. types I and III, glycosaminoglycans (GAGs), glycoproteins	Peripheral odontoblastic zone, cell free zone, cell rich zone and pulp core	Highly cellular, vascularized and innervated connective tissue filling tooth core
TMJ disc	Coll., GAGs, proteoglycans and elastin	Fibrocartilage: fibroblasts and fibrochondrocytes	Cartilaginous structure between head of condyle and glenoid fossa

ALP, alkaline phosphatase; PDL, periodontal ligament.

15.3   Engineering mineralized collagenous craniofacial structures

Bone is a complex composite of mineralized collagen fibrils, units which are the very building blocks of mineralized hard tissues. Regulation of the intra-fibrillar mineralization process is achieved through interaction between the collagen matrix, non-collagenous extracellular proteins and hard inorganic nanocrystalline hydroxyapatite (HA) (20–80 nm long and 2–5 nm thick).[21] During bone regeneration, HA serves as a chelating agent for mineralizing osteoblasts while the collagen provides mechanical support, and promotes adhesion and proliferation.[7]

In recent years, the science of nanotechnology has been fine-tuning the emulation of collagenous organic and/or the HA mineral phases of bone. The goals are two-fold: to achieve osteoconductivity along with the desired mechanical properties.[22,23] Strategies in bone regeneration have indeed advanced through the synthesis of scaffolds with a pattern of highly mineralized collagen fibrils identical to those of natural bone – specifically, with nano-apatite crystallites preferentially aligned along the collagen fibril axes.[22] Incorporation of HA nanoparticles into a bone engineering scaffold has demonstrated an enhancement in compressive mechanical properties and stiffness. Such constructs also boast improvement in *in vitro* bioactivity,[24] while the nanostructure of HA particles can be designed as a biomimetic surface to promote osteogenic differentiation.[25]

Biomimetic bone tissue engineering technologies have progressed in response to specific challenges. Among them are the mechanical limitations of calcium phosphate and polymeric scaffolds, challenges in nutrient transfer, and the need for adequate support for vascularization and mineralization critical to craniofacial load-bearing bones. Notable were the introduction of nanoparticle-composite scaffolds for increasing mechanical strength in bone grafts, the development of novel genetic delivery and targeting systems (encoding osteogenic growth factors), and the fabrication of nanofibrous scaffolds which support cell growth and differentiation through morphologically favored architectures.[26] In addition, polymeric nanofiber matrices have been shown to be biologically similar to native bone ECM architecture, and when incorporated onto biodegradable microscopically porous polymeric three-dimensional (3D) scaffolds, and have the potential to be used as bone biomimetic regeneration substrates. The nano-scale topography of such scaffolds enhances cellular adhesion and mesenchymal stem cell stimulation, thereby enhancing bone mineral deposition, while the open geometry and porosity of the scaffold promote cell penetration and nutrient transport.[27]

Another regenerative nano-tool that has garnered recent attention is the carbon nanotube. Boasting excellent mechanical strength, nanotubes are also amenable to promoting cell attachment and proliferation. They are pro-osteogenic – showing support to osteoblastic growth and modulation to the osteoblastic phenotype. In addition, they can be readily incorporated as reinforcing agents into the 3D architectures of polymeric scaffolds[28–30] that would not otherwise perform as efficiently.[31]

Furthermore, nano-scale strategies are being developed to incorporate functional motif sequences of complex biomimetic materials or short peptides promoting cellular adherence and osteogenic differentiation into scaffolds.[32–34] Modifications of the polymer surface with BMP-related peptide,[35,36] full length BMP[37] or osteocalcin cross-linked to nano-HA[38,39] have been employed to mimic ECM signaling and have been shown to enhance osteoblastic cell attachment and bone matrix synthesis. When compared to conventional polymers, the adsorption and conformation of proteins that regulate osteoblast adhesion and function (e.g., fibronectin and vitronectin) were enhanced on nanophase surfaces.[40] Finally, in a yet-limited application for dental tissue, scaffold coating with RGD integrin recognition sequences resulted in more mineralized osteodentin-like tissue.[41]

In the dental implant arena, nano-scaling of textured titanium fixtures greatly improved osseointegration, evidenced by higher osteoblasts adhesion, alkaline phosphate activity and calcium deposition.[42] Nanophase ceramics, with their unique osteoconductivity and vitronectin-rich surface (osteoblast adhesion protein), can improve bone implant efficacy by enhancing bonding of orthopedic/dental implants to juxtaposed bone.[16,40] With similarly mineralized collagenous tissues, such nano-engineering approaches could be applied for dentin and cementum, however, potential avenues and directions have yet to be fully explored.[23]

15.3.1 Vascularization

Bone is a dynamic tissue that forms only within an adequately vascularized site. For a certain critical-sized defect, conventional bone engineering constructs may achieve bone formation at the periphery of the scaffold, while the viability of inner cells is compromised by slow and inadequate vascularization at the center.[6] As nanotechnology can increase control of scaffold features at the micro-and nano-levels, define the cellular microenvironment and direct cell fate, the opportunities it offers for guiding the formation of micro-engineered capillary beds and vascular ingress while arborizing the hierarchical vascular networks are clear.[43-47] High-resolution modification of scaffold properties by incorporation of growth factors, molecules and cell ligands can also provide avenues for vascularization control. Indeed, vascular cell adhesion and proliferation were greatly improved on nanostructured titanium compared to conventional implants.[48] Furthermore, polymers with 200 nm structures promoted vascular cell responses and greater fibronectin interconnectivity than smooth micro-featured polymers.[49]

15.3.2 Enamel regeneration

Enamel, a complex structure produced by specialized cells, is not only the hardest tissue in the body but is also primarily non-regenerative. The enamel forming cells, or ameloblasts, terminate their function and cease to exist concomitant with tooth eruption, condemning enamel to be non-vital.[6] Despite a few studies distinguishing ameloblastic differentiation from other cell lineages,[2,50] attempts to engineer this noncellular tissue have focused mainly on its nano-structure and chemical synthesis. Indeed, a nonbiological, acellular, enamel prism-like structure of fluorapatite crystals has been engineered by hydroxyapatite nanorod self-assembly.[51,52] Researchers have postulated that the production of mineralized films could be used to restore decayed teeth.

15.3.3 Other craniofacial tissues

Despite steady advancement, nano-scale applications are as yet under-explored in other craniofacial tissues. Dental stem cells loaded on peptide nanostructured hydrogels (eminently applicable in small defects as periodontal pockets) maintained high proliferation and collagen production, expressing osteoblast gene markers and showing mineral deposits.[53] In the treatment of periodontal disease, nanotubes or wires on the outer surfaces of 3D scaffolds were reported to modulate macrophage adhesion and viability, possibly also serving to reduce inflammatory responses in transplanted engineered tissues. Zinc oxide nanorods (50 nm in diameter and around 500 nm long) sputtered on polyethylene terephthalate discs, for example, inhibited the formation of the typical foreign body capsule occurring after subcutaneous transplantation in mice.[54] For a potential temporo mandibular joint (TMJ) regeneration application, nanofibrous osteochondral scaffold with HA nanoparticles[55] or with dual factor conjugation (insulin for chondrogenesis and α-glycerophosphate for osteogenic differentiation),[56] were reported to regenerate the osteochondral interface.

15.3.4   Whole tooth regeneration

Motivated by growing awareness of the importance of conserving tooth structure, tooth tissue engineering is an area of regenerative medicine which enjoys great momentum. As the knowledge of developmental, cellular and molecular tooth biology expands, the previously unthinkable paradigm of tooth regeneration appears increasingly imaginable.[6] Conceptually, there is little doubt that the best substitute for lost tooth structure is a biologically functional regenerated tooth, and nanotechnology could conceivably be leveraged in myriad ways toward that end.

Ongoing research utilizes two strategies to achieve whole tooth regeneration: either combining cells, biofactors and scaffolds[41,42,57–62] or by recapitulating organogenesis.[63–65] Even though the latter approach is fundamentally superior, tremendous challenges remain in providing the microenvironmental spatiotemporal cues that orchestrate tooth morphogenesis and differentiation. An example of such cues could include translating in an adult animal/individual the presence of the same combination of cryptic factors present in that specific location along the dental lamina for a specific crown morphology during a definite stage of embryonic development.

The TE approach appears to be more 'clinical-friendly', yet challenged by the ability to engineer the complex interfaces between biologically and mechanically distinct tissues, namely the enamel–dentin junction, the periodontal ligament and the junctional epithelium.[6]

Addressing the 'bottleneck', providing blood supply to an engineered tooth or to necrotic pulp (regenerative endodontics), is challenged not only by the pulp being encased in an impermeable shell, with consequent prevention of large-scale diffusion of nutrients or metabolites, but also by the presence of only one point of vascular access at the apex of the tooth root.[66,67] The incorporation of microvasculature into tissue-engineered constructs is a promising advancement toward tooth regeneration and has been reported so far in combination with microengineered polymers with the potential for rapid vascularization and fluid exchange.[37,68] Perfused agarose hydrogels with microfluidic channels have also been reported to support cell metabolism and function by delivering oxygen and nutrients to cells encapsulated within 200 μm from the hydrogels surface, which appeals for the potential of assisting the ingress of microvascular networks to the interior of the tooth and dental pulp.[69] Nano-scale technologies providing biomimetic matrices for escorting vascular entry through the apex to an engineered dental pulp are of particular benefit.

The engineering of complex tissues composed of several cell types organized in specialized niches, for example, the tooth, brings about unique challenges. In one new approach, assembly of 3D tissues capitalizes on the cellular compatibility and subsequent magnetic levitation of a hydrogel composed of phage, magnetic iron oxide and gold nanoparticles. By spatially controlling the magnetic field while cells divide and grow, the system enables the manipulation of cell mass geometry into specific locations for the engineering of 3D structures.[70] Indeed, nanotechnology systems appear to have the potential for applying nature-inspired biomimetics for engineering 3D bioactive matrices that provide necessary biological signals; this promise may pave the course for an engineered cap-stage tooth that could be engrafted into a host.[71]

15.4   Nano-scale scaffolds with integrated delivery systems

Since nano-scale scaffolds can be taken into cells directly by endocytosis, they can be used for novel gene delivery system without viral transfection. Also, the nano-scale structures of scaffold allow us to control degradation rate to consistently release drugs incorporated to scaffold, and to construct ideal structure for cell migration and angiogenesis. These gene and drug delivery systems using nano-scale scaffolds have a strong potential to better control dental and craniofacial tissue engineering.

15.4.1   Nano-scale gene delivery systems

Also directed toward advancing dental and craniofacial tissue regeneration via nanotechnology are nano-scale scaffolds with integrated gene or drug delivery systems. These nano-scale carriers are considered to be viable alternatives to viral gene therapy. They include, among others, cationic polymers,[72] cationic lipids,[73] electroporation,[74] poly(D,L-lactide-co-glycolide) (PLGA) microparticles[75,76] and calcium phosphate (CaP).[77–79] Although the mechanisms of these carriers are varied, the method of gene delivery into cells is similar. The carriers are combined with DNA, and are taken up by target cells through endocytosis. Through the endocytic pathway, DNA is released from the endosome before digestion with lysosomal enzymes. Then DNA is translocated into the nucleus via nuclear localizing sequences,[80] and undergoes transcription and translation of desired proteins.[81] Gene delivery carriers can also be leveraged in tissue engineering strategies by embedding the carriers in the scaffolds. For example, Bonadio *et al.* proposed the concept of such a delivery system as a gene-activated matrix (GAM).[82] The GAM consists of two ingredients: plasmid DNA and a biodegradable scaffold. The scaffold holds the DNA *in situ* until the fibroblasts have proliferated at the wound area. Fibroblasts are subsequently transfected with DNA, and start to secrete plasmid-encoded protein to enhance proper tissue regeneration.[82]

The gene delivery carriers known as calcium phosphates (CaP) are made by DNA-calcium phosphate co-precipitation[83,84] and range optimally from 25 to 50 nm in size.[85] CaP has proven to be one of the favored non-viral gene delivery carriers because of its biocompatibility, biodegradability and absorptive capacity for plasmid DNA (pDNA).[77-79] In *in vivo* experiments using mice, CaP showed higher transfection efficiency compared to that achieved via injection of naked DNA.[78] The mechanism responsible for the high efficiency of CaP has been hypothesized as follows: the CaP promotes pDNA to escape from the endosome into the cytosol during endocytosis; as the pH drops and hydrolytic enzyme is activated in the endosome, the CaP is dissolved into Ca^{2+} and PO_4^{3-}. These ions then increase the osmotic pressure in endosomes, and lead to the bursting of endosomal vesicles. As a result, the pDNA is released into the cytosol before being degraded (Fig. 15.2 from Ref. 81). CaP is the major mineral component of hard tissues such as enamel, dentin and bone, and has been studied as an important material needed for hard tissue regeneration.[37,86]

Other major materials used for gene delivery systems are cationic polymers and cationic lipids. Cationic polymers such as poly(L-lysine),[87] poly(ethyleneimine) (PEI)[88] and poly(beta-amino ester)[89] are commonly used. Moreover, successful gene delivery into mesenchymal stem cells using cationic polymers has been reported,[90] promising potential use in dental and craniofacial hard tissue regeneration. However, cationic polymers are prone to have cytotoxicity,[91,92] and have low efficacy *in vivo* due to an association with negatively charged serum proteins.[93] Further studies on improving the efficacy of cationic polymer carriers, such as coating the nano-particles with negatively charged peptides,[94] or electrostatic ligands,[95] are required.

Cationic lipids consist of a positively charged moiety, hydrophobic lipid anchor and a linker that bridges these two components.[96] Although widely used for *in vitro* gene transfection (i.e., Lipofectamin™), cytotoxicity remains an issue.[96] Moreover, the concentration of lipid and lipid/DNA ratios need to be carefully controlled during application.[97] Overall, nano-scale gene delivery carriers are powerful, yet largely underutilized, tools to establish more efficient and tissue-specific tissue engineering systems.

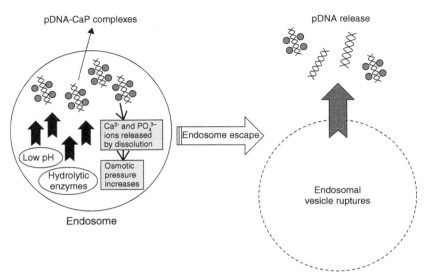

15.2 Schematic of pDNA-CaP complex endosomal escape by the increase of osmotic pressure. As the endocytic vesicle matures, pH starts to drop and hydrolytic enzymes start to activate. As a result, calcium phosphate begins to dissolve concomitant with osmotic pressure inside the vesicle increasing (due to Ca^{2+} and PO_4^{3-} ion release into the inner-vesicular environment). Water fluxes into the vesicle continuously while calcium phosphate continuously dissolves. Eventually, the vesicle ruptures and pDNA is released into the cytosol.[81]

15.4.2   Nano-scale drug delivery systems

Controlled release of protein for tissue engineering has been widely studied,[98] and nano-scale scaffolding is attractive due to its biocompatibility and ability to enhance cell attachment, proliferation and differentiation.[26,99] For periodontal tissue engineering, Feng *et al.* engineered nanofibrous poly lactic acid (PLLA) scaffolds releasing doxycycline targeted against bacterial infection of the periodontia.[100] After incorporating doxycycline into PLGA nanospheres (300–730 nm diameter) with a modified water-in-oil-in-oil emulsion method, prefabricated nanofibrous PLLA scaffolds were created. The release profile of doxycycline was stable, and the scaffold inhibited common bacterial growth (*S. aureus* and *E. coli*) *in vitro*.[100] In the same group, Wei *et al.* loaded bone morphogenetic protein 7 (BMP7) in the PLGA nanospheres, incorporated them into PLLA nanofibrous scaffolds,[101] and implanted them subcutaneously into rats. There was significant bone formation on BMP7-containing scaffolds but not on the scaffold without BMP7, or the same scaffold soaked in BMP7 solution. These results indicated that the biological activity of BMP7 was prolonged by encapsulation into the PGLA nanospheres.[101] This scaffold can be used for dental and craniofacial bone regeneration.

15.5 Micro/nano-arrays as libraries for high-throughput characterization

Another noteworthy innovation has been the introduction of micro/nano-arrays employing 3D micro/nano-engineered platforms.[102] These arrays employ mathematical algorithms to design or print homogeneous micro/nano-scaled libraries of topographies or protein coatings on chips where the culture conditions and function/differentiation of cultured cells are evaluated. The potential for creating a high-throughput characterization of conditions and disease biomarkers with high sensitivity, using less time and sample, make this technology appealing. Massive amounts of data are collected from the nano-textured, information-rich matrices, and this knowledge would function as a repository for conditions that control the biochemical and biomechanical environment in cell culture, and for proteins that could manipulate cell signal pathways.[71,103,104] Recent studies have reported the ability of mesenchymal cells from bone marrow to 'Braille read' varying topographies on which they were cultured, and thus alter their proliferation and osteogenic differentiation behavior.[105,106] Similar techniques could be used to study dental/craniofacial stem cells from various sources, with a goal of selecting the optimum biomimetic approach for regeneration of a certain craniofacial structure.

15.6 Conclusion

Nano-scale technologies offer compelling benefits for control of scaffold architecture, biomechanics, growth factor delivery, vascularity, cellular spatial orientation and temporal signaling. Although scientific strides in the field are rapid, they are still overshadowed by a paucity of knowledge of the actual signaling events occurring in the nanoenvironment. Through the application of micro/nano-scale technologies and the generalization of high-throughput characterization, it is hoped that knowledge and technology can continue to grow in concerted achievement of craniofacial/dental tissue regenerative therapeutic benefit.

15.7 References

1. Feynman RP (1960) There's plenty of room at the bottom. An invitation to enter a new field of physics. *Engineering and Science Magazine*, **XXIII**(5):22–36.
2. Yuan J, Cao Y and Liu W (2012) Biomimetic scaffolds: Implications for craniofacial regeneration. *J Craniofac Surg*, **23**(1):294–7.

3. Langer R and Vacanti JP (1993) Tissue engineering. *Science*, **260**(5110):920–6.
4. Costello BJ, Shah G, Kumta P and Sfeir CS (2010) Regenerative medicine for craniomaxillofacial surgery. *Oral Maxillofac Surg Clin North Am*, **22**(1):33–42.
5. Mao JJ, Giannobile WV, Helms JA, Hollister SJ, Krebsbach PH, Longaker MT and Shi S (2006) Craniofacial tissue engineering by stem cells. *J Dent Res*, **85**(11):966–79.
6. Zaky SH and Cancedda R (2009) Engineering craniofacial structures: Facing the challenge. *J Dent Res*, **88**(12):1077–91.
7. Dvir T, Timko BP, Kohane DS and Langer R (2010) Nanotechnological strategies for engineering complex tissues. *Nat Nanotechnol*, **6**(1):13–22.
8. Stevens MM and George JH (2005) Exploring and engineering the cell surface interface. *Science*, **310**(5751):1135–8.
9. Taipale J and Keski-Oja J (1997) Growth factors in the extracellular matrix. *FASEB J*, **11**(1):51–9.
10. Buxboim A, Ivanovska IL and Discher DE (2010) Matrix elasticity, cytoskeletal forces and physics of the nucleus: how deeply do cells 'feel' outside and in? *J Cell Sci*, **123**(Pt 3):297–308.
11. Hacking SA, Harvey E, Roughley P, Tanzer M and Bobyn J (2008) The response of mineralizing culture systems to microtextured and polished titanium surfaces. *J Orthop Res*, **26**(10):1347–54.
12. Curtis A and Wilkinson C (1999) New depths in cell behaviour: Reactions of cells to nanotopography. *Biochem Soc Symp*, **65**:15–26.
13. Webster TJ, Ergun C, Doremus RH, Siegel RW and Bizios R (2000) Specific proteins mediate enhanced osteoblast adhesion on nanophase ceramics. *J Biomed Mater Res*, **51**(3):475–83.
14. Colon G, Ward BC and Webster TJ (2006) Increased osteoblast and decreased Staphylococcus epidermidis functions on nanophase ZnO and TiO_2. *J Biomed Mater Res A*, **78**(3):595–604.
15. Zhang L and Webster TJ (2009) Nanotechnology and nanomaterials: Promises for improved tissue regeneration. *Nano Today*, **4**:66–80.
16. Tran N and Webster TJ (2009) Nanotechnology for bone materials. *Wiley Interdiscip Rev Nanomed Nanobiotechnol*, **1**(3):336–51.
17. Bettinger CJ, Langer R and Borenstein JT (2009) Engineering substrate topography at the micro- and nanoscale to control cell function. *Angew Chem Int Ed Engl*, **48**(30):5406–15.
18. Goldberg M, Langer R and Jia X (2007) Nanostructured materials for applications in drug delivery and tissue engineering. *J Biomater Sci Polym Ed*, **18**(3):241–68.
19. Shi J, Votruba AR, Farokhzad OC and Langer R (2010) Nanotechnology in drug delivery and tissue engineering: From discovery to applications. *Nano Lett*, **10**(9):3223–30.
20. Nanci A (2008) *Ten Cate's Oral Histology. Development, Structure and Function*. Seventh Edition. Moby, Elsevier.
21. Boskey AL (2003) Biomineralization: An overview. *Connect Tissue Res*, **44**(1):5–9.
22. Kim YK, Gu LS, Bryan TE, Kim JR, Chen L, Liu Y, Yoon JC, Breschi L, Pashley DH and Tay FR (2010) Mineralisation of reconstituted collagen using polyvinylphosphonic acid/polyacrylic acid templating matrix protein analogues in the presence of calcium, phosphate and hydroxyl ions. *Biomaterials*, **31**(25):6618–27.

23. Sachlos E, Gotora D and Czernuszka JT (2006) Collagen scaffolds reinforced with biomimetic composite nano-sized carbonate-substituted hydroxyapatite crystals and shaped by rapid prototyping to contain internal microchannels. *Tissue Eng*, **12**(9):2479–87.

24. Jack KS, Velayudhan S, Luckman P, Trau M, Grondahl L and Cooper-White J (2009) The fabrication and characterization of biodegradable HA/PHBV nanoparticle-polymer composite scaffolds. *Acta Biomater*, **5**(7):2657–67.

25. Wang G, Zheng L, Zhao H, Miao J, Sun C, Ren N, Wang J, Liu H and Tao X (2011) In vitro assessment of the differentiation potential of bone marrow-derived mesenchymal stem cells on genipin-chitosan conjugation scaffold with surface hydroxyapatite nanostructure for bone tissue engineering. *Tissue Eng Part A*, **17**(9–10):1341–9.

26. Woo KM, Jun JH, Chen VJ, Seo J, Baek JH, Ryoo HM, Kim GS, Somerman MJ and Ma PX (2007) Nano-fibrous scaffolding promotes osteoblast differentiation and biomineralization. *Biomaterials*, **28**(2):335–43.

27. Dalby MJ, Gadegaard N, Tare R, Andar A, Riehle MO, Herzyk P, Wilkinson CD and Oreffo RO (2007) The control of human mesenchymal cell differentiation using nanoscale symmetry and disorder. *Nat Mater*, **6**(12):997–1003.

28. Edwards SL, Werkmeister JA and Ramshaw JA (2009) Carbon nanotubes in scaffolds for tissue engineering. *Expert Rev Med Dev*, **6**(5):499–505.

29. Zanello LP, Zhao B, Hu H and Haddon RC (2006) Bone cell proliferation on carbon nanotubes. *Nano Lett*, **6**(3):562–7.

30. Shi X, Sitharaman B, Pham QP, Liang F, Wu K, Edward Billups W, Wilson LJ and Mikos AG (2007) Fabrication of porous ultra-short single-walled carbon nanotube nanocomposite scaffolds for bone tissue engineering. *Biomaterials*, **28**(28):4078–90.

31. Sitharaman B, Shi X, Walboomers XF, Liao H, Cuijpers V, Wilson LJ, Mikos AG and Jansen JA (2008) In vivo biocompatibility of ultra-short single-walled carbon nanotube/biodegradable polymer nanocomposites for bone tissue engineering. *Bone*, **43**(2):362–70.

32. Gelain F, Bottai D, Vescovi A and Zhang S (2006) Designer self-assembling peptide nanofiber scaffolds for adult mouse neural stem cell 3-dimensional cultures. *PLoS One*, **1**:e119.

33. Xu J, Zhou X, Ge H, Xu H, He J, Hao Z and Jiang X (2008) Endothelial cells anchoring by functionalized yeast polypeptide. *J Biomed Mater Res A*, **87**(3):819–24.

34. Rexeisen EL, Fan W, Pangburn TO, Taribagil RR, Bates FS, Lodge TP, Tsapatsis M and Kokkoli E (2009) Self-assembly of fibronectin mimetic peptide-amphiphile nanofibers. *Langmuir*, **26**(3):1953–9.

35. Xu HH, Weir MD and Simon CG (2008) Injectable and strong nano-apatite scaffolds for cell/growth factor delivery and bone regeneration. *Dent Mater*, **24**(9):1212–22.

36. Lin ZY, Duan ZX, Guo XD, Li JF, Lu HW, Zheng QX, Quan DP and Yang SH (2010) Bone induction by biomimetic PLGA-(PEG-ASP)n copolymer loaded with a novel synthetic BMP-2-related peptide in vitro and in vivo. *J Control Release*, **144**(2):190–5.

37. Ryu HS, Youn HJ, Hong KS, Chang BS, Lee CK and Chung SS (2002) An improvement in sintering property of beta-tricalcium phosphate by addition of calcium pyrophosphate. *Biomaterials*, **23**(3):909–914.

38. Sarvestani AS, He X and Jabbari E (2007) Effect of osteonectin-derived peptide on the viscoelasticity of hydrogel/apatite nanocomposite scaffolds. *Biopolymers*, **85**(4):370–8.

39. Sarvestani AS, He X and Jabbari E (2008) Osteonectin-derived peptide increases the modulus of a bone-mimetic nanocomposite. *Eur Biophys J*, **37**(2):229–34.

40. Webster TJ, Schadler LS, Siegel RW and Bizios R (2001) Mechanisms of enhanced osteoblast adhesion on nanophase alumina involve vitronectin. *Tissue Eng*, **7**(3):291–301.

41. Xu WP, Zhang W, Asrican R, Kim HJ, Kaplan DL and Yelick PC (2008) Accurately shaped tooth bud cell-derived mineralized tissue formation on silk scaffolds. *Tissue Eng Part A*, **14**(4):549–57.

42. Yao ZQ, Yang WS, Zhang WB, Chen Y and Zhou YX (1992) Hepatic stimulator substance from human fetal liver for treatment of experimental hepatic failure. *Chin Med J (Engl)*, **105**(8):676–83.

43. Kane RS, Takayama S, Ostuni E, Ingber DE and Whitesides GM (1999) Patterning proteins and cells using soft lithography. *Biomaterials*, **20**(23–24):2363–76.

44. Rozkiewicz DI, Kraan Y, Werten MW, de Wolf FA, Subramaniam V, Ravoo BJ and Reinhoudt DN (2006) Covalent microcontact printing of proteins for cell patterning. *Chemistry*, **12**(24):6290–7.

45. Kim P, Jeong HE, Khademhosseini A and Suh KY (2006) Fabrication of non-biofouling polyethylene glycol micro- and nanochannels by ultraviolet-assisted irreversible sealing. *Lab Chip*, **6**(11):1432–7.

46. Borenstein JT, Weinberg EJ, Orrick BK, Sundback C, Kaazempur-Mofrad MR and Vacanti JP (2007) Microfabrication of three-dimensional engineered scaffolds. *Tissue Eng*, **13**(8):1837–44.

47. Wong AP, Perez-Castillejos R, Christopher Love J and Whitesides GM (2008) Partitioning microfluidic channels with hydrogel to construct tunable 3-D cellular microenvironments. *Biomaterials*, **29**(12):1853–61.

48. Choudhary S, Haberstroh KM and Webster TJ (2007) Enhanced functions of vascular cells on nanostructured Ti for improved stent applications. *Tissue Eng*, **13**(7):1421–30.

49. Miller DC, Haberstroh KM and Webster TJ (2007) PLGA nanometer surface features manipulate fibronectin interactions for improved vascular cell adhesion. *J Biomed Mater Res A*, **81**(3):678–84.

50. Hu B, Unda F, Bopp-Kuchler S, Jimenez L, Wang XJ, Haikel Y, Wang SL and Lesot H (2006) Bone marrow cells can give rise to ameloblast-like cells. *J Dent Res*, **85**(5):416–21.

51. Chen H, Clarkson BH, Sun K and Mansfield JF (2005) Self-assembly of synthetic hydroxyapatite nanorods into an enamel prism-like structure. *J Colloid Interface Sci*, **288**(1):97–103.

52. Yin Y, Yun S, Fang J and Chen H (2009) Chemical regeneration of human tooth enamel under near-physiological conditions. *Chem Commun (Camb)*, **39**:5892–4.

53. Galler KM, Cavender A, Yuwono V, Dong H, Shi S, Schmalz G, Hartgerink JD and D'Souza RN (2008) Self-assembling peptide amphiphile nanofibers as a scaffold for dental stem cells. *Tissue Eng Part A*, **14**(12):2051–8.

54. Zaveri TD, Dolgova NV, Chu BH, Lee J, Wong J, Lele TP, Ren F and Keselowsky BG (2010) Contributions of surface topography and cytotoxicity to the macrophage response to zinc oxide nanorods. *Biomaterials*, **31**(11):2999–3007.

55. Bini P, Gabay JE, Teitel A, Melchior M, Zhou JL and Elkon KB (1992) Antineutrophil cytoplasmic autoantibodies in Wegener's granulomatosis recognize conformational epitope(s) on proteinase 3. *J Immunol*, **149**(4):1409–15.

56. Erisken C, Kalyon DM, Wang H, Ornek-Ballanco C and Xu J (2011) Osteochondral tissue formation through adipose-derived stromal cell differentiation on biomimetic polycaprolactone nanofibrous scaffolds with graded insulin and Beta-glycerophosphate concentrations. *Tissue Eng Part A*, **17**(9–10):1239–52.

57. Thesleff I and Tummers M (2003) Stem cells and tissue engineering: prospects for regenerating tissues in dental practice. *Med Princ Pract*, **12**(1):43–50.

58. Duailibi MT, Duailibi SE, Young CS, Bartlett JD, Vacanti JP and Yelick PC (2004) Bioengineered teeth from cultured rat tooth bud cells. *J Dent Res*, **83**(7):523–8.

59. Duailibi SE, Duailibi MT, Zhang W, Asrican R, Vacanti JP and Yelick PC (2008) Bioengineered dental tissues grown in the rat jaw. *J Dent Res*, **87**(8):745–50.

60. Modino SA and Sharpe PT (2005) Tissue engineering of teeth using adult stem cells. *Arch Oral Biol*, **50**(2):255–8.

61. Yelick PC and Vacanti JP (2006) Bioengineered teeth from tooth bud cells. *Dent Clin North Am*, **50**(2):191–203, viii.

62. Yen AH and Sharpe PT (2008) Stem cells and tooth tissue engineering. *Cell Tissue Res*, **331**(1):359–72.

63. Sharpe PT and Young CS (2005) Test-tube teeth. *Sci Am*, **293**(2):34–41.

64. Sartaj R and Sharpe P (2006) Biological tooth replacement. *J Anat*, **209**(4):503–9.

65. Ikeda E, Morita R, Nakao K, Ishida K, Nakamura T, Takano-Yamamoto T, Ogawa M, Mizuno M, Kasugai S and Tsuji T (2009) Fully functional bioengineered tooth replacement as an organ replacement therapy. *Proc Natl Acad Sci U S A*, **106**(32):13475–80.

66. Nor JE (2006) Tooth regeneration in operative dentistry. *Oper Dent*, **31**(6):633–42.

67. Hacking SA and Khademhosseini A (2009) Applications of microscale technologies for regenerative dentistry. *J Dent Res*, **88**(5):409–21.

68. Fidkowski C, Kaazempur-Mofrad MR, Borenstein J, Vacanti JP, Langer R and Wang Y (2005) Endothelialized microvasculature based on a biodegradable elastomer. *Tissue Eng*, **11**(1–2):302–9.

69. Ling Y, Rubin J, Deng Y, Huang C, Demirci U, Karp JM and Khademhosseini A (2007) A cell-laden microfluidic hydrogel. *Lab Chip*, **7**(6):756–62.

70. Souza GR, Molina JR, Raphael RM, Ozawa MG, Stark DJ, Levin CS, Bronk LF, Ananta JS, Mandelin J, Georgescu MM, Bankson JA, Gelovani JG, Killian TC, Arap W and Pasqualini R (2010) Three-dimensional tissue culture based on magnetic cell levitation. *Nat Nanotechnol*, **5**(4):291–6.

71. Snead ML (2008) Whole-tooth regeneration: It takes a village of scientists, clinicians, and patients. *J Dent Educ*, **72**(8):903–11.

72. Davis ME (2002) Non-viral gene delivery systems. *Curr Opin Biotechnol*, **13**(2):128–131.

73. Li B, Li S, Tan Y, Stolz DB, Watkins SC, Block LH and Huang L (2000) Lyophilization of cationic lipid-protamine-DNA (LPD) complexes. *J Pharm Sci*, **89**(3):355–64.

74. Somiari S, Glasspool-Malone J, Drabick JJ, Gilbert RA, Heller R, Jaroszeski MJ and Malone RW (2000) Theory and in vivo application of electroporative gene delivery. *Mol Ther* **2**(3):178–87.

75. Li Y, Ogris M, Pelisek J and Roedl W (2004) Stability and release characteristics of poly(D,L-lactide-co-glycolide) encapsulated CaPi-DNA coprecipitation. *Int J Pharm*, **269**(1):61–70.
76. Oster CG, Kim N, Grode L, Barbu-Tudoran L, Schaper AK, Kaufmann SH and Kissel T (2005) Cationic microparticles consisting of poly(lactide-co-glycolide) and polyethylenimine as carriers systems for parental DNA vaccination. *J Control Release*, **104**(2):359–77.
77. Wilson SP, Liu F, Wilson RE and Housley PR (1995) Optimization of calcium phosphate transfection for bovine chromaffin cells: Relationship to calcium phosphate precipitate formation. *Anal Biochem*, **226**(2):212–20.
78. Roy I, Mitra S, Maitra A and Mozumdar S (2003) Calcium phosphate nanoparticles as novel non-viral vectors for targeted gene delivery. *Int J Pharm*, **250**(1):25–33.
79. Bisht S, Bhakta G, Mitra S and Maitra A (2005) pDNA loaded calcium phosphate nanoparticles: highly efficient non-viral vector for gene delivery. *Int J Pharm*, **288**(1):157–68.
80. Fernandez CA and Rice KG (2009) Engineered nanoscaled polyplex gene delivery systems. *Mol Pharm*, **6**(5):1277–89.
81. Lee D, Upadhye K and Kumta PN (2012) Nano-sized calcium phosphate (CaP) carriers for non-viral gene delivery. *Materials Science and Engineering B-Advanced Functional Solid-State Materials*, **177**(3):289–302.
82. Bonadio J (2000) Tissue engineering via local gene delivery. *J Mol Med (Berl)*, **78**(6):303–11.
83. Jordan M, Schallhorn A and Wurm FM (1996) Transfecting mammalian cells: optimization of critical parameters affecting calcium-phosphate precipitate formation. *Nucleic Acids Res*, **24**(4):596–601.
84. Jordan M and Wurm F (2004) Transfection of adherent and suspended cells by calcium phosphate. *Methods*, **33**(2):136–43.
85. Olton D, Li J, Wilson ME, Rogers T, Close J, Huang L, Kumta PN and Sfeir C (2007) Nanostructured calcium phosphates (NanoCaPs) for non-viral gene delivery: Influence of the synthesis parameters on transfection efficiency. *Biomaterials*, **28**(6):1267–79.
86. Lee D and Kumta PN (2010) Chemical synthesis and characterization of magnesium substituted amorphous calcium phosphate (MG-ACP). *Mat Sci Eng C-Mater Biol Appl*, **30**(8):1313–17.
87. Kwoh DY, Coffin CC, Lollo CP, Jovenal J, Banaszczyk MG, Mullen P, Phillips A, Amini A, Fabrycki J, Bartholomew RM, Brostoff SW and Carlo DJ (1999) Stabilization of poly-L-lysine/DNA polyplexes for in vivo gene delivery to the liver. *Biochim Biophys Acta*, **1444**(2):171–90.
88. Kircheis R, Wightman L, Schreiber A, Robitza B, Rossler V, Kursa M and Wagner E (2001) Polyethylenimine/DNA complexes shielded by transferrin target gene expression to tumors after systemic application. *Gene Ther*, **8**(1):28–40.
89. Zugates GT, Peng W, Zumbuehl A, Jhunjhunwala S, Huang YH, Langer R, Sawicki JA and Anderson DG (2007) Rapid optimization of gene delivery by parallel end-modification of poly(beta-amino ester)s. *Mol Ther*, **15**(7):1306–12.
90. Pandita D, Santos JL, Rodrigues J, Pego AP, Granja PL and Tomas H (2011) Gene delivery into mesenchymal stem cells: A biomimetic approach using RGD nanoclusters based on poly(amidoamine) dendrimers. *Biomacromolecules*, **12**(2):472–81.

91. Hill IRC, Garnett MC, Bignotti F and Davis SS (1999) In vitro cytotoxicity of poly(amidoamine)s: relevance to DNA delivery. *Biochim Biophys Acta*, **1427**(2):161–74.

92. Choksakulnimitr S, Masuda S, Tokuda H, Takakura Y and Hashida M (1995) In-vitro cytotoxicity of macromolecules in different cell-culture systems. *J Control Release*, **34**(3):233–41.

93. Alexis F, Pridgen E, Molnar LK and Farokhzad OC (2008) Factors affecting the clearance and biodistribution of polymeric nanoparticles. *Mol Pharm*, **5**(4):505–15.

94. Harris TJ, Green JJ, Fung PW, Langer R, Anderson DG and Bhatia SN (2010) Tissue-specific gene delivery via nanoparticle coating. *Biomaterials*, **31**(5):998–1006.

95. Green JJ, Chiu E, Leshchiner ES, Shi J, Langer R and Anderson DG (2007) Electrostatic ligand coatings of nanoparticles enable ligand-specific gene delivery to human primary cells. *Nano Lett*, **7**(4):874–9.

96. Lonez C, Vandenbranden M and Ruysschaert JM (2008) Cationic liposomal lipids: From gene carriers to cell signaling. *Prog Lipid Res*, **47**(5):340–7.

97. Percot A, Briane D, Coudert R, Reynier P, Bouchemal N, Lievre N, Hantz E, Salzmann JL and Cao A (2004) A hydroxyethylated cholesterol-based cationic lipid for DNA delivery: Effect of conditioning. *Int J Pharm*, **278**(1):143–63.

98. Biondi M, Ungaro F, Quaglia F and Netti PA (2008) Controlled drug delivery in tissue engineering. *Adv Drug Deliv Rev*, **60**(2):229–42.

99. Woo KM, Chen VJ and Ma PX (2003) Nano-fibrous scaffolding architecture selectively enhances protein adsorption contributing to cell attachment. *J Biomed Mater Res A*, **67**(2):531–7.

100. Feng K, Sun H, Bradley MA, Dupler EJ, Giannobile WV and Ma PX (2010) Novel antibacterial nanofibrous PLLA scaffolds. *J Control Release*, **146**(3):363–9.

101. Wei G, Jin Q, Giannobile WV and Ma PX (2007) The enhancement of osteogenesis by nano-fibrous scaffolds incorporating rhBMP-7 nanospheres. *Biomaterials*, **28**(12):2087–96.

102. Zorlutuna P, Annabi N, Camci-Unal G, Nikkhah M, Cha JM, Nichol JW, Manbachi A, Bae H, Chen S and Khademhosseini A (2012) Microfabricated biomaterials for engineering 3D tissues. *Adv Mater*, **24**(14):1782–804.

103. Flaim CJ, Chien S and Bhatia SN (2005) An extracellular matrix microarray for probing cellular differentiation. *Nat Methods*, **2**(2):119–25.

104. Flaim CJ, Teng D, Chien S and Bhatia SN (2008) Combinatorial signaling microenvironments for studying stem cell fate. *Stem Cells Dev*, **17**(1):29–39.

105. Unadkat HV, Hulsman M, Cornelissen K, Papenburg BJ, Truckenmuller RK, Carpenter AE, Wessling M, Post GF, Uetz M, Reinders MJ, Stamatialis D, van Blitterswijk CA and de Boer J (2011) An algorithm-based topographical biomaterials library to instruct cell fate. *Proc Natl Acad Sci U S A*, **108**(40):16565–70.

106. Truckenmuller R, Giselbrecht S, Escalante-Marun M, Groenendijk M, Papenburg B, Rivron N, Unadkat H, Saile V, Subramaniam V, van den Berg A, van Blitterswijk C, Wessling M, de Boer J and Stamatialis D (2012) Fabrication of cell container arrays with overlaid surface topographies. *Biomed Microdevices*, **14**(1):95–107.

Index

433

CPSIA information can be obtained at www.ICGtesting.com
Printed in the USA
LVOW02*1806070813

346777LV00002B/2/P